F. Weinberg International Symposium on Solidification Processing

Titles of Related Interest—

Ashby ENGINEERING MATERIALS 1
Ashby ENGINEERING MATERIALS 2
Brook IMPACT OF NON-DESTRUCTIVE TESTING
Koppel AUTOMATION IN MINING, MINERAL AND METAL PROCESSING 1989
Ruhle METAL-CERAMIC INTERFACES
Taya METAL MATRIX COMPOSITES

Other CIM Proceedings Published by Pergamon

Bergman FERROUS AND NON-FERROUS ALLOY PROCESSES
Bickert REDUCTION AND CASTING OF ALUMINUM
Bouchard PRODUCTION, REFINING, FABRICATION AND RECYCLING OF LIGHT METALS
Chalkley TAILING AND EFFLUENT MANAGEMENT
Closset PRODUCTION AND ELECTROLYSIS OF LIGHT METALS
Dobby PROCESSING OF COMPLEX ORES
Embury HIGH TEMPERATURE OXIDATION AND SULPHIDATION PROCESSES
Jaeck PRIMARY AND SECONDARY LEAD PROCESSING
Jonas DIRECT ROLLING AND HOT CHARGING OF STRAND CAST BILLETS
Kachaniwsky IMPACT OF OXYGEN ON THE PRODUCTIVITY OF NON-FERROUS METALLURGICAL PROCESSES
Macmillan QUALITY AND PROCESS CONTROL IN REDUCTION AND CASTING OF ALUMINUM AND OTHER LIGHT METALS
Mostaghaci PROCESSING OF CERAMIC AND METAL MATRIX COMPOSITES
Plumpton PRODUCTION AND PROCESSING OF FINE PARTICLES
Purdy FUNDAMENTALS AND APPLICATIONS OF TERNARY DIFFUSION
Rigaud ADVANCES IN REFRACTORIES FOR THE METALLURGICAL INDUSTRIES
Ruddle ACCELERATED COOLING OF ROLLED STEEL
Salter GOLD METALLURGY
Thompson COMPUTER SOFTWARE IN CHEMICAL AND EXTRACTIVE METALLURGY
Twigge-Molecey MATERIALS HANDLING IN PYROMETALLURGY
Twigge-Molecey PROCESS GAS HANDLING AND CLEANING
Tyson FRACTURE MECHANICS
Wilkinson ADVANCED STRUCTURAL MATERIALS

Related Journals

(Free sample copies available upon request)

ACTA METALLURGICA
CANADIAN METALLURGICAL QUARTERLY
MATERIALS RESEARCH BULLETIN
MINERALS ENGINEERING
SCRIPTA METALLURGICA

PROCEEDINGS OF THE F. WEINBERG INTERNATIONAL SYMPOSIUM ON SOLIDIFICATION PROCESSING, HAMILTON, ONTARIO, AUGUST 27-29, 1990

F. Weinberg International Symposium on Solidification Processing

Editors

J.E. Lait
Research and Development
Stelco Steel
Hamilton, Ontario

I.V. Samarasekera
The Centre for Metallurgical
Process Engineering
The University of British Columbia
Vancouver, British Columbia

Symposium organized by the Basic Sciences Section of The Metallurgical Society of CIM

29th ANNUAL CONFERENCE OF METALLURGISTS OF CIM
29e CONFÉRENCE ANNUELLE DES MÉTALLURGISTES DE L'ICM

Pergamon Press
Member of Maxwell Macmillan Pergamon Publishing Corporation
New York Oxford Beijing Frankfurt São Paulo Sydney Tokyo Toronto

Pergamon Press Offices:

U.S.A.	Pergamon Press, Inc., Maxwell House, Fairview Park, Elmsford, New York 10523, U.S.A.
U.K.	Pergamon Press plc, Headington Hill Hall, Oxford OX3 0BW, England
PEOPLE'S REPUBLIC OF CHINA	Pergamon Press, 0909 China World Tower, No. 1 Jian Guo Men Wai Avenue, Beijing 1000004, People's Republic of China
FEDERAL REPUBLIC OF GERMANY	Pergamon Press GmbH, Hammerweg 6, D-6242 Kronberg, Federal Republic of Germany
BRAZIL	Pergamon Editora Ltda, Rua Eça de Queiros, 346 CEP 04011, Paraiso, São Paulo, Brazil
AUSTRALIA	Pergamon Press Australia Pty Ltd., P.O. Box 544, Potts Point, NSW 2011, Australia
JAPAN	Pergamon Press, 8th Floor, Matsuoka Central Building, 1-7-1 Nishishinjuku, Shinjuku-ku, Tokyo 160, Japan
CANADA	Pergamon Press Canada Ltd., Suite 271, 253 College Street, Toronto, Ontario M5T 1R5 Canada

Library of Congress Cataloging in Publication Data

ISBN 0-08-040413-8

Printing: 1 2 3 4 5 6 7 8 9 Year: 0 1 2 3 4 5 6 7 8 9

Printed in the United States of America

The paper used in this publication meets the minimum requirements of American National Standard for Information Sciences-Permanence of Paper for Printed Library Materials, ANSI Z 39.48-1984

Foreword

This international symposium, to be held August 27-29, 1990 in Hamilton, Ontario, is in honour of Prof. Fred Weinberg, who retired from The University of British Columbia in June 1990, following a distinguished career. The symposium is an integral part of the 29th Annual Conference of Metallurgists of The Metallurgical Society of The Canadian Institute of Mining and Metallurgy.

Professor Fred Weinberg is a pioneer in research on solidification processing and has made seminal contributions to the fundamentals of solidification. He has gained international recognition for his work on continuous casting of steel, casting of non-ferrous metals and cast iron, semiconductor and optoelectronic crystal growth, as well as for his research on the high-temperature low ductility of steels, grain boundary studies and interdendritic fluid flow, to mention but a few. He was a lecturer of the Institute of Metals and recipient of the Robert H. Mehl medal of the AIME in 1975. For his outstanding career contributions he has also received the Alcan Award of The Metallurgical Society of CIM in 1988, and was made a Fellow of the Metallurgical Society of the AIME in 1988, and a Fellow of The Canadian Institute of Mining and Metallurgy in 1990. Professor Weinberg has published over 100 papers and has received recognition for his work through several best paper awards, including the Charles H. Herty Award of the AIME in 1974, Robert W. Hunt Award of the ISS-AIME in 1980 and the Howe Medal of the AIME in 1979.

Professor Weinberg graduated from the University of Toronto in 1947 with a degree in engineering physics, and was awarded an M.Sc. in 1948. He received a Ph.D. in 1951 for his work on "Dendritic Growth of Metals" with the late Prof. Bruce Chalmers who was renowned for his reserarch on solidification. Following his Ph.D., Dr. Weinberg joined the Physical Metallurgy Research Laboratory of the Department of Energy, Mines and Resources in Ottawa and was head of the Metal Physics Section between 1961 and 1967. He has been a professor with the Department of Metals and Materials Engineering at The University of British Columbia since 1967, and was head from 1980 to 1985. Professor Weinberg's dedication to research and commitment to excellence has been a source of inspiration to all who have worked with him. He has challenged and guided many graduate students contributing immeasurably to their professional growth and has mentored numerous others. This symposium is a tribute to Professor Weinberg—an outstanding educator and researcher.

The symposium consists of six sessions — Fundamentals of Solidification, Non-ferrous Casting, Continuous Casting of Steel, Static Casting of Cast Iron, Novel Solidification Studies, and Semiconductor and Optoelectronic Crystal Growth — each a reflection of Professor Weinberg's research interests. This world-class symposium owes its success to the quality of the contributions, and we wish to express our appreciation to all the authors especially to the invited speakers, Dr. N. Bryson, Prof. M. Flemings, Prof. H. Fredricksson, Prof. I. Minkoff and Prof. R. Brown for their contributions. This distinguished group will enhance the event with their knowledge of the subject and outstanding work. Thanks are also due to the session chairmen, Prof. M.E. Glicksman, Prof. J.E. Gruzleski, Prof. J.K. Brimacombe, Dr. H. Biloni, Prof. D. Apelian and Dr. G. Elliot for their efforts in soliciting papers and organizing each session. Without their commitment this symposium would not have come to fruition.

The financial sponsorship of The Metallurgical Society of CIM which enabled promotion of the symposium is gratefully acknowledged. The sponsorship of the Basic Sciences Section of The Metallurgical Society is also recognized. We would further like to express our gratitude to Prof. G.A. Irons, Technical Program Chairman for the 29th Conference of Metallurgists, for his efforts in integrating the symposium with the conference program, to Prof. D.A.R. Kay, Conference Chairman, and to the Board of The Metallurgical Society who encouraged and supported our endeavour to honour an eminent Canadian metallurgist.

J.E. Lait
Research and Development
Stelco Steel

I.V. Samarasekera
The Centre for Metallurgical Process Engineering
The University of British Columbia

June 1990

TABLE OF CONTENTS

IV. NOVEL SOLIDIFICATION STUDIES

V. STATIC CASTING OF CAST IRON

VI. SEMICONDUCTOR AND OPTOELECTRONIC CRYSTAL GROWTH

I. Fundamentals of solidification

Chairman: M.E. Glicksman
Rensselaer School of Engineering,
Troy, New York, U.S.A.

Recent research in solidification (Invited Paper)

F. Weinberg
Metals and Materials Engineering Department, The University of British Columbia, Vancouver, British Columbia, Canada, V6T 1W5

Abstract

A number of areas in the general field of solidification have been recently examined by the present author and his associates. These investigations will be discussed with particular reference to the solidification problems they present and, in some cases, possible solutions. The areas include the columnar to equiaxed transition, large spangles in galvanized sheet, solidification and segregation of nodular iron, magnetic fields in semiconductor crystal growth, Cu in GaAs, corrosion at bicrystal grain boundaries in AlCu, and foamed metals.

The present author and his associates have recently been investigating a range of topics in solidification. In this presentation the topics will be considered in turn. The basic problems in solidification encountered will be outlined, and the problems discussed.

Columnar to Equiaxed Transition

The columnar to equiaxed transition is a classic problem, important in both ferrous and nonferrous solidification. The transition involves many of the basic features of solidification - nucleation, dendritic growth, solute segregation, heat flow, fluid flow and others. If the columnar to equiaxed transition can be made precise and reproducible, and a model developed which quantitatively predicts when the transition occurs, using known parameters, then experimental measurements can be made to quantitatively verify the overall model and the constituent mechanisms.

Models of the columnar to equiaxed transition have been proposed, (Hunt (1)) which predict when the transition occurs. However, they have not been verified experimentally in castings, and they require input parameters which can only be estimated. In a series of experiments the present author and his associates have examined the columnar to equiaxed transition in Sn-Pb (2), Al-Cu (3) and Pb-Sn (4).

In the Sn-Pb alloys, for directional solidification from a water cooled copper chill, columnar to equiaxed transitions were observed which occurred sharply and reproducibly, as shown in Figure 1. The position of the transition could be moved toward the chill by reducing the rate of heat extraction from the chill. Let us assume that nuclei are present in the melt during solidification, and that the transition occurs when a nucleus is ahead of a dendrite. This requires the liquid ahead of the dendrite tip to be thermally or constitutionally supercooled. Since pure metals do not exhibit a transition, and thermal supercooling is not generally possible where a positive thermal gradient exits, the supercooling can be attributed to constitutional supercooling ahead of the dendrite tip. Such supercooling has been measured by Burden and Hunt (5) in Al-Cu alloys, the amount of supercooling being in the order of several degrees C for shallow thermal gradients and normal freezing rates. As the thermal gradient ahead of a dendrite tip decreases, the width of the supercooled region in which a nucleus can grow increases. At some point there could be a high probability that a nucleus is present in the supercooled region which grows and results in the transition occurring.

To determine whether this concept could account for the position of the transition, at least as a first step, the temperature gradient ahead of the dendrite tip in the Sn-Pb alloys, determined by interpolation of measured temperatures, was plotted as a function of distance from the chill and related to the transition position. The results are shown in Figure 2.

The transition occurs when the temperature gradient in the melt at the liquidus temperature is 0.11°C/mm for Sn-10 wt pct Pb and 0.10 and 0.13°C/mm for 5 and 15 wt pct. Pb respectively (2). In Al-3 wt pct Cu the transition occurs near 0.66°C/mm (3). When TiB_2 is added to Al-3 wt pct Cu a fine grained structure is obtained with gradients much larger than 0.06°C/mm. The amount of TiB_2 required is small, about 171 ppm, below which columnar growth occurs. It is proposed that a very large density of nuclei are produced by the TiB_2 in the melt. The probability of nuclei being present in the supercooled layer ahead of an advancing tip is high. These nuclei grow blocking further dendritic growth. This continues as the new grains start to grow as columnar grains. Accordingly the critical parameters when a nucleant is added to a melt are the density of nuclei, and their ability to grow at small supercoolings.

Since the temperature and solute gradients being dealt with are small, a heat transfer mathematical model of the solidification system offers a means of establishing accurate temperatures, temperature gradients, and growth velocities in the non-steady state conditions being considered. This was done in the Sn-Pb and Al-Cu systems but the measurements could not be fitted to the model predictions. Further modelling and experiments have been done with care in the Pb-Sn system aiming for an accuracy of 0.2°C in predicting temperatures during solidification. It was found that the boundary conditions related to heat flux across the alloy surfaces could not be defined well enough to produce the accuracy required.

If the assumption is made that the columnar to equiaxed transition occurs when a critical small gradient is reached ahead of the dendrite tip, and heterogeneous nuclei are present in the melt, what is the effect of fluid flow produced by electromagnetic stirring or buoyancy flow on the transition? Increasing fluid flow would tend to decrease the thermal gradients in the liquid and produce a transition sooner in a given thermal field, as is observed. This differs from the normal assumption that fluid flow generates nuclei by a dendritic secondary branch remelting process, producing the transition.

Hot Dipped Galvanized Steel Sheet

Although hot dipped galvanizing has been used industrially for many years, the solidification of the galvanized layer - particularly the growth of large spangles - is not clearly understood.

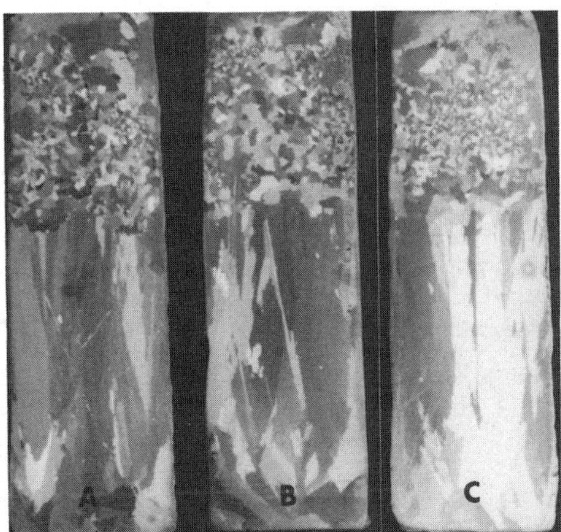

Fig.1. Etched vertical sections of ingots showing the columnar to equiaxed transition with 3 superheats: (a) 19°C, (b) 31°C, (c) 36°C. Sample width 35 mm.

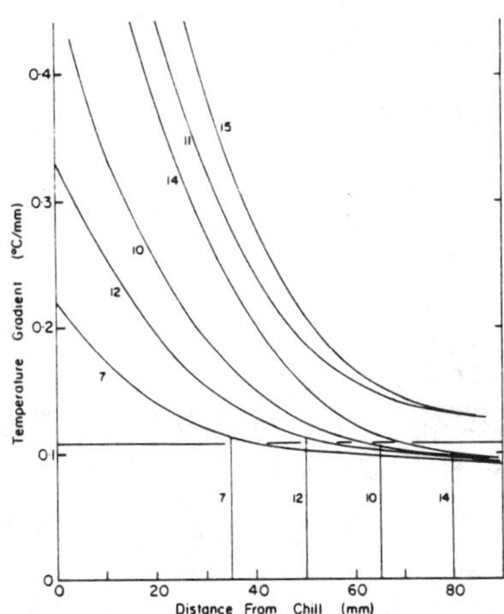

Fig.2. Temperature gradient vs distance from the chill in Sn 10 pct Pb with different heat transfer coefficients. Vertical lines are CET positions for tests 7,12,10 & 14.

Spangles are very large, faceted, dendritic grains. An example of a spangle of 13mm diameter is shown in Figure 3. Much larger spangles can be obtained in commercially galvanized sheets. In an investigation by the present author and his associates (6) a number of features concerning spangle growth were observed.

1) Spangle growth is dependent on bath composition. Spangles from when at least 0.04 wt pct of Pb, Bi or Sb is added to the galvanizing bath. They do not form when the same amount of Mg, Sn or Cd are added.
2) The growth of the large dendrites producing the spangle is not due to an increase in melt supercooling. Melt supercooling is small - less than 1°C generally.
3) Dendrites in hexagonal zinc grow in the basal plane. If dendritic growth in the spangle is confined to the basal plane, the spangle surface will have to be closely aligned to the basal plane since the galvanized layer is about 40 μm thick. This is not the case. The basal planes of spangles are tilted at large angles with respect to the spangle surface.
4) When a galvanizing bath which produces spangles on a galvanized sheet is solidified in bulk, spangles do not form.

One can speculate on the mechanism of spangle formation.

The size of the spangle is determined by the velocity of the primary dendrite branches forming the skeleton of the grain. Large spangles result from high dendrite velocities. Adding a small amount of Pb, Bi or Sb to the melt increases the dendrite velocity by decreasing the dendrite tip radius. Each of the added elements have a very small segregation coefficient in Zn (less and 0.01) which would result in a high solute concentration ahead of the dendrite tip. The decrease in dendrite tip radius results from the relatively low surface tension of the additions. The relationship between grain diameter and surface tension of the six additions considered is shown in Figure 4.

The large title angle of the basal plane with respect to the spangle surface, combined with the sharply linear primary dendrite spikes in the basal spangle, indicate growth occurs in crystallographic planes but not in crystallographic directions. The directions are confined by the melt surface. The mechanism by which this occurs is unclear.

The observation that large spangles form in the galvanized layer, and not in the bulk, for a given melt could be accounted for by assuming dendrites have higher velocities along melt surfaces. This was observed to be the case for water (7) where the dendrite velocity increased from 0.3 cm/s in free growth to 1.2 cm/s on a glass rod and 3.0 cm/s on a brass plate. Whether a similar effect occurs for Zn dendrites growing at the melt surface is not clear. We note that with the basal plane tilted with respect to the melt surface, some primary dendrites will grow along the melt/air interface, others along the melt/steel interface.

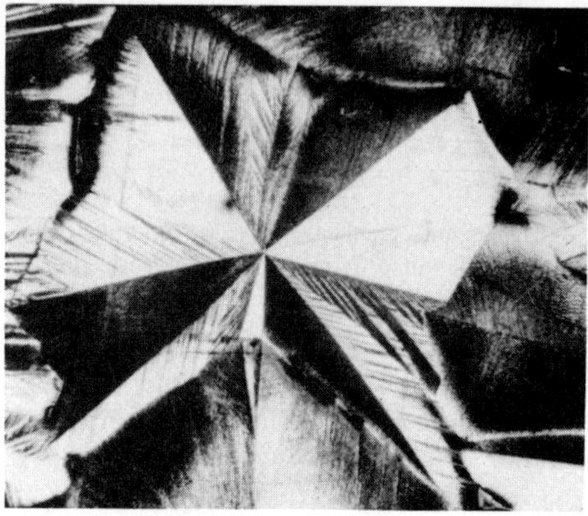

Fig.3. Galvanized sheet steel spangle. Diameter 13 mm.

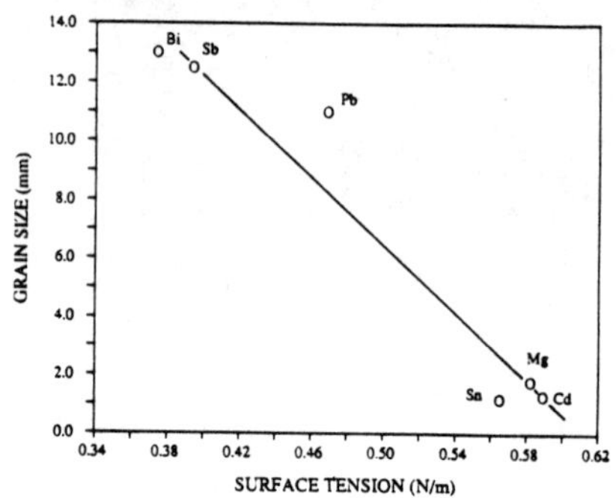

Fig.4. Grain size vs element surface tension. Zn-0.2 wt pct Pb bath.

Nodular Cast Iron

A. The nucleation and growth mechanism for nodules in cast iron is still not clearly defined. The role of magnesium added to the melt could be to act as a getter for elements which contribute to the flake form of the graphite in grey cast iron, or itself could directly affect the growing graphite particle. If it is the latter, then one might anticipate that Mg would be distributed throughout a nodule since it would be continually acting on the nodule surface.

Experiments using secondary ion mass spectrometry (SIMS) have shown that Mg, at low levels, is uniformly distributed throughout the nodule (8) indicating the prime role of Mg is not gettering. This does not clarify how Mg atoms modify the growth morphology of graphite to produce spheres instead of flakes. The effect of impurity elements present in the melt on growth morphologies in metals is not clearly understood.

B. The segregation of alloying elements in nodular iron has been examined, primarily using microprobe analysis. The results show a number of interesting features (9).

1) Defining an effective segregation coefficient as the ratio of solute concentration at the centre of an initial primary dendrite branch, to that of the solute melt concentration in which the dendrite is growing, the effective segregation coefficients differ markedly from the equilibrium values taken from Fe based binary diagrams. Cu, Ni and Si have equilibrium coefficients of 0.62, 0.75 and 0.87 which increase to

effective coefficients of 1.37, 1.23 and 1.09 respectively. On the other hand Cr and Mo with equilibrium coefficients of 0.99 and 0.53 decrease to effective coefficients of 0.60 and 0.26 respectively. Mn shows little change, from 0.64 to 0.70 for equilibrium to effective segregation coefficient. The results clearly indicate that solutes cannot be considered individually in a melt component system when considering segregation during solidification. It is not clear how the various solutes interact, or how to determine the interaction from first principles for such a complex system.

2) Using the effective segregation coefficients, and the Scheil equation, a reasonable correlation could be obtained for the solute distribution of each element, measured by microproble analysis and the calculated distribution. It thus appears segregation can be determined on an individual basis.

Crystal Growth of GaAs

A) LEC GaAs has a high dislocation density, 10^4 to $10^5/cm^2$ being typical values. Much effort is being directed towards reducing the dislocation density by varying the growth conditions.

In metals, the dislocation density can be markedly reduced by high temperature annealing, or by annealing under an applied stress. This does not appear to be the case for GaAs.

An investigation was undertaken (10) to determine to what extent as-grown dislocations moved and annihilated in GaAs as a result of annealing and stress/annealing. The dislocations were observed by cathodoluminescence. The results showed the following:

1) At annealing temperatures of 1000°C (0.81 of the melting temperature) the as-grown dislocations did not move.
2) With a small stress, both static and cyclic, the as-grown dislocations did not move.
3) The applied stress results in slip occurring in the GaAs producing slip lines on the surface consistent with normal slip in an F.C.C. system. The as-grown dislocations disappeared when local slip occurred.
4) The inability of the as-grown dislocations to move cannot be readily attributed to impurity locking, since the residual impurity level in the GaAs wafers is very low. It is postulated that movement is inhibited because dislocation climb is inhibited by the structure. Edge dislocations terminate on a Ga or an As plane. As a result climb would require the edge to move through alternate Ga and As planes which appears unlikely.

B) Crystal Growth - Magnetic Fields

Efforts are continuing by producers of LEC GaAs crystals to improve crystal yield, and crystal quality. This is being done by modifying the growth conditions. Because there are many subtle growth variables contributing to crystal quality, and growth is done under high pressure, it is difficult and expensive to conduct extensive experimental programmes. As a result much effort has been directed to mathematical modelling of the process, both related to the growth process, and the generation of dislocations during growth (11). The modelling of dislocation generation indicates where and when dislocations are generated during growth, and how this varies with the most important growth variables.

There are relatively few major changes which can be made to the LEC growth process and still grow single crystals. One change currently being proposed and investigated is applying a magnetic field to the bulk liquid. The field will reduce or stop buoyancy driven fluid flow in the melt which could lead to better quality crystals. It is not clear how fluid flow specifically affects crystal growth and therefore crystal quality.
An investigation is underway in which the effect of a large magnetic field on crystal growth is being investigated. The project is being examined experimentally and by fluid flow mathematical modelling (12). We note the fluid flow is complex since forced convection is produced by the rotation of the crucible, and the counter rotation of the crystal. Buoyancy driven convection occurs since the heat to the melt comes from the crucible walls, and top surface convective cells may be anticipated since heat is being lost from the top surface.

Questions

1) What temperature distribution exists in a melt under normal growth conditions?

2) What affect does a magnetic field have on this distribution?
   a) as a function of strength of field
   b) as a function of direction of field
   c) bulk flow
   d) oscillations

3) How does the magnetic field change the liquid under the crystal?
   a) thickness of δ layer
   b) shape of solid/liquid interface
   c) impurity segregation and stoichiometry
   d) velocity of growth - banding
   e) Taylor-Proudman cells

Observations

Temperature measurements were made in melts of Hg and Ge with and without axial magnetic fields of about 0.1 Tor.

1) On applying a magnetic field the upper central region of the melt dropped in temperature due to stopping of flow from the crucible walls. The drop was as much as 30°C in Ge.
2) The oscillations were sharply reduced, in agreement with previous reports.
3) The large, long wavelength oscillations under the crystal due to Taylor-Proudman flow in Ge were strongly damped.
4) A mathematical model of the effect of a radial magnetic field on fluid flow has been developed and results for assumed imput variables calculated (12).
5) At present the effects on the magnetic field on the liquid under the crystal and the interface shape have not been examined.

Copper in Gallium Arsenide

An investigation has been undertaken to examine the preferential diffusion and concentration of Cu at dislocations, grain boundaries, and the free surface of GaAs (13). Preferential diffusion along dislocations generally occurs at low temperatures with respect to the melting temperature. Cu diffuses very rapidly in GaAs, with a diffusion coefficient near $10^{-5}cm^2s^{-1}$.

Accordingly, appreciable diffusion of Cu should occur at low temperatures enabling preferential diffusion to be detected.

A number of techniques were used to look for preferential diffusion along dislocation pipes, the most successful being with tracer experiments using radioactive Cu. The results differed from what was anticipated.

1) Only a very small part of the Cu deposited on the surface of a GaAs wafer diffused at a high rate in the bulk material - less than 1 ppm. The remainder diffused at normal substitutional rates. I find it intuitively difficult to envisage an element with a very high diffusion rate in a material, having hardly any solubility in the material.
2) There was no evidence of preferential diffusion down dislocation pipes on grain boundaries at diffusion temperatures of 500 or 600°C (0.4 and 0.5 of melting temperature).
3) There was no evidence of segregation of Cu at the surface of the GaAs wafer.
4) At temperatures above 800°C the Cu alloyed with the Ga and As forming liquid compounds. This is consistent with what might be anticipated from the phase diagrams.
5) The formation of liquid on the wafer surface resulted in the formation of Cu rich pipes which penetrated through the wafer. In some cases the liquid drained from the pipe, leaving a hole.

In another investigation of GaAs (14) the effect of annealing was examined. Cathodoluminescence observations were made on the surfaces and cross sections of GaAs wafers, as a function of annealing times and temperatures. It was found that anomalous bright bands were present adjacent to the surface of the wafer, the thickness of the bands being related to the annealing treatment. The bright bands have been shown to be clearly related to the presence of very low concentrations of Cu which diffuse into the GaAs from the outer surface. This was shown with photoluminescence spectra of cross-sections of the annealed wafers obtained at Simon Fraser University by Dr. M. Thewalt. Cu peaks were present in the spectra in the bright band regions, and not present in the dark region in the centre of the wafer.

Corrosion of Al-Cu Bicrystals

A study has been undertaken to examine corrosion - particularly pitting corrosion - in Al-Cu bicrystals (15). The question of interest from a solidification point of view is, to what extent segregation of Cu at the grain boundary influences the corrosion in this system? This includes segregation in the as-grown bicrystal, in the fully annealed bicrystal, and at the precipitates on the grain boundary.

The results showed that, in general, there was no preferential pitting corrosion at either small or large angle grain boundaries in the bicrystals. This was the case for both the as-grown and fully annealed samples. When precipitates formed on the boundary, these sites were often sites for preferential pitting. It was observed that particular crystallographic planes in the alloy pitted at different rates and exhibited pits of different morphologies.

Foamed Metals

An investigation is underway, supported by the Canada Space Agency to examine the generation, movement, and interaction of gas bubbles in metals. If small bubbles can be generated, distributed, and kept separate, then it should be possible to produce foamed metals. In general, buoyancy forces move bubbles very rapidly to the melt surface where they are lost. In a microgravity environment buoyancy forces are very small, and bubble behaviour can be investigated.

The project has been considered in two parts, one dealing with bubbles in liquid metals by the present author, and the second dealing with mathematically modelling the flowing liquid containing bubbles and testing the model using transparent liquids, by M. Salcudean of Mechanical Engineering, University of British Columbia.

Results to Date

1) Attempts to introduce a large number of small bubbles into liquid tin in a microgravity environment have not been successful. Gas passed through porous stainless steel sheets or through individual holes in sheets produced bubbles in the melt under normal gravity conditions, but not in microgravity.
2) Tests on a water/glycerine system in microgravity by M. Salcudean with fluid flow introduced in the melt by electromagnetic means, showed that small bubbles moved with the flowing liquid in a manner consistent with the predictions of the mathematical model.
3) The behaviour of gas bubbles in liquid metals is normally very strongly influenced by gravity forces. Without these forces surface tension plays a major role in bubble behaviour. Since the surface tension of liquid tin (530 dynes/cm) is much higher than water (73 dynes/cm) using water as a modelling system for liquid metals can be misleading. At present it is unclear to what extent gas bubbles will stick to the wall of the container or react with one another in a microgravity environment.

References

1. J.D. Hunt, Matl. Sc. and Eng. 65 (1984) pp. 75-83.
2. R.B. Mahapatra and F. Weinberg, Met. Trans. B, 18B (1987) pp. 425-431.
3. I. Ziv and F. Weinberg, Met. Trans. B, 20B (1989) p. 731.
4. G.T. Lowe, "The Columnar to Equiaxed Transition in Pb-Sn and Sn-Pb Alloys", M.A.Sc. Thesis, University of British Columbia, 1990.
5. M.D. Burden and J.D. Hunt, J. Crystal Growth, 22 (1974) pp. 99-108.
6. F.A. Fasoyinu, "The Solidification of Hot Dipped Galvanized Coatings in Steel", Ph.D. Thesis, University of British Columbia, 1989.
   F.A. Fasoyinu and F. Weinberg, "Spangle Formation in Galvanized Sheet Steel Coatings", Met. Trans. B, in press, 1990.
7. G.S. Lindenmeyer, G.T. Orak and B. Chalmers, J. of Chemical Physics, 27 (1957), p. 822.
8. G. Torga, R. Boeri and F. Weinberg, Cast Metals, 2 (1989) p. 169.
9. R. Boeri, "The Solidification of Ductile Cast Iron", Ph.D. Thesis, University of British Columbia, 1989.
   R. Boeri and F. Weinberg, Trans. AFS, 106 (1989) pp. 179-184.
10. P. Gallagher and F. Weinberg, J. of Crystal Growth, 94 (1989) pp. 299-310.

11. C. Schvezov, I.V. Samarasekera and F. Weinberg, J. of Crystal Growth, 97 (1989) pp. 146-151; 92 (1988) pp. 479-488; 92 (1988) pp. 489-497; 84 (1987) pp. 212-218; 84 (1987) pp. 219-230; 85 (1987) pp. 142-147.
12. P. Sabhapathy and M.E. Salcudean, Journal of Crystal Growth (1990) in press.
13. D. Macquistan and F. Weinberg, "The Behaviour of Cu on GaAs", M.A.Sc. Thesis, University of British Columbia, 1989.
14. C. Third, F. Weinberg and L. Young, Appl. Phys. Lett 54 (1989) pp.2671-6273.
C. Third - Ph.D. Thesis, University of British Columbia - in progress.
15. M. Yasuda, F. Weinberg and D. Tromans, J. Electrochemical Society (1990) - in press, "Pitting Corrosion of Al and Al-Cu Single Crystals" "Pitting Corrosion of Al and Al-Cu Bicrystals."

Fluid flow and microstructure development

J.A. Dantzig and L.-S. Chao
Department of Mechanical and Industrial Engineering, University of Illinois, Urbana, Illinois 61801, U.S.A

Introduction

Metal and semiconductor manufacturing represents an important and highly contested fraction of the United States economy. The typical production route for these materials begins with melting and alloying raw materials to a predetermined composition, followed by controlled solidification. The solidification process produces a *microstructure* and this microstructure provides the materials with the properties that they will have in service. Once all of the material is solidified it is very difficult to change any of the gross features of the microstructure, because the transport processes which change the structure occur much more rapidly when one of the phases is liquid. Thus it is desirable to control the microstructure during freezing, which necessitates that one have a good understanding of the processes by which the structure forms.

In this article we will examine the interaction of macroscopic processing conditions with microstructure evolution, paying particular attention to the role of fluid flow during solidification. Most theories describing microstructure development during solidification are based on diffusive transport of heat and mass, and the presence of fluid flow acts to change the solute field, rather than fundamentally altering the phase change process. This makes the microgravity environment attractive for studying these phenomena because it represents an opportunity to study phase change processes with much less convective transport.

Solidification begins with *nucleation* of the parent crystal followed by continued growth. Nucleation is affected most by the purity of the solidifying material, and is not strongly affected by fluid flow. The microgravity environment can be very important for studying nucleation phenomena, however, because it affords the possibility of observing nucleation in experiments done without a container, which is a possible source of contamination. These phenomena are discussed in other contributions in this volume, so this article will focus on *growth* processes, emphasizing the use of analysis to study and control structure development.

Metals and semiconductors can be considered to be incompressible. For that reason, we may write the following equations representing the balance of mass, momentum, energy, and solute in the solidifying material: For the fluid

$$\nabla \cdot \mathbf{u} = 0 \quad [1]$$

$$\rho_l \left(\frac{\partial \mathbf{u}}{\partial t} + \mathbf{u} \cdot \nabla \mathbf{u} \right) = - \nabla p + \nabla \cdot (\mu \mathbf{D}) + \rho_l \mathbf{F} + \rho_l \mathbf{g} \left\{ 1 - \beta_T (T - T_{ref}) - \beta_S (c - c_{ref}) \right\} \quad [2]$$

$$\rho_l c_{pl} \left(\frac{\partial T}{\partial t} + \mathbf{u} \cdot \nabla T \right) = \nabla \cdot (k_l \nabla T) + Q \quad [3]$$

$$\frac{\partial c_l}{\partial t} + \mathbf{u} \cdot \nabla c_l = \nabla \cdot (D_l \nabla c_l) \quad [4]$$

and for the solid

$$\rho_s c_{ps} \frac{\partial T}{\partial t} = \nabla \cdot (k_s \nabla T) + Q \quad [5]$$

$$\frac{\partial c_s}{\partial t} = \nabla \cdot (D_s \nabla c_s) \quad [6]$$

where it has been implicitly assumed that the solid cannot move (**u**=0). The momentum balance equation for the fluid includes terms on the right-hand-side representing the variation in density of the fluid as a function of its temperature and composition. These terms represent driving forces for convection due to these buoyant effects. Note that each of these terms is proportional to the magnitude of the gravity vector, so the microgravity environment provides a unique means to suppress convection. It is not sufficient for most purposes to simply take an experiment into low earth orbit and assume that convection will be suppressed. After examining the effects of fluid flow on microstructure, conditions under which buoyant forces produce significant convection will be studied.

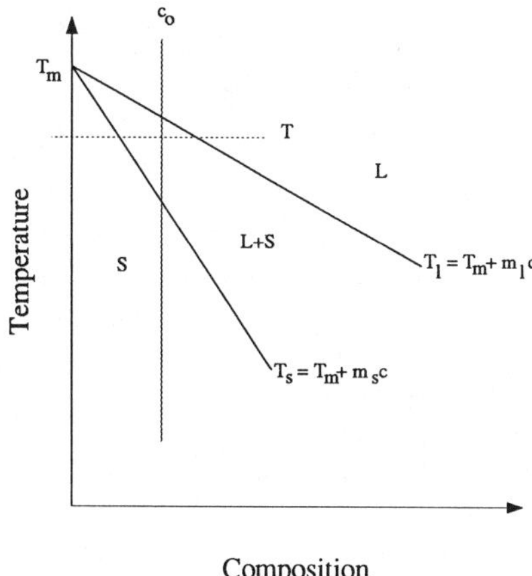

Figure 1: Portion of a typical phase diagram for a binary system.

The phenomena of interest in this article relate to solidification, and thus the coupling conditions for heat, mass and momentum transport at the liquid-solid interface are also required. Figure 1 shows a typical phase diagram for a dilute binary alloy. A vertical line in this diagram represents a single composition and the various regions in the diagram indicate which phases are present at various temperatures. The equilibrium thermodynamics implied by the phase diagram usually holds true at least at the liquid-solid interface, so that the following boundary conditions apply at the interface: Thermodynamic equilibrium yields a relation between temperature, composition and the local interface curvature

$$T_s = T_l = T_m + m_l c_l - \frac{\gamma_{sl}}{L_f} K \quad [7]$$

and the compositions of the two phases are related by the phase diagram

$$c_s = k_o c_l \quad [8]$$

The mechanical and thermal boundary conditions which apply at the interface specify that mass is conserved and there is no slip of the fluid on the solid

$$\rho_l \mathbf{u} \cdot \mathbf{n} = (\rho_l - \rho_s) \mathbf{V} \cdot \mathbf{n} \quad [9]$$

where **V** is the velocity of the liquid-solid interface and **u** is the fluid velocity. Heat is also conserved at the interface

$$k_s \nabla T_s \cdot \mathbf{n} - k_l \nabla T_l \cdot \mathbf{n} = L_f \mathbf{V} \cdot \mathbf{n} \qquad [10]$$

Taken together, Equations [1]-[10] completely specify the solidification field problem.

The process of setting the microstructure is essentially one of the interaction of the thermal and solute fields. In metallic systems, a typical boundary layer thickness for the thermal field ($\frac{\alpha_{s,l}}{V}$) is several millimeters, while for the solute ($\frac{D_{s,l}}{V}$) it is just a few microns. For this reason, solidification problems can be thought of as ones where the thermal field is set by the processing conditions and the solute field follows through diffusion.

When fluid flow is present, the same concepts apply. Flow may affect the morphological stability and scaling of the microstructure, and it can alter the existing thermal and solute fields to alter the microstructure. This article will consider several problems where the altered thermal and solute field cause microstructure changes. The flow can have effects on both the local (microscopic) and the global (macroscopic) scale. One important exception to this viewpoint, however, is the area of coupled instabilities for solidification and fluid flow. A recent review by Glicksman, *et al*(1) provides an excellent summary of recent advances in the coupling of fluid flow and morphological stability.

The Role Of Fluid Flow

Microscopic Scale

The effect of fluid flow on the morphological stability of the liquid-solid interface has been considered by several authors. In each case, a linear stability analysis was performed similar to that of Mullins and Sekerka(2), but also considering a basic state including some fluid flow ahead of the interface. Delves(3) considered a simple shear flow parallel to the advancing interface and found that the processing conditions, *i.e.*, the interface velocity and temperature gradient in the liquid which produced morphological instability were essentially the same as those when no flow was present. However, the interface instability in this case came as a set of traveling waves across the interface. Solutions admitted waves which traveled parallel or anti-parallel to the fluid flow.

Coriell, *et al*(4) considered the stability of a binary alloy solidified anti-parallel to the gravity vector, for the case where the rejected solute was less dense than the parent alloy, admitting the possibility of double diffusive instabilities. They found extended regions of instability, but almost no coupling between the morphological and double diffusive modes. Davis, *et al*(5) examined the coupling of the natural convection with a solidifying interface by considering an ice layer frozen from above, *i.e.* parallel to the gravity vector. The liquid-solid interface served as a record for the Rayleigh-Bénard cells present in the fluid.

Davis and Brattkus(6) considered a stagnation flow coming approaching the interface. They found traveling waves similar to those found by Delves, but they also found stationary waves which were not quite sinusoidal. The interface was distorted from an initially sinusoidal disturbance to one which became more like a sawtooth. The mechanism for this distortion was that the advancing flow compressed the solute boundary layer on the leading edge of the interface disturbance and expanded it behind the disturbance. This led to a higher concentration gradient on the front side of the interface disturbance than on the back, making the bump asymmetric.

There has been relatively less attention paid to the interaction of fluid flow with fully developed cellular and/or dendritic interfaces. Dantzig and Chao(7) considered the case of a shear flow perpendicular to an advancing cellular array. The physical model for this work is shown in Figure 2. Several dilute alloys of lead in tin were considered, under several different processing conditions. Assuming the process to be steady and using the following scales:

Length: λ

Velocity: $\dot{\gamma}\lambda$
Concentration: c_o
Temperature: T_o

reduces the governing equations to

$$\begin{aligned}
\left(\frac{\dot{\gamma}\lambda}{\nu}\right) \mathbf{u}^* \cdot \nabla^* \mathbf{u}^* &= -\nabla^* p^* + \nabla^{*^2} \mathbf{u}^* \\
\left(\frac{\dot{\gamma}\lambda}{D}\right) \mathbf{u}^* \cdot \nabla^* c &= \nabla^{*^2} c \\
\left(\frac{\dot{\gamma}\lambda}{\alpha}\right) \mathbf{u}^* \cdot \nabla^* \theta &= \nabla^{*^2} \theta
\end{aligned} \qquad [11]$$

Note that the buoyancy terms were neglected because the experimental configuration is stably stratified with respect to both thermal and solutal disturbances. For typical values of the shear rate, $\dot{\gamma}$ =10 s^{-1} and λ = 30 μm, the dimensionless premultipliers of the advective terms are

$$\begin{aligned}
\left(\frac{\dot{\gamma}\lambda}{\nu}\right) &= Re = 0.045 \\
\left(\frac{\dot{\gamma}\lambda}{D}\right) &= Pe_m = 15 \\
\left(\frac{\dot{\gamma}\lambda}{\alpha}\right) &= Pe_h = 0.0025
\end{aligned} \qquad [12]$$

This indicates that the effect of the flow on the thermal field can be neglected, but that the flow has an important effect on the solute field.

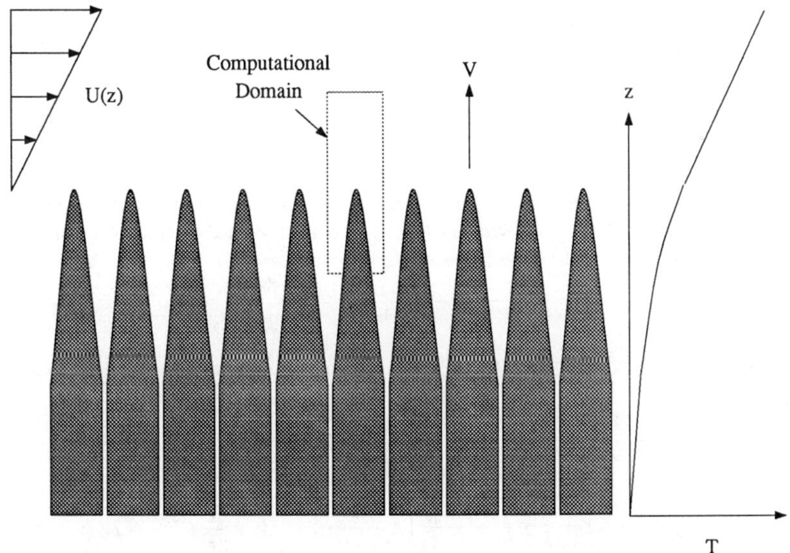

Figure 2: Schematic view of physical model of shear flow near the interface

The velocity and concentration fields were solved in the domain within ±3D/V of the cell tip using the finite element method, implemented in FIDAP(8). Owing to the symmetry of the interface, a single cell was modeled, using a finite element mesh such as that shown in Figure 3. The mesh consisted of 9x21x25 nodes, joined into 8–noded trilinear brick elements. See FIDAP manuals(8) or standard textbooks for details of the theory of the numerical methods.

The biggest difficulty in this problem is that the shape of the liquid-solid interface is not known *a priori*, so that the interface shape must be found as a consequence of the solution to the field problem for solute and temperature. The boundary conditions which must be satisfied at the interface were given in the previous section, Equations [7]-[10]. For a fixed linear temperature profile, as suggested from the small value of Pe_h, only two of these remain, Equations [7] and [8].

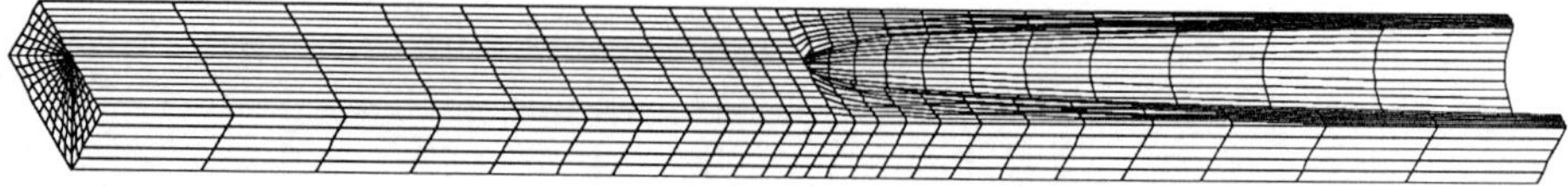

Figure 3: FEM mesh for the cell

There are several strategies available in the literature for constructing solutions to free boundary problems. Most of these involve first selecting a shape for the interface and constructing a solution using *one* of these boundary conditions, then updating the shape based on the error in satisfying the other boundary condition. In this work an approximate technique was used to select a shape for the cell. Bower, *et al* (9) considered directional solidification and noted that deep within the cells, the remaining liquid is in such a small area that it could be assumed to be completely mixed. For thermodynamic equilibrium to hold at the interface, the liquid composition must be $\frac{G}{m_l}$ deep within the cells. By assuming that this condition held all the way up to the cell tips, they were able to obtain an equation for the fraction solid as a function of alloy concentration and temperature.

For a linear temperature profile this reduces to

$$f_s = 1 - \left[\frac{(k_o - 1)(T_{base} + Gz - T_m) - ac_o}{c_o(k_o - 1 - ak_o)}\right]^{\frac{1}{k_o - 1}}$$
$$a = \frac{DG}{m_l V c_o} \qquad [13]$$

Considering a round cell inside a square array, the radius is related to the fraction solid by

$$r(z) = \lambda\sqrt{\frac{f_s(z)}{\pi}} \qquad [14]$$

where λ is the cell spacing. Combining Equations [13] and [14] yields a shape for the interface. This shape was then corrected to include the undercooling associated with curvature. Computing the mean curvature, K, as

$$K = \frac{1}{r(z)\left[1 + r'(z)^2\right]^{1/2}} + \frac{r''(z)}{\left[1 + r'(z)^2\right]^{3/2}} \qquad [15]$$

then combining the four preceding equations yields a nonlinear ordinary differential equation for the cell shape:

$$\frac{d^2r}{dz^2} = \frac{2}{G}\left[f(r) - Gz\right]\left[1 + \left(\frac{dr}{dz}\right)^2\right]^{3/2}$$
$$f(r) = m_l c_o \left[\frac{a}{k_o - 1} + \frac{1 - ak_o}{k_o - 1}\left(1 - \frac{\pi r^2}{\lambda^2}\right)^{k_o - 1}\right] - \frac{G}{2r} + T_m - T_{base} \qquad [16]$$

Integration of this equation by a Runge-Kutta scheme was performed using the unmodified cell shape far from the cell tip as a boundary condition. Example cell shapes computed in this way is shown in Figure 4, compared to an unmodified cell shape.

The boundary conditions on the computational domain were as follows:

Vertical boundaries: $\frac{\partial c}{\partial n} = 0$
Bottom: $\frac{\partial c}{\partial z} = \frac{G}{m_l}$
Top: $-D_l \frac{\partial c}{\partial z} = V(c - c_o)$

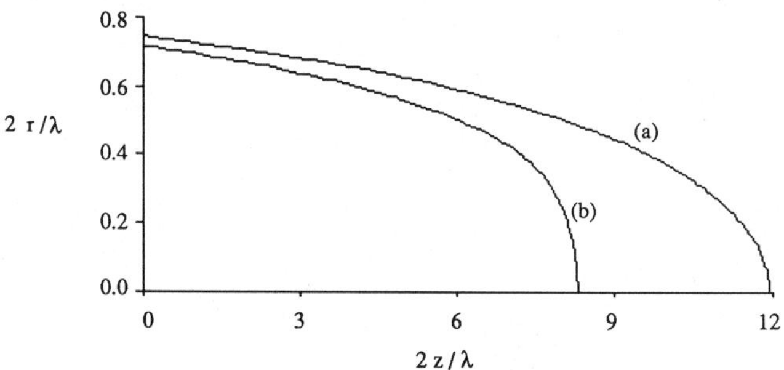

Figure 4: Cell shapes near the cell tip:
(a) Bower, *et al* (9) solution
(b) Corrected for curvature by Equation [16].

The boundary condition explicitly satisfied on the interface was that of local equilibrium, and the solute balance equation was used to update the cell shape and microstructure as described below.

In the absence of any shear flow, the concentration field will be symmetric with respect to the center-line of the cell, and the lateral sides will be symmetry planes. When the shear flow exists, solute is convected from the leading edge, around the cell and appears on the trailing edge. This leads to asymmetry in the concentration field, and in particular makes the concentration gradient on the leading edge greater than on the trailing edge. The solute balance equation for the interface tells us, then, that the normal velocity on the front face of the cell must be greater than on the trailing face, *i.e.* the cell translates into the flow. The resulting microstructure will consist of cell tilted toward the flow direction, with an inclination angle given by

$$\psi = \tan^{-1}\frac{U_{tr}}{V} \qquad [17]$$

The translation velocity, U_{tr}, was applied as a boundary condition on the cell. A series of processing conditions was examined using dilute alloys of 0.5 and 1.0 w/o Pb in Sn. Detailed results are included for just one case to demonstrate the phenomena, and the remainder are summarized in Figure 5.

Note that in steady growth, the lateral sides must be *both* periodic and no-flux boundaries,so the no-flux boundary condition was used to reduce the computational load. The procedure used to determine ψ was to use trial values until a concentration field was found that was again symmetric with respect to the cell centerline. This was done by integrating the solute flux over the entire leading edge and over the entire trailing edge and comparing them. If the flux on the leading edge exceeded that on the trailing edge, the angle was increased, while if the flux on the leading edge was smaller than on the trailing edge, the angle was decreased.

The results are summarized in Figure 5. The trends in the data can be interpreted as showing that the primary indicator for the angle is the "openness" of the structure. For fixed concentration, increasing the temperature gradient or the interface velocity results in a less porous structure, and thus smaller inclination angles. Similarly, increasing the concentration of the alloy produces less solidification per unit volume for the same processing conditions, so that these alloys show larger inclinations. This interpretation was integrated into the representation in Figure 5 by correlating the inclination angle with the product $\lambda c_o^{0.25}$ which is a measure of the openness of the structure. There is no special significance to the exponent 0.25.

The broadest conclusion from this study is that fluid flow can affect the solidification microstructure, but that it does so mainly by altering the solute fields near the interface. For this reason, it is important that the causes of fluid flow be well understood and controlled. Most of these flows originate on the macroscopic scale, and the next section will consider these problems.

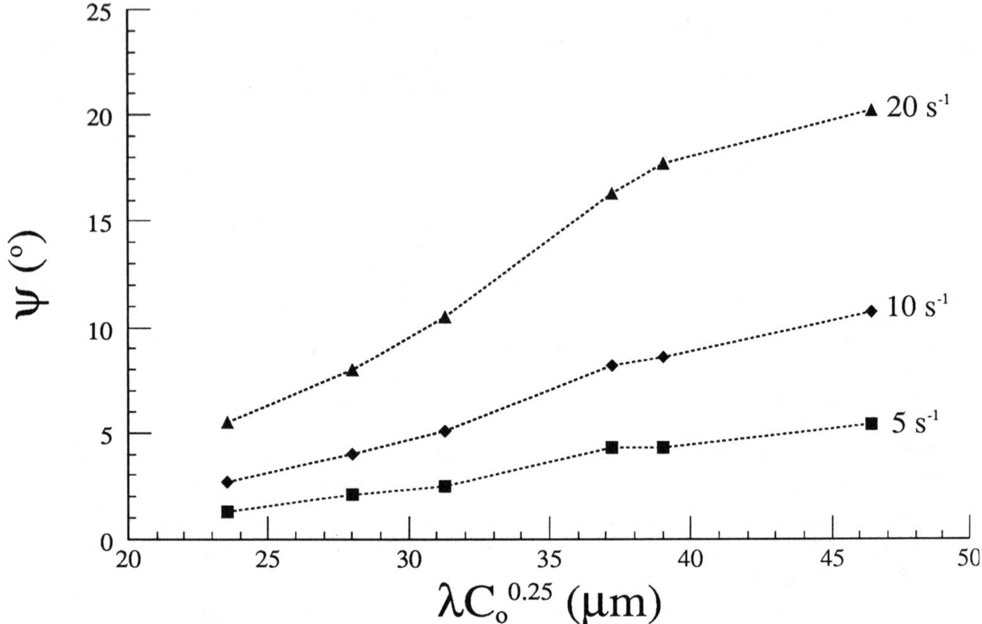

Figure 5: Computed inclination angle of advancing cells for various processing conditions.

Macroscopic Scale

Having shown that the microstructure can be strongly affected by flow in the melt, we will now consider means for determining and controlling buoyant flows during solidification. Note that the reduced gravity to be expected under space processing conditions is not sufficient to completely eliminate this convection. As in the previous sections, we will consider Bridgman growth experiments, this time for GaAs crystals.

To properly analyze samples produced in directional solidification, the investigator must know the existing temperature, solute and velocity fields throughout the sample during the entire experiment. What is usually measured, however, is a relatively small number of temperature histories on thermocouples, most of which reside in the furnace or on the outside of the ampoule. There is really no way to avoid this, because embedding large numbers of thermocouples within the samples would undoubtedly contaminate the sample and overly influence the experiment, obscuring the effects that the experiment was intended to observe. Thus the investigator must try to infer from the limited amount of recorded data, and the sample, just what actually occurred in the experiment. This arrangement usually provides far less information than is needed to properly interpret the resulting sample. Perhaps more important is the fact that because there is no sufficiently accurate means for predicting what the experimental conditions will be before performing the experiment, it is difficult to properly design the experiment so that it will be performed under the conditions which the experimenter wishes to investigate.

In our early simulations of the crystal growth process, we found that the liquid-solid interface was highly curved. The curvature of the interface induces natural convection in the melt. Mixing of the fluid by the convection causes segregation. Since the interface is often an isoconcentrate, this situation causes radial segregation in the final crystal. The following analysis demonstrates that the primary cause of non-planarity of the interface is the difference in thermal conductivity of the solid and liquid GaAs, and this fact also suggests means for improving the planarity of the interface by altering the applied furnace temperature distribution. Under steady-state conditions, the energy balance at the liquid-solid interface given in Equation [10] becomes

$$k_s \nabla T_s \cdot \mathbf{n} - k_l \nabla T_l \cdot \mathbf{n} = 0 \qquad [18]$$

When the interface is flat and perpendicular to the gravity vector in our models

$$\mathbf{n} = \hat{\mathbf{z}} \qquad [19]$$

where $\hat{\mathbf{z}}$ is the unit vector along the crystal growth axis. Consider first the simplified model shown in Figure 6, with insulated ends and a temperature profile specified on the outer diameter. For a constant temperature gradient imposed on the boundary, the temperature distribution near the interface can be expressed as

$$T_s = T_l = T_{ref} + Gz \qquad [20]$$

where G is the temperature gradient on the boundary, then combining Equations [18]-[20] yields

$$k_s G - k_l G = 0 \qquad [21]$$

which implies that

$$k_s = k_l \qquad [22]$$

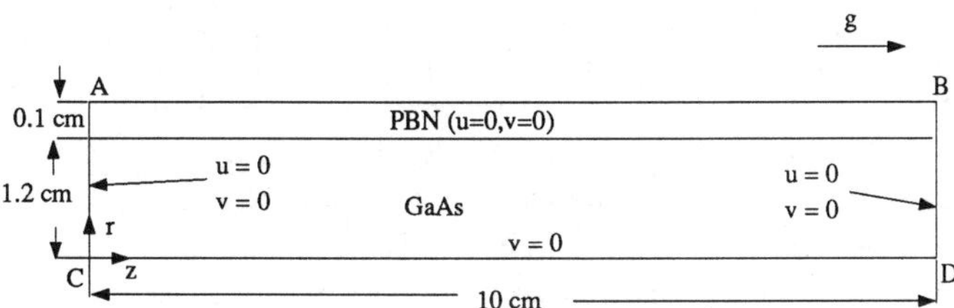

Figure 6: Simplified model of growth ampoule with temperature distribution imposed on the boundary

However, in GaAs, k_l is not equal to k_s, so that Equations [18]-[20] cannot be satisfied simultaneously. If Equation [20] is imposed as a processing condition, and the material ensures that Equation [18] is satisfied, then the radial component of the unit normal vector of the interface cannot be zero and the interface is not flat. If we wish to have a flat interface when k_l is not equal to k_s, by combining Equations [18] and [20], the energy balance at the interface becomes

$$k_s G_s = k_l G_l \qquad [23]$$

with G_s not equal to G_l. This suggests that the interface could be flattened by applying a bilinear temperature profile with gradients satisfying Equation [23] and joined at the melting temperature on the boundary of the crystal.

Because of the presence of the PBN container, there is a small difference between the temperature distribution near the interface and that applied on the boundary. Trial-and-error was used to obtain the temperature profiles on the boundary leading to a maximum out-of-plane error in the interface less than 1.0 x 10^{-3}cm. In a typical application, either G_s or G_l was fixed, while changing the other one to achieve a flat interface.

In all of the calculations which follow, G_l was set to be 5 K/cm and G_s was varied to achieve a flat interface. The dimensionless parameter, R, is defined as the ratio of the two gradients. From Equation [23], an estimate for R which would yield a flat interface is

$$R \approx \frac{G_s}{G_l} = \frac{k_l}{k_s} = 2.06 \qquad [24]$$

The results are summarized in Figure 7, where three trial temperature gradient ratios were applied on the boundary producing these steady temperature solutions. These results were obtained considering conduction only. The figure shows that for R = 1 (constant gradient in both phases) the interface has strong concave curvature. When $R = \frac{k_l}{k_s}$ it can be seen that the interface shape has been overcorrected, and is significantly convex. By using trial-and-error methods to vary R, a flat interface was obtained when R = 1.91.

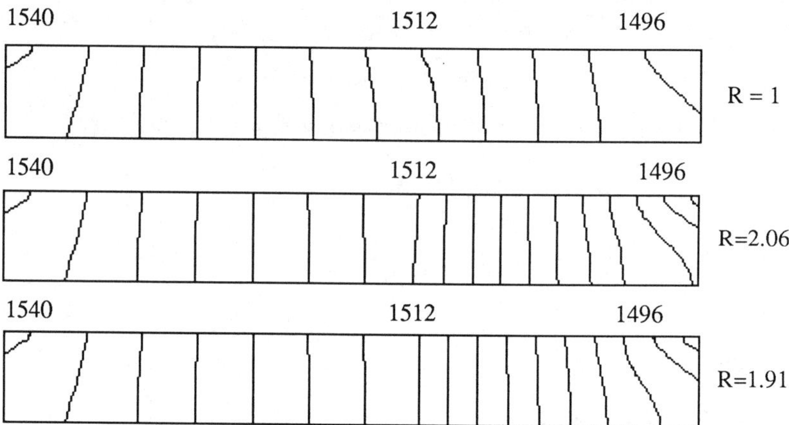

Figure 7: Steady state temperature distributions computed for various trial imposed temperature distributions (G_l = 5 K/cm)

Figure 8 shows the results from an analysis of a similar problem, but now also including natural convection (g = 0.0981 m/sec^2). When R = 1, there are two circulations, but the circulation near the interface disappears for the flat interface. Note that convection is eliminated only near the interface, while the end effect of the insulated boundary still induces some buoyant flow.

Figure 9 shows a schematic of the real apparatus used in the experiments to grow GaAs in space. A radiative heat transfer boundary condition, instead of an imposed temperature, applies on the ampoule boundary facing the furnace wall. This was implemented in FIDAP by modifying the radiation heat transfer element so that for any time and for any location on the boundary, the reference temperature could be calculated from measured temperature data. All the other boundaries were treated as though they were insulated. The temperature contours and streamlines of the steady solutions for R=1 and R=2.67 are shown in the Figure 10, in which the inner rectangle is the boundary of the GaAs. The same streamlines are plotted in both figures. When R = 1, the interface is not flat and there is a strong clockwise circulation. When R = 2.67, somewhat above the value predicted on the basis of the thermal conductivity ratios, the interface is flat and there is a slightly weaker circulation. Note that the primary reason for the difference between the value of R to obtain the flat interface for the earlier cases and that for the present one is not the different boundary conditions for temperature; rather it is the different surrounding materials and geometry, which cause different boundary effects, especially at the ends.

Figure 11 shows the value of R needed to produce a flat interface under steady state conditions at different locations in the crystal. The range of z-coordinate corresponding to GaAs is 4.4 cm to 13.2 cm. Because of end effects, it was not possible to find a value of R to produce a flat interface when the interface was near either end. From the figure, it can be seen that increasing **g** necessitates a larger value for R to flatten the interface. This is because larger gravitational accelerations induce stronger convective

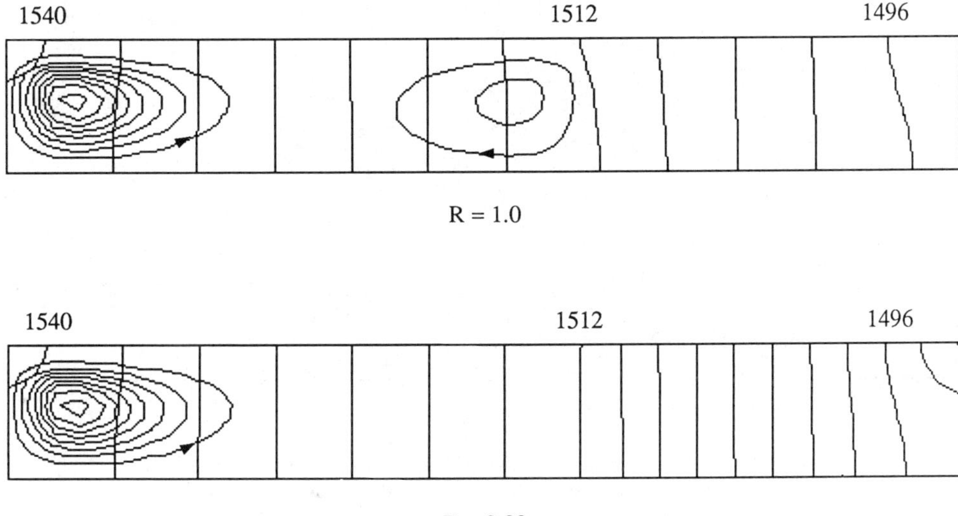

Figure 8: Computed results for idealized geometry when convection is included in the calculation (G_l = 5 K/cm, g = 0.0981 cm/s^2)

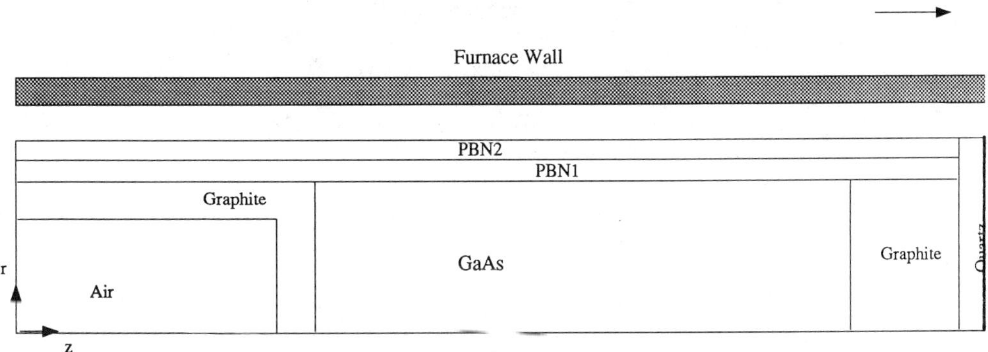

Figure 9: Schematic view of the model for the experiment used to grow GaAs by GTE

heat transfer in the liquid, which in turn induce larger values for G_l in Equation [24], therefore a larger value of R is needed to compensate.

Consider next the case where the crystal is being solidified over time. The initial condition for the simulation was taken to be the steady state solution, with the radiation reference temperatures given by those at time equal to zero, corresponding to approximately half of the GaAs being molten. The temperature profile on the furnace was then slid up the crystal to simulate transient growth.

Solutions for this model using a single temperature gradient, *i.e.* R=1, are illustrated in Figure 12, showing the computed streamlines and successive interface positions under reduced gravity (g = 9.8 $\times$ 10^{-4} m/s^2). It is clear that the interface is far from flat, and this situation will lead to radial segregation on slices cut perpendicular to the growth axis. To extend the earlier steady state analysis to transient problems, the heat balance at the interface is that given in Equation [10]. If the interface is flat, using

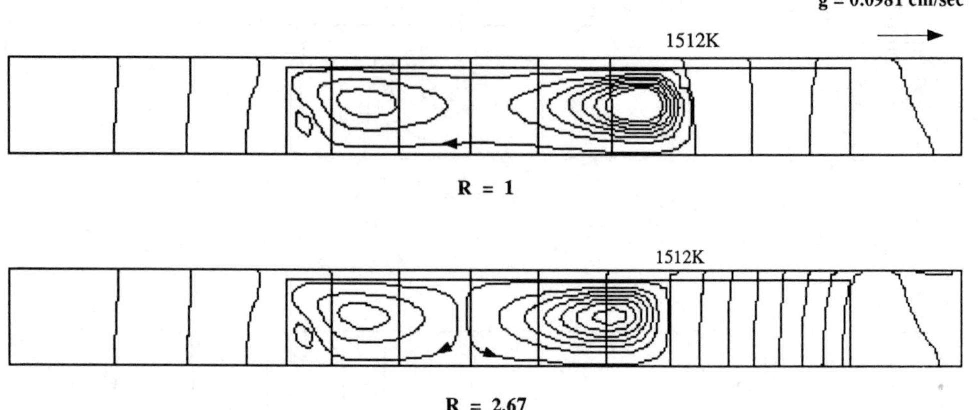

Figure 10: Computed isotherms and streamlines for GaAs in the GTE space experiment apparatus. Note that the maximum absolute value of the stream function was 1.7×10^{-7}.

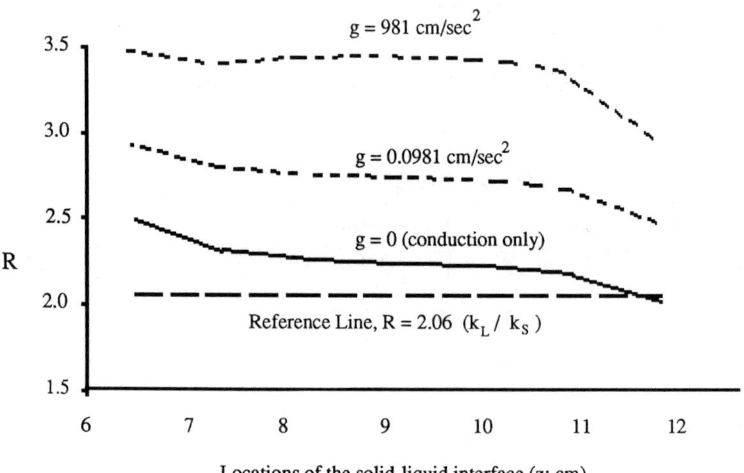

Figure 11: Values of R required to produce a flat interface under steady state conditions at various locations in the crystal and under several different gravitational accelerations

Equation [19] yields

$$k_s G_s = k_l G_l + \rho L_f |\mathbf{V}| \qquad [25]$$

In a similar manner to that described in the preceding paragraphs, we may use this equation to adjust the temperature profile on the boundary to obtain a flat interface. The estimated value for R corresponding to the flat interface for a given G_l is

$$R \approx \frac{k_l}{k_s} + \frac{\rho |\mathbf{V}| L_f}{k_s G_l} \qquad [26]$$

The first term on the right hand side of this equation corresponds to steady state while the second represents the heat liberated by the moving interface. Simple trial-and-error methods for calculating R to produce a flat interface in transient problems proved to be difficult. The best results were obtained using

$$R = R_{steady} + \frac{\rho |\mathbf{V}| L_f}{k_s G_l} \quad [27]$$

Note that if $|\mathbf{V}|$ is 10 μm/sec and G_l is 5 K/cm, the second term on the right hand side is 11.66, compared to R_{steady}, which was 2.67. Figure 13 shows that this method improves the shape of the moving interface very significantly. However, the interface near the initial point (between A and B in Figure 13–b) was not very flat. The initial condition used was the steady solution with the heat source at the initial location, so R = 2.67. As soon as the heat source leaves the initial point, R jumps up to 14.33. This inconsistency caused some curvature in the interface near the starting point. (See Figure 13–b.)

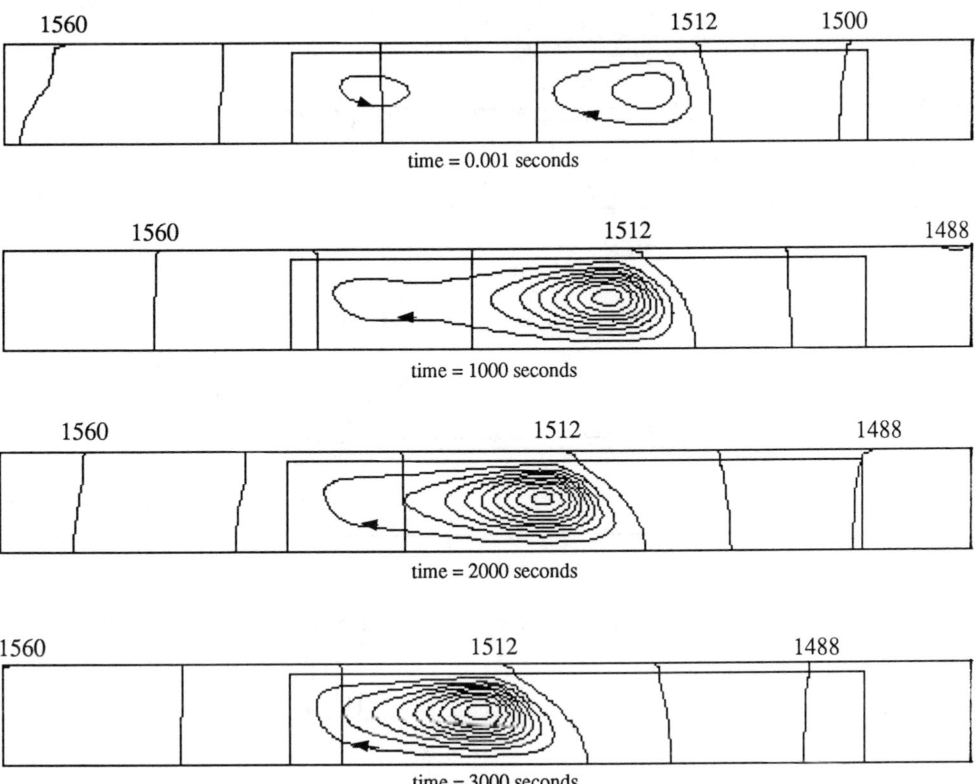

Figure 12: Streamlines and successive interface positions for GaAs crystal growth under 10^{-4} X earth gravity.

This situation was improved by ramping the speed of the heat source from zero to 10 μm/sec (with constant acceleration) in the beginning of processing. During this period, the value of R was changed according to Equation [26]. The corresponding result (Figure 14–b) is much improved over the result with a sudden change in speed (Figure 14–a). The final results are illustrated in Figure 15. The streamlines in this figure and Figure 12 correspond to the same values of the stream function, so it can be seen that application of this method not only flattens the interface considerably, but also decreases the intensity of the natural convection near the interface. This model could also be applied to processing on earth. Figure 16 shows the result for g = 981 cm/s^2. The resulting interface could be made as flat as in 10^{-4} g_e, but there remains significantly more convection. Please note that these streamlines are *different* from those used in the previous figure. At 981 cm/s^2, the velocities were approximately 1000 times larger than those computed for 10^{-4} g_e.

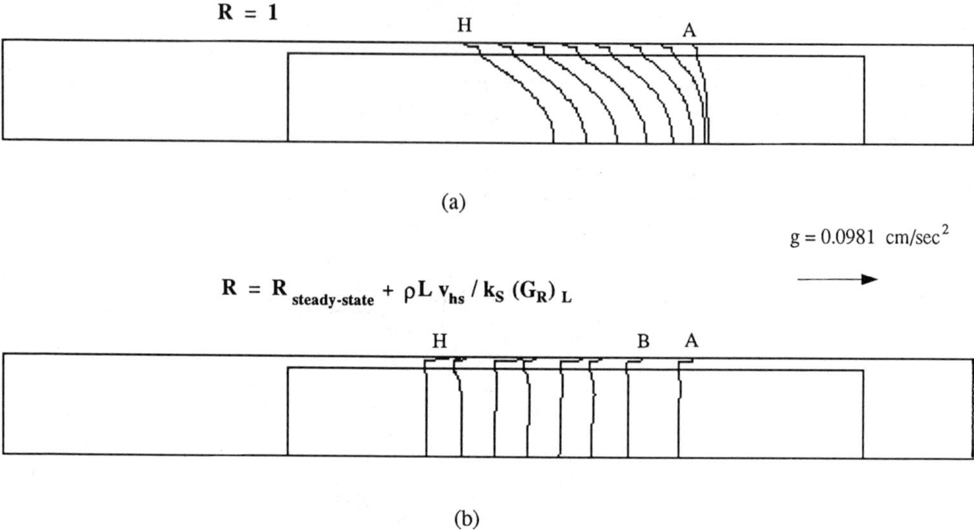

Figure 13: Computed locations of liquid-solid interface and streamlines during transient simulation

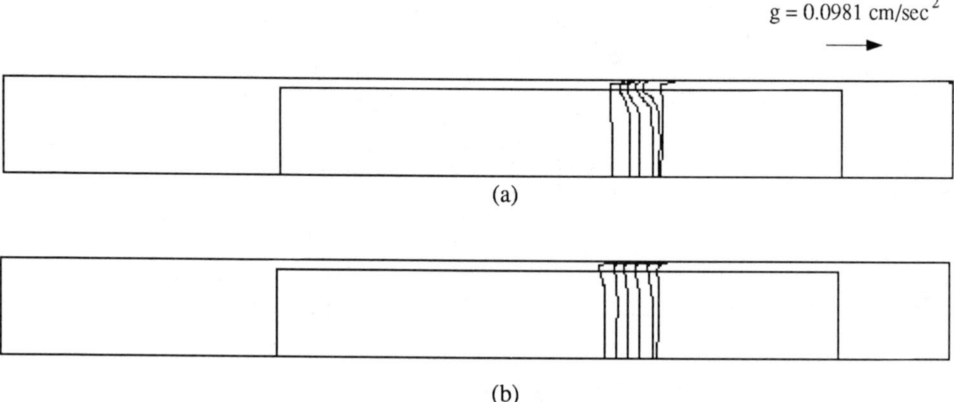

Figure 14: Computed locations of liquid-solid interface and streamlines in the early stages of transient simulation using a ramp-up in speed. (a) Interface positions with no ramp-up. (b) Interface positions with ramp-up.

The most important conclusion from this work is that a model considering fluid flow and heat transfer can be used to specify processing conditions for controlling fluid flow and solidification near the liquid-solid interface in GaAs Bridgman growth. The model is based on the energy balance at the liquid-solid interface, and uses this condition to force the radial component of temperature gradient at the interface to be zero by adjusting the temperature profile of the furnace.

Discussion

The results of the preceding section should be interpreted in the broader context of the importance of fluid flow in the development of solidification structures, as well as the role that analysis can play in studying these phenomena. Fluid flow plays an important part in determining the solute field, and thus

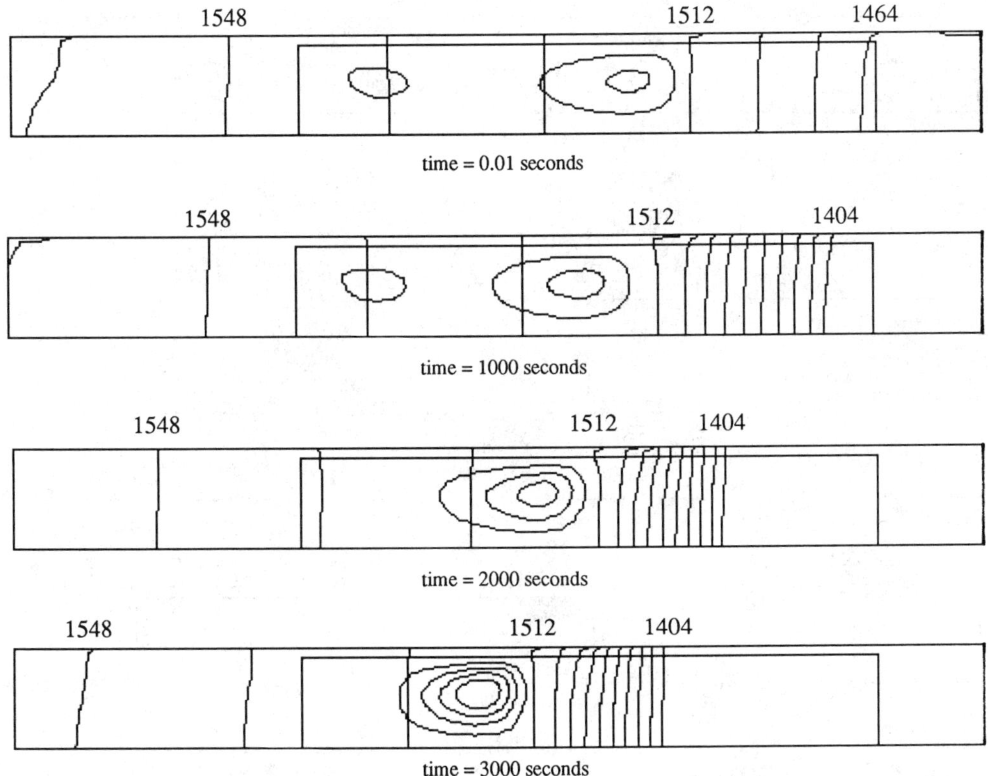

Figure 15: Streamlines and successive interface positions for GaAs crystal grown at $10^{-4}g_e$ using controlled temperature profiles.

the final properties of the solidified material. Depending on the objectives of the experiment, fluid flow may be a help or a hindrance.

The importance of the analytical methods is that they allow experimenters to separate the phenomena into diffusive and convective effects. They also allow one to better design and interpret experiments under a variety of processing conditions. It was readily apparent in the second example that significant convection can be observed even at $10^{-4}g_e$ if care is not taken to carefully control the experimental conditions. At the same time, the analysis showed that if the goal of the crystal grower was simply to achieve a flat growth interface, this could be done on earth.

Acknowledgment

The authors would like to thank the National Science Foundation, which supported the work on cellular microstructures under Grant DMR 83–03917, NASA, which supported the work on furnace modeling under Grants NAGW-1683 and NAG 3–947, and the National Center for Supercomputing Applications at the University of Illinois for grants of computer time for both problems.

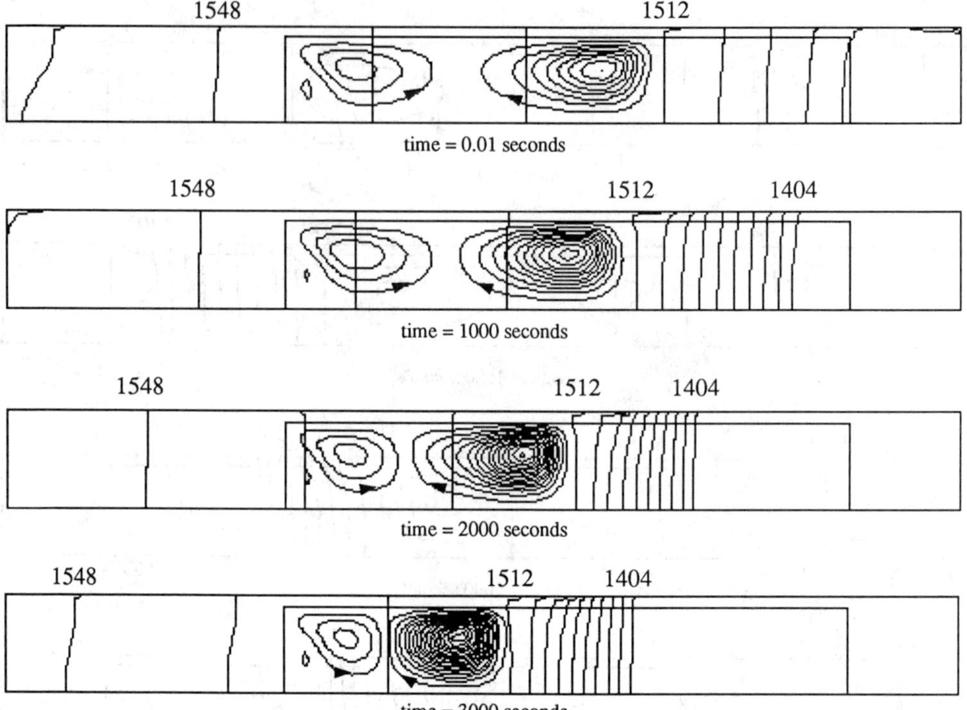

Figure 16: Streamlines and successive interface positions for GaAs crystal grown at $1g_e$ using controlled temperature profiles.

References

[1] M. E. Glicksman, S. R. Coriell, and G. B. McFadden. Interaction of Flows with the Crystal-Melt Interface. *Ann. Rev. Fluid Mech.*, 18:307–335, 1986.

[2] W. W. Mullins and R. F. Sekerka. Morphological Stability of a Particle Growing by Diffusion or Heat Flow. *J. Appl. Physics*, 34:323–329, 1963.

[3] R. T. Delves. In B. R. Pamplin, editor, *Crystal Growth*, pages 40–103. Pergamon, 1974.

[4] S. R. Coriell, F. S. Biancanello, and W. J. Boettinger. Solutal Convection Induced Macrosegregation and the Dendrite to Composite Transition in Off-Eutectic Alloys. *Met. Trans.*, 12A:321–327, 1981.

[5] S. H. Davis, U. Müller, and C. Dietsche. Pattern Selection in Single-Component Systems Coupling Benard Convection and Solidification. *J. Fluid Mech.*, 144:133–151, 1984.

[6] K. Brattkus and S. H. Davis. Flow Induced Morphological Instabilities: Stagnation Point Flows. *J. Crystal Growth*, 89:423, 1988.

[7] J. A. Dantzig and L. S. Chao. The Effect of Shear Flows on Solidification Microstructure. In J. P. Lamb, editor, *Proc. Tenth U. S. Natl. Cong. Appl. Mech.*, pages 249–255. ASME, 1986.

[8] M. S. Engelman. *FIDAP Theoretical Manual.* Fluid Dynamics International, Evanston, IL, 1987.

[9] T. F. Bower, H. D. Brody, and M. C. Flemings. Measurements of Solute Redistribution in Dendritic Solidification. *Trans. AIME*, 236:624, 1966.

On understanding cellular, dentritic and Widmanstätten patterns

J.S. Kirkaldy
Institute for Materials Research, McMaster University, Hamilton, Ontario, Canada L8S 4M1

Abstract

A retrospective and current review on theories of dendrites is presented followed by a parallel treatment of solidification cells. The free dendrite problem is first defined and the Ivantsov free boundary model is introduced. An equivalent approximate derivation of the paraboloidal shape solution due to Hillert is outlined. The extensions of Horvay and Cahn, Trivedi, Nash and Glicksman are described, the last-named in relation to Zener's maximum velocity criterion. The alternate marginal stability criterion of Langer and Muller-Krumbhaar is introduced and its successful application to succinonitrile thermal dendrites by Huang and Glicksman is described. Our own modification of this approach is applied to data on succinonitrile (SN), Xe and Kr with salutary results. A brief summary of the more recent solvability approach of Langer and others is presented and its possible bearing on the dendrite to Widmanstätten transition is discussed.

The current observations on thin film cells are reviewed. The linear stability theory of Mullins and Sekerka and the non-linear theory of Wollkind and Segel are referenced. These are examined in the light of Fourier analysis and a coefficient counting argument and found to possess underdetermination and convergence problems. Some of the conclusions are accordingly in doubt. Our own complete analysis favours the constitutional supercooling criterion with an oversaturation, finite amplitude onset of cells via nucleation as currently observed. Analytic scaling relations for stable single-valued deep-rooted cells based on Zener-Hillert arguments and dissipation optimization, which is favoured by Hamilton's principle in classical mechanics, are successfully tested against the data for succinonitrile-salol (S-S). By contrast, current computer simulations are non-quantitative. We synthesize the offering with a discussion of cellular dendrites.

Introduction

Cells, dendrites and Widmanstätten figures are all variants of a diffusion-controlled first-order phase transition from a supersaturated solution and as a consequence possess a morphological commonality. Figs. 1 and 2 offer some comparative microstructures (1-3). The prevalence of a cooperative branching mode and the strong involvement of anisotropy of surface tension stand out. On the theoretical side they are all self-organizing systems which share an asymptotic steady state free boundary degeneracy. For dendrites and Widmanstätten figures, the process driving the supersaturation can be applied identically so there exists an identical dimensionless theoretical formulation for the two processes. Although steady state features generally appear asymptotically, the configurations are inherently non-steady. This means that morphological differences can only arise from differences in initial conditions and from differing strengths of surface tension anisotropy and interface kinetics (if applicable). Furthermore, the degeneracy is removed if one computationally integrates up from a realistic initial condition.

The cell problem differs in that the front velocity is forced at a true steady state whereby the supersaturation arises indirectly in a morphology controlled configuration. This acts to generate a constrained steady state inverse Widmanstätten reaction at the cell root controlled by capillarity and solid state diffusion (4; Fig. 3). Because the thermal length m/G (where m is the absolute value of the liquidus slope and G is the temperature gradient) is added to the others the scaling laws must reflect this. It is particularly helpful to a synthesis that deep-rooted cells possess Widmanstätten features at the root (Fig. 2) and dendritic features near the tips (Fig. 4).

We begin our exposition with a treatment of the irreversible thermodynamics of surface reactions which informs us in a general way as to why anisotropy and interface kinetics are important considerations in problems of this type. Dendritic and Widmanstätten theory will then be reviewed, followed independently by the theory of thin film cell growth. A synthesis will then be undertaken through the vehicle of cellular dendrites.

Thermodynamics and Physics of the Boundary Value Problem

We consider a binary alloy and write the Gibbs-Duhem equations for the respective L and S sides of a generally moving interface assuming pressure of surface tension on the S phase only

$$dA^L = X_1^L \, d\mu_1^L + X_2^L \, d\mu_2^I = -S^L \, dT \tag{1}$$

and

$$dA^S = X_1^S \, d\mu_1^S + X_2^S \, d\mu_2^S = V^S \, dP - S^S \, dT \tag{2}$$

For the thermal dendrite case, the left-hand sides are differentials of the molar Gibbs Free Energy μ of the pure system. The differential equation which relates to the kinetics is

$$d(A^L - A^S) = -V^S \, dP - (S^L - S^S) \, dT \tag{3}$$

where at constant L and S compositions the left hand side is proportional to the differential of de Donder's affinity A introduced by Prigogine (5) and is in general not an exact differential since

$$\frac{\partial V^S}{\partial T} \neq \frac{\partial (S^L - S^S)}{\partial P} \tag{4}$$

Noting that the dissipation (or entropy production rate) is A multiplied by the reaction rate or velocity, and fixing the compositions, we can integrate Eq. (3) up from the planar state of thermal equilibrium ($T = T_m$) and zero local entropy production rate where

$$A^L - A^S = A = 0 \tag{5}$$

to yield

$$A = -V^S \sigma K - \ell(T - T_m)/T_m \tag{6}$$

where σ is the surface tension, K is the curvature, ℓ is the molar latent heat and T_m is the melting point of the alloy. Note that $A = 0$ does not necessarily imply chemical equilibrium. This is not even necessarily the case for pure material ($A_L = \mu_L = \mu_S = A_S$), since the terms on the right of Eq. (6) are normally of opposite sign, so can cancel with $T \neq T_m$.

We now introduce the kinetics for a dilute alloy via the linear or Fickian interface reaction equation

$$v_n \rho = MA/\delta \tag{7}$$

where v_n is the normal velocity, ρ is the mass density, M is the single independent mobility and δ is the thickness of the Gibbsian interface phase. Combining Eqs. (6) and (7) we obtain the expression for the undercooling with drives the surface reaction

$$T_m - T = \frac{T_m}{\ell} [V^S \sigma K + v_n \rho \, \delta/M] \tag{8}$$

We must now recognize the existence of a subsidiary free boundary problem (compare Eq. 12) since anisotropic σ and M, through the shared effect of surface roughness, are mutually monotone functions and this monotonicity we represent formally by

$$\sigma = \eta M^N; \qquad \eta, N = \text{constants} \tag{9}$$

In the simplest atomic model both σ and M would be proportional to the number of dangling bonds so $N \rightarrow 1$. In lieu of a transparent form for the dissipation function we choose to hypothesize the trend to minimize the undercooling as a surrogate, yielding the stable values at the minimum

$$M = [\rho \, \delta v_n / V^S K \eta \, N]^{1/(1+N)} \quad \text{or} \quad \sigma = [\rho \, \delta v_n / V^S K \eta \, N]^{N/(1+N)} \tag{10}$$

Substituting the optimal M in Eq. (8) and re-evaluating Eqs. (6) and (7), we obtain

$$A = N V^S \sigma K = v_n \rho \, \delta / M \tag{11}$$

This indicates that for the particular integral of Eq. (3) chosen that irrespective of the detailed relation between M and σ the interface velocity tends to vanish with the curvature at the optimum. Furthermore, the interface dissipation (through A) and the undercooling (cf. Eq. (8)) vanish together at an absolute minimum in a planar ($K = 0$) orientation even though the alloy system is not in chemical equilibrium at the planar wall. This means that where anisotropy is strong enough through local free energy minimization to abet facetting (reduced σ and M; $K \rightarrow 0$) the interface velocity will tend to vanish. These considerations bear on the immobile facetting observed in Figs. 1 and 2 and upon the perceived role of anisotropy in the more advanced models of dendrite stability. We shall also be able to use them as the basis for quantifying the relative importance of surface kinetics.

Free Dendrites and Widmanstätten Figures

In 1946, Zener (6) addressed the problem of growth of individual Widmanstätten plates in steel (Figs. 1 and 2). With the explicit inclusion of the Gibbs-Thomson capillarity effect in the constitutional boundary conditions he set the requirement for all subsequent quantitative studies of fine scale precipitates in solids and liquids. Zener's dimensional arguments were placed on a more rigorous basis by Hillert (7) in 1957. A different and incomplete train of thought was initiated by Ivantsov (8) in 1947 which demonstrated that in paraboloidal coordinates there exist shape-preserving solutions of the diffusion equation under the assumption of constant concentration or temperature at the interface between solid and liquid (thus excluding capillarity effects). In 1960, Temkin (9) expressed diffusion (but not shape) corrections to the Ivantsov theory due to capillarity and interface drag and noted how to include chemical dendrites rigorously within the thermal dendrite theory. Horvay and Cahn (10) in 1961 reported further shape-preserving parabolic solutions in the absence of capillarity which subsumes plates as well as needles. Trivedi (11) expanded the scope of the Temkin theory in 1970.

Hillert's treatment of plates and needles with the inclusion of capillarity is the most transparent and quite sufficient to an experimental test where interface drag and dendrite side-branching are not factors. He solved the Fick differential equations in planar and cylindrical coordinates with correctly formulated interface thermodynamics including capillarity, obtaining in his first approximation the Zener tip velocity for plates of tip radius r

$$v = \frac{D}{a}\,\frac{\Delta C}{\Delta C_0}\cdot\frac{1}{r}\left(1-\frac{r_c}{r}\right); \qquad r_c \simeq \frac{\sigma}{RT\,\Delta C} \tag{12}$$

where ΔC is the driving concentration difference in moles/unit volume, $\Delta C/\Delta C_0$ is the relative supersaturation, r_c is the cutoff or nucleation radius and a = 2. The generic problem of non-uniqueness or degeneracy recognized by Ivantsov and Zener is here evident. In explanation of the degeneracy, Nash and Glicksman (12) remark, and we agree, that "the steady-state treatment is in fact an attempt to determine the long time asymptotic solution to a time-dependent problem without ascertaining the entire time-history of the growth process ... a price is paid in the form of lost information ...". Zener conjectured that nature chooses the state of maximum growth rate, which Nash and Glicksman identify as surrogate to the lost information. Until recently most workers including Hillert have accepted this plausible conjecture as near to the truth. Hillert's application to plates yielded an optimal velocity (7)

$$v_m \simeq \frac{DRT}{12\sigma}\,\frac{(\Delta C)^2}{\Delta C_0} \tag{13}$$

and near parabolic shape near the tip

$$u_0 \simeq -\frac{3\pi D}{16\,v_m r_0^2}\,y^2; \qquad r_0^2 \simeq \left(\frac{3\sqrt{\pi}\,\sigma}{RT}\right)^2 \frac{\Delta C_0}{(\Delta C)^3} \tag{14}$$

and similar relations, including a paraboloidal tip shape were offered for needles.

Townsend and Kirkaldy (1) successfully tested this Zener-Hillert theory against their velocity observations in Fe-C for Widmanstätten plate growth without side-branching (Fig. 5). Although there are no systematic observations of tip shapes the predicted scale was found to be of the right magnitude ($\sim 10^{-2}$μm). As we shall see the application of variants of this theory to the thermal analogue, which always involves side-branching, has been less successful, and this dichotomy will have to be explained.

To underline the last point, we refer the reader to Fig. 6 abridged from Langer (13). The so-called "spherical approximation" is the same as that of Hillert for a paraboloidal needle. In contrast to the Ivantsov solution (8) it subsumes capillarity, and due to the approximation lies a factor of 2 above Sekerka et al's version of the Ivantsov theory modified for capillarity (14). The Trivedi result mentioned above is similar to that of Temkin (9) while the rigorous computational treatment of Nash and Glicksman, which allows the shape to deviate from paraboloidal, lies above the Temkin result. Nash and Glicksman have emphasized the failure of the maximum velocity criterion applied to all theories interpreted for thermal dendrites. Furthermore, they make clear that the errors are in the direction of very low curvatures where the direct capillarity effect tends to be negligible. Significantly, the Ivantsov and Hillert continuum of solutions remains intact within the approximations and uncertainties (Fig. 6).

Solidification specialists were apparently uninformed about the successful Widmanstätten result so they were not led automatically to the possibility that side-branching interactions associated with weak anisotropy might be the source of the discrepancies. Notwithstanding, Oldfield (15), and later, Langer and Muller-Krumbhaar (LM; 16) were led to a viable resolution, albeit on the basis of a premise which attempts to discount branch-tip interactions. This is now known as the marginality hypothesis pertaining to the application of a formula from the successful stability analysis of a growing sphere in a supersaturated environment due to Mullins and Sekerka (MS, 17). It is eminently reasonable that one should undertake to directly transform the theory of destabilization of the spherical set of shape-preserving solutions to an equivalent one for the destabilization of the Ivantsov or Hillert paraboloidal set of shape-preserving solutions. While the LM procedure, modified by Huang and Glicksman (18) has proven to be an excellent predictor of all the scaling relations for SN over wide ranges of the undercooling, it in our view misrepresents for apparent philosophical reasons the true nature of the transformation as just expounded. We here present an alternative construction which recognizes the morphological match between thermal dendrites and Widmanstätten figures (Fig. 1) and the successful theory of the latter phenomenon given by Townsend and Kirkaldy (1). This theoretical construction traded upon the same quasi-steady MS theory with salutary experimental comparisons (Fig. 7) but also qualitatively incorporated some important time-dependent calculations of Sekerka (19) which explicitly demonstrated selection of waves to favour the fastest growing wavenumber (Fig. 8). This suggested the model of the dendrite-like Widmanstätten manifold of Fig. 9 and implies that in understanding a dendrite causally one should focus not on the tip in a moving frame, but on an aged deep branch in the laboratory frame and the side-bands which it has continuously projected outward in time, the tip in the liquid being merely the current limit of the side-bands (Fig. 9) and an instantaneous frontal wave maximum.

To make this model semi-quantitative, in accord with the foregoing, we impose a symmetric two-dimensional, decreasing amplitude wave array of fastest growing wavelength (17; C_p is heat capacity, D_{th} is thermal diffusivity, L is latent heat per unit volume)

$$\lambda_m = 2\pi\sqrt{6T_m \sigma C_p D_{th} / L^2 v_n} \tag{15}$$

on the surface of a cylinder near the tip, and to incorporate four-fold anisotropy we require that for a cylindrical, tip trailing radius r_c

$$4\lambda_m = 2\pi r_c \tag{16}$$

The normal velocity of the cylindrical surface is given by the Laplacian or Hillert approximation

$$v_{nc} \simeq \frac{1}{2} \frac{D_{th}}{r_c} \frac{\Delta T}{T_u} \tag{17}$$

where D_{th} is the thermal diffusivity and T_u is the unit undercooling. Since the tip represents only a partial wave, we set

$$r_t = \lambda_m/3 \tag{18}$$

The normal velocity at the tip to the same approximation as Eq. (17) is

$$v_{nt} \simeq \frac{D_{th}}{r_t} \frac{\Delta T}{T_u} \tag{19}$$

Combining Eq. (15)–(18) consistently we obtain the scaling laws of Table 1 based on the data of Table 2 and the salutary low undercooling comparisons with the experiments on SN, Kr and Xe (2, 18, 23). The magnitudes are in excellent agreement with the predictions, but the scaling law exponents are imperfect due to the imperfections of the experiments, approximations and the likely involvement of convection and (or) interface kinetics. If we combine the hemispherical velocity (19) with Eqs. (7) and (8), neglecting capillarity, we obtain the estimate

$$\frac{\Delta T_i}{\Delta T_e} = \frac{R T_m^2}{T_u \ell} \frac{(D_{th}/r_c)}{(D_L/\delta)} \tag{20}$$

where ΔT_e is the effective undercooling, ΔT_i is the part of the actual undercooling sacrificed to driving the interface reaction, D_L is the liquid self-diffusion coefficient, δ is the interface width (~1 nm) and ℓ is the molar latent heat. This ratio can be significant (~0.2) within the range of undercooling to which Table 1 applies. Despite the discrepancies we can confidently claim a theoretical understanding of the process from the metallurgical and pedagogical points-of-view. However, a number of mathematicians and theoretical physicists would not agree that these are acceptable criteria, proceeding with an alternative called solvability theory.

Table 1: Scaling Relations

Invariants		$vr/\Delta T$ (mm^2/Ks)	$^*\lambda_{opt}\Delta T$ (mm/K)	$^*r\Delta T$ (mm/K)	$^*v/(\Delta T^2)$ (mm/K^2s)
SN	Predicted	5×10^{-3}	1.9×10^{-2}	0.63×10^{-2}	0.79
	Experimental	3×10^{-3}	3.0×10^{-2}	1.0×10^{-2}	0.30
Kr	Predicted	1.7×10^{-3}	6.8×10^{-3}	2.2×10^{-3}	0.77
	Experimental	1.9×10^{-3}	6.3×10^{-3}	2.1×10^{-3}	0.90
Xe	Predicted	1.3×10^{-3}	10.3×10^{-3}	3.4×10^{-3}	0.39
	Experimental	1.1×10^{-3}	8.4×10^{-3}	2.5×10^{-3}	0.44

* These are middle range, low supersaturation relations which primarily test magnitudes. Bilgram et al. find variations of $\Delta T^{0.80}$ in column 2, $\Delta T^{0.68}$ in column 3 and $\Delta T^{1.73}$ in column 4. The data of Huang and Glicksman do not deviate significantly from the tabulated power laws in the low supersaturation range.

Table 2: Selected Constitutive Properties (18,23)

Property	SN	Kr	Xe	Units
Molecular wt.	0.080	0.083	0.131	Kg/mol
Density of Liquid	970	2440	2960	Kg/m^3
Melting point	331.24	115.8	161.4	K
Heat of fusion	3688.3	1638.8	2291.0	joule/mole
Heat Capacity of Liquid	160.1	42.00	44.58	joule/mole K
Thermal Conductivity Liquid	2.23×10^{-2}	2.06×10^{-2}	47.60×10^{-2}	joule/μm s K
Thermal Diffusivity Liquid	1.16×10^{-8}	7.42×10^{-8}	7.29×10^{-8}	m^2/s
Intrefacial Energy	8.96×10^{-3}	9.00×10^{-3}	11.0×10^{-3}	$joule/m^2$
Unit Supercooling	23.1	43.6	59.2	K
Capillary length, d	2.77	0.44	0.49	nm

The Significance of Solvability Theory

Since 1986, it has been the view of the more mathematically minded that there exists a very fundamental principle of physics governing the unique dendrite tip growth velocity perceived as a strict asymptotic constant, the ubiquitous trailing side-branches representing nothing more than a bothersome artifact. "Solvability" is jargon pertaining to a self-consistency requirement within eigenvalue expansions associated with perturbation theory and used by Wollkind and Segel (20) and Langer (21). The first theoretical element is that the imposition of the Gibbs-Thomson effect on the Ivantsov steady state paraboloidal manifold of tip shapes overdetermines the system and destroys the steady states. The second discovery, which is more significant, is that when the surface tension (or interface kinetics) is recognized as sinusoidally anisotropic, steady states are recovered but not in a continuum. There is a discrete spectrum of steady velocity states, the highest of which is the only one which is perturbation stable. Ergo, nature selects for the only stable maximum velocity state.

It has not escaped the attention of Langer (21) and others that this outcome of deep and extended theoretical research fails, as many would have liked (22), to completely dislodge the involvement of "ad hoc" principles in wave number selection. Indeed, we have been brought back to Zener's maximum velocity criterion (6) in a quantitatively different form. This now becomes the solvability hypothesis described by Langer (21) as "A natural guess .... that the selected dendrite is the one for which the stable solution exists." Regretting the loss of the intuitive appeal of the marginality hypothesis, he goes on to say that "there remains the possibility ... that the solvability theory might be mathematically correct but physically irrelevant – that real systems might simply ignore these steady-state solutions and find other, perhaps oscillatory or even irregular, states of motion " In view of the observations of Bilgram et al (23) that side-branch activity extends right to the dendrite tips in Xe and Kr it is not difficult to imagine that the near tip of a dendrite lies in a limit cycle with a slight, generally undetectable velocity oscillation, rather than in a strict asymptotic steady state (see also, Karma and Pelcé, 24). Langer concludes in effect that the matter will have to be settled by experiment, which seems to undermine the entire enterprise.

Current refinements of the theory have steered the quantitative predictions towards convergence with the aforementioned MS based structures. This means that there exists a theoretical correspondence between the discrete spectrum of stable and unstable tip velocity eigenvalues and the continuum of side-branch Fourier wave numbers subject to amplitude velocity selection. Seemingly in this general convergence a divergence remains in the non-paralleled solvability prediction that the stable tip velocity increases, and the tip radius decreases rather strongly with the strength of the anisotropy. This disagreement could, however, be simply a defect due to approximations (15)–(18) or their marginality equivalent. While there exists no experimental evidence for the strong anisotropy effect (21), the solvability-independent computer simulations (Fig. 10) of Saito et al (25) for dendrite development up to a unique asymptote are significant. While generally supporting the approximate MS-based structures these results parallel the solvability magnitudes and directions of the anisotropy effects (cf. Fig. 10). Note in passing that in the computer simulation wave number selection takes place implicitly in accord with the fastest growing wave number as in Sekerka's earlier (19) calculation.

To take the most positive view on solvability theory the extrapolation of the low anisotropy trends could help provide a non-computational theoretical bridge between the "stability point" of weakly anisotropic branched thermal dendrites and strongly anisotropic smooth-walled, maximum velocity Fe-C Widmanstätten figures (Fig. 2). A supporting diagnostic on the solvability structure would be a predicted increase in the density of steady velocity states with increase in the anisotropy strength. Such a correspondence between structures involving widely varying scales of length would suggest a homomorphism with critical point structures within the renormalization group, irreversible thermodynamic context (26). Solvability theory could therefore turn out after all to be another way of expressing a non-linear thermodynamic stability principle. If the proponents are serious about their oft-repeated claim to an impact on materials science they are duty-bound to establish the limits of equivalence between MS-based selection, asymptotic computational solutions and solvability, and this in a transparent way. We turn next to the theory of thin film cells where the forced rather than asymptotic steady state is central.

General Theory of Thin Film Cells

The extra degree of freedom in the forced velocity cell problem provided by the non-zero temperature gradient has a number of complicating effects which have hindered the advance towards full understanding. Firstly, there are two control parameters rather than one at fixed composition, velocity v and gradient G, rather than undercooling. This introduces the new thermal length m/G which ultimately acts to change the exponents in the scaling laws. Furthermore, the parameters to be scaled increase from two (R_t and λ_c) to at least three (ℓ, R_t and λ) where ℓ is the length of the cells, R_t is the tip radius and λ is the spacing. The degree of degeneracy has been deemed to rise from one to two, which can be identified in free values of R_t and λ. It has been claimed through analogy with the dendrite tip problem by Kessler and Levine (27) that the degeneracy on R_t can be removed via solvability principles (see below). On the other hand, there is agreement that this theoretical method is inapplicable to wave length selection (27). There is an MS marginality condition (Mullins and Sekerka 28; de Cheveigné et al, 29) for the cellular front analogous to that implicated in dendrite side-branch stability, viz.,

$$\lambda_m = \left\{ \frac{16 \pi^3 T_m}{m X_0 (1-k)} \cdot \frac{\sigma}{L} \cdot \frac{D}{v} \cdot \frac{D}{v} \right\}^{1/3} \tag{21}$$

where k is the partition coefficient, m is the absolute value of the liquidus slope and X_0 is the alloy mole fraction, and this defines a point of marginal interface instability which is different than that given by the classical constitutional supercooling (CS) criterion (Tiller et al, 30). However, comparing with

$$\frac{v}{G} \geq \frac{kD}{X_0(1-k)m} \tag{22}$$

and thin film cells at λ_m according to Eq. (21) and subsidiary relations the difference corresponds to as little as 1% in G which is not experimentally significant. Note that the CS criterion contains the ratio of the characteristic lengths (D/v)/(m/G) whereas the MS critical wavelength is independent of m/G.

This in itself is a reason for being suspicious of Eq. (21). Furthermore, one might expect that a marginal prediction would at least qualitatively reflect non-marginal scaling. The empirical scaling of λ and ℓ for stable deep-rooted S-S cells is given by (31)

$$\lambda v^{1/3} G^{1/3} X_0^{1/6} = 0.180 \pm .018\,(\mathrm{mm})(\mu\mathrm{m/s})^{1/3}(\mathrm{K/mm})^{1/3} \tag{23}$$

$$\ell v^{1/3} G^{1/3} X_0^{-1/6} = 7.71 \pm .20\,(\mathrm{mm})(\mu\mathrm{m/s})^{1/3}(\mathrm{K/mm})^{1/3} \tag{24}$$

so the MS formula fares poorly. We shall later argue that like solvability theory (27), it is inapplicable to the wave number selection problem.

A number of interesting computer simulations based on the assumption of strict Gibbs-Thomson equilibrium and methods similar to those for dendrites have been presented for marginal and not-so-marginal conditions (27, 32, 33) leading to cell shapes as in Fig. 11 There seems to be computational agreement that deep-rooted reentrant forms, which imply remelting, are ubiquitous. While similar unstable forms have been observed near the margin (Fig. 12) there is little positive data pertaining to their general existence as a stable form. The comprehensive data set of Venugopalan and Kirkaldy (3, 31) pertained to deep-rooted cells whose walls approached parallelism and lay in a facetted orientation chosen parallel to the growth direction. When observable, roots were always of submicroscopic dimensions approached by converging flat walls (Figs. 2 and 13). The systematics of the computational forms as offered by Ramprasad et al (33) seem to bear little relation to the empirical Eqs. (23) and (24). In contrast to the dendrite case, computer simulation of cells seems still to be in a immature quantitative state. The modellers (27, 32, 33) agree, however, that the question of wave number selection remains unsettled. We will return to this all-important consideration as part of a closed form development of the diffusion problem and estimation of the scaling laws (31).

For our theoretical development we assume that there are no surface tension singularities so that the cell manifold shape is smoothly periodic and subject to the single-valued convergent Fourier representation

$$\phi(y) = \sum_{n=0}^{\infty} \phi_i \cos 2n\pi/\lambda \tag{25}$$

and the diffusion equation

$$\frac{\partial^2 X}{\partial x^2} + \frac{\partial^2 X}{\partial y^2} + \frac{v}{D}\frac{\partial X}{\partial x} = 0 \tag{26}$$

with the everywhere n-convergent general solution form (3, 31)

$$X = X_0 + A_0 + b\exp(-vx/D) + \sum_{n=1}^{\infty} A_n \exp(\mu_n x)\cos\omega_n y\,; \quad x < 0 \tag{27}$$

$$X = X_0 + B_0 \exp(-vx/D) + \sum_{n=1}^{\infty} B_n \exp(-\eta_n x)\cos\omega_n y\,; \quad x > 0 \tag{28}$$

where

$$\omega_n = \frac{2n\pi}{\lambda}\,; \qquad \mu_n = \frac{1}{2}\left\{\sqrt{\left(\frac{v}{D}\right)^2 + 4\omega_n^2} - \frac{v}{D}\right\} \tag{29}$$

$$\eta_n = \frac{1}{2}\left\{\sqrt{\left(\frac{v}{D}\right)^2 + 4\omega_n^2} + \frac{v}{D}\right\}\,; \qquad n = 1, 2, \ldots \tag{30}$$

Note the typographical error in the versions of Eqs. (27) and (28) given in the foregoing reference (31). The matching of the two solutions in value and in gradient at the origin, which has been made possible by the addition of the b term to the A series, implies

$$B_0 = A_0 + b\,; \text{ and } A_n = B_n\,; \quad n = 0, 1, 2, \ldots \tag{31}$$

$$-\frac{v}{D} A_0 = \sum_{n=1}^{\infty} A_n(\mu_n + \eta_n)\cos\omega_n y \tag{32}$$

and by induction on Eq. (26) we can show that this assures differentiability to all orders at the origin and therefore uniqueness. This procedure for making the n-convergence explicit, involving the same number of independent coefficients as the A or B series alone, has been adopted in analogous heat transfer problems (Arpaci, 35). The common practice of identifying the cell diffusion solution as the B series alone applied for all x requires for convergence that the B_n decay exponentially with some negative power of n and it can easily be demonstrated via form (25) that the decay is in fact via negative powers of n so convergence does not exist. The local mass balance

$$(1-k)\, v_n(\phi')\, X(\phi(y)) = \partial X(\phi)/\partial N \tag{33}$$

and the Gibbs-Thomson relation (K is the curvature and N is the unit normal)

$$X(\phi(y)) = X_0/k - G\,\phi/m + \sigma VK/RT_0(1-k) \tag{34}$$

together with the conditions $X \rightarrow X_0$ at $x = \infty$ and an infinite extent of interface provide two independent Dirichlet constraints upon the ϕ and A coefficients. Now if we express the steady state global mass balance as an integral over the Gibbs-Thomson relation (34) we can prove that $\phi_0 = 0$, which defines a unique location for the x-origin and eliminates one unknown (equal enclosed areas for $x = \pm 0$). The necessarily equivalent pair of Dirichlet conditions operating upon the independent Fourier coefficients impose n+1 constraints upon the n+3 independent variables (ϕ_i, v, G, λ). Eliminating the n non-zero shape coefficients we are left with the feasible existence of stable states in the manifold

$$f_1(\lambda, v, G) = 0 \tag{35}$$

or alternatively, retaining ϕ_1 and eliminating λ, we have

$$f_2(\phi_1, v, G) = 0 \tag{36}$$

These can be deemed consistent with the marginal (M) conditions from relation (22)

$$\frac{D}{v_M} \Big/ \frac{m X_0 (1-k)}{G_M k} = \alpha = 1 \tag{37}$$

for if there exists such a marginal relation of the form

$$f_3(\phi_M, \lambda_M) = 0 \tag{38}$$

there will also exist via (35) and (36) a corresponding marginal set (λ_M, ϕ_{1M}). If we exclude the feasible case $f_3(0, \infty) = 0$ as unrepresentative then the marginal first harmonic amplitude ϕ_1 will in general be non-zero in agreement with scaling theory which predicts λ increasing with v decreasing towards the margin (3,4), and with current experiments of de Cheveigné et al (36) and Eshelman and Trivedi (37). All of this is in conflict with the smoothly vanishing amplitude MS marginal stability theory and the related computational modelling of Brown and coworkers (32,33). Indeed, this latter theoretical approach is usually interpreted with an extra non-existent degree of freedom wherein

$$f_4(\phi_1, \lambda, v, G) = 0 \tag{39}$$

Thus, at fixed G and assumed marginal $\phi_1 = 0$, there exists a neutral stability curve of the form

$$f_5(v, G) = 0 \tag{40}$$

There is now an undertaking to rescue the foundations of the failed first order linear theory leading to Eq. (21) by invoking Wollkind and Segel's non-linear third order theory (20) which encompasses what has been designated as sub-critical (finite amplitude) instability. It appears to us to represent an unnecessary and perhaps misleading undertaking. It is sufficient to consider finite amplitude instability within currently accepted nucleation concepts.

Closed Form Scaling Relations for Near Critical and Facetted Deep-Rooted Cells

Venugopalan and Kirkaldy first reported the existence of a distinct near-marginal cellular regime which is qualitatively different than the well-known deep-rooted stable cell manifestation (3,34). This has been recently confirmed and to some extent quantified by Trivedi and coworkers (Fig. 14). As can be seen the cell amplitude in both regimes increases in the direction of the margin. Kirkaldy and Venugopalan (31) have presented two closed form, quantitative scaling theories which can now be associated one-to-one with the two cellular regimes. We designate these as the loose-coupling and the tight-coupling theories, the former being associated with the near-marginal regime and the latter with the stable facetted regime. Both theories are based on successful Zener-Hillert structures developed originally for solid state Widmanstätten and eutectoid structures. The loose-coupling theory, trading upon the connection defined by Fig. 2, considers the cell length and root problem as a Widmanstätten analogue which is independent of the eutectoid-like cell spacing problem (3,4). The resulting scaling laws (cf (23) and (24)) are:

$$\lambda = \left\{ \frac{16}{(1-k)X_0} \cdot \frac{G_0}{G} \cdot \frac{\sigma_e}{L} \cdot \frac{D}{v} \right\}^{1/2} \tag{41}$$

and

$$\ell = \left\{ \frac{(1-k)X_0}{ke} \cdot \frac{m}{G} \cdot \frac{D}{v} \right\}^{1/2} \qquad (42)$$

and are of the typical $1/v^{1/2}$ form where $G_0/G \simeq \ell/\lambda$ represents the relative supersaturation and $\sigma_e > \sigma$ is the solid free energy segregation parameter. Quite appropriately λ is the geometric mean of the diffusion and a modified capillarity length while ℓ is the geometric mean of a thermal and the diffusion length. Note that if G_0 and σ_e can be regarded as constants the aspect ratio ℓ/λ is independent of v and G and proportional to X_0. Indeed, for non-singular near marginal S-S cells (compare Figs 12 and 13) we estimate $\sigma_e \simeq 10\sigma$ and $G_0 \simeq 100$ K/mm (see below) whence λ and ℓ both prove according to the S-S data of Table 3 to be of the right magnitude for marginal cells. However, there is as yet insufficient systematic data for a thorough test of these laws There are three arguments which associate the weak coupling theory with near marginal states. Firstly, the characteristic lengths in (42) are defined explicitly and uniquely within the marginal CS condition (37) as a simple ratio. Secondly, the smaller is v the closer to equilibrium is the diffusion-reaction manifold, and therefore to constraint decoupling. Finally, the direct dependence of the aspect ratio ℓ/λ on X_0 is highly appropriate to a near equilibrium configuration.

Table 3: Constitutive Parameters for S-S

Equilibrium melting (T_M) 331.24°K	Latent heat of fusion (ΔH) 4.7×10^7 J/m^3
Partition coefficient (k) 0.20	Diffusion coefficient in liquid (D) 1×10^{-9}m^2/s
Liquidus slope (m) 180°K/mole fraction	Solid-liquid surface tension (σ) 9mJ/m^2

In developing the tight-coupling theory for conditions further removed from equilibrium we recognized that the thermal length m/G must exhibit an effect in the cell spacing as well as the length, and analogously to the phase boundary diffusive short-circuiting coupling that occurs in some eutectoid reactions through the boundary width we argued that one should introduce the scaling transformation

$$\frac{D}{v} \rightarrow \left\{ \frac{D}{v} \cdot \frac{m}{G_0} \right\} \qquad (43)$$

with a decease in the power index from 1/2 to 1/3. Furthermore we have argued that the augmented and strongly coupled concentration segregation is to be expressed by the substitution $X_0 \rightarrow X_0^{1/2}$ ($X_0 < 1$). These modifications of Eqs. 41 and 42 yield the scaling relations

$$\lambda = \left\{ \frac{24T_m}{(1-k)X_0^{1/2} m} \cdot \frac{\sigma_e}{L} \cdot \frac{D}{v} \cdot \frac{m}{G} \right\}^{1/3} \qquad (44)$$

and

$$\ell = \left\{ \frac{3(1-k)X_0^{1/2} G}{2k\,e\,G_0} \cdot \frac{D}{v} \cdot \frac{m}{G} \cdot \frac{m}{G} \right\}^{1/3} \qquad (45)$$

where σ_e must now be made explicit as an average free energy storage solution parameter which is in excess of the surface tension σ in rough proportion to the aspect ratio and G_0 is an average relative supersaturation solution parameter in excess of the experimental G-range defined such that average G_0/G is also close to the aspect ratio. For the S-S data set with G=10K/mm we can iterate trial values of σ_e and G_0 against the calculated ℓ/λ finding convergence at:

$$\sigma_e/\sigma \simeq \overline{G_0/G} \propto \overline{\ell/\lambda} \simeq 10 \qquad (46)$$

Note the very plausible weighting of the three characteristic lengths in Eq. (44) in contrast to that in the MS marginal Eq. (21). With $\sigma_e/\sigma \simeq 10$ and $\overline{G_0/G} \simeq 10$, these formulas generate very close to the correct values of the empirical coefficients in Eqs. (23) and (24) (31). Furthermore, Table 4 provides a comparison between formulas (41), (42) and (44), (45) for $X_0 = 0.01$, G=10K/mm and mutual relations (46) subject to the observed transition v≃7μm/s. The increase in λ is roughly in accord with Fig. 14, and the overall trends reported by Eshelman et al (38) for pivalic acid-ethanol are reproduced.

The transition from the marginal to the facetted mode can be understood in terms of a dissipation transition to a lower interfacial reaction value in accord with the considerations pertaining

Table 4: Typical Predicted Cell Dimensions

Dimensions	Facetted Cells	Marginal Cells
λ	140 μm	76 μm
ℓ	990 μm	195 μm
ℓ/λ	7.0	2.5

to the irreversible thermodynamic affinity relation (11). The transition data for S-S and stability considerations suggest the criterion

$$\frac{vX_o}{G} \simeq \frac{D}{m}\left(\frac{3ek}{1-k}\right)^{2/3} \tag{47}$$

and this is tested in Fig. 15 for S-S.

There remains a question first raised by the author as to whether near marginal cells in the absence of strong surface tension anisotropy possess true steady states and this has been recently been repeated by Kurz and Trivedi (39). While such in accord with relation (35) is feasible, there is no proof that such feasible states are stable. The experiments on S-S are negatively disposed (3,34) towards this disposition. As with the weak coupling relations (41) and (42) relations (44) and (45) define an aspect ratio which depends only upon X_0. This trend is clearly evident in Fig. 13 where spatial uniformity is evolving at constant aspect ratio by a soliton mechanism (see below).

Cellular Dendrites

Because the temperature in cell formation is essentially uniform in the lateral direction ($D_{th} >> D$) there will always be CS in the near frontal grooves and its side-branching effect will be augmented by increasing concentration and by a decrease in the tip radius relative to the spacing. This is to be combined with the suggestion from the marginal stability condition (22) that general instability will be favoured by higher v relative to G. Proceeding more quantitatively we note that the tip radius as for free dendrites should vary approximately as $1/v^{1/2}$. That is to say, the loose coupling formula (41) for $\lambda/2$ should give a good approximation to the tip radius, viz.,

$$R_t = \lambda/2 \simeq \left\{\frac{4}{(1-k)X_o} \cdot \frac{G_o}{G} \cdot \frac{\sigma_e}{L} \cdot \frac{D}{v}\right\}^{1/2} \tag{48}$$

and this has a scaling behaviour similar to dendrite tip formulas based on the MS expression (15). On the other hand the spacing of deep facetted cells varies as $1/v^{1/3}$ (Eq. (41)) so the lateral instability is again on this basis favoured by high v. We have succeeded in combining these trends analytically yielding an instability relation for the onset of side-branching with units as in Eqs. (23) and (24)

$$v\,X_0^{1/2}/G \geq \gamma(0.5)10^{-5} \tag{49}$$

where the empirical value of $\gamma \simeq 6$ for S-S represents the relative excess of CS as measured by relation (22) over that to initiate the onset of stable cells (Eq. 47; γ was evaluated from Fig. 17). It would be interesting to ascertain whether the cell side-branching model of Karma and Pelcé (24) can generate this empirically tested formula. We note that this construction ties in further with thermal dendrites in that the propensity for dendritic morphologies increases with the approach to isothermal conditions (G small).

As already mentioned, the cell root configuration can be perceived in an inverse reaction context (Fig. 2) as the analogue of a two-dimensional dendrite or Widmanstätten plate. Within this context it is important to recognize that the configuration defines the quintessential dendrite problem, for according to Fig. 3 a two-sided model is essential and a true, forced velocity steady state, with attendant pattern degeneracy is defined. For this model we obtained the root radius expression (4)

$$R_r = \left\{\frac{\gamma_s D_s V\sigma k/RTX_o(1-k)^2}{v\,[1-\alpha(v)\exp(v\ell/D)]}\right\}^{1/2} \tag{50}$$

defining the characteristic free boundary degeneracy $R = R(v)$ with $\ell = \Delta T/G$ constant and $\gamma_s(\sim 4)$ a known geometric constant. When $R_r(v)$ is minimized as an expression of minimum dissipation (maximum segregation) we are led to the stable length ℓ which according to Eq. 42 is independent of the root parameters. Were we justified in characterizing the cell tips as growing independently in a symmetric model the same derivation yields, with $\ell \sim 0$,

$$R_t = \left\{ \frac{\gamma_L DV \sigma k / RTX_0 (1-k)^2}{v[1-\alpha(v)]} \right\}^{1/2} \tag{51}$$

and this is to be compared with a recent solvability formula for dendrite tips due to Langer (40,41) as quoted by Kurz and Trivedi (39), which with the exception of the γ value is identical. Since $\alpha \sim 1/v$ this has no optimum so one might be tempted with Kessler and Levin (27) to apply the solvability condition. However, it is difficult to justify the decoupling of the cell tips. If this is not done than an optimizable free boundary problem is recovered leading to formulas (41) and (44). Indeed with the conjecture that the the tip radius $r_t \sim \lambda/2$ in the weak coupling approximation we have obtained Eq. 48 which is to be compared with the solvability result with $\alpha << 1$

$$R_t \simeq \left\{ \frac{40k}{(1-k)^2 X_0} \cdot \frac{V\sigma}{RT} \cdot \frac{D}{v} \right\}^{1/2} \tag{52}$$

Although the magnitudes are similar the latter is independent of G, which we believe, reflects the unjustified decoupling of cell tips. This is a general defect of MS formulas applied to cells (cf. Eq. 21).

The optimized Eq. 50 yields

$$R_r \simeq \left\{ \frac{k}{X_0(1-k)^2} \cdot \frac{V\sigma}{RT} \cdot \frac{D_s}{v} \right\}^{1/2} \tag{53}$$

In contrast to R_t it is reasonable that this should not depend explicitly on G when $\alpha << 1$. On comparison of Eqs. (51) and (53) we conclude that the ratio R_t/R_r scales like $\sqrt{D/D_s}$. The observations (Figs. 2 and 13) therefore imply that for SN based solutions that $D/D_s > 10^3$ a consideration which is unfavorable to the development of reentrant grooves (37; Figs. 2, 13 and 12). This is a matter which we will not pursue further here. In any case, all of this affirms the intimate relationship between cell, dendrite and Widmanstätten morphologies and gives more than a hint that solvability is old wine in a new bottle.

Discussion

As regards dendrite morphologies the simulations of Sekerka (19) and Saito et al (25) demonstrate that the asymptotic forms can be predicted by integration up from a realistic initial condition (Figs. 9 and 11) in agreement with early perceptions of Sekerka (19) and Nash and Glicksman (12). Clearly, tip noise can be nothing more than a weak perturbation to the evolution of side branches (25). While as a corollary, pattern selection proceeds in accord with the fastest growing wavelength (19), this is not to be taken as identifying a free boundary problem.

As a surrogate to the information contained in the ignored temporal evolution, solvability theory attempts to predict a tip stability point without reference to the asymptotic reality thus confronting an artificial free boundary degeneracy, which as usual can be removed only by invoking a hypothesis concerning pattern selection. In this case it is selection for the tip velocity eigenvalue which is both maximal and uniquely stable against tip-splitting (21,37). This offers the prediction that the stable tip velocity increases (or the tip radius decreases) strongly with the strength of the anisotropy, an effect which is verified by the Saito et al simulation but which has not been observed (21,39). We are attracted to this idea for it is capable of converging at very high anisotropy with the successful maximum velocity free boundary criterion for Widmanstätten plates and needles. One of the missing links in this line of reasoning is represented by the need to show that spatial selection over a velocity spectrum is equivalent to a temporal selection over a wavenumber spectrum. Such a ergodicity concept is not unattractive. The increasing evidence for a role of stochastically formed solitons in the temporal change of pattern is relevant to the ergodicity generalization (42-44, 45, 46; cf. Fig. 14).

The problem of cellular morphology, while overlapping with the foregoing in several strong respects has a true free boundary degeneracy, as evidenced by the fact that there exists a feasible steady state planar interface solution for the full spectrum of unstable conditions. The anisotropy dependent phase space of morphologies is a very complex one with a finite amplitude marginal bifurcation to a weakly turbulent cellular state (Fig. 13), a second finite amplitude bifurcation to a stable anisotropic cellular state (Figs. 1 and 14), and a third bifurcation to the anisotropic dendritic state which seems to be continuous (Fig. 4). With such complexity, hysteresis may be expected to be a ubiquitous phenomenon, as indeed observed.

We can anticipate that there will emerge objections to the newly introduced non-interfacial cellular solution parameters σ_e and G_0 as adjuncts to a dissipation optimization procedures. However, these considerations state nothing more nor less than that a full accounting of molecular trajectories in the classical mechanical sense, including those which pertain to solution thermodynamics, should remove all degeneracies. On a more quantitative basis we have long ago argued that Hamilton's Principle for multiparticle mechanical systems (48) is the ultimate foundation of dissipation principles (47). In recent research we have demonstrated that this mechanical variational principle of the general form

$$\delta \int_{t_1}^{t_2} (T - V + W)\,dt = 0 \tag{53}$$

where t is the time, T is the total kinetic energy, V is the potential energy and W is the cumulative energy made unavailable by internal friction, when subject to the Second Law in isolation, generates the corollary principles

$$\delta \int_{t_1}^{t_2} W\,dt = 0; \quad \delta \int_{t_1}^{t_2} T\,dt = 0; \quad \delta \int_{t_1}^{t_2} V\,dt = 0; \tag{54}$$

For the steady state asymptote of mechanical trajectories in isolation the first of these implies stability at a maximum or minimum in the integral rate of dissipation, the second implies a maximum or minimum in the undercooling and the third translates to a maximum or minimum in interface curvature and (or) solute unmixing. Thus all of the principles previously regarded as "ad hoc" find a firm foundation within mechanics and can be substituted for dissipation principles in estimates which strive for precision to within a factor of about 2.

Acknowledgements

The author is grateful to Professors D. Venugopalan and Y. Brechet for discussions.

References

1. R.D. Townsend and J.S. Kirkaldy, Trans. Quart 61, 605 (1968).
2. S.-C. Huang and M.E. Glicksman, Acta Met. 29, 717 (1981).
3. D. Venugopalan and J.S. Kirkaldy, Acta Met. 32, 893 (1984).
4. J.S. Kirkaldy, Scripta Met. 14, 739 (1980).
5. I. Prigogine, Introduction to Thermodynamics of Irreversible Processes, p. 23 et seq. Charles C. Thomas, Springfield, IL, (1955).
6. C. Zener, Trans. AIME 167, 550 (1946).
7. M. Hillert, Jernkontorets Ann. 141, 757 (1957).
8. G.P. Ivantsov, Dokl. Akad. Nauk. SSSR 58, 567 (1947).
9. D.E. Temkin, Dokl. Akad. Nauk SSSR 132, 1307 (1960).
10 G. Horvay and J.W. Cahn, Acta Met. 9, 695 (1961).
11. R. Trivedi, Acta Met. 18, 287 (1970).
12 G.E. Nash and M.E. Glicksman, Acta Met. 22, 1283 (1974).
13. J.S. Langer, Rev. Mod. Phys. 52, 1 (1980).
14. R.F. Sekerka, R.G. Seidensticker, D.R. Hamilton and J.D. Harrison, Investigation of Desalination by Freezing, Westinghouse Research Laboratory Report (1967) Chapter 3.
15. W. Oldfield, W., Mat. Sci. Eng. 11, 211 (1973).
16. J.S. Langer and H. Müller-Krumbhaar, Acta Met. 26, 1681 (1978).
17. W.W. Mullins and R.F. Sekerka, J. Appl. Phys. 34, 323 (1963).
18. S.-C. Huang and M.E. Glicksman, Acta Met. 29, 701 (1981).
19. R.F. Sekerka, J. Phys. Chem. Solids 28, 983 (1967).
20. D.J. Wollkind and L.A. Segel Proc. Roy. Soc. (London), 268, 351 (1970).
21. J.S. Langer, Science 243 1150 (1989).
22. F. Liu and N. Goldenfeld, Phys. Rev. A 38, 407 (1988).
23. J.H. Bilgram, M. Firmann and E. Hurliman, J. Cryst. Growth, 96, 175 (1989).
24. A. Karma and P. Pelcé, Phys. Rev. A 39, 4162 (1989).
25. Y. Saito, G. Goldbeck-Wood and H. Muller-Krumbhaar, Phys. Rev. A, 38, 2148 (1988).
26. K.G. Wilson, Sci. Am. 241, 158 (1979).
27. D.A. Kessler and H. Levine, Phys. Rev. A 39, 3041, 3208 (1989).
28. W.W. Mullins and R.F. Sekerka, J.Appl. Phys. 35, 444 (1964).
29. S. de Cheveigné, C. Guthmann and M.-M. Lebrun, J. Cryst. Growth 73, 242 (1985).
30. W.A. Tiller, K.A. Jackson, J.W. Rutter and B. Chalmers, Acta Met. 1, 428 (1953).
31. J.S. Kirkaldy and D. Venugopalan, Scripta Met. 23, 1603 (1989).

32. L.H. Ungar and R.A. Brown, Phys. Rev. B 31, 5931 (1985).
33. N. Ramprasad, M.J. Bennett and R.A. Brown, Phys. Rev. B 38, 583 (1988).
34. D. Venugopalan, Ph.D. Dissertation, McMaster University (1982).
35. V.S. Arpaci, Conduction Heat Transfer, p. 211 et seq. Addison-Wesley, Reading, MA (1966).
36. S. de Cheveigné, C. Guthmann and M.-M. Lebrun, J. Physique 47, 2095 (1986)
37. M.A. Eshelmann and R. Trivedi, Acta Met. 35, 2443 (1986).
38. M.A. Eshelman, V. Seetharaman and R. Trivedi, Acta Met. 36, 1165 (1988)
39. W. Kurz and R.Trivedi, Acta Met. 38, 1 (1990).
40. J.S. Langer, Phys. Rev. A33, 435 (1986).
41. M. Benamar and P. Pelcé, Phys. Rev. A 39, 4263 (1989).
42. J.S. Kirkaldy and R.C. Sharma, Acta Met. 28, 1009 (1980).
43. J.S. Kirkaldy, Scripta Met. 15, 1255 (1981).
44. D. Venugopalan and J.S. Kirkaldy, Acta Met. 32, 893 (1984).
45. A.J. Simon, J. Backhofer and A. Libchaber, Phys. Rev. Letts. 61, 2574 (1988).
46. G. Faivre, S. deCheveigné, G. Guthmann and P. Kurowski, Preprint, 1989.
47. J.S. Kirkaldy, Can. J. Phys. 37, 739 (1959).

Fig. 1 Illustration of morphological similarity between Widmanstätten colonies (a) and thermal dendrites (b). After Townsend and Kirkaldy (1) and Huang and Glicksman (2).

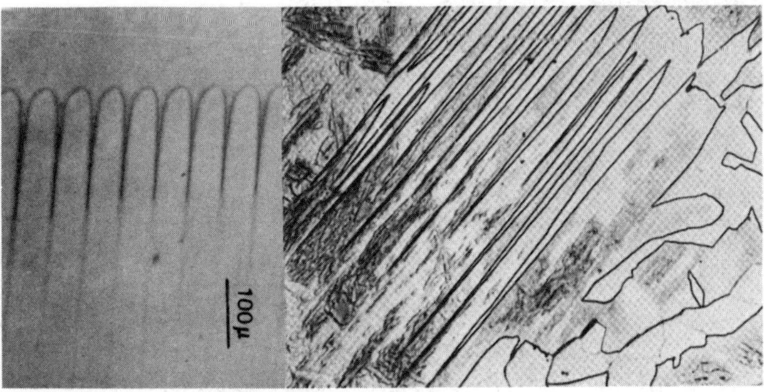

Fig. 2 (a) Stable cellular interface in succinonitrile-salol. The cell root radii are not resolved. (b) The conventional Widmanstätten structure in a hypoeutectoid steel. After Venugopalan and Kirkaldy (44) and Townsend and Kirkaldy (1).

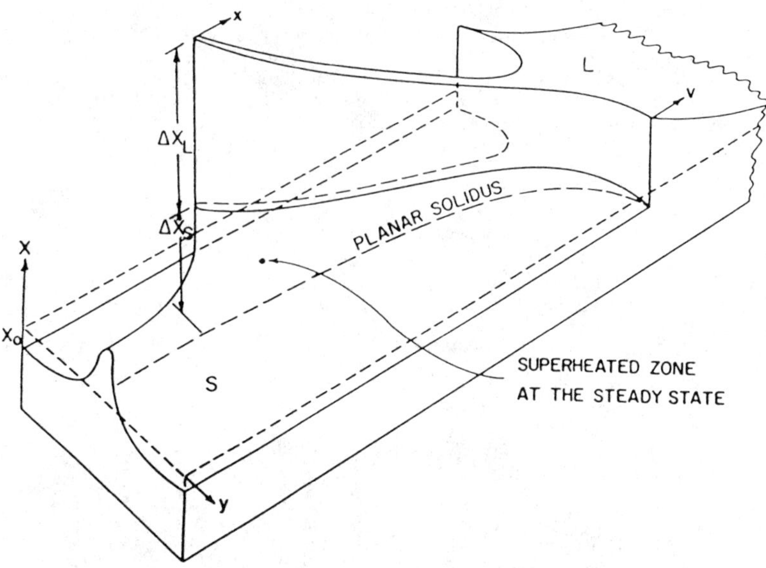

Fig 3 Solid-liquid concentration distributions corresponding to a liquid "Widmanstätten" spike or plate at the steady state. After Kirkaldy (4).

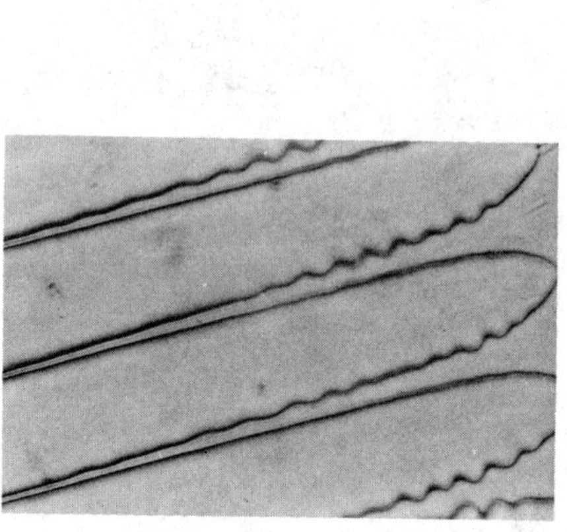

Fig. 4 Cell wall instability as the precursor of cellular dendrites. Courtesy of D. Venugopalan.

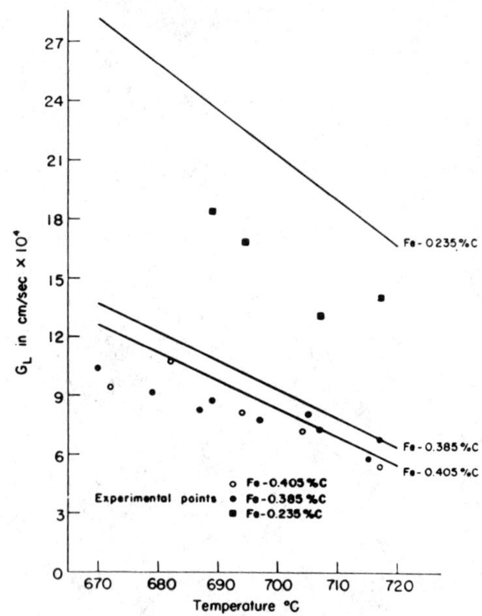

Fig. 5 Comparison of experimental and theoretical velocity (G_L) values on a linear scale. Analogous curves for dendrites are invariably plotted logarithmically concealing discrepancies of the magnitude seen here. After Townsend and Kirkaldy (1).

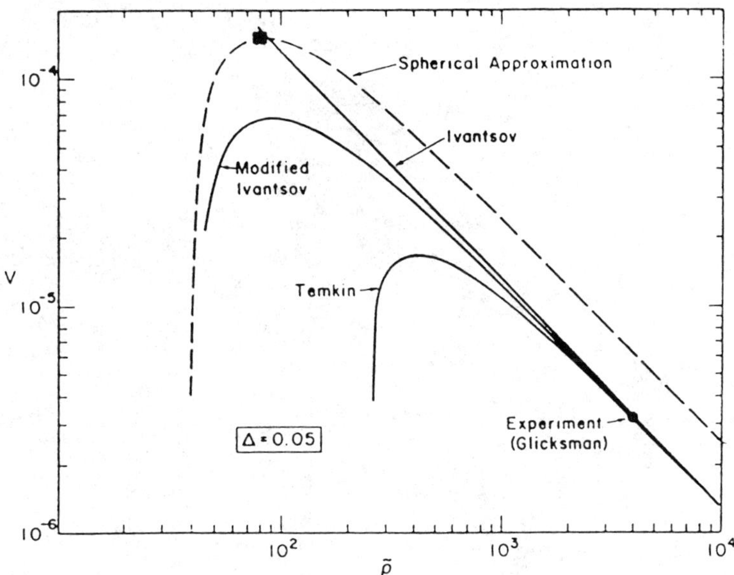

Fig. 6 Dimensionless dendritic growth velocity as a function of tip radius for a dimensionless under cooling $\Delta = 0.05$, determined by the four steady-state approximations indicated. The experimental point (solid circle) has been provided by Glicksman from measurements on succinonitrile. The solid square point is for a Widmanstätten plate in steel. After Townsend and Kirkaldy (1). Abridged from Langer (13).

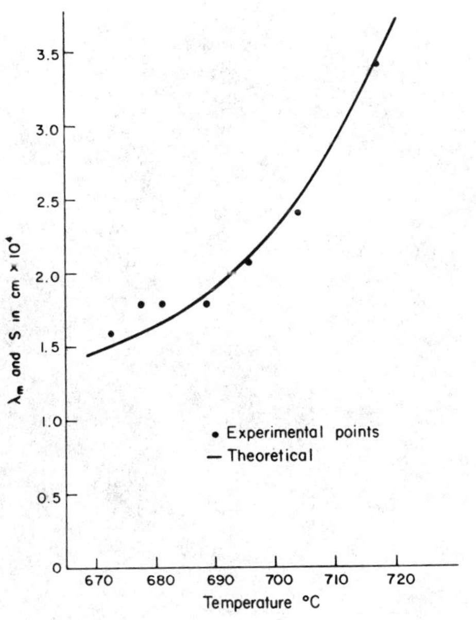

Fig. 7 Comparison of experimental and theoretical Widmanstätten plate spacings; Fe-0.405%C. Theory based on Mullins and Saekerka fastest growing wavelength. After Townsend and Kirkaldy (10.

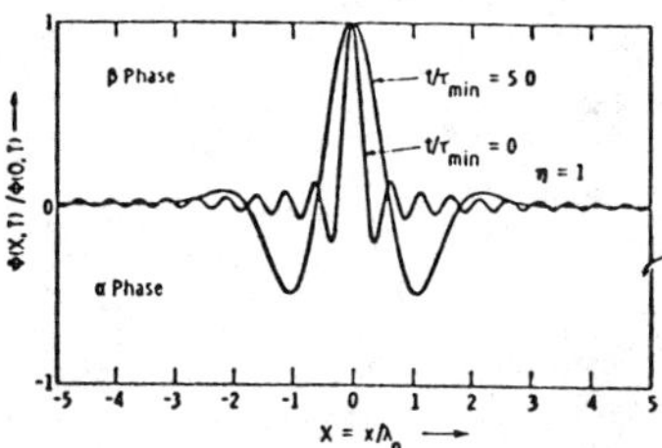

Normalized shape of the two-phase interface at times $t/\tau_{min} = 0$ and $t/\tau_{min} = 5{\cdot}0$.

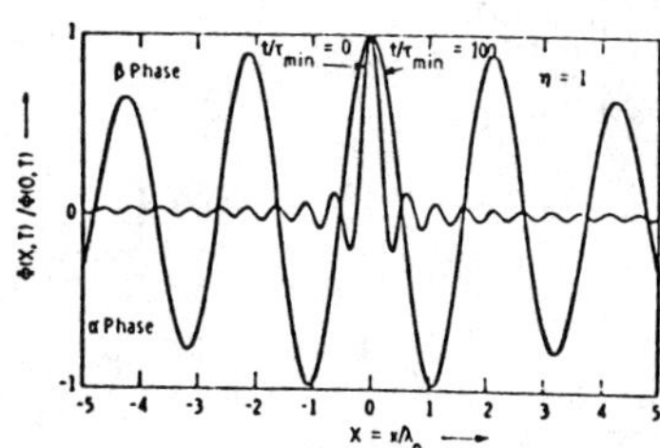

Normalized shape of the two-phase interface at times $t/\tau_{min} = 0$ and $t/\tau_{min} = 100$.

Fig. 8 Nonsteady state growth by side-band emission of a perturbation which selects the fastest growing wavelength from the Fourier manifold of the initial condition. After Sekerka (19).

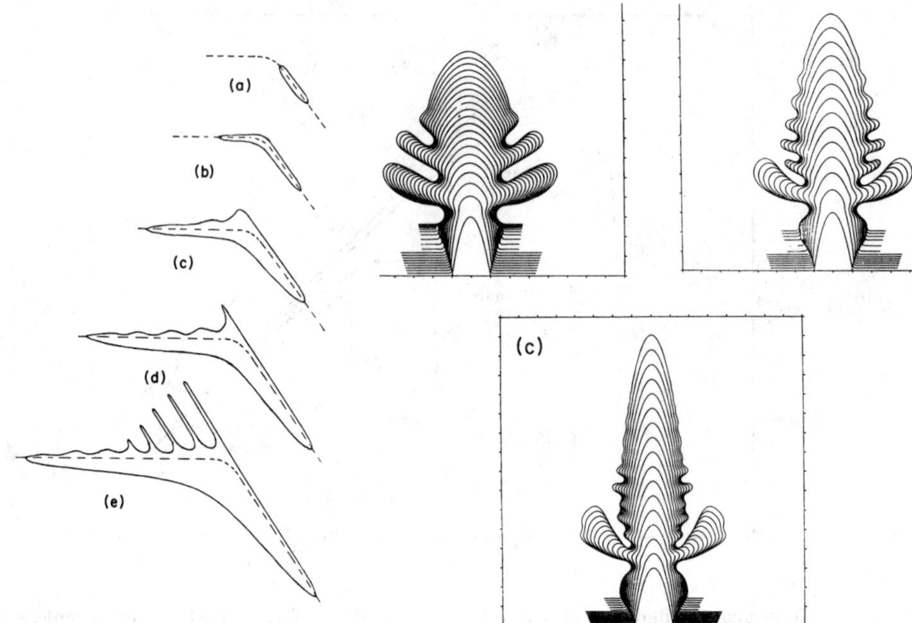

Fig. 9 Schematic of Widmanstätten side-branch colony in steel. After Townsend and Kirkaldy (1).

Fig. 10 Computer simulation of growth of thermal dendrites. After Saito et al. (25).

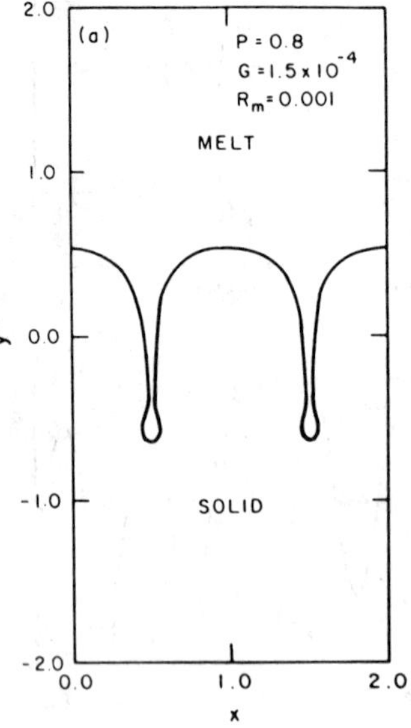

Fig. 11 Computer simulation of reentrant cells which are claimed to be stable. After Ungar and Brown (32).

Fig. 12 Unstable reentrant cells near the margin in succinonitrile-salol. Courtesy of Venugopalan.

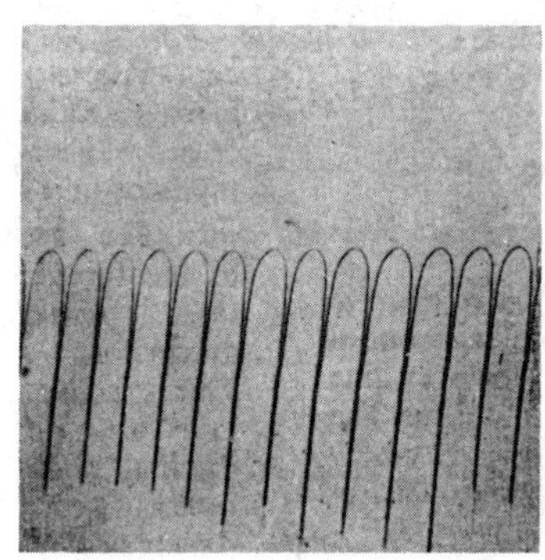

Fig. 13 Non-reentrant deep-rooted cells illustrating invariance of the aspect ratio and soliton mechanism of spacing change Anonymous.

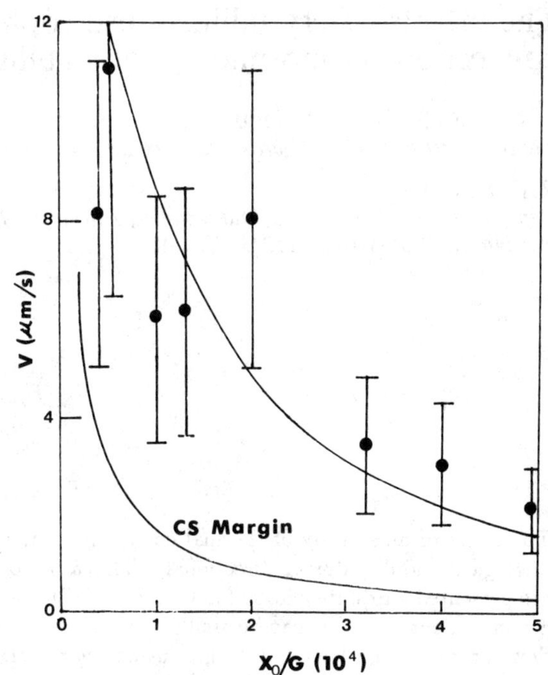

Fig. 15 Theoretical correlation for weak-coupling to strong-coupling transition in succinonitrile-salol compared to CS margin. Data after Venugopalan and Kirkaldy (3).

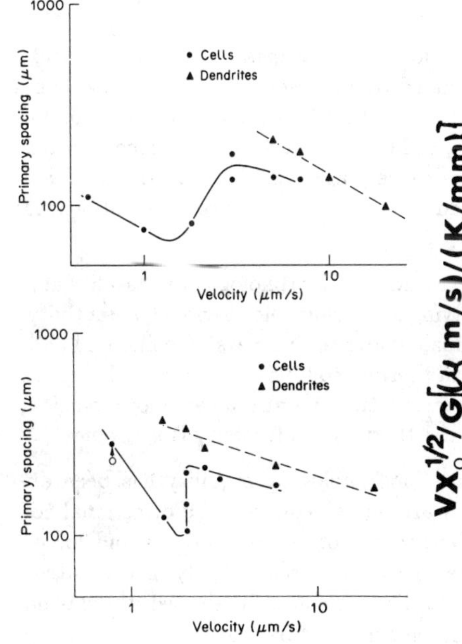

Fig. 14. Observation of the weak-coupling to the strong-coupling transition in pivalic acid-acetone (top) and succinonitrile-acetone. After Eshelman et al. (38).

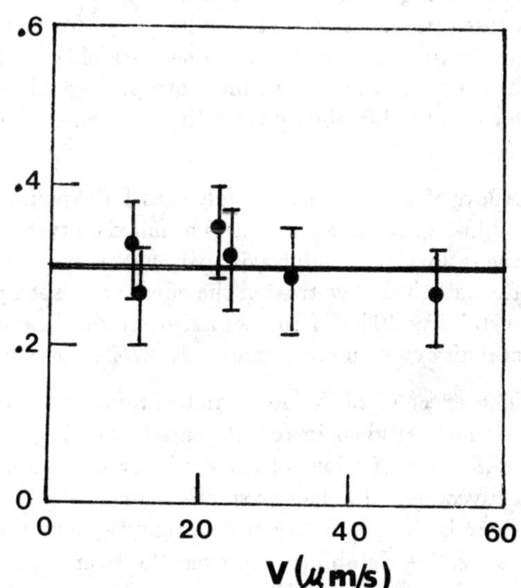

Fig. 16 Theoretical correlation for cell to side-branching transition in succinonitrile-salol. Data after Venugopalan and Kirkaldy (3).

The effects of crystalline anisotrophy and buoyancy-driven convection on morphological stability

S.R. Coriell, G.B. McFadden
National Institute of Standards and Technology, Gaithersburg, Maryland, 20899, U.S.A.

R.F. Sekerka
Departments of Physics and Mathematics, Carnegie Mellon University, Pittsburgh, Pennsylvania 15213, U.S.A.

The effect of anisotropy of thermal conductivity and buoyancy-driven convection on morphological stability during directional solidification of a binary alloy is considered. Results are presented for dilute alloys of indium in tin. The onset of instability depends on crystallographic orientation, and can be oscillatory in time under certain conditions. Buoyancy-driven flow can result in a long wavelength instability that enhances morphological instability.

Introduction

During directional solidification of a binary alloy, morphological instability can occur, often leading to cellular or dendritic growth accompanied by microsegregation of solute. The constitutional supercooling criterion [1] for instability was based on the thermal and solutal gradients in the liquid. The linear stability analysis of Mullins and Sekerka [2] incorporated heat flow in the crystal and isotropic crystal-melt surface tension in the analysis. Morphological stability theory is continually being extended to encompass more complex situations [3–8].

Alloys of tin have been widely used for experimental studies of morphological instability and cellular growth [9–19]. Tin is a uniaxial crystal, having an anisotropic thermal conductivity, the ratio of the conductivity perpendicular to the uniaxial axis to that parallel being 1.44 [20]. Recently, we have treated the effect of anisotropy in thermal conductivity on morphological instability [21]. Of course, anisotropy in interface properties can also affect morphological stability even in cubic materials [22–24], in which the thermal conductivity is isotropic.

The effect of both forced and natural convection on morphological stability has been extensively studied in recent years [3,25–42]. In this article, we consider the directional solidification of alloys of tin vertically upward in the absence of radial gradients and forced convection. If a lighter solute is rejected (or a heavier solute preferentially incorporated) there is the possibility of a convective instability. If a heavier solute is rejected, there is no convective instability; however, the lateral gradients in temperature and solute induced by morphological instability can give rise to fluid flow that affects the solute and temperature distributions, which in turn alters the morphological instability. The tin-indium system is particularly interesting since the liquid densities of pure tin and pure indium are almost the same at the melting point of tin, namely, 6.981 and 6.977 g/cm^3, respectively [43]. However,

TABLE I

Thermophysical Properties of Sn-In Used in Numerical Computations.

liquid diffusion coefficient	D	$2.3 \cdot 10^{-5} \text{cm}^2/\text{s}$
liquid thermal conductivity	k_L	$0.303\, J/(\text{cm} \cdot K \cdot s)$
solid thermal conductivities	$k_{\parallel}$	$0.460\, J/(\text{cm} \cdot K \cdot s)$
	$k_{\perp}$	$0.662\, J/(\text{cm} \cdot K \cdot s)$
distribution coefficient	k	0.4
liquidus slope	m	$-1.83\, K/\text{at}\%$
capillary parameter	$T_M\Gamma$	$8.6 \cdot 10^{-6} \text{cm} \cdot K$
heat of fusion	L_V	$418.0\, J/\text{cm}^3$
kinematic viscosity	ν	$2.6 \cdot 10^{-3}\, \text{cm}^2/\text{s}$
liquid thermal diffusivity	κ	$0.17\ \text{cm}^2/\text{s}$
solid specific heat	C_p	$1.85\ J/(\text{cm}^3\text{K})$
density change with temperature	α	$1.0 \cdot 10^{-4}\ \text{K}^{-1}$
density change with composition	β	$0\ \text{at}\%^{-1}$
fractional density change	ϵ	0

there is disagreement over whether the density of tin increases or decreases with the addition of indium [43, 44]. For purposes of calculation in this article we will assume that the addition of indium to tin does not change the density.

Onset of Morphological Stability

Recently, we carried out a linear stability analysis of an initially planar crystal-melt interface for a crystal with anisotropic thermal conductivity [21]. The planar crystal-melt interface is assumed to be moving in the z direction at constant velocity V with respect to the crystal. We choose a Cartesian coordinate system (x, y, z) which is fixed with respect to the unperturbed planar interface located at $z = 0$. The perturbed interface, $z = h(x, y, t)$, is assumed to have the form

$$h(x, y, t) = \delta \exp(\sigma t + i\omega_x x + i\omega_y y). \tag{1}$$

The interface is unstable if the real part of σ is positive for any values of ω_x and ω_y.

For a uniaxial crystal, such as tin, the thermal conductivity tensor in a coordinate system aligned with the principal axes of the crystal can be written in the form

$$(k_{ij}) = \begin{pmatrix} k_{\perp} & 0 & 0 \\ 0 & k_{\perp} & 0 \\ 0 & 0 & k_{\parallel} \end{pmatrix},$$

where $k_{\perp}$ and $k_{\parallel}$ are generally unequal. For our (x, y, z) coordinate system at least one principal crystallographic axis corresponding to $k_{\perp}$ can always be taken to lie along the x axis because of degeneracy in the plane perpendicular to the uniaxial principal axis. The orientation of the crystal with respect to the growth direction z can be described by the angle θ between the growth direction and the uniaxial axis. It is convenient to write the wave vector of the perturbed interface in the form $(\omega_x, \omega_y) = \omega(\cos\alpha, \sin\alpha)$. In the absence of anisotropy, the dispersion relation for σ does not depend on α or on θ. Anisotropy of thermal conductivity for an uniaxial crystal introduces the two additional variables α and θ

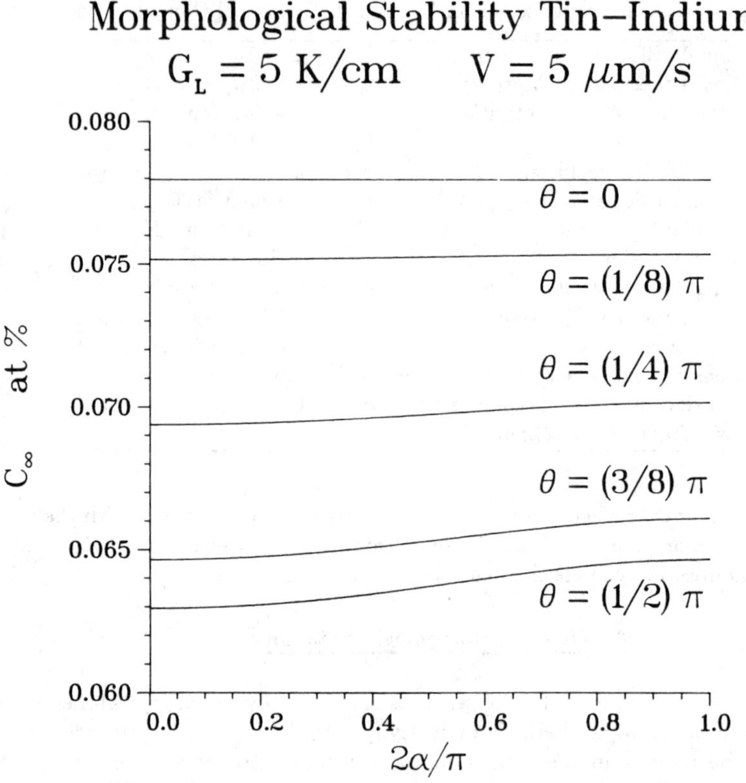

FIG. 1. Marginal values of the solute concentration above which morphological instability occurs as a function of α for various growth directions θ for alloys of indium in tin for a growth velocity of 5 μm/s and a temperature gradient in the liquid of 5 K/cm.

in the dispersion relation. For growth of a single crystal, the value of θ can be fixed, and stability theory predicts the value of α for which instability first occurs.

The marginal values of the indium concentration in tin for which the real part of σ vanishes are plotted in Fig. 1 as a function of the angle α for various values of the crystallographic orientation angle, θ, for a temperature gradient, G_L, in the melt of 5 K/cm and a growth velocity, V, of 5 μm/s. These marginal values are calculated by minimizing c_∞ over ω for fixed α. The thermophysical properties used in the calculations are given in Table I. In Fig. 2, we plot $\sigma_i = \text{Im}(\sigma)$ as a function of α for the same conditions as in Fig. 1. For non-zero values of σ_i , the perturbation moves laterally along the interface (travelling waves). From Fig. 1 for any θ, the onset of instability occurs at $\alpha = 0$ with $\sigma_i = 0$. For $\theta = 0$ and $\theta = \pi/2$, $\sigma_i = 0$ for all values of α; these values of θ correspond to growth along a principal crystallographic axis. However, for other values of the temperature gradient in the liquid and the growth velocity, the onset of instability can occur for $\alpha = \pi/2$ and correspond to travelling waves [21]. The variation of marginal concentration with the angle α, which

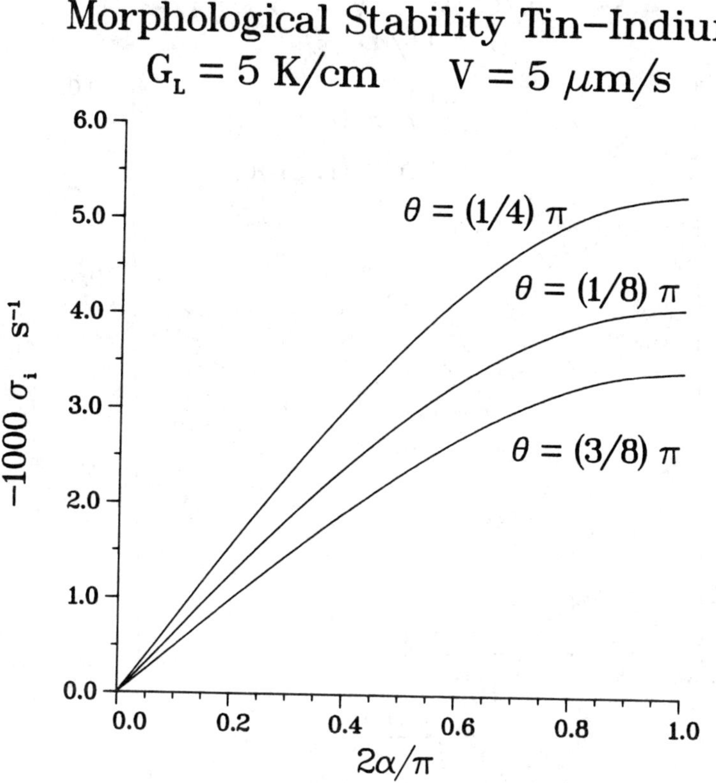

FIG. 2. Imaginary part of σ for the same conditions as Fig. 1 For $\theta = 0$ and $\theta = \pi/2$, which are principal crystallographic directions, $\sigma_i = 0$. Note also that $\sigma_i = 0$ at the onset of instability, which takes place at $\alpha = 0$.

measures the direction of the perturbation, is very small: for $\theta = \pi/2$, the marginal concentration changes from 0.0629 to 0.0647 at.% as α ranges from 0 to $\pi/2$. The critical concentration at the onset of instability varies from 0.078 to 0.063 at.% as the crystallographic orientation varies from $\theta = 0$ to $\pi/2$, respectively. From the viewpoint of linear theory, for all crystallographic orientations other than growth along the uniaxial direction ($\theta = 0$), two-dimensional planforms are preferred over hexagonal planforms. Very few of the experimental articles on interface instability and cellular growth of tin report the crystallographic orientation with respect to the growth direction, an exception being the paper by Weinberg [16], who investigated the morphology of tin crystals doped with Tl^{204} by both microscopic observation of decanted interfaces and radiographic analysis of solute segregation. His growth direction corresponded to our $\theta = \pi/2$, and he reported morphologies that were two-dimensional (elongated cells) at the onset of instability, as well as three-dimensional cells

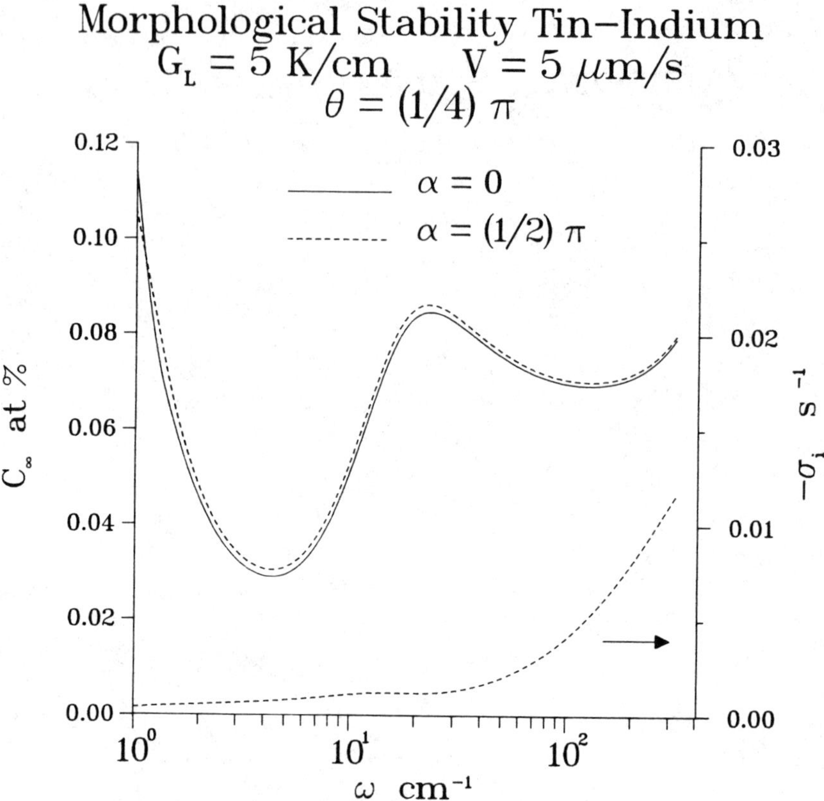

FIG. 3. The concentration, c_∞, of indium in tin at the onset of instability during directional solidification at 5 μm/s and a temperature gradient in the liquid of 5 K/cm as a function of the wave number of a sinusoidal perturbation for two values of the angle, α, specifying the direction of the wave vector. For $\alpha = \pi/2$, the onset of instability is oscillatory, and σ_i is shown, while for $\alpha = 0$, the onset of instability is non-oscillatory.

for larger growth rates and solute concentrations.

The above calculations assume that there is no fluid flow in the melt. Since we have assumed that for tin-indium alloys, the liquid density is independent of concentration, any fluid flow is due to temperature gradients (assuming also the absence of forced convection). In a real furnace, radial gradients will give rise to fluid flows with length scales typically of the container size. At present, there is no detailed understanding of the effect of these flows on morphological stability. Since the scale of the flow field is usually orders of magnitude larger than the wavelength of the morphological instability, a rigorous treatment appears extremely difficult at present. For simple flow fields, there are a number of results, but the effect on morphological stability can depend sensitively on the detailed structure of the flow field [3, 7, 26, 31, 37, 38, 40, 42].

Even in the absence of radial gradients and forced flows, the effects of flow on morphological stability can be complex. The large number of dimensionless parameters in this problem makes a full investigation intractable, and there may be interesting regimes that have not been investigated. We therefore confine ourselves to an examination of the cases shown in Figs. 1 and 2, except that fluid flow is now allowed to occur.

In Fig. 3, we show the concentration of indium in tin above which instability occurs and the imaginary part of σ as a function of the wave number, ω, for $\alpha = 0$ and $\pi/2$ for $\theta = \pi/4$. For $\alpha = 0$, the imaginary part of σ vanishes and so is not shown in Fig. 3. The methods used in the linear stability calculations have been previously described [25, 41]; the anisotropy of thermal conductivity was incorporated into the previous analysis. In Fig. 3, the concentration exhibits two relative minima as a function of wave number; at each relative minimum the curve for $\alpha = 0$ lies below the curve for $\alpha = \pi/2$ so that the instability first occurs for $\alpha = 0$, although the concentration difference between the two curves is very small. We therefore confine our discussion to the $\alpha = 0$ curve. The two relative minima occur at ω = 4.3 and 130 cm^{-1} with c_∞ = 0.029 and 0.069 at.%, respectively.

In the absence of flow (Fig. 1), the critical concentration and wave number are in excellent agreement with the relative minimum that occurs at the larger value of ω in Fig. 3. The absolute minimum occurs at the lower value of ω and arises from the flow. When the interface deforms, radial gradients are produced that give rise to a buoyancy-driven flow. For the tin alloy considered, the thermal conductivity of the crystal and the melt are such that the flow sweeps solute into the depressions in the interface, lowering its melting point, and enhancing morphological instability [41]. This enhancement of morphological instability by the thermally-driven flow is less pronounced at higher growth velocities and lower thermal gradients. The wavelength of the instability is of the order of the sample size, and there may be interactions between this flow and flows due to radial gradients. The maximum enhancement occurs for solutes that are neutrally buoyant, as shown in [41], and thus the tin-indium system is particularly interesting.

Discussion

Our understanding of morphological stability has advanced in the years since the constitutional supercooling criterion and morphological stability theory were developed. It is now possible to calculate deep cells in two dimensions [45], the interaction of shallow multiple cells in two dimensions [46], and shallow cells in three dimensions [47]. Linear and weakly nonlinear theory are being refined to treat more complex situations. In this article we have discussed the effect of the anisotropy of thermal conductivity on the onset of morphological stability in alloys of tin containing indium. Under certain conditions the onset of instability can be oscillatory in time corresponding to travelling waves. The onset is likely to be oscillatory when growth is not along a principal axis and the temperature gradient in the solid exceeds that in the liquid [21]. For the calculations shown here, the temperature gradient in the liquid exceeds that in the solid, and the onset is non-oscillatory. The experiments of Davis and Fryzuk [17] on alloys of tin containing indium used growth rates of 7.5, 26.5, and 53.0 μm/s with temperature gradients in the liquid of 3.5, 2.8, and 1.0 K/cm. For the lowest velocity, the temperature gradient in the liquid exceeds that in the solid, while the reverse is true at the highest velocity. However, the experimental observations of the onset of instability did not agree with morphological stability theory; the planar interface was less stable than predicted by the theory using available thermophysical properties. The fluid flow effects we have just described do not resolve this discrepancy.

An oscillatory phenomena has been observed during the solidification of tin containing lead by ultrasonics[19]; however, it is unclear that this is related to the travelling wave instability. The anisotropy of interface kinetics can also give rise to oscillatory instabilities [22], as can fluid flow [3, 31]. However, the values of the kinetic coefficient and its anisotropy are unknown, so it is impossible to calculate the magnitude of the effect.

Acknowledgments

This work was conducted with the support of the Microgravity Science and Applications Division of the National Aeronautics and Space Administration, and the Applied and Computational Mathematics Program of the Defense Advanced Research Projects Agency. One of us (RFS) received partial support from the National Science Foundation under Grant DMR89-12752.

References

1. W. A. Tiller, K. A. Jackson, J. W. Rutter, and B. Chalmers, Acta Met. 1, 428 (1953).

2. W. W. Mullins and R. F. Sekerka, J. Appl. Phys. 35, 444 (1964).

3. R. T. Delves, Crystal Growth, p. 40, B. R. Pamplin, ed., Pergamon, Oxford (1974).

4. D. J. Wollkind, Preparation and Properties of Solid State Materials, p. 111, W. R. Wilcox, ed., Vol. 4, Marcel Dekker, New York (1974).

5. J. S. Langer, Rev. Mod. Phys. 52, 1 (1980).

6. S. R. Coriell, G. B. McFadden, and R. F. Sekerka, Ann. Rev. Mater. Sci. 15, 119 (1985).

7. M. E. Glicksman, S. R. Coriell, and G. B. McFadden, Ann. Rev. Fluid Mech. 18, 307 (1986).

8. R. A. Brown, AIChE J. 34, 881 (1988).

9. B. Chalmers, Principles of Solidification, Wiley, New York (1964).

10. D. Walton, W. A. Tiller, J. W. Rutter, and W. C. Winegard, Trans. AIME 203, 1023 (1955).

11. F. Weinberg, Trans. AIME 224, 628 (1962).

12. J. J. Kramer, G. F. Bolling, and W. A. Tiller, Trans. AIME 227, 374 (1963).

13. H. Biloni and G. F. Bolling, Trans. AIME 227, 1351 (1963).

14. H. Biloni, G. F. Bolling, and G. S. Cole, Trans. AIME 236, 930 (1966).

15. H. Biloni, R. DiBella, and G. F. Bolling, Trans. AIME 239, 2012 (1967).

16. F. Weinberg, Crystal Growth, p. 639, H. S. Peiser, ed., Pergamon, Oxford (1967).

17. K. G. Davis and P. Fryzuk, J. Crystal Growth 8, 57 (1971).

18. J. C. Warner and J. D. Verhoeven, Metall. Trans. 4, 1255 (1973).

19. K. Suzuki, A. Hikata, and C. Elbaum, Phys. Rev. Lett. 59, 2686 (1987).

20. Y. S. Touloukian, R. W. Powell, C. Y. Ho, and P. G. Klemens, Thermophysical Properities of Matter. Vol. 1. Thermal Conductivity. Metallic Elements and Alloys, p.408, IFI/Plenum, New York (1970).

21. S. R. Coriell, G. B. McFadden, and R. F. Sekerka, J. Crystal Growth, in press.

22. S. R. Coriell and R. F. Sekerka, J. Crystal Growth 34, 157 (1976).

23. G. W. Young, S. H. Davis, and K. Brattkus, J. Crystal Growth 83, 560 (1987).

24. G. B. McFadden, S. R. Coriell, and R. F. Sekerka, J. Crystal Growth 91, 180 (1988).

25. S. R. Coriell, M. R. Cordes, W. J. Boettinger, and R. F. Sekerka, J. Crystal Growth 49, 13 (1980).

26. S. R. Coriell and R. F. Sekerka, PhysicoChemical Hydrodynamics 2, 281 (1981).

27. S. R. Coriell, M. R. Cordes, W. J. Boettinger, and R. F. Sekerka, Adv. Space Res. 1, 5 (1981).

28. D. T. J. Hurle, E. Jakeman, and A. A. Wheeler, J. Crystal Growth 58, 163 (1982).

29. D. T. J. Hurle, E. Jakeman, and A. A. Wheeler, Phys. Fluids 26, 624 (1983).

30. R. J. Schaefer and S. R. Coriell, Metall. Trans. 15A, 2109 (1984).

31. S. R. Coriell, G. B. McFadden, R. F. Boisvert, and R. F. Sekerka, J. Crystal Growth 69, 15 (1984).

32. B. Caroli, C. Caroli, C. Misbah, and B. Roulet, J. Phys. (Paris) 46, 401 (1985).

33. B. Caroli, C. Caroli, C. Misbah, and B. Roulet, J. Phys. (Paris) 46, 1657 (1985).

34. D. R. Jenkins, PhysicoChemical Hydrodynamics 6, 521 (1985).

35. D. R. Jenkins, IMA J. of Appl. Math. 35, 145 (1985).

36. G. W. Young and S. H. Davis, Phys. Rev. B34, 3388 (1986).

37. M. Hennenberg, A., Rouzaud, J. J. Favier, and D. Camel, J. Phys. (Paris) 48, 173 (1987).

38. K. Brattkus and S. H. Davis, J. Crystal Growth 89, 423 (1988).

39. K. Brattkus and S. H. Davis, J. Crystal Growth 91, 538 (1988).

40. G. B. McFadden, S. R. Coriell, and J. I. D. Alexander, Comm. Pure and Appl. Math. 41, 683 (1988).

41. S. R. Coriell and G. B. McFadden, J. Crystal Growth 94, 513 (1989).

42. S. A. Forth and A. A. Wheeler, J. Fluid Mech. 202, 339 (1989).

43. P.-E. Berthou and R. Tougas, Metall. Trans. 1, 2978 (1970).

44. B. Predel and A. Emam, J. Less-Common metals 18, 385 (1969).

45. N. Ramprasad, M. J. Bennett, and R. A. Brown, Phys. Rev. B 38, 583 (1988).

46. M. J. Bennett and R. A. Brown, Phys. Rev. B 39, 11705 (1989).

47. G. B. McFadden, R. F. Boisvert, and S. R. Coriell, J. Crystal Growth 84, 371 (1987).

Micro- and macro-segregation during alloy solidification*

A. Hellawell
Department of Metallurgical and Materials Engineering, Michigan Technological University, Houghton, Michigan, U.S.A.

Abstract

Formal expressions for fractional crystallization or 'coring' are briefly reviewed with attention to their limitations and modifications to account for partial stirring in the liquid or back diffusion in the solid. The application of these approaches is considered for the solidification of a casting, especially the extent to which they may be used to accurately predict the form of a cooling curve. Hence, the practice of curve fitting thermal analysis data and the possibilities of the predictive modelling of microstructural development are discussed.

Various examples of macro-segregation - normal, inverse and those arising from gravitational forces - are reviewed, and the possibilities of realistically modelling these are also considered. In particular, the convective paths, rates and mechanisms for relatively long range transport are considered for different geometrical configurations. It is noted, in conclusion, that in most systems there can be micro-segregation with little or no macro-segregation, but rarely the reverse.

*Full paper on p. 395.

Processing-microstructure relationships in advanced cast aluminum alloys

S. Shivkumar and D. Apelian
Aluminum Casting Research Laboratory, Department of Materials Engineering, Drexel University, Philadelphia, Pennsylvania 19104, U.S.A.

ABSTRACT

Premium quality aluminum alloy castings are used extensively in various applications requiring a high strength to weight ratio such as in aerospace, automotive and other structural components. The mechanical properties in these structure-sensitive alloys are determined primarily by the secondary dendrite arm spacing and the morphology of interdendritic phases. In addition, the amount of porosity in the casting and the inclusion concentration have strong influence on fracture, fatigue and impact properties. During the production of the casting, various molten metal processing techniques can be implemented to control the microstructural parameters. These melt treatments include grain refinement with Ti-B, eutectic modification with Sr or Na, degassing with purge gases and filtration of inclusions. Recent developments in these areas are reviewed and salient results of ongoing investigations in our laboratory are presented.

KEY WORDS

Casting, Aluminum Base Alloys, Molten Metal Processing, Modification, Filtration

INTRODUCTION

The majority of the cast aluminum alloys are primarily hypoeutectic Al-Si alloys which have widespread applications especially in the aerospace and automotive industries. These foundry alloys possess excellent tensile and fatigue properties and good corrosion resistance. The addition of Si imparts excellent castability and resistance to hot-tearing in these alloys. Also, since silicon increases in volume during solidification, the susceptibility of the castings to shrinkage defects is reduced. Consequently, Al-Si alloys are ideally suited for high volume production in the aluminum foundry. The presence of magnesium in the alloy (356 and 357) offers the ability to heat treat Al-Si castings to high strength levels. Magnesium contents are typically less than about 0.75%, because increased additions impair fluidity and feeding. In many cases, copper is added (319) to enhance as-cast and heat treated strength properties. These hypoeutectic Al-Si alloys are used for investment, sand, lost foam and permanent mold castings.

The mechanical properties in these alloys are determined primarily by the melt quality, the efficiency of feeding systems, and the subsequent heat treatment of the casting. Because of their commercial and technological importance, these alloys have been the subject of extensive research. As a result, a number of innovative molten metal processing techniques aimed at improving the casting quality have emerged in recent years. The primary purpose of this contribution is to review some of the developments pertaining to the processing of molten aluminum alloys and to examine their effects on mechanical properties. In particular, recent results generated at our laboratory in the area of molten metal processing will be discussed.

SOLIDIFICATION OF CASTINGS

In hypoeutectic Al-Si alloys, solidification begins with the development of a dendritic network of primary (α) aluminum. The secondary dendrite arm spacing (DAS) is essentially determined by the following parameters: alloy composition, cooling rate, local solidification time and temperature gradient. Among the primary alloying elements, Si has been reported to have an influence on DAS [1]. Increasing the Si content lowers the secondary dendrite spacing. It has been shown that DAS varies with cooling rate according to the relation [2]:

$$\log (dT/dt) = - (\log (DAS) - 2.37)/0.4 \qquad (1)$$

where dT/dt, the cooling rate is in C/min and DAS is in μm. At any particular location in the casting, the local solidification time determines DAS and, hence, the mechanical properties. The secondary dendrite arm spacing controls the size and the distribution of porosity and intermetallic particles in the casting. As DAS becomes smaller, porosity and second phase constituents are dispersed more finely and evenly. This refinement of the microstructure leads to a substantial improvement in mechanical properties [3,4]. The tensile properties most affected by variations in DAS are ultimate tensile strength (UTS) and %elongation. In Al-Si alloys, both strength and fracture properties increase with solidification rate (Fig. 1) [5]. This behavior is in contrast to cast Al-Cu alloys where fracture toughness varies inversely with strength properties. Furthermore, in Al-Si alloys, there is only a marginal decrease in toughness with increasing yield strength. Thus, the Si particle morphology and distribution is much more important than the matrix strength in these alloys.

As solidification progresses, various non-equilibrium reactions may occur in the interdendritic liquid depending on cooling rate and on the amount of impurities (primarily iron and Manganese) in the liquid metal. A comprehensive study by Backerud and coworkers [6] reveals the following reactions for A356 alloys:

Liq. $\rightarrow \alpha + Si$
Liq. $\rightarrow \alpha + Al_5FeSi$
Liq. $\rightarrow \alpha + Si + Al_5FeSi$
Liq. $+ Al_5FeSi \rightarrow \alpha + Si + Al_8Mg_3FeSi_6$
Liq. $\rightarrow \alpha + Si + Mg_2Si$
Liq. $\rightarrow \alpha + Si + Mg_2Si + Al_8FeMg_3Si_6$

Similar reactions have also been identified for A357 and 319 family of alloys. The optical photograph in Fig. 2 shows the morphology of typical interdendritic phases.

INFLUENCE OF MELT CHEMISTRY

The chemical composition of the melt has a significant influence on microstructure and mechanical properties in the cast component. In Al-Si alloys of commercial importance, there are a number of naturally occurring impurities and great care must be exercised in minimizing their concentration in order to produce premium quality castings. Iron is probably one of the most important elements that needs to be controlled because it forms brittle intermetallic compounds which seriously impair ductility of the alloy. Several iron-rich compounds may form during solidification of the casting (Fig. 2) including large platelets of β ($FeSiAl_5$), chinese script $(Fe,Mn)_3Si_2Al_{15}$ particles and π ($Al_8FeMg_3Si_6$) phases. The reduction in ductility, fracture toughness and fatigue properties in Fe containing alloys has been attributed to the precipitation of thin, coarse needles of plate-shaped $FeSiAl_5$.

The deleterious effect of Fe can be countered by starting with a low iron charge, by minimizing iron pickup during melting and holding, by increasing the cooling rate and by making suitable metallic additions to the melt. At low cooling rates, $FeSiAl_5$ needles are coarse and concentrated at grain boundaries, where it can be expected that they will promote a brittle fracture. At higher cooling rates, $FeSiAl_5$ particles are quite small and interspersed more evenly. The most common addition to neutralize the effect of iron has been Mn, and to a lesser extent Co, Be and Cr. The addition of these elements changes the morphology of the Fe phase from a plate like crystals to chinese script particles. Recent studies by Awano and Shimizu [7] indicate that even in the absence of iron neutralizers the morphology of the Fe phase may also be altered by superheating the melt appropriately. For example, their results indicate that the Fe phase crystallizes predominantly as chinese script particles instead of Fe platelets when the melt is superheated in the temperature range 850 to 900C for short periods. The deleterious effects of Fe can be also overcome to a large extent by Sr modification. It has been reported that the formation of large, brittle iron intermetallic phases can be suppressed in a well modified alloy [8].

MELT DEGASSING

Aluminum alloy melts are highly susceptible to the absorption of hydrogen during melting and holding. The solubility of hydrogen in liquid aluminum is a function of temperature and decreases rapidly at the solidification temperature. The hydrogen rejected from the liquid metal during solidification leads to porosity in the casting. The relatively large solidification shrinkage in aluminum alloys exacerbates this problem. Depending on the processing conditions, porosity levels of 0.5 to 3% may be detected in commercial castings.

The amount of porosity that may be tolerated excessive amounts of porosity begin to form depends on several factors including alloy composition, local solidification time, thermal gradients, design of casting and inclusion concentration. Extensive mathematical and experimental analysis of microporosity formation has been conducted at our laboratory. Our results indicate that depending on cooling rate, there may exist a critical hydrogen concentration below which no porosity may form in the casting (Fig. 3). This critical hydrogen concentration is of the order of 0.05 to 0.1 cc/100 g in untreated A356.2 alloys. This value is in agreement with the results of Thomas and Gruzleski [9] who have demonstrated that if the H_2 content in the melt is maintained below approximately 0.08 cc (STP)/100 g of Al, pore free castings can be obtained under common casting conditions.

The mechanism of hydrogen induced porosity may be analysed by considering the partitioning of dissolved hydrogen at the solid/liquid interface. During solidification, the interdendritic liquid is gradually enriched with hydrogen as the fraction solid increases, since most of the hydrogen is rejected at the solid-liquid interface. Solubility of hydrogen in liquid and solid aluminum is in the order of 0.65 and 0.035 ml/100 g alloy respectively [10]. The variation of hydrogen concentration in the liquid during solidification as calculated by our model is shown in Fig. 4. As solidification progresses, the hydrogen content in the liquid increases and eventually exceeds the solubility limit. Ideally, a gas pore should nucleate at this point. However, the creation of a new pore requires the establishment of a new surface. Because of this surface barrier, the hydrogen concentration in the liquid will continue to increase above the solubility limit until it reaches a maximum value at which pores can form, stage I in Fig. 4 [11]. At this point, pores begin to nucleate, stage II. This nucleation occurs predominantly at the root of dendrites or at other heterogeneous sites such as inclusions. Once the pore is nucleated, there is rapid growth during the initial stages and the hydrogen concentration in the liquid decreases, stage III. Subsequently the pore detaches from the dendrite arm and grows at a relatively slow rate, stage IV.

Inclusions in the melt offer heterogeneous sites for the nucleation of the pore and thus promote pore formation. Even in the presence of small quantities of inclusions the threshold hydrogen content may be lowered significantly. The original work of Blondyke and Hess [12] illustrated that for the same hydrogen content varying amounts of porosity may form depending on the inclusion concentration. In the absence of inclusions, extremely high gas pressures are required to overcome surface tension forces when the first molecules of hydrogen coalesce to form the bubble. Thus, the critical hydrogen content varies inversely with the inclusion content and can be expected to be greater than 0.3 cc/100 g at low inclusion levels.

Depending on the application, a variety of methods are being used to degas aluminum alloy melts including addition of fluxes to the melt, bubbling of purge gases through the molten metal, vacuum degassing and ultrasonic vibration of the melt. Many in-line commercial processes such as MINT (Melt in-line treatment system), SNIFF (Spinning nozzle inert floatation) and Alcoa 622 process are also available for this purpose [13]. Most often, however, a purge gas is bubbled through the molten aluminum for extended periods to accomplish degassing. Various gas or gas mixtures, which include nitrogen , chlorine and nitrogen-freon mixture, have been employed. Factors such as the size and surface area of the gas bubble, the rate of gas flow, the depth of the container holding the melt and the bath temperature exert an influence on the degassing process. Commercially available tablets which release chlorine when added the melt are also used to degas the metal. The efficiency of the various techniques used to degas the melt have been the subject of some controversy, largely because of the difficulties associated with the sampling of aluminum melts and analysis for dissolved hydrogen.

The results of a recent mathematical model indicate that the degassing process is essentially controlled by three parameters: metal temperature, purge gas bubble size and purge gas composition [14,15]. The solubility of H_2 in aluminum increases exponentially with temperature and so the time to degas to a certain level doubles for each 70C increase in melt temperature. It becomes imperative therefore that the melt temperature should be maintained as low as possible. It has also been shown that the degassing process operates at almost 100% efficiency when the purge gas bubble diameter is less than about 5 mm. The efficiency decreases rapidly as the bubble size increases. These results form the basis of the rotary impellor degasser which is increasingly being employed in aluminum foundries to maximize the degassing efficiency.

INCLUSION REMOVAL

The term "inclusion" refers to any exogeneous solid or liquid phase particle present above the liquidus temperature of the alloy. Several types of inclusions may be present in the melt including furnace dross, salts, oxides, and unmelted elements. Nonmetallic and intermetallic particulate suspensions may be introduced into the melt through the charge material or during various processing operations. Inclusion concentrations in unprocessed aluminum melts range from 0.005 to 0.02% (volume fraction) and have been observed in a variety of morphologies. Even at very low inclusion concentration (by volume) a substantial number of inclusions may be present in the melt [16]. For example, if the average inclusion size is about 40 µm, then at an inclusion concentration of 1 ppm, 1 Kg of metal will contain approximately 11,000 particles. The critical size of inclusions that may be tolerated depends on the end application. In most cast components, inclusions with particle sizes greater than 10 to 20 µm may have a drastic effect on the quality of the part. Inclusions in the melt may lead to several problems such as reduction of mechanical properties in the casting, poor machinability, loss of fluidity, increased porosity in the casting, poor surface finish and lack of pressure tightness.

The particles present in molten metal prior to casting are an inevitable feature of the production and can broadly be classified as [17]:

- *Solid Inclusions:* isolated, rigid particles or agglomerates of different phases, textures and morphologies which may be introduced in the melt through extraneous sources (such as refractory brick, grain refiner, etc.) or may form during processing.
- *Liquid Inclusions:* particles which are deformable and may coalesce into globules.
- *Fluxes and Salts* are suspended in the melt as a result of prior melt treatment process.

Solid inclusions in aluminum castings may be metallic or non-metallic and may originate from a variety of sources. The most common type of solid inclusions are oxides, nitrides, carbides and borides. Because of the strong affinity for oxygen, aluminum can form more than 15 different oxides, ranging in specific gravity from 2.3 to 4 for corundum. The concentration of oxygen in commercial aluminum is in the range 0.1 to 100 ppm. In aluminum alloys, oxide films and clusters of oxide particles are often observed in the casting. Fortunately, the Pilling-Bedworth ratio for Al is 1.27 indicating that the oxide film is dense, continuous and protective. Oxides may form at different stages in the melting process [18]. They may originate from electrolytic cells, from reaction between the melt and the atmosphere, and from the interaction between the melt and the refractory linings in the furnace. Other solid inclusions such as nitrides (AlN), carbides (Al_4C_3), borides (TiB_2, VB_2), $CaSO_4$ CaS and Al_3Zr have also been detected in aluminum melts [19-21].

Liquid inclusions are predominantly chlorides and flourides. Chlorine gas and chloride fluxes are often used with aluminum alloys for degassing the melt. Hence chloride conclusions may form in melt, especially in Mg containing alloys [22]. The compound $MgCl_2$ may form during the chlorination process. Aluminum alloys may normally contain less than 1 ppm of cryolite, but 10-30 ppm of cryolite may be found in some commercial alloys. The source of the cryolite particles is the alumina electrolyte. NaF may also be present in large inclusions.

Excessive temperatures during melting accelerate the oxidation of aluminum especially when fluxes are used to help facilitate the removal of dross. If the flux material is not removed properly at the correct temperature, separation of the salts occur, resulting in contamination of the melt and in salt-wetted oxide inclusions.

A number of methods have been adopted to reduce the incidence of inclusions in the casting including sedimentation, floatation, adjustments in the gating system design and the use of a variety of filter materials in the mold cavity [22]. Gravity sedimentation techniques based on Stoke's law are practically limited to inclusion sizes greater than 90-100 μm. Floatation devices may effective up to about 30 to 40 μm. Removal of inclusions less than about 40 μm in size may be accomplished by filtration. Melt filtration is a two-step serial transport process. The inclusions are transported to the filter surface as a result of bulk fluid flow and subsequently, the inclusions are captured in the filter material due to interfacial or surface forces. Materials such as fiber-glass cloths, steel gauze screens and wire wool have been used in foundries for inclusion removal. In recent years, the introduction of ceramic foam filters in the foundry has largely overcome many problems experienced with filtration. The ceramic foam filter is highly effective in reducing the turbulence of the molten metal and does not cause contamination of the melt. Ceramic foam filters can now be produced in a wide variety of sizes and shapes and can be used for filtering bulk metal in the ladle or within the mold cavity. They are applied extensively in sand, pressure, investment and die casting processes. Ceramic filters in dipant wells or crucibles, and small ceramic foam filters [22] and fine mesh steel screens in the ingots of the mold are commonly used in the production of aerospace castings.

Two types of filters which have wide scale industrial application are being studied at our laboratory: deep-bed filtration and cake filtration. The deep bed filter consists of a packed bed of refractory particles (usually tabular alumina) through which the molten aluminum flows. The inclusions deposit onto the grains of the filter medium due to diffusion, direct interception, gravity, and/or surface forces. Mechanical entrapment has been observed to be responsible for filtration of inclusions larger than 30 μm; whereas, it is believed that surface forces are responsible for the retention of inclusions smaller than 30 μm. The primary mode of transport of inclusions smaller than 30 μm to the grain surface of the filter medium is due to flow dynamics, while surface forces are responsible for retention of the inclusion. In a depth filter the inclusions are dispersed through part or all of its volume (depth). It thus has the advantage of having a large surface area for entrapment and can trap particles much smaller than the pores present in the filter bed. In cake filtration, the inclusions are deposited at the upper surface of the filter. The latter aggregate of inclusions acts as a "second" filter above the original medium rendering an additional resistance to flow and thus causing separation of the particulates from the melt.

Ceramic foam filters are generally used in many commercial applications to minimize inclusion concentration. The ceramic foam filter consists of a sheet of open pore ceramic foam approximately 5 cm thick. Typically, it would have 85% volume void fraction, an average pore size of 1 mm, with approximately 30 pores per linear inch. A ceramic foam filter can be installed prior to the casting station with minimal capital expenditure. However, it only has a limited life because with use the inclusions form a filter cake which increases the flow resistance. Thus, the ceramic foam is a good example of a filter medium which at the onset operates as a depth filter and at the end of its "useful life" operates as a cake filter.

Other techniques of inclusion removal include fluxing treatments and flushing with reactive gases. These techniques are useful on a limited scale and are not as efficient and consistent as the filtration systems. It has been reported recently that extremely fine purge gas bubbles used with rotary impellor degassers may "float out" oxides present in the melt [22]. This flotation effect is accentuated by adding small amounts of chlorine or foam to the inert purge gas [23] since the presence of the halogen changes the surface tension between the oxide film and the ascending gas bubble.

GRAIN REFINEMENT

Aluminum alloy castings are grain refined with Ti or Ti-B master alloys in order to control the grain structure in the casting. The grain refiner provides nucleating sites for the formation of primary α-aluminum dendrites and facilitates the production of a casting with a large number of small, uniform and equiaxed grains. The results of Apelian and Cheng [24] indicate that the addition of grain refiner to the melt may have a beneficial effect irrespective of the cooling rate during solidification. Grain refining the casting has several advantages including reduction of the amount of porosity, improved resistance to hot tearing and enhancement of feeding properties.

Typical microstructures in samples that are not grain refined and in well grain refined samples are shown in Fig. 5. When no grain refiner is added, the structure is very coarse and the average grain size is greater than 3000 μm. Upon addition of the master alloy, grain size decreases rapidly and is measured to be less than 300 μm . Typical grain size values for sand, permanent mold and lost foam 319 castings produced under similar conditions are plotted in Fig. 6 as a function of Ti concentration [25]. As can be expected, lowest grain sizes are obtained with permanent mold castings for any particular Ti concentration. At most locations in lost foam specimens, the grain size is finer than in green sand castings. A minimum Ti content of 0.15% is necessary to obtain substantial grain refinement. This alloy contains appreciable amounts of Cu and Zn in the solid solution. Apparently, the presence of these elements in the solid solution interferes with the performance of the grain refiner. Consequently, optimum grain refiner additions in alloy 319 are greater than in alloy 356 which contains less than about 0.2% Cu and no Zn [26]. The effects of various alloying elements on the grain refining efficiency have recently been reviewed by Abdel-Hamid [27].

Grain refinement of the casting may have a significant effect on the porosity levels in the casting. The average amount of porosity in 319 castings is plotted as a function of Ti concentration in Fig. 7 [25]. It is evident that for lost foam, green sand and permanent mold castings the amount of porosity varies inversely with the Ti concentration. Addition of 0.2% Ti lowers the amount of porosity from about 1.5% to 0.5%. This reduction in the fraction of porosity is directly related to the decrease in average pore diameter upon grain refinement [28]. It can be seen from Fig. 8 that the pore diameter decreases from 55 μm in untreated alloys to 35 μm in alloys containing 0.2% Ti. Since the radius of interdendritic liquid pool decreases with the addition of Ti, the average pore size is lowered. Under these conditions, the gas pressure in the liquid is not strong enough to overcome the contribution of interfacial energies and pore formation becomes difficult. Furthermore, grain refinement leads to improved feeding of the alloy. The dendritic network is broken down into small equiaxed grains. These fragmented dendrites are transported within the casting and contribute significantly to the mass feeding that occurs during solidification. The improved feeding characteristics in grain refined alloys minimize the incidence of shrinkage voids in the casting.

MODIFICATION

An irregular eutectic structure is observed in Al-Si alloys with non-faceted aluminum growing together with faceted Si crystals. Thus, depending on the growth conditions (interface velocity and temperature gradient) and the melt concentration, the minor Si phase can adopt a variety of morphologies. At low cooling rates, Si is present as coarse, acicular and interconnected flakes in the unmodified state. These flakes act as crack initiators and consequently the material fractures in a brittle mode. A flake to fiber transition occurs at growth rates greater than about 1 mm/s [29]. The fibrous form of Si minimizes stress concentration effects and therefore improves ductility substantially is samples that are rapidly cooled during solidification. The flake to fiber transition can also be attained chemically by the addition of small amounts of reactive elements such as Na or Sr to the melt. This process which is referred to as "modification" in the trade literature improves mechanical properties of cast Al-Si alloys significantly [30]. Modification with Na & Sr results in a fine and fibrous eutectic Si as shown in Fig. 9. Sr is used extensively in North America in the form various Al-Sr (sometimes Al-Si-Sr) master alloys. A partial modification effect is also obtained by the addition of Sb to the melt, which produces a lamellar Si structure. The beneficial effects of modification on the microstructure and mechanical properties of Al-Si alloys have been well documented [31].

The amount of modifier necessary to achieve modification generally depends on the type of modifier and the concentration of impurity elements which "poison" the modifier. About 0.005 to 0.01% Na is sufficient to obtain complete modification while 0.01 to 0.02% Sr may be necessary to produce the same effect. Increasing the Na or Sr concentrations beyond the values indicated above may have detrimental effect on microstructure and mechanical properties. Overmodification effects with Na have been well established [32]. Recently, it has been observed that increasing the Sr content beyond about 0.015% results in a coarsening of Si particles [25]. In 319 alloys, it has been observed that both particle diameter and aspect ratio begin to increase at Sr levels greater than about 0.015-0.02% (Fig. 10). DasGupta et al [33] have observed a similar coarsening effect in sand cast A356 alloys. This coarsening of Si particles may explain the results of Closset and Gruzleski [34] and Ozelton et al [35] who report a reduction in mechanical properties at Sr contents greater than 0.01 to 0.015%Sr (Fig. 11). It should be noted that the Sr content at which coarsening of

Si particles begins may depend on the cooling rate.

One of the major problems associated with chemical modification is the susceptibility of castings to porosity formation. Our work with 319 alloy castings indicates that there is a significant increase in porosity upon Sr modification. The porosity in simple strip castings increases from about from 1.8% to about 2.8% after addition of 0.005% Sr to the melt (Fig. 12). Subsequent increases in Sr content result in a marginal increase of porosity and at 0.03% Sr, porosity levels are of the order of 3.4%. The severity of this problem can be reduced substantially by using a combination of grain refinement and modification. For example, in samples containing 0.2% Ti and various amounts of Sr, the maximum amount of porosity was less than 1.5%.

It has also been observed that the average pore size increases upon modification [25]. The variation of average pore size in 319 alloy castings with Sr content is plotted in Fig. 8. The pore size increases by a factor of 2 (from 35 μm to 70 μm) when 0.01% Sr is added to the melt. Increasing the Sr content beyond 0.01% does not have any significant effect on pore size. The pore size in grain refined and modified melts is smaller than the preceding value and is measured to be about 54 μm at 0.01% Sr. Similar results have been reported by Fang and Granger [28] for A356 alloys.

The exact mechanism by which modification increases the amount of porosity has not been clearly established. Some authors suggest that modification exacerbates porosity problems by increasing the H_2 content of the melt and/or the rate of regassing of the melt [30,36]. Others contend that the addition of Sr does not increase the hydrogen concentration or the rate of regassing [37,38]. It is clear, however, that several factors may contribute to increased porosity in modified castings. Because of the depression of the eutectic temperature in modified alloys, the freezing range increases and pores may form over a longer period. Thus, pores may grow to larger sizes in modified alloys. Argo and Gruzleski [39] indicate that large pores may be observed in modified alloys because of problems associated with interdendritic feeding. The reduction of eutectic temperature in modified alloys increases the length of the mushy zone and therefore large pockets of interdendritic liquid may become isolated. Solidification of this liquid may result in the formation of large pores. The addition of reactive elements such as Na and Sr to the melt reduces the surface tension and may facilitate the nucleation of the pore. Fang and Granger [35] suggest that Sr may also increase the inclusion content of the melt and thereby provide additional sites for the nucleation of the pore.

Chemical modification of the melt has a strong effect on the structural changes occurring during heat treatment [40]. Al-Si castings are generally solutionized for relatively long periods in order to change the Si particle morphology. The primary objective during solution treatment is spherodize the Si particles through "thermal" modification processes. Plate-like eutectic structures observed in unmodified alloys are resistant to shape perturbations and hence spherodize very slowly. It can be seen from Fig. 13 that even after 26 hr of solution treatment the unmodified structure is not completely spherodized. By comparison, spherodization occurs very rapidly in Sr modified alloys which undergo fragmentation readily. Consequently, chemical modification with Sr or Na can be used to reduce solution times significantly. The data shown in Fig. 14 show that optimum solution times for maximizing mechanical properties in the casting are significantly lower in Sr modified than in unmodified samples. Thus, chemical modification can be used effectively to reduce the overall heat treatment costs.

SUMMARY

Optimization of microstructural parameters in cast aluminum alloys entails proper control of molten metal processing operations. Degassing and filtration techniques can be utilized to control melt cleanliness. Our understanding of the factors controlling melt cleanliness is not at an advanced stage primarily because of the lack of a quantitative and inexpensive technique to monitor melt cleanliness. Clearly, more work is needed to alleviate this problem. Although the mechanism of grain refinement has not been determined unambiguously from a scientific point of view, the technology is well established. Foundrymen can use Ti-B master alloys to lower grain size and improve feeding properties. Grain refinement also leads to a reduction in the amount amount and the average pore size. The whole area of chemical modification is a subject of controversy. It is clear that the addition of Sr or Na to the melt produces the Si particles in the desired morphology and also leads to a redistribution of the shrinkage. In addition, it has been shown that chemical modification with Sr may also be used to lower substantially solution heat treatment times and hence reduce heat treatment costs. However, the increase in porosity because of the addition of Na or Sr to the melt may reduce the overall beneficial effects of chemical modification. The exact mechanism of this increase in porosity is not clear although several theories have been proposed by various researchers. It is imperative that additional work be conducted in this area so that a clear understanding of the fundamental mechanisms may be obtained.

REFERENCES

1. M. Bamberger, B.Z. Weiss and M.M. Stupel, *Materials Science and Technology*, 1987, **3**, 49-56
2. M. Drouzy and M. Richard, *Fonderie* , 1969, **285**, 500-511
3. J.C. Jaquet and H.J. Huber, *Geissereiforschung*, 1985, **38**(1), 11-20
4. K.J. Oswalt and M.S. Misra, *Trans AFS* , 1980, **88**, 845-862
5. G.D. Scott, B.A. Cheney and D.A. Granger: in "Technology for Premium Quality Castings", (ed. E. Dunn and D.R. Durham), 123-149; 1988, Warrendale, PA, The Metallurgical Society
6. L. Backerud, G. Chai and J. Tamminen, "Solidification Characteristics of Aluminum Alloys: Vol. 2 Foundry Alloys", AFS publication, 1990
7. Y. Awano and and Y. Shimizu, 56th World Foundry Congress, Dusseldorf, May 19-23, 1989, paper #27
8. G. Sigworth, *Modern Casting* , 1987, **77**(7), 23-25
9. P.M Thomas and J.E. Gruzleski, *Met Trans B*, 1978, **9B**, 139-141.
10. D. E. Talbot: *Inter. Metall. Rev.*, **20**, 1975, 166-183
11. J. Zou, S. Shivkumar and D. Apelian: *AFS Trans.*, **98**, 1990, in press
12. K.J. Brondyke and P.D. Hess, *Trans AIME* , 1964, **230**, 1542-1546
13. D.A. Anderson, D.A. Granger and J.G. Stevens: in "Production and electrolysis of light metals", 28th Annual Conference of Metallurgists, Halifax, Canada, August 20-24, 1989, Canadian Institute of Mining and Metallurgy, 163-172
14. G.K. Sigworth and T.A. Engh, *Met. Trans. B*, 1982, **13B**, 447-460
15. G.K. Sigworth, *Trans AFS* , 1987, **95**, 73-78
16. J.P. Martin, G. Dube, D. Frayce and R. Guthrie, *Light Metals*, 1988, 445-455
17. D. Apelian, Proceedings of the 46th Electric Furnace Conference, Pittsburgh, December 1988
18. C.J. Siemensen and G. Berg, *Aluminium*, 1980, **56**, 1-7
19. C.J. Siemensen and G. Strand, *Z. Anal. Chem.*, 1981, **308**, 11-16
20. C.J. Siemensen, *Met Trans B*, 1981, **12B**, 733-743
21. C.J. Siemensen, *Z. Anal. Chem.*, 1978, **292**, 207-212
22. Metals Handbook, Volume 15, Ninth edition, American Society for Metals, 1988
23. T. Pedersen: "Purge gas treatment of Aluminum to remove hydrogen, dissolved alkali metals and inclusions", Ph.D. Thesis, 1984, Metallurgical Institute of the Norwegian Technical University, Trondheim, Norway
24. D. Apelian and J.J.A. Cheng, *Trans AFS*, 1986, **94**, 797-808
25. L. Wang, S. Shivkumar and D. Apelian: in "2nd International Conference on Molten Aluminum Processing", Orlando, Florida, Nov. 6-7, 1989, AFS, paper # 5
26. G. Sigworth, "International Molten Aluminum Processing", City of Industry, California, Feb 17-18, 1896, 75-101
27. A.A. Abdel-Hamid, *Z. Metallkde* , 1989, **80**, 566-569
28. Q.T. Fang and D.A. Granger, *Trans AFS*, 1989, **97**, in press
29. L. Hogan, Shu-Zu Lu and A. Hellawell, *Met. Trans A* , 1987, **18**, 1721-1739
30. J.R. Denton and J.A. Spittle, *Materials Science & Technology* , 1985, **1**, 305-311
31. G.K. Sigworth, *Trans AFS* , 1983, **91**, 7-16
32. L.F. Mondolfo, "Aluminum Alloys: Structure & Properties", 2nd Edition, 1976, Butterworth's & Co. Ltd.
33. R. DasGupta, C.G. Brown and S. Marek, *Trans AFS* , 1988, **96**, 297-310
34. B. Closset and J. E. Gruzleski, *Trans AFS* , 1982, **90**, 453-464
35. M.W. Ozelton, G.R. Turk and P.G. Porter: same as Reference 5, 81-106
36. T. Hurley and R. Atkinson, *Trans AFS* , 1985, **93**, 291-296
37. F.C. Dimayuga, N. Handiak and J.E. Gruzleski, *Trans AFS* , 1988, **96**, 83-88
38. H. Shahani, *Scandinavian Journal of Metallurgy* , 1985, **14**, 306-312
39. D. Argo and J.E. Gruzleski, *Trans AFS* , 1988, **96**, 65-74
40. S. Shivkumar, S. Ricci, Jr., B. Steenhoff, D. Apelian and G. Sigworth, *Trans AFS*, 1987, **97**, 791-810

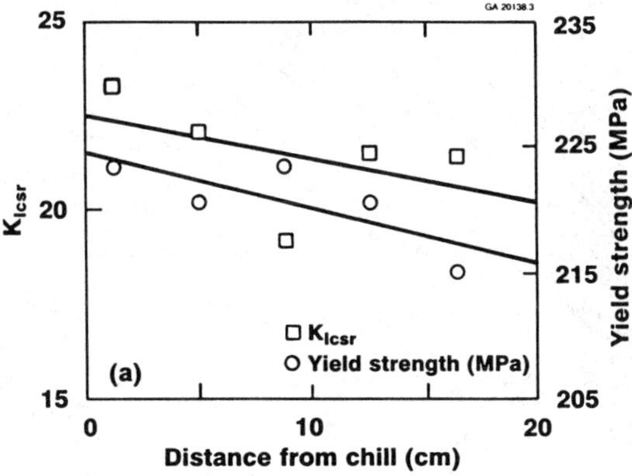

Fig. 1 Variation of short rod fracture toughness (K_{ICSR}) and yield strength (YS) with distance from chill in unmodified A356-T6 castings [7].

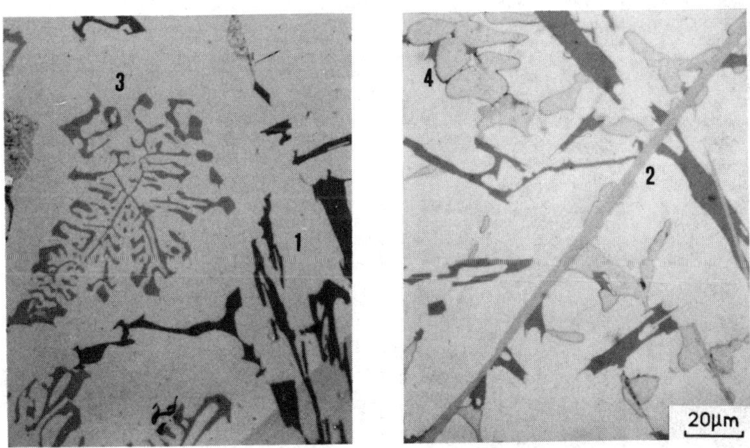

Fig. 2 Typical microstructures in an unmodified alloy showing various interdendritic phases: 1) Si 2) $FeSiAl_5$ 3) $(Fe,Mn)_3Si_2Al_{15}$ 4) $CuAl_2$.

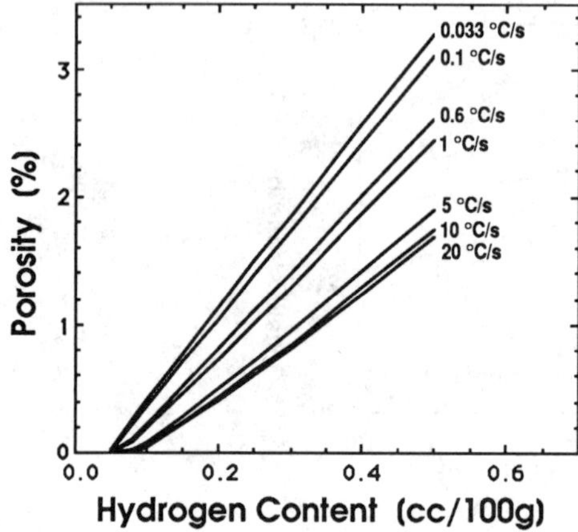

Fig. 3 Calculated amount of porosity in an unmodified A356 alloy as a function of hydrogen content and cooling rate.

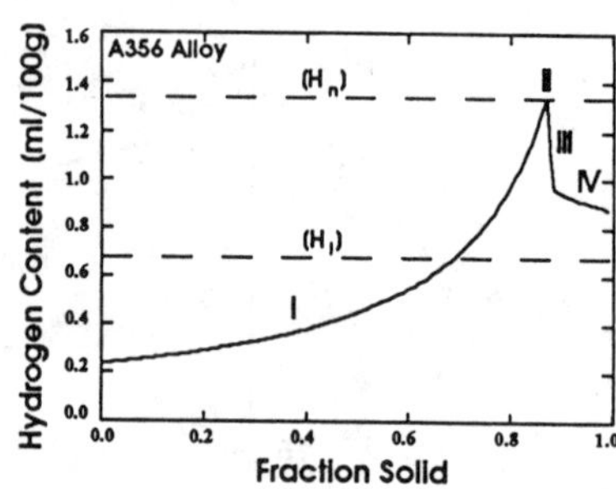

Fig. 4 Calculated variation of hydrogen concentration in the interdendritic liquid as a function of fraction solid.

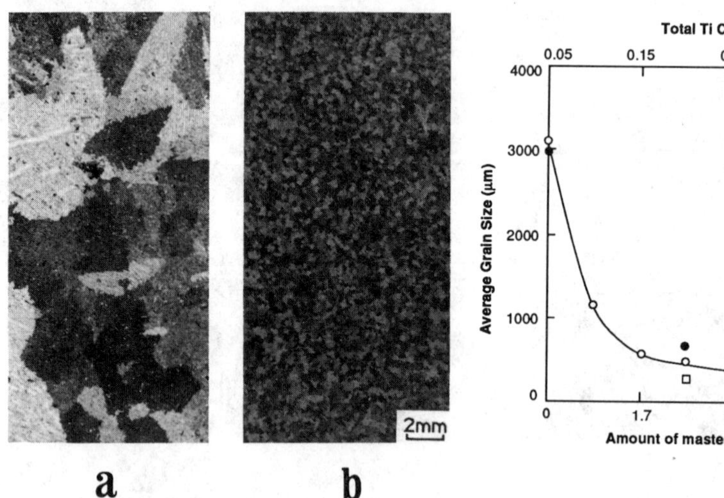

Fig. 5 Grain structures in alloy 319 castings: a) No grain refinement b) Grain refined with Ti-B (0.2%Ti).

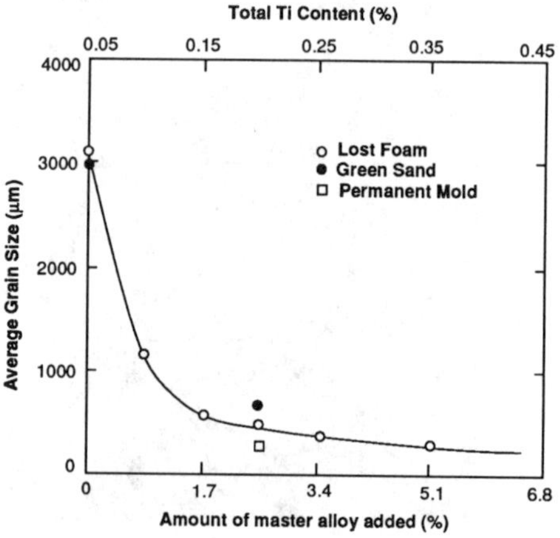

Fig. 6 Variation of average grain size in alloy 319 castings with Ti concentration. The data for three different casting processes are plotted.

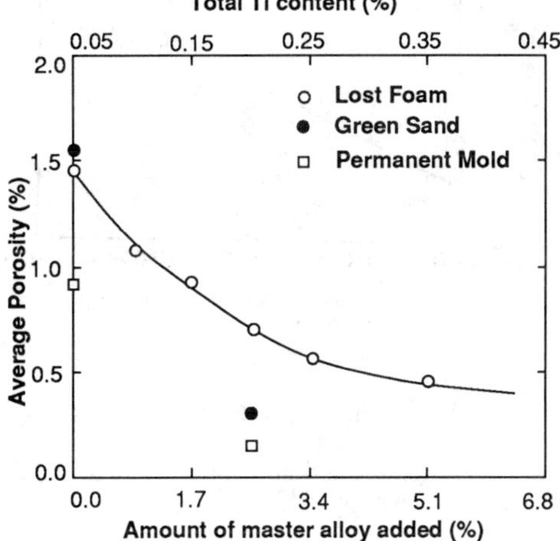

Fig. 7 Average porosity in alloy 319 castings as a function of Ti concentration.

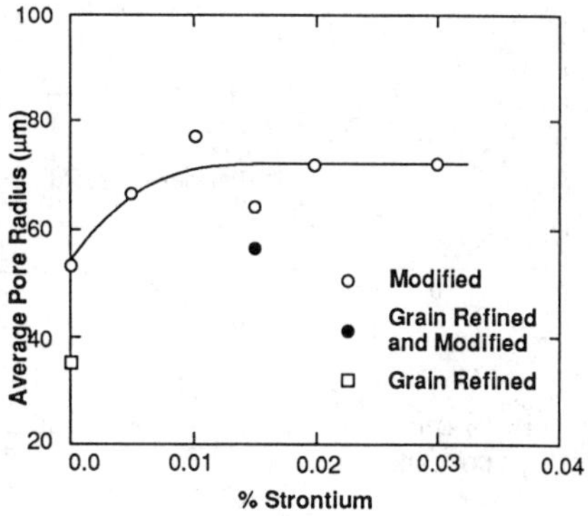

Fig. 8 Variation of average pore size in alloy 319 castings.

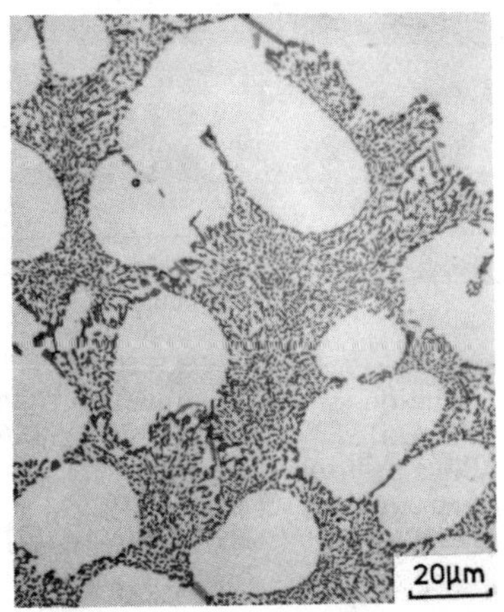

Fig. 9 Photograph showing Si particle morphology in modified A356 alloys.

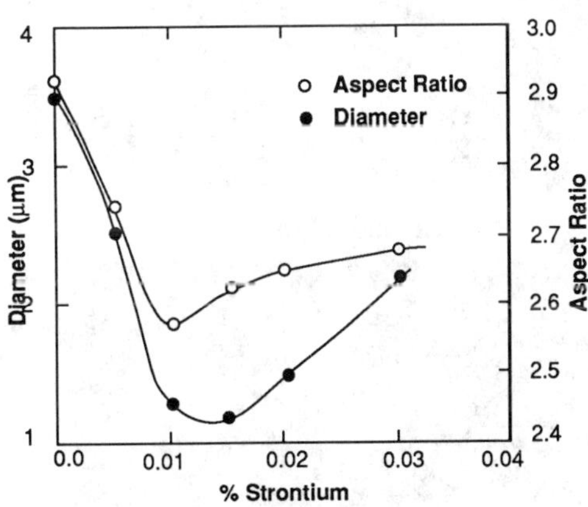

Fig. 10 Variation of Si particle average diameter and aspect ratio with Sr concentration.

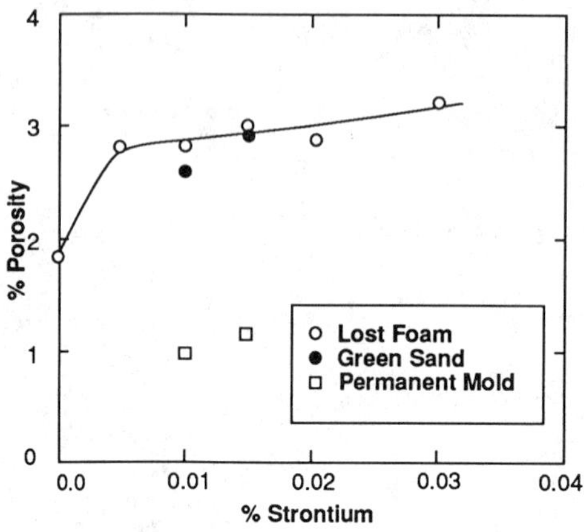

Fig. 11 % porosity as a function of Sr concentration.

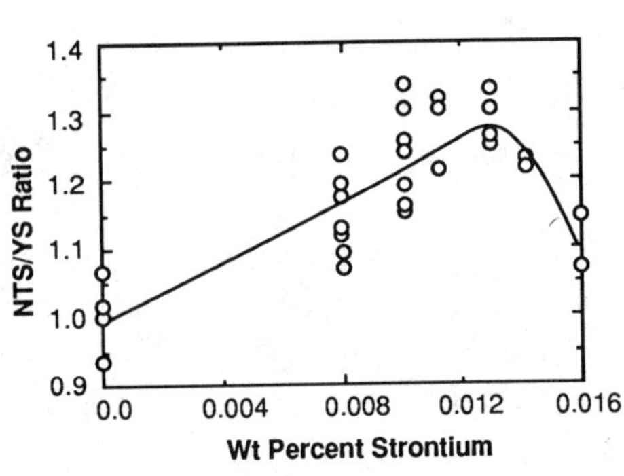

Fig. 12 Variation of notched tensile strength (NTS) and Yield strength (YS) ratio as a function of Sr content in A357-T6 alloys [43].

Unmodified

Sr modified

20μm

1 hr **24 hr**

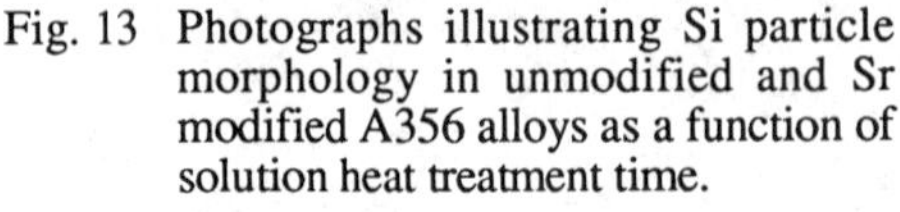

Fig. 13 Photographs illustrating Si particle morphology in unmodified and Sr modified A356 alloys as a function of solution heat treatment time.

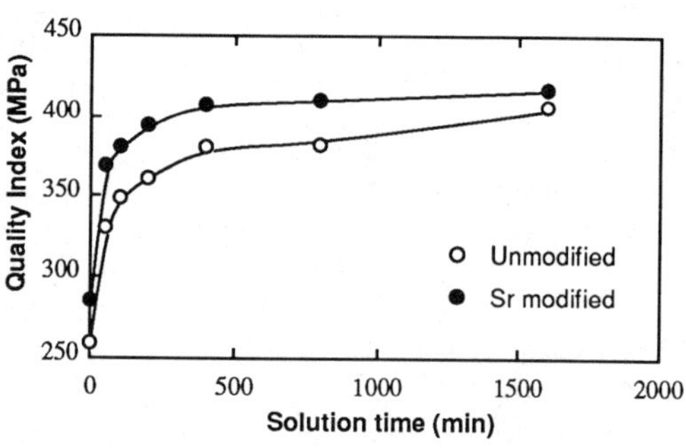

Fig. 14 Variation of quality index (Q) in unmodified and Sr modified A356 alloys as a function of solution heat treatment time. Q is defined as:
Q (MPa)=UTS (MPa)+150(log(%El)

Directional solidification microstructures

R. Trivedi
Ames Laboratory, U.S. Department of Materials Science and Engineering,
Iowa State University, Iowa State University, Ames, Iowa, U.S.A

Abstract

Directional solidification technique is important both in understanding fundamental phenomena of solidification in technological applications where highly controlled microstructures are desired for close control of properties. One of the critical aspects of directional solidification is to characterize specific microstructure and its length scale as a function of processing conditions, e.g. growth rate, temperature gradient and composition. Our current understanding of the basic ideas which control microstructural scales and which characterize microstructural transitions will be discussed. Specific attention will be given to the role of anisotropic interface properties on microstructural scale selection process, and on the microstructural transitions which occur under rapid solidification conditions.

II. Non-ferrous casting processes

Chairman: J.E. Gruzleski
Department of Mining and Metallurgical Engineering,
McGill University, Montreal, Quebec

Direct chill ingot casting: progress in control (Keynote Paper)

N. B. Bryson
Alcan International Ltd., Kingston, Ontario, Canada

Introduction

From an outsider's viewpoint, the direct chill casting process appears to be a relatively unsophisticated method of casting large blocks of aluminum for subsequent working into semi-fabricated products. Not surprisingly, the DC ingot may exhibit a wide spectrum of solidification defects, from cracking through porosity to segregations. Despite thermal treatments and mechanical working, many ingot defects may persist through to the end product. Clearly, metallurgical defects must be eliminated at the ingot casting stage if high quality end products are to be produced.

The DC process is the predominant process used today by the aluminum industry to cast ingots in a wide range of alloys, shapes and sizes for semi-fabrication processing.

As depicted in (Figure 1), a shallow, water-cooled mould, of shape and size corresponding to the cross-section of the desired ingot, is initially closed by a starting block. Casting is begun by feeding molten metal into the open top of the mould, while slowly lowering the starting block downward by a ram.

A thin solid shell is solidified by the mould, but the major heat flow is rejected to the cooling water flowing directly against the sides of the ingot - hence the name "direct chill process".

The depth of the molten pool may be many times deeper than the mould length, depending upon ingot thickness (up to 75 cm) and the casting rate (up to 150 mm/min.).

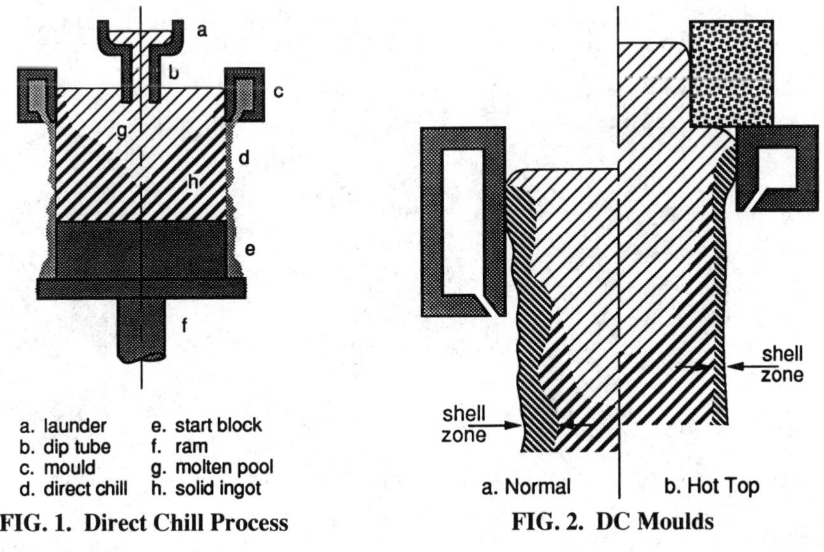

FIG. 1. Direct Chill Process

FIG. 2. DC Moulds

When the desired ingot length has been produced by the travel of the ram (up to 8 metres), withdrawal is stopped, the ingot removed, the starting block re-entered in the mould, and the process is repeated. Typically, from six very large ingots to ninety-six small ingots are cast simultaneously. In a few plants, the process has been operated fully continuously in the horizontal direction.

Examples of Metallurgical Control

Considerable progress has been made during more than 40 years of DC casting development in controlling process and metallurgical problems. In recent years, a number of sophisticated techniques have been developed. Three such examples are described:

1. Problem: On cylindrical ingot for the extrusion process, rough surfaces and deep sub-surface segregations are common on ingot cast through conventional moulds. During extrusion, the ingot surface may become part of the extrusion surface. Hence, a thick, segregated ingot shell zone leads to excessive scrap during extrusion.

Cause: The conventional, ''long DC mould'' (Figure 2a), is known to generate non-uniform heat transfer due to air-gap formation. A highly segregated rough surface and a thick, segregated shell is typical with conventionally cast DC extrusion ingot.

Control: By minimizing the heat removed by the mould and maximizing the heat rejected to the direct chill by means of a ''short mould'', thin shell, virtually segregation-free ingot may be cast. In practice, short moulds are difficult to operate because the molten metal level must be closely controlled to prevent molten metal bleed-out. However, the level control problem has been overcome by placing an insulated ''hot top'' above the mould (Figure 2b). This technique eliminates the need for precise level control, and makes short moulds feasible for production use.

Three types of hot-top mould designs are currently in widespread production use (Figure 3): (a) Porous graphite mould with gas infiltration[1], (b) Normal mould with gas injection[2], (c) Very short, ''critical length'' mould[3]. All are capable of casting thin shell ingot (2-3 mm), with very shallow surface segregation (<200 μm), by reducing mould heat transfer to a low level.

The porous mould design (Figure 3a) generates an insulating gas film between ingot and mould to reduce mould heat flow.

The gas injection mould design (Figure 3b) uses gas pressure to depress the molten metal/mould contact level so that the effective length of the mould is reduced.

The ''critical length'' mould design, (Figure 3c) makes use of a conventional, precisely calculated very short mould (ca 25 mm) so that the cooling effect of the direct chill reaches up inside the mould.

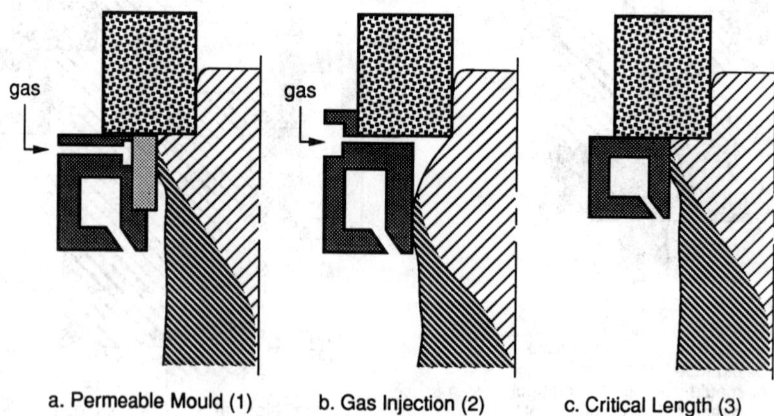

FIG 3. Advanced Mould Designs

2. **Problem:** During the starting phase of casting very large rectangular DC ingot for rolling into sheet, the foot (butt) of the ingot warps away from the starting block. This allows molten metal to escape from the mould, creating a safety hazard and generating scrap.

Cause: Until the growing ingot attains a substantial length, the foot has insufficient stiffness to resist the thermal stresses generated by the strong direct chill cooling. As a result, the ingot foot curls upward (Figure 4b).

Control: The thermal stresses causing the ingot to warp are reduced during starting by reducing (temporarily) the direct chill heat transfer (Figure 4a). Three different techniques are in commercial use: (a) Pulsed water cooling[4], (b) Carbon dioxide injection[5], and (c) Air injection[6].

The pulsed water process has been widely used for many years. By cyclically varying water flow to near zero every few seconds, the direct chill heat transfer can be reduced in a controllable manner.

The carbon dioxide injection process is also very effective, and simpler in equipment. Carbon dioxide gas is dissolved in the cooling water supplied to the mould. Because of the inverse solubility of CO_2 gas in water, the gas remains in solution until the direct chill coolant leaves the mould and strikes the hot ingot surface. The sudden reduction in pressure and increase in temperature of the direct chill causes the CO_2 to precipitate as fine bubbles. The bubbles of gas form an insulating film on the ingot surface and reduce the chilling rate. Butt curl is greatly reduced.

The air injection process utilizes compressed air, which is injected as fine bubbles into the cooling water just before exiting the mould. The air bubbles impede the direct chill cooling, reduce thermal stresses and hence reduce butt curl.

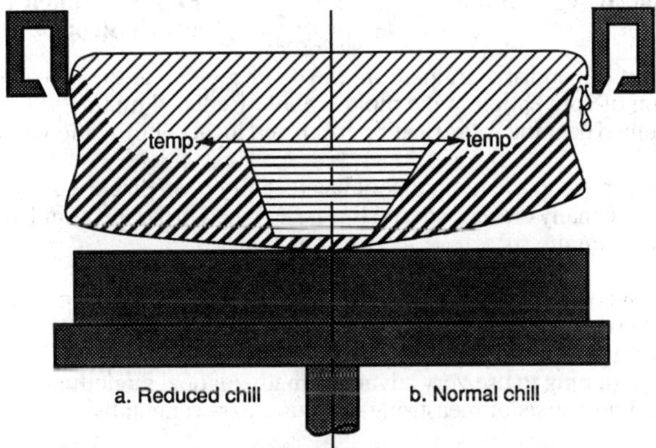

FIG. 4. Ingot Butt Thermal Distortion

3. **Problem:** The final example of control concerns the macro and microstructure of very large DC rolling ingot. The average thickness of rolling ingot has increased from 500 mm to 750 mm. The object is to improve rolling efficiency by increasing the coil weight. Unfortunately, the increased ingot thickness results in an undesireable coarsening of the dendrite structure and a severe increase in macrosegregation in the centre of the ingot.

Cause: As ingot thickness is increased, the molten metal pool deepens rapidly (parabolic rate increase). A recent paper by Yu and Granger[7] postulates that centre macrosegregation in thick DC ingot is the result of flow in the molten pool. Currents generated by thermal convection are thought to break off solute depleted dendrites from the outer regions of the ingot, and convey the dendrite pieces to the central region, where they become entrapped by the solidification front. A high concentration of solute depleted dendrites in the central region can result in a solute depletion as much as 15% below nominal (Figure 5).

In addition to the macrosegregation, microstructural coarsening of the dendrites due to Ostwald ripening becomes significant in the centre of thick DC ingot. Local solidification times become extended due to the rapid loss of superheat resulting from strong convective mixing. Copious nucleation of fine equi-axed grains, promoted by the use of efficient grain refiners, occurs throughout the entire molten pool. The grains move about in the pool for extended times, growing and coarsening all the while. As a result, the microstructure in the central zone of the ingot becomes very coarse, leading to reduced fabricability during ingot processing.

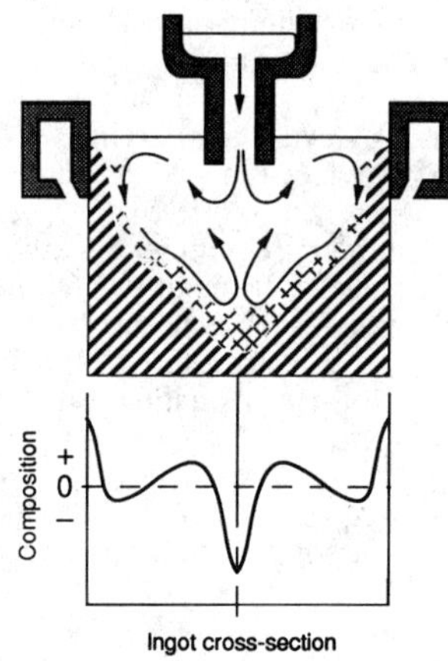

FIG. 5. Macro Segregation

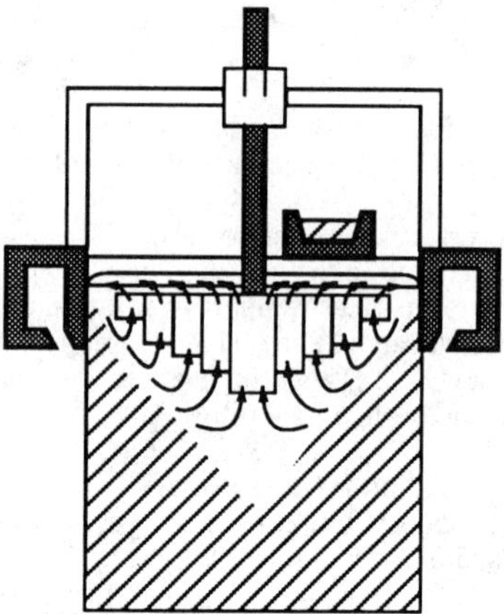

FIG. 6. Molten Pool Flow Control Device [8]

Control: The transport of dendritic grains from the outer regions of the ingot to the centre may be reduced by attenuation of convection in the molten pool. Reduced mixing will also permit thermal gradients to be maintained near the solidification front, and local solidification times may be reduced.

A flow control device for use in the molten pool has recently been disclosed by Yu and Granger[8]. The grid-like device, with many closely-spaced fins is designed to dampen fluid motion in the pool by increasing fluid friction (Figure 6).

Many more examples of control in the DC process could be described. I have intentionally concluded this paper with a problem involving the molten pool in order to emphasize that our knowledge of what is going on in the molten pool (of any casting process) is extremely limited. We can readily measure temperatures, we are beginning to use very advanced mathematical modelling techniques to simulate flows, but we have had no means of measuring the flows experimentally.

A possible explanation for the lack of concern paid to molten metal flow control by the casting industry may have been the lack of a tool to measure molten metal flow in "real time".

Fortunately, there is now available, at least for metals up to the melting range of aluminum, a molten metal flow transducer. It has been invented by Professor Charles Vivès at the University of Avignon[9]. The new device permits the accurate determination of both magnitude and direction of fluid flow in the molten pool.

Conclusions

The use of sophisticated control techniques in all phases of the DC casting process for aluminum ingot is increasing in response to the critical metallurgical specifications of today's aluminum end products.

Theoretical evidence is strong that fluid flow in the molten pool during solidification is extremely important to ingot structure. In 1977, Caldwell, Campagna, Flemings and Mehrabian[10] published a keynote paper on heat flow during solidification. They concluded that the local solidification time decreased, and hence the dendrite arm spacing also decreased, as: (a) heat transfer rate increased, (b) superheat increased, (c) gradient at the freezing front increased, and (d) the multi-dimensionality of the heat flow increased.

Because heat flow in thick DC ingot is limited by the thermal resistance of the solidified ingot, it is only by the manipulation of thermal conditions in the molten pool that cast structure may be controlled. It is primarily for this reason that future developments in DC casting will most likely be related to the molten pool.

References

1. U.S. Patent #4 598 763 (1986)
2. U.S. Patent #4 157 728 (1979)
3. U.S. Patent #4 709 744 (1987)
4. U.S. Patent #3 411 079 (1969)
5. U.S. Patent #4 166 495 (1979)
6. U.S. Patent #4 693 298 (1987)
7. H. Yu and D.A. Granger, Symposium NASA Lewis Research Center, Cleveland, U.S.A., 1984, pp. 157-168.
8. U.S. Patent #4 709 747 (1987)
9. Ch. Vivès and Ch. Perry, Int J. Heat Mass Transfer, 1986, Vol. 29(1), pp. 21-33.
10. T.W. Caldwell, A.J. Campagna, M.C. Flemings and R. Mehrabian, Met Trans B, Vol. 88, June 1977, pp. 261-270.

Gravity casting of zinc-aluminum (ZA) alloys — from theory to practice

R.J. Barnhurst
Energy and Product Applications Laboratory, Noranda Technology Centre, Pointe Claire, Quebec, Canada

ABSTRACT

Zinc-Aluminum (ZA) alloys are now widely used in foundries worldwide. However, within the long history of the casting industry the development of ZA alloys has taken place only very recently. In the last few years much has been learnt about the solidification behaviour of these alloys and our understanding continues to evolve.

The paper will discuss the development of ZA alloys, their solidification characteristics and, especially, the development of foundry parameters for gravity casting ZA alloys. This information will be presented with the practicing foundryman in mind and with direct reference to the contents of Noranda's recently published ZA alloy foundry practices manual.

Introduction

Hundreds of articles and dozens of textbook have been written dealing with the fundamentals of solidification and casting of metals and alloys. What is of academic interest however, may be very much less so to the average foundrymen who actually produces castings. Their needs are for an easily digestible, concise treatment of how to successfully cast the alloys of direct interest to them in simple terms that can be easily understood.

At Noranda Technology Centre we, like other applied research centres, need to provide this sort of information for the benefit of our foundry customers. We take the excellent teachings of our colleagues in academia, such as those of the distinguished Fred Weinburg, add to it practical research data and, it is hoped, package it in a usable form of direct benefit to the foundry.

The objective of the paper today will be a discussion of how the family of ZA alloys was developed and why, the applied research that was conducted to identify their casting parameters, and how this information has been packaged.

Development of ZA Alloys

During the years 1959 to 1962, The International Lead Zinc Research Organization (ILZRO) sponsored a zinc die casting alloy development program at the New Jersey Zinc Co. which culminated in the development of an alloy containing 12%Al-0.75% Cu-0.02%Mg which became ILZRO 12. Later, adjustments were made in the aluminum content, bringing it to 11%. This alloy, developed as a prototyping alloy for zinc pressure die cast alloy No. 3, was eventually recognized in its own right as a foundry alloy with exceptional properties.

Early in the 1970's, it was recognized that a family of zinc-aluminum alloys with a wide range of properties would be a useful addition to the nonferrous casting alloys. Early work by St. Joe Minerals Corp. identified a short freezing range permanent moulding alloy which was later modified by Noranda Technology Centre (hereafter NTC) to overcome various shortcomings with the original formulation. At the same time a wide freezing range sand casting alloy was developed at NTC (1). These alloys are now known throughout the world as ZA-12, ZA-8 and ZA-27 respectively.

Metallurgical Considerations

Avenues to development of zinc based alloys were limited. A review of the literature and the early experience gained with ZA-12 clearly pointed to the binary zinc-aluminum system as offering the best possibility for alloy development (1). The system (figure 1) showed no intermetallic phases and unlimited liquid solubility. Further, the range of solidification was large enough to allow the flexibility required for gravity casting. The system features a peritectic reaction at 443°C and approximately 28.4% Al, an eutectic reaction at 380°C and approximately 5% Al, a monotectoid reaction at 340°C and approximately 30% Al, and a eutectoid reaction at 275°C and approximately 22% Al. Solid solution hardening and the precipitation of β and η phases are characteristics of the system. The additions of Cu and Mg were investigated and found to benefit the strength of zinc-aluminum alloys by precipitation of intermediate phases, by influencing the solid state reactions and by providing for enhanced corrosion resistance.

The properties of ZA-12 include a tensile strength of approximately 300 MPa combined with 1 to 3% elongation. A target tensile strength of 240 MPa was set for the development of the (ZA-8) permanent moulding alloy and a target tensile strength of 380 MPa, combined with a 10% elongation, for the (ZA-27) sand casting alloy.

Exhaustive experimentation with various chemistries eventually isolated the ZA-27 alloy composition of 25 to 28% Al, 2 to 2.5% Cu and 0.01 to 0.02% Mg. The equivalent permanent mould alloy composition was eventually identified at 8 to 8.8% Al, 0.8 to 1.3% Cu and 0.015 to 0.03% Mg. The full castings chemistries of the ZA family are given in Table 1. Figures 2 to 4 demonstrate the results of the work on the ZA-27 alloy, while figures 5 to 7 show similar results for the ZA-8 alloy.

TABLE 1

Castings Specification (ASTM B791-88) for Zinc-Aluminum Alloys

	ZA-8	ZA-12	ZA-27
Aluminum	8.0-8.8	10.5-11.5	25.0-28.0
Copper	0.8-1.3	0.5-1.2	2.0-2.5
Magnesium	0.015-0.03	0.02-0.03	0.01-0.02

For the ZA-27 alloy, an aluminum content of 26% and a copper content of 2.5% gave the best combination of strength and ductility. The addition of magnesium further strengthened the alloy but at a loss in ductility. By reference to other properties, including creep, impact resistance and long term ageing, a Mg level of 0.01 to 0.02% was eventually chosen.

For the ZA-8 alloy, approximately 50 compositions were investigated in the range of 7 to 10% Al, 0 to 5% Cu and 0 to 0.01% Mg. Based on these property data plus fluidity and freezing range considerations the composition of 8.4% Al, 1.0% Cu and 0.02% Mg was chosen. Freezing range played a very important part in the selection of the chemistry of this alloy because it was demonstrated in foundry trials that a freezing range greater than 15 °C would be most acceptable in terms of foundry performance.

Microstructure and Solidification

The microstructure of the alloys is classical dendritic/eutectic typical of many intermediate and long range freezing alloys, viz., primary aluminum rich dendrites (α for ZA-27, β for ZA-8 and ZA-12), surrounded by a eutectic of varying amount, consisting of $\alpha + \eta$ phases. Copper rich $\epsilon(CuZn_4)$ particles are present interdendritically following solidification (2). Photomicrographs of the ZA-8 alloy in the permanent moulded condition and the ZA-12 and ZA-27 alloys in the sand cast and permanent moulded conditions are shown in Figure 8. The ZA-27 alloy undergoes a peritectic reaction at 443 °C, yielding a coating of β phase on the original α dendrites. Both α and β phases are metastable and undergo the eutectoid reaction at 275 °C. The ϵ phase is also metastable and undergoes a long term reaction with α to give a lower density T' phase, thereby causing a very slight increase in dimensions in ZA-27 castings, where the copper content is above 2%. When cast under a higher solidification rate (e.g. pressure die cast), the finer the microstructure and dispersion of precipitates becomes. These factors are important because of their influence on the performance of the alloys, including creep and wear resistance.

As with all dendritic/eutectic structures, particularly in wide freezing range alloys such as ZA-12 and ZA-27, directional solidification under high thermal gradient conditions is critical to the production of sound castings. This aspect is discussed below.

Casting Parameters

Having identified the chemistry of the alloys, work began in earnest in identifying the foundry parameters needed to cast ZA alloys, whether gravity or pressure die cast, and thereby produce premium quality castings. Although these alloys are now used extensively in pressure die casting, the discussion here will focus on gravity casting of ZA alloys, including sand casting and permanent mould casting. These data are presented in reference to the gravity casting manual published by Noranda in 1989 (3).

Fluidity (Castability)

The fluidity of the ZA alloys is very high. As measured by the standard spiral test the ZA alloys have fluidity which exceeds that of A356 and which is equal to that of the high silicon A413 alloy at high superheats as shown in (Figure 9) (4). All three alloys have about equal fluidity. Compared to A356, ZA alloys will flow in thinner walls. For example ZA-12 has been sand cast successfully in mould wall thicknesses of 1.7-3.8mm. Because of the high fluidity of ZA alloys heavy sections can be poured at low temperatures without misruns.

Sand Casting

The family of ZA alloys, particularly the ZA-27 alloy, have wide freezing ranges which render them susceptible to shrinkage porosity upon casting. In addition, gravity segregation of the solidifying phases, leading to underside shrinkage, can occur in heavy section slowly cooled castings (4). For these reasons, care must be taken with the design of the pattern and running and risering systems. When sand casting ZA alloys, chills have been found to be very important for obtaining sound castings. Filtering of ZA alloys prior to or during casting has been found to be unnecessary except at high Fe levels exceeding 0.075%. When present, the removal of large intermetallic particles greatly increases the casting performance, particularly in terms of ductility and impact strength (5).

(i) Risering

As with most foundry alloys, where high quality castings with high soundness are required, risers must be incorporated in order to feed the castings during solidification. The results of experimentation to develop recommended riser volumes and feeding distances are given below (6). Figures 10 and 11 show the casting designed used in the experiments. In each case, castings were sectioned and radiographed to obtain measurements of relative soundness.

a) Height to Diameter Ratio

The optimal geometry of a riser for ZA-12 and ZA-27 alloys was found to be a cylindrical riser with a height to diameter ratio of 0.5-1.0. The use of a H/D ratio of less than 0.5 and more than 2.0 resulted in increased porosity in the casting (figure 12). A high H/D ratio, although increasing the metallostatic head, also reduces the efficiency of the riser.

b) Feeding Distance

The effective feeding distance of risers for both chilled and unchilled ZA alloy castings increases with casting thickness up to 2.5-3.8cm, reaching a maximum of 30cm and 50cm respectively for chilled ZA-27 and ZA-12 castings (Table 2) (3). At a wall thickness of 5cm or more, the feeding distance falls by 20-25% in ZA-12 and by 50% in ZA-27. At wall thicknesses of 2.5-3.8cm, ZA-12 has 50-80% greater feeding capability than ZA-27. The feeding distance in unchilled castings was found to be half that of chilled castings.

TABLE 2

Effective Riser Feeding Distance, cm

	ZA-12		ZA-27	
Wall Thickness, cm	Unchilled	Chilled	Unchilled	Chilled
1.3	15	20	15	20
2.5	36	50	20	30
3.8	38	46	20	30
5.1	30	41	10	15

c) Riser Volume

As a general guide to selection of riser size, figure 13 gives an NRL type risering curve for both ZA-12 and ZA-27 in the unchilled and chilled conditions(4). The NRL (7) type curves are one of two ways of presenting riser volume information and these were developed under controlled conditions using cylindrical blind risers. The solidification of blind risers falls between that for open top risers and insulated risers and the position of the risering curves reflect this fact. For insulated risers 25% less volume, and 25-50% more when using open top risers, should be used. In all cases the modulus, V/SA (Volume/Surface Area), of the riser must exceed that of the casting by a factor of 1.3.

In the NRL curves, a shape factor (L + W/T), where L is the length of the casting, W is the width and T is the thickness, represents the surface area to volume ratio of the casting to be risered. This replaces the calculation of a modulus (V/SA) for both the casting and the riser and allows for a quicker and less error prone estimation of the minimum riser volume required to achieve a sound casting. In complex castings where approximated values of L, W and T could lead to incorrect estimation of the riser volume castings can be broken down into a number of elements or sections and each risered separately based on the shape factor estimated for each element.

By comparing the curves for chilled and unchilled castings it can be quickly appreciated that the volume of riser required to feed an unchilled casting is 50-100% more than a chilled casting, both for ZA-12 and ZA-27. Of the two alloys, the risering curves for ZA-12 are much lower than for ZA-27, due mainly to the lower freezing range of the ZA-12. For all the curves it can be appreciated that a chunky casting with a low shape factor value (low surface area to volume ratio) requires a higher volume of riser to achieve a sound casting than a long thin casting, such as a plate, with a high value of the shape factor (high surface area to volume ratio).

Figure 14 shows the risering requirements of the ZA alloys in comparison to several commonly used nonferrous casting alloys. The behaviour displayed by unchilled ZA alloys was found to fall between that of of the unchilled bronze castings produced with and without hot topping (8,9). The risering requirements of ZA alloys produced using chills is equivalent to that of bronze alloys produced with risering aids and to aluminum alloys cast with end chills (10).

(ii) Chilling

It is generally recommended that, for ZA-12 and ZA-27 castings, copper, iron/steel or even graphite chills be used to impart directional solidification (that is, freezing from the chill towards the riser) which thereby reduces shrinkage porosity (4). A reduction in porosity is critical to improved properties as shown for impact toughness in figure 15 (5).

Directional solidification results directly from thermal or temperature gradients in the casting, expressed as temperature change per unit length such as °C per cm. The chills are placed so as to promote directional solidification towards the riser(s), which can be further assisted by tapering the section towards the riser. Tapered chills have also been used successfully to promote directional solidification in ZA alloys.

As a general guide the following sizes can be used for ZA castings (Table 3). It has been determined by this work that ZA castings in sections of 19mm or less in thickness can be over chilled resulting in unsoundness, as found for underchilled castings. For sections with wall thicknesses of 1.3cm or less graphite chills may prevent over chilling.

TABLE 3

Recommended Chill Size and Casting Superheat as a Function of Thickness

Casting Thickness, cm	Chill Volume*, cm^3	Casting Spht, °C
up to 1.3	65	≈110
1.3-2.5	65-130	≈55-80
2.5-5.1	130-195	≈55
5.1	>195	≈55

* based on iron chills - for copper chills reduce by 25%., for graphite chills add 25%

(iii) Shake Out (Demoulding)

Shakeout time was also investigated (6). It was found that to achieve premium properties, shake out of ZA castings should take place no later than 30 min. following pouring. Experiment has shown that marked loss in impact toughness may result in ZA-12 castings if a demoulding time longer than 30 min. is used (figure 16). Other property loss will occur in both ZA-12 and ZA-27 castings but to a much lesser extent.

Permanent Mould Casting

The ZA alloys are particularly well adapted to permanent mould casting using iron or graphite moulds. Determination of risering requirements is one of several areas in which experimental work has been conducted in some detail. In addition to laboratory work, plant work on gate size, pour tilt time and mould temperature was also conducted at Stahl Specialty Co., Kingsville, Mo.

(i) Risering

As for sand cast alloys, where high quality castings with high soundness are required, risers must be incorporated to feed the casting during solidification. For the permanent mould process experimentation led to the development of recommended riser volumes and feeding distances as given below (11). A custom designed and built, fully interchangeable, iron mould was used in the experiments with a riser/casting pattern similar to that used in the sand casting experiments (see figure 10a).

a) Height to Diameter Ratio

For unchilled ZA-27, the minimum riser height to diameter ratio was found to be between 0.6 and 0.8. For ZA-12 castings the minimum riser height to diameter ratio was found to be between 0.5 and 0.7. In permanent mould casting risers tend to be taller than in typical sand castings due to the physical constraints of the mould.

b) Feeding Distance

Investigations have shown that effective feeding distance is often much less than that for sand castings and decreases as the alloy freezing range increases. Very thin section feeding remains difficult. Effective feeding distances was found to be increased markedly with an increase in wall thickness of the castings. The maximum effective feeding distance of ZA-8 was, on average, 1.5 times that of the ZA-27 at all casting thicknesses, resulting chiefly from the much narrower solidification range of the ZA-8 and less dendritic mass. When compared to unchilled castings, the improvement in effective feeding distance varied from 30 to 90%. Here, the overall improvement for ZA-8 was minimal, especially in greater wall thicknesses, due to the already high feeding distance possible in unchilled castings. The effective feeding distance data in iron permanent moulds is as follows (Table 4):

TABLE 4

Effective Riser Feeding Distance, cm

Casting geometry	ZA-8		ZA-12		ZA-27	
W x T, cm	Unchilled	Chilled	Unchilled	Chilled	Unchilled	Chilled
1.3 x 1.3	5	10	5	7.5	5	6.5
1.9 x 1.9	15	15[1]	8	15	6	10
2.5 x 2.5	20[2]	20[2]	11.5	18	8	13
3.2 x 3.2	15	20[2]	15	18	10	15

[1] Preliminary results
[2] Maximum distance evaluated

c) Riser Volume

As a general guide to selection of riser size, figure 17 gives an NRL type risering curve for both ZA-12 and ZA-27 permanent moulded castings in the unchilled and chilled conditions. A similar curve is currently being prepared for ZA-8.

In common with the results for sand cast alloys, chilling lowers the risering curve to lower ratios of V_R/V_C for both ZA-12 and ZA-27. Again, the risering curve for unchilled ZA-12 castings is similar to that of the wider freezing range ZA-27 in the chilled state. In all cases the risering curves for permanent mould ZA-12 and ZA-27 was found to lie above that of the sand cast alloys (figure 13) at an equivalent shape factor value, whether cast with or without chills present in the mould. This translates into an increase in the risering requirements of permanent mould cast ZA alloys when compared to sand cast alloys and which can be ascribed to differences in the solidification pattern in the permanent mould castings. Progressive solidification becomes a much more important factor relative to directional solidification, particularly in the thinner castings typically found in permanent mould casting. Larger risers, and the associated increase in heat content, are therefore needed to ensure that the casting is properly fed. Plant trials conducted at the Stahl Specialty Co. foundry have shown that ZA alloys require similar or slightly more risering than A356 (LM 25) and about 25% more than A319 (LM 4) for castings of equal quality. Such a large difference is due, in part, to a difference in riser geometry, i.e. rectangular versus round, the former being more commonly found in permanent mould casting due to mould constraints.

ii) Gate size

The effect of gate cross-sectional area was investigated using a brake piston mould which produced two sets of six castings each (12). One set had a gate cross-sectional area of 1.56 cm^2 and the other a cross-sectional area of 3.23 cm^2. Each casting weighed approximately 200g and had a riser volume to casting volume ratio of 0.6. The riser height to diameter ratio was approximately 1.3. The results showed the larger gate cross-sectional area to be very beneficial in producing high quality castings, especially at low to medium mould temperatures. Data obtained for a mould temperature of 305°C is given in Table 5 as follows:

TABLE 5

Effect of Gate Cross-Sectional Area on ZA-27 Casting Quality

Gate Area	UTS	Elongation (2.5 cm G.L.)	Impact Strength	Vol. Percent Porosity
cm^2	MPa	%	J	
1.56	212	<1	5	2.7
3.23	425	3.2	56	<0.1

At higher mould temperatures of approximately 340°C the measured differences in properties caused by lowering gate cross-sectional area are less. Experience has shown that 10-15% smaller gates than for A356 may be required because of the high fluidity of the ZA alloys, as long as the gate remains open until the mould is full.

(iii) Mould Tilt Time

A tilt pour casting procedure offers the best results when casting the wide freezing range

ZA alloys by virtue of the low turbulent conditions and improved feeding made possible with this process. The effect of mould tilt time was therefore determined (12). In general, longer tilt times promote directional solidification plus feeding of the casting by the incoming metal as the casting solidifies. This results in sounder castings with improved properties at medium to high mould temperatures. An example of the data for a ZA-27 commercial casting poured at a mould temperature of 340 °C is shown in Table 6. The exception is where casting design necessitates rapid filling to prevent premature freezing of the gates. Although general recommendations can be formulated, each casting design has an optimum tilt time which must be determined by trial and error.

TABLE 6

Effect of Mould Tilt Time on the Properties of a ZA-27 Casting at a Mould Temperature of 340 °C.

Tilt Time (s)	UTS MPa	0.2% Yield Strength MPa	Elongation %	Impact Strength J (ft lb)
12	123	376	4	12
48	444	395	6	41

(iv) Mould Temperature

Experimentation has shown that, in general, high mould temperatures of approximately 315 °C result in castings with higher tensile strength properties but lower toughness and elongation (13). Conversely, a low mould temperature of approximately 150 °C can result in both poor tensile and impact properties. At intermediate mould temperatures of 175-260 °C both higher tensile and impact toughness are more readily achieved, as long as a high thermal gradient is maintained. Figure 18 shows a typical example of the properties measured on a commercial casting.

Extreme mould temperatures for ZA alloys fall in the range of 150-340 °C with typical temperatures running at 170-250 °C but with some U.S. experience achieving best results with 300-320 °C, a somewhat unexpectedly high temperature. Despite this, the emphasis should clearly be on as low a mould temperature as possible with a pour cup temperature of 465-545 °C for ZA-8, 490-570 °C for ZA-12 and 540-620 °C for ZA-27. The exact mould temperature is dependent primarily on section thickness, thinner walls requiring the highest mould temperatures to ensure mould filling and a minimum of misruns. The number of cavities in the mould, as well as the complexity of the casting and mould material, also clearly play a role. The temperature of the mould (cavity) nearest the riser (normally the top of the mould) should be at least 50 °C or approximately 4 °C/cm higher than at the base of the mould (cavity), particularly for ZA-27, which has the highest freezing range.

Both ZA-8 and ZA-12 were found to also benefit from a high thermal gradient in the mould. A high thermal gradient rather than just a high chilling rate is clearly more important in achieving casting soundness, a factor often overlooked by foundrymen.

The final selection of in-mould temperatures will, in addition to the properties required and the quality of the casting achieved, depend almost equally on a number of important economic factors including cycle time, size of casting machines and the complexity and number of cavities per mould.

Acknowledgments

The author wishes to thank the technical staff of the Energy and Product Applications Laboratory at the Noranda Technology Centre, particularly D. Argo, R. Bouchard, D. Jacob and M. Lefebvre, for conducting the foundry work, sample preparation, mechanical testing and metallographic work over the last decade. Thanks are due to F. DeHart and R. Andriano for their efforts in the permanent mould work at the Stahl Specialty Co. foundry. The author gratefully acknowledges the financial support of Canadian Electrolytic Zinc Limited, Valleyfield, Quebec.

References

1. E. Gervais, H. Levert and M. Bess, Trans. American Foundrymen's Society (AFS),p. 68, (1980)

2. R.J. Barnhurst, and L. Mongeon, Metals Handbook, Vol. 9, Ninth Edition, American Society for Metals, Metals Park, Ohio, p. 488 (1985).

3. R.J. Barnhurst, Gravity Casting Manual for Zinc-Aluminum Alloys, Published by Noranda Inc., June 1989

4. R.J. Barnhurst, E. Gervais, and F.D. Bayles, Trans. AFS, Vol 91, p. 569, (1983).

5. R.J. Barnhurst and E. Gervais, Trans. AFS, Vol. 93, p. 591, (1985).

6. R.J. Barnhurst, and D. Jacob, D., Trans. AFS Vol. 96, p. 321, (1988).

7. H.F. Bishop, E.T. Myskowski and W.S. Pellini, Trans. AFS p. 271, (1955).

8. P.J. Guichelar, M.J. Weins and R.A. Flinn, Trans. AFS, p. 355, (1965)

9. R.A. Flinn, R.E. Rote and P.J. Guichelar, Trans. AFS, p. 380, (1966)

10. R.M. Pillai, V. Panchanathan and U.D. Mallaya, Trans. AFS, p. 103, (1978)

11. D. Argo, R.J. Barnhurst, Unpublished work, Noranda Technology Centre, Pointe Claire , Quebec

12. D. Argo, R.J. Barnhurst, F. DeHart, R. Andriano, Unpublished work, Noranda Technology Centre and Stahl Specialty Co.

13. D. Argo, R.J. Barnhurst, F. DeHart and R. Andriano, Trans. AFS, Vol 97, p. 757, (1989).

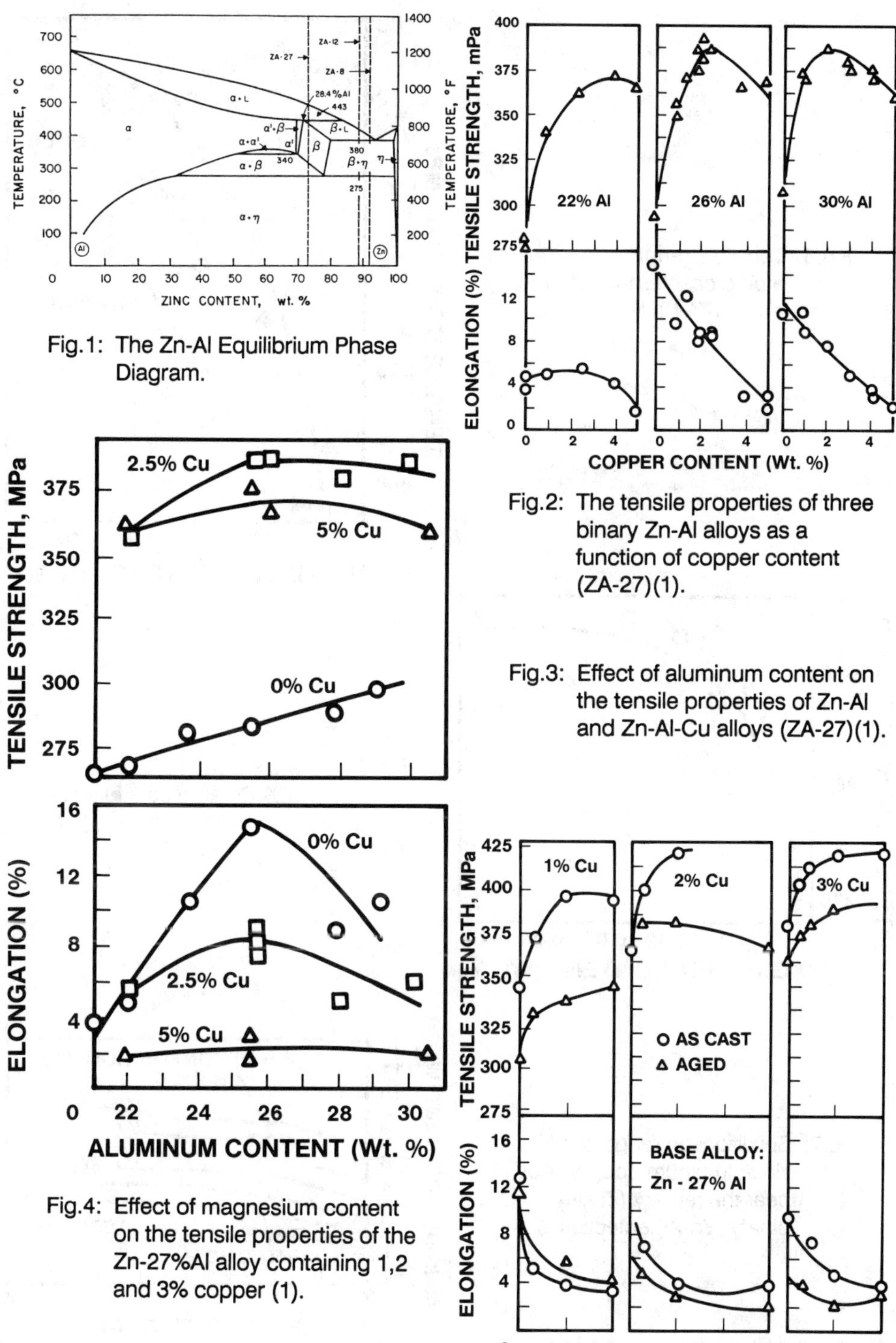

Fig.1: The Zn-Al Equilibrium Phase Diagram.

Fig.2: The tensile properties of three binary Zn-Al alloys as a function of copper content (ZA-27)(1).

Fig.3: Effect of aluminum content on the tensile properties of Zn-Al and Zn-Al-Cu alloys (ZA-27)(1).

Fig.4: Effect of magnesium content on the tensile properties of the Zn-27%Al alloy containing 1,2 and 3% copper (1).

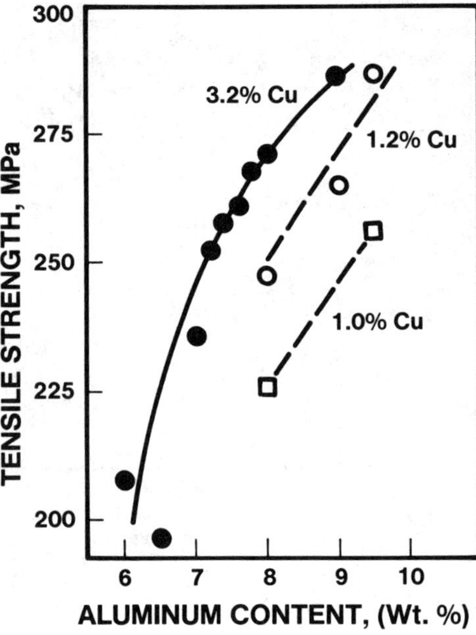

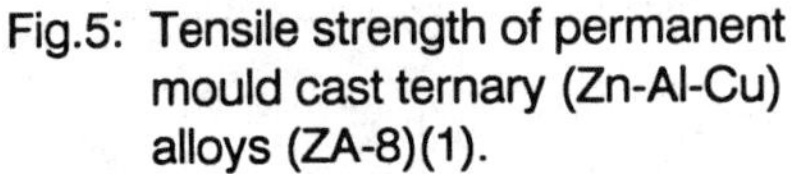
Fig.5: Tensile strength of permanent mould cast ternary (Zn-Al-Cu) alloys (ZA-8)(1).

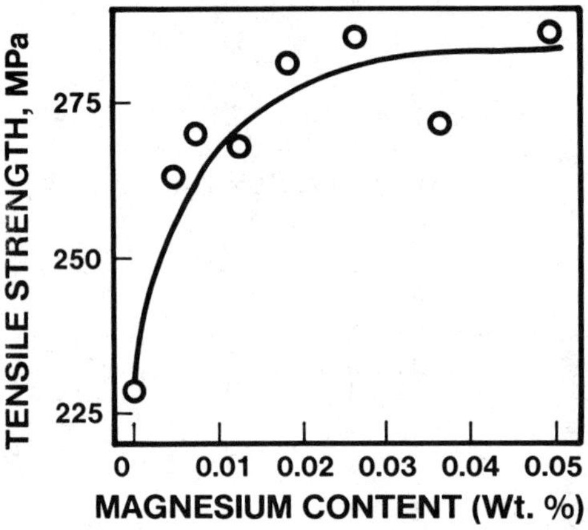

Fig.6: Tensile strength of Zn-9% Al-1.7% Cu alloy as a function of magnesium content.

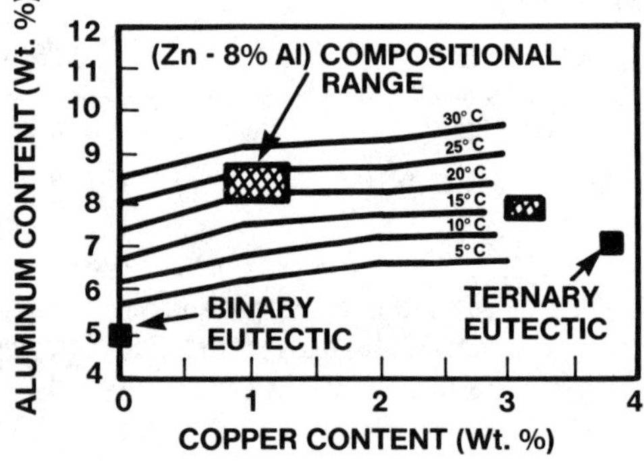

Fig.7: Solidification range of zinc-aluminum-copper alloys near the ternary (Zn-Al-Cu) and binary (Zn-Al) eutectics.

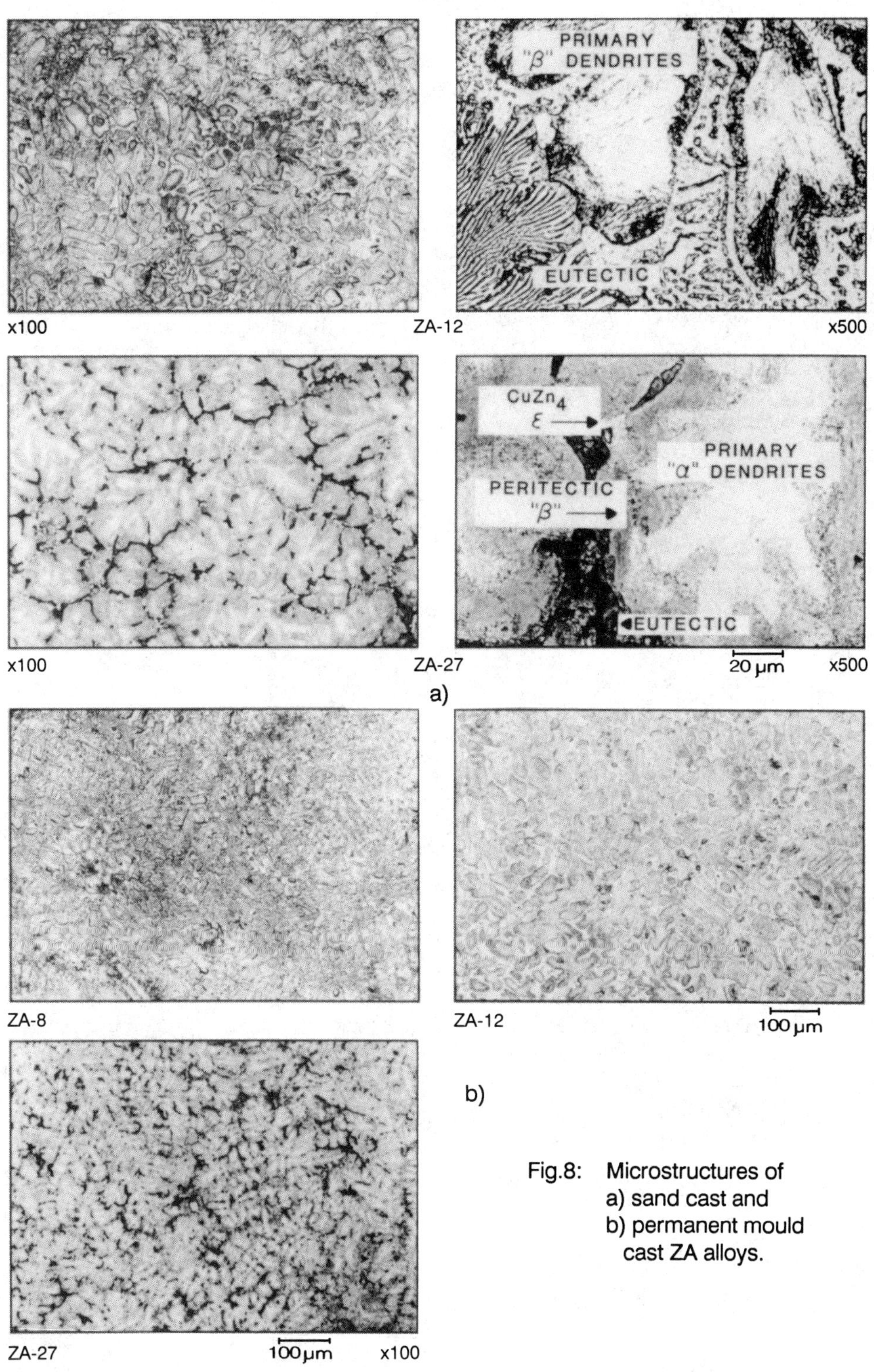

Fig.8: Microstructures of
a) sand cast and
b) permanent mould cast ZA alloys.

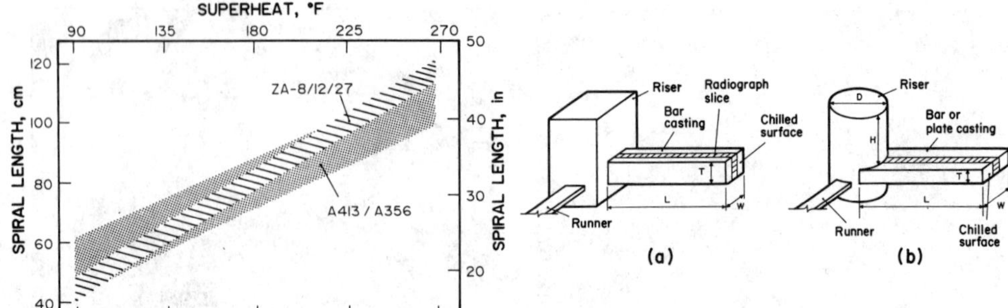

Fig.9: Fluidity of the ZA alloys in comparison to aluminum alloys(3,4).

Fig.10: Schematic of a) feeding distance and b) riser H/D-V_R/V_C patterns

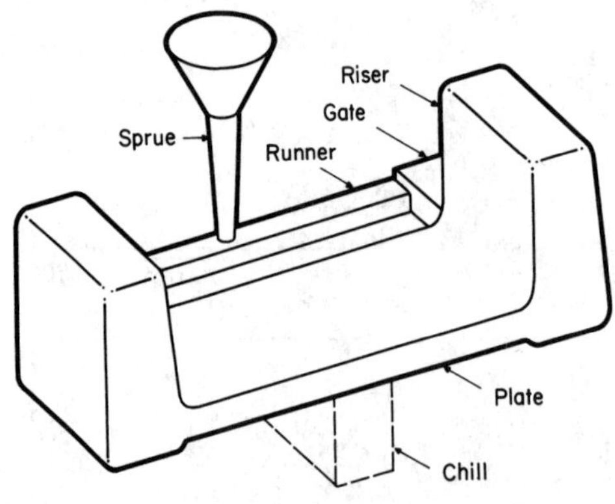

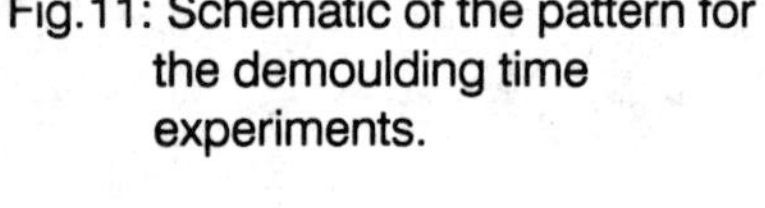

Fig.11: Schematic of the pattern for the demoulding time experiments.

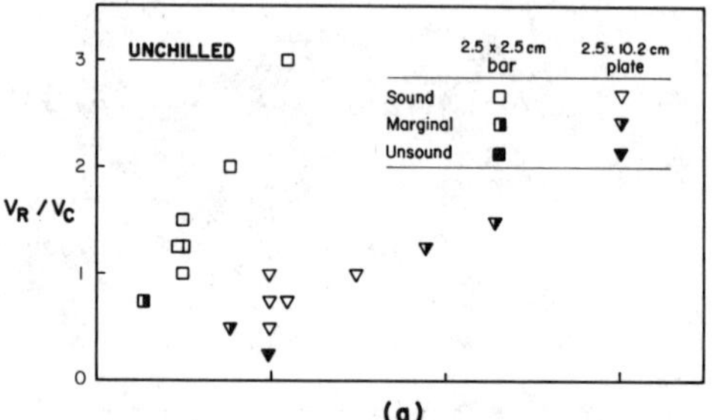

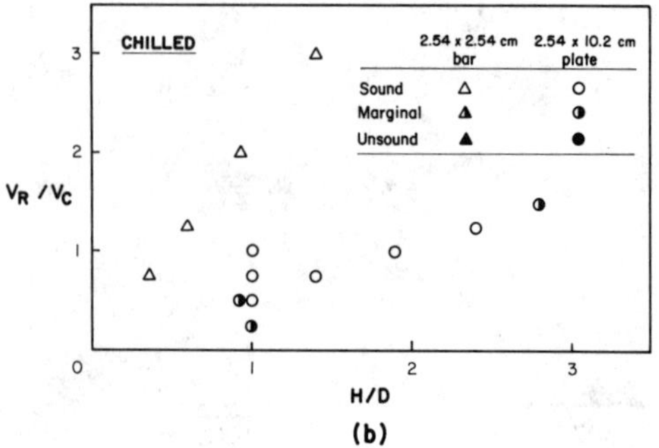

Fig.12: Determination of most efficient H/D ratio for a) unchilled and b) chilled ZA-27 castings, based on radiographic analysis(4).

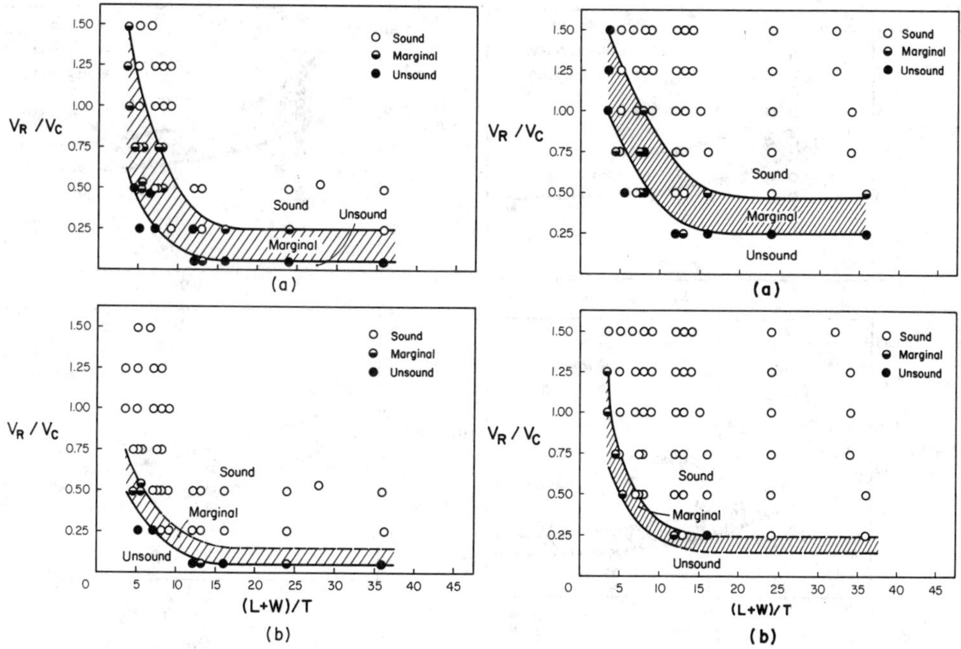

Fig.13: NRL type risering curves for a) unchilled and b) chilled, ZA-12 (left) and ZA-27 (right) castings based on radiographic analysis(4).

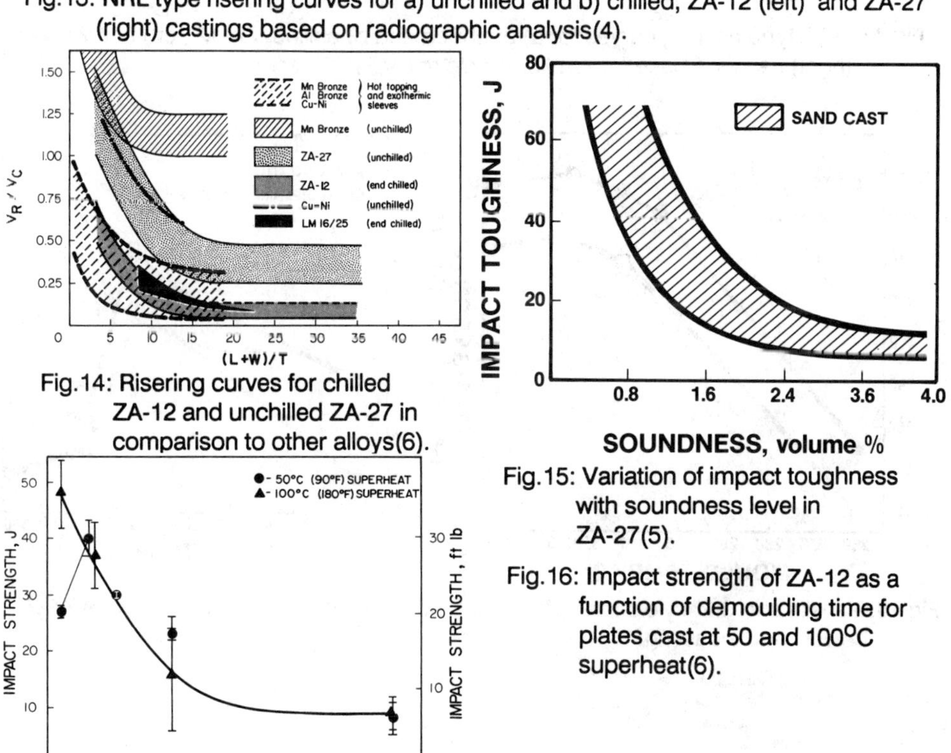

Fig.14: Risering curves for chilled ZA-12 and unchilled ZA-27 in comparison to other alloys(6).

Fig.15: Variation of impact toughness with soundness level in ZA-27(5).

Fig.16: Impact strength of ZA-12 as a function of demoulding time for plates cast at 50 and 100°C superheat(6).

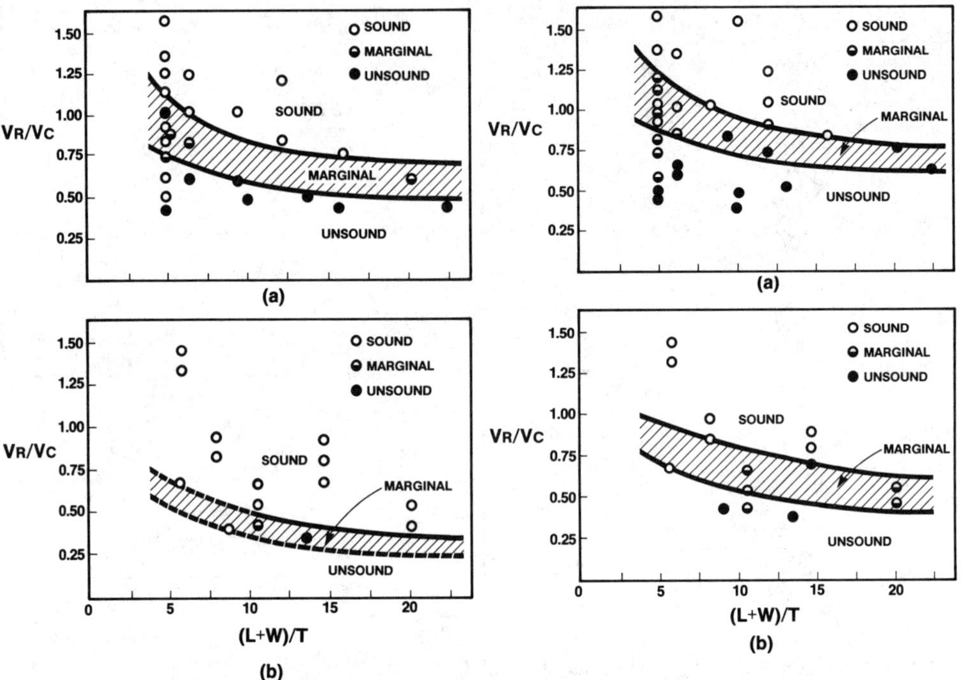

Fig.17: NRL type risering curves for a) unchilled and b) chilled, ZA-12 (left) and ZA-27 (right) castings based on radiographic analysis(11).

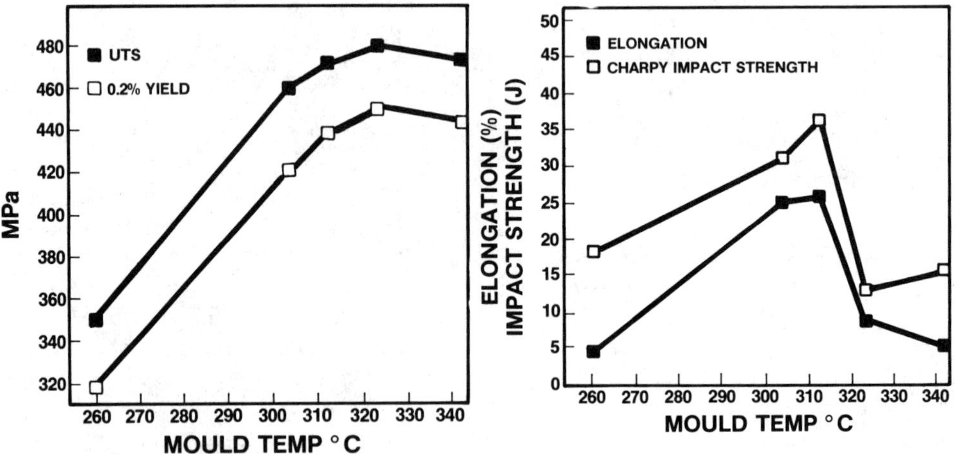

Fig.18: Tensile and 0.2% yield strengths as a function of mould temperature for a ZA-27 trowel handle casting(13).

Effect of casting techniques on the properties of cast Al-Si alloys

B. Closset
Timminco Metals, Toronto, Ontario, Canada

Abstract

Samples and parts have been cast in sand, permanent, and investment casting moulds. The effect of strontium modification and cooling rates on the eutectic silicon morphology has been investigated. It has been shown that the addition of strontium to a level comprised from 0.01% to 0.03% is usually necessary to obtain a finely dispensed eutectic silicon. As a consequence, improvements in tensile and impact strength have been observed for each moulding technique.

Nondestructive techniques such a thermal analysis and electrical conductivity have been used to characterize microstructural changes.

Vacuum pouring

J.L. Dion, R.K. Buhr and M. Sahoo
Metals Technology Laboratories, CANMET, Ottawa, Ontario, Canada

Abstract

Oxidation of the molten steam of metal during tapping, pouring and filling of a mould results in the formation of inclusions in the solidifying metal. A possible remedy is to fill the mould by applying vacuum, sucking the molten metal up a hollow tube. The method minimizes the exposure of the molten metal to air and also draws the metal from beneath the surface of the molten metal, thus ensuring that the cleanest metal is transferred to the mould cavity. This technique has been applied to pour castings in a wide variety of alloys including steel, monel, leaded red brass and Al-Si-Cu alloy using both green sand and CO_2 bonded sand moulds. The casting parameters included the degree of vacuum, type of sand, melt temperature and cast time. Castings were also produced by the conventional gravity-pouring practice in these two types of sand moulds. The problems associated with the development of this process and the quality of the castings produced are described.

Introduction

The development of the vacuum pouring system stems from Metals Technology Laboratories' (MTL's) in-house work on low-pressure die casting and evaporative pattern casting processes, where the low-pressure unit was used to fill a mould containing an evaporative pattern (1,2). As the mould was filled by pressurizing from under, vacuum was used to hold the sand. Upon measuring mould filling time, it was discovered that vacuum was filling the mould before there was enough pressure in the low-pressure unit for the metal to reach the bottom of the mould. The initial results led to the development of the vacuum pulling of full mould castings with disposable tubes. At a later stage, the project was extended to vacuum pouring of green sand and CO_2 bonded sand moulds to produce ferrous and non-ferrous castings. To this end, a large container where regular sand moulds could be inserted was designed and studied for vacuum pouring through disposable tubes. Although the project started in 1988, it is only recently that a review paper was published on the commercialization of vacuum casting (3).

This paper describes the quality of the castings produced in a variety of alloys such as monels, steels, leaded red brass and Al-Si-Cu alloys and compares them with those produced by the gravity casting process. The influence of casting parameters such as type of sand, degree of vacuum, melt temperature, etc., on the quality of the castings is also discussed.

Equipment

A schematic of the vacuum pouring unit is shown in Fig. 1. The disposable feed tube extends from below the molten metal surface to the sprue of the mould. A fiberfax gasket is inserted between the flange of the tube and the flange of the container. The container is 92 cm (36 in.) × 92 cm (36 in.) and 76 cm (30 in.) deep. A restraining bar is used to prevent movement of the mould during pouring or handling of the container. The top of the container which rests on a rubber gasket is held in place by vacuum during pouring. The vacuum was set in the container by opening a large valve on the right side. This valve was connected to a large vacuum furnace by a 9 cm (3.5 in.) hose. This assembly, consisting of the container, valve, hose, etc., is seen better in Fig. 2. The large vacuum furnace was used as a vacuum reservoir as the regular vacuum pump was unusable because of the size of the container. The degree of vacuum in the casting unit was controlled by the degree of vacuum in the reservoir (larger furnace).

Experimental

Process Development

In the initial stage, CO_2 sand moulds were used with Al alloys, Cu alloys and grey iron. A steel snorkle was used for these experiments. Figure 3 shows the casting used in this study. The dimensions of the elbow fitting are shown in Fig. 4. The runners were 19 mm (3/4 in.) x 13 mm (1/2 in.) and the tapered sprue 19 mm (3/4 in.) x 19 mm (3/4 in.) at the base with the ingate 19 mm (3.4 in.) x 13 mm (1/2 in.). Two problems emerged from these runs. First, the cast time, which is defined here as the interval of time the vacuum is applied was much too long to permit the risers to act. A blind riser cannot act under vacuum, and 60-90 s was needed to prevent the molten metal from leaking down the tube into the ladle. In the literature where this casting system is reported to be used for chemically-bonded sand, a mechanical device is used to pinch the tube after mould filling (4). Based on data obtained from the work on evaporative patterns, we replaced the mechanical device by freezing spots on the runner. This is shown in Fig. 5. The minimum thickness 5 mm (3/16 in.) was determined by successive trials. The width at that spot was 5 cm (2 in.). It was calculated in order to maintain the same cross-sectional area to eliminate turbulence. This cut down the cast time to 10-15 s depending on the particular alloy and the pouring temperature.

The second problem relates to the steel tube as it was inadequate for steel casting, and attempts were made to develop a sand snorkle. The ASTM standard permeability test was done on different combinations of sand such as silica 45, McConnellsville AFS 120 fineness, Olivine 180 with additions such as iron oxides, cement and zircon flour. The CO_2 process was used to bind these mixtures. The regular height of 50 mm (2 in.) for the permeability test samples was reduced to 25 mm (1 in.) for these tests. To further seal the surface, different coatings or washes were also evaluated. Then, a pattern was designed to fabricate these tubes. The sand tube along with the steel tube are seen in Fig. 6. Both tubes are 30.5 cm (12 in.) long under the flange. The steel tube is made of 2.5 cm black pipe welded to the flange. The sand tube is 19 mm (3/4 in.) thick. A vacuum tester (Fig. 7) for the sand snorkle was designed and built to test the most promising mixture. Combinations of fine sand produced by the CO_2 process such as olivine 180 with 10% zircon flour coated with a zircon wash on the outside appeared fine. However, it was seen that under casting trial, their resistance to the thermal shock in steel was unreliable.

Finally, the following mixture was adopted:

2 parts Sairbone (clay)
1 part Grain of Mullite
0.5 parts water
0.5 parts water glass (CO_2 binder)

These tubes had to be oven dried. They proved adequate in casting trials. However to eliminate a possible variable during this study, a steel tube was inserted inside the sand tube.

No difficulties were experienced in producing castings with the CO_2 sand mould. However, there were two major problems with the green sand mould. The first problem was that the mould collapsed very fast. Vacuum testing in a smaller vacuum chamber (Fig. 7) showed that it was caused by two factors. The mould was drying rapidly under vacuum, and this action was even faster with a leak in the system such as the one provided by the feed tube. Also, with a leak, very rapid erosion of the mould occurred. Under static conditions, collapse of the mould was not so fast. This problem was solved by introducing the feed tube inside the melt before setting up the vacuum. Such an erosion problem was also noted with the CO_2 mould. However it was much slower. The second problem with green sand mould was the pulling of the tube inside the mould under vacuum. This was due to the lower compressive strength of the green sand. A perforated steel plate was then interposed between the mould and the tube.

Production of Castings

After an initial period of familiarization which was needed to learn the operational characteristics of that casting process, the following casting procedures were adopted. First, the feed tube was plunged in the melt, then the vacuum set in by opening the valve. At the end of the cast time, the valve was closed, the relief valve opened, and the casting unit lifted. Although it was needed only for steel, the sand tube was used throughout. Casting trials using the elbow fitting were made in CO_2 sand, pep set and green sand. Figure 8 shows the two types of riser used with the elbow: a cylindrical riser on the runner and a trapezoidal riser on top of the flange. As the elbow was symmetrical, in some cases both risers were used at the same time for comparison. Casting trials were run with Al-alloy 319, Cu-alloy C83600 (85 Cu-5 Sn-5 Pb-5 Zn), monel (weldable grade), grey iron and low-carbon steel. However to study the castings parameters such as degree of vacuum, pouring temperature, type of sand, etc., the emphasis was on steel, monel and alloy 319. For steel and monel, a step block to provide tensile test specimens where section thickness effect could also be evaluated was used. It is shown in Fig. 9. The overall dimensions were 20 cm (8 in.) long by 15 cm (6 in.) wide by 2.5 cm (1 in.) and 6 cm (2 1/2 in.) thick. The pouring sprue was 16 mm (5/8 in.) in diameter. The freezing spot connected to the riser was 25 mm (2 1//2 in.) wide and 3 mm (1/8 in.) thick. Every vacuum-pulled casting was accompanied by a gravity-poured casting. For gravity pouring, the same feeding system was used, except that the sprue was inverted. This is shown in Fig. 10. The same applies to the elbow fitting when gravity poured.

All melts were prepared by standard melting procedures in both the tilt-type and push-up induction furnaces, the latter being used for aluminum- and copper-base alloys.

Results and Discussion

Risers

Cylindrical and trapezoidal risers respectively with insulating sleeves and fibrefax were used for both gravity and vacuum pouring as shown in Fig. 8. The advantage of the side riser over the ingate is that it reduces the amount of sand in the mould. In addition, relatively clean metal is expected to enter the mould cavity as the molten metal passes through the riser. However, in the present work, a top riser is to be preferred since occasionally some gas could get trapped at the top of the flange with a side riser. An example of the gas defect is shown in Fig. 11 for the side riser only. This defect could be attributed to air being sucked in with the molten metal. An improperly degassed melt could also cause this defect.

Another defect was a shrinkage on the bottom half of the flange. It could also occur on the top part to a lesser extent. An example is shown in Fig. 12 for a red brass casting. Corrective measures such as an increase in riser size, use of an insulating sleeve and a chill on the bottom part of the flange were the same for both vacuum and gravity pouring.

Type of Sand Moulds

Sand penetration was observed with the vacuum pouring of CO_2 bonded sand (Silica AFS 56) moulds. Use of a zircon wash on the mould cavity eliminated this defect for all the alloys except for the aluminum alloy where the pouring temperature and degree of vacuum had to be lowered to 680°C and 380 mm, respectively, to overcome this problem.

There was no sand penetration in the green sand moulds with the fine sand used (i.e., McConnellsville AFS 126 and silica AFS 128) for all the alloys except the aluminum alloy 319 where the pouring temperature had to be lowered to 665-675°C and the degree of vacuum to 500 mm. However, there was an expansion or swelling problem on the flange of the elbow fitting in the green sand mould, especially in steel castings (see Fig. 13). Lowering the steel pouring temperature from 1610°C to 1575°C had no noticeable effect on the swelling. Extra ramming in the flange area or lowering the moisture content of sand to 2% aggravated the situation. Similarly, increasing the moisture content to more than 3% had no further effect on swelling. However, lowering the degree of vacuum to 380 mm and removing the vents on the mould (cope half) for vacuum pouring produced some encouraging results. Although, a better casting could be produced with such modifications, it was felt to be a borderline situation. This problem was not observed with the step-block casting in steel.

The swelling problem was observed with the aluminum alloy at high pouring temperatures as well as a high degree of vacuum. Sand penetration problems are also encountered under such casting conditions.

Mechanical Properties

Table 1 shows the properties of steel in CO_2 and green sand moulds for both vacuum and gravity pouring. The following heat treatments were carried out on the castings before tensile testing for the 25 mm (1 in.) sections:

1 h at 927°C (1700°F) and air-cooled

1 h at 885°C (1625°F) and water-quenched

1 h at 607°C (1125°F) and air-cooled

The heat treatment time was increased to 2 h for the 63 mm (2.5 in.) thick sections.

As shown in Table 1, there was no significant difference in the mechanical properties of the specimens produced by either gravity or vacuum pouring. Microstructures of the specimens produced by both casting processes were also similar. However, the non-metallic inclusion content of the specimens produced by vacuum pouring were less than those produced by gravity pouring.

The mechanical properties of monels cast in CO_2 bonded sand moulds employing both gravity and vacuum pouring are given in Table 2. For each section thickness, there is definite improvement in the UTS and YS values by using vacuum pouring, although such increases may be within the experimental error.

Chemical compositions of the steel and monel melts are given in Tables 3 and 4, respectively.

TABLE 1. Mechanical Properties* of 1025 Steel Castings (step block) Produced by both Gravity and Vacuum Pouring

	Sand	Process	Section thickness (mm)	Young's modulus MPa	0.2% offset YS (MPa)	UTS MPa	% Elong	% RA	Impact toughness (J)
3090	Green sand	Vacuum	25	202 704	394	593	29	60	125
		380 mm	63	207 875	353	560	32	63	123
	Silica 135	Gravity	25	213 736	389	588	30	63	124
			63	205 462	351	558	31	60	128
3094	Green sand	Vacuum	25	207 521	415	644	27	55	92
		380 mm	63	208 254	368	599	30	57	103
	Silica 135	Gravity	25	204 428	419	644	27	54	93
			63	191 327	377	612	30	56	85
3096	CO_2 sand	Vacuum	25	214 883	458	666	24	51	96
		380 mm	63	204 083	405	639	29	55	116
	Silica 45	Gravity	25	203 394	462	671	25	55	90
			63	198 263	407	642	29	56	96
3108	CO_2 sand	Vacuum	25	206 840	422	619	28	55	112
		380 mm	63	200 635	375	590	31	60	120
	Silica 45	Gravity	25	208 910	420	619	28	59	112
			63	174 781	381	594	30	59	114

*Average of two tensile test specimens

TABLE 2. Mechanical Properties* of Cast Monels in CO_2 Moulds Employing both Gravity and Vacuum Pouring

Melt No.	Process	Section thickness (mm)	UTS MPa	0.2% offset YS MPa	0.5% YS MPa	% Elong	% RA
56	Vacuum	25	492	198	212	32	28
	200 mm	63	490	203	215	29	28
	Gravity	25	485	197	208	31	30
		63	471	196	208	29	27
57	Vacuum	25	511	196	209	39	37
	355 mm	63	499	197	209	36	34
	Gravity	25	470	185	195	41	42
		63	480	193	206	37	34

*Average of two tensile test specimens.

TABLE 3. Spectrographic Analysis of Steel

Melt No.	Element %				
	C	Si	Mn	S	P
3090	0.21	0.53	1.2	0.0075	0.018
3094	0.29	0.47	1.2	0.0079	0.015
3096	0.28	0.58	1.2	0.0080	0.016
3108	0.21	0.54	1.3	0.0069	0.014

TABLE 4. Analysis of monels

Melt No.	Element %											
	Ni	Cu	Fe	Mn	Si	Nb	C	Mg	Ti	Pb	S	P
56	65.6	29.6	1.49	0.68	1.40	0.94	0.12	0.018	0.075	<0.005	0.0026	<0.024
57	65.4	29.6	1.83	0.81	1.68	0.75	0.096	0.019	0.074	<0.005	0.0028	<0.024

Casting Soundness

The surface finish of the elbow fitting and the step-block was better in vacuum pouring especially for the steel where trapped slag inclusions and other surface defects were eliminated to a great extent.

It was observed with monel and Al-alloys that there was an improvement in the porosity level with vacuum pouring when using gassy melts. Despite such improvements, the castings were still judged to be unacceptable.

Mould Filling Time

The mould filling time was faster with vacuum pouring for both the lower density aluminum alloy and the higher-density monels. It decreased further with increases in the degree of vacuum. These features are demonstrated in Table 5.

TABLE 5. Mould filling time for Monel & Al-319

Alloy	Process	Degree of vacuum (mm)	Time (s)
Monel	Vacuum	205	5.4
	Vacuum	290	5.5
	Gravity	--	8.3
Al-319	Vacuum	210	1.9
	Vacuum	380	2.8
	Vacuum	510	3.1
	Vacuum	550	3.0
	Gravity	--	6.9

Fluidity

Figure 14 shows the data obtained by vacuum and gravity casting of fluidity spirals with monel, where the temperature varied between 1320 and 1500°C. These were cast as follows.

Gravity casting of two series of five spirals with the same melt in green and CO_2-bonded sand moulds, gravity and vacuum pouring of three series of four spirals with the same melt in pep set sand and gravity and vacuum pouring of two series of four spirals with the same melt in green sand.

Figure 14 shows that there is practically no difference in fluidity between the two processes. However, it must be remembered that, in vacuum pouring, the molten metal has to travel up 320 mm (12 3/4 in.) the feed tube before it gets to the sprue entrance. Thus, the temperature of the molten metal at the sprue entrance is less than that measured in the ladle. It can then be assumed that an increase in fluidity can be expected during vacuum pouring. It may be noted that the mould filling time during vacuum pouring is less than that during gravity pouring.

Conclusions

1. A vacuum pouring process has been developed to use green sand mould along with a chemically-bonded sand mould.
2. With vacuum pouring, a wash was needed to prevent sand penetration in the CO_2 mould. In green sand, this problem was solved by the use of finer sand. Another casting defect related to vacuum techniques was the swelling of the flanges on the elbow fitting in steel in green sand.
3. There was a slight improvement in mechanical properties for monel in vacuum pouring, whereas there was no difference for steel despite the fact that fewer non-metallic inclusions were produced in vacuum pouring.
4. Vacuum pouring decreased the porosity in castings when using gassy melt of monel or Al-alloys.
5. Mould filling time is decreased with vacuum. It also decreased in proportion with the amount of vacuum.
6. The increase in fluidity observed with monel would be an asset in thin-wall casting.

Acknowledgements

The authors gratefully acknowledge the technical assistance provided by the foundry support staff and Messrs. J.R. Emmett, D. Cousineau and R. Matte.

References

1. Dion, J.L, Buhr, R.K., Warda, R.D., Emmett, J.R. and Sahoo, M. "Progress report on low-pressure permanent mould casting process"; Metals Technology Laboratories, CANMET, Energy, Mines and Resources Canada, Report MTL 89-28(TR); 1989.
2. Dion, J.L, Warda, R.D., Buhr, R.K., Emmett, J.R. and Sahoo, M. "Production of aluminum alloy castings using evaporative pattern and vacuum techniques"; Trans AFS 98; 1990.
3. Blackburn, R.D. "Vacuum casting goes commercial"; Advanced Materials and Processes 137:2:17-21; Feb. 1990.
4. Metal Handbook 9th Edition; Casting 15:319; 1988.

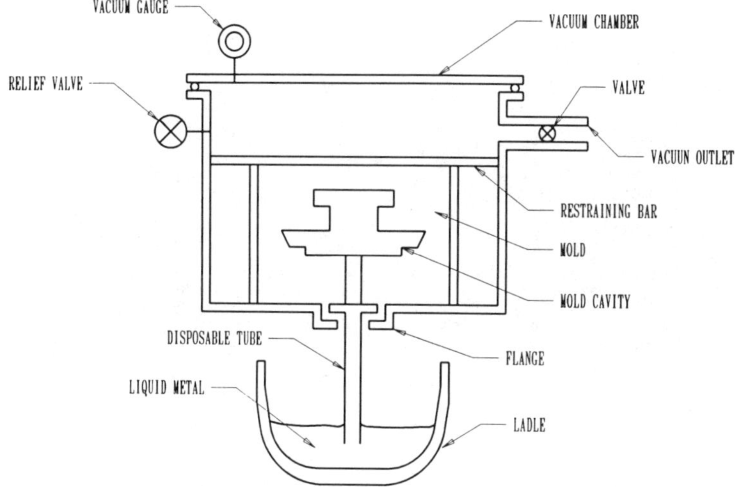

FIG. 1. Schematic of the container for vacuum pouring of regular mould.

FIG. 2. Container used for vacuum pouring of regular mould.

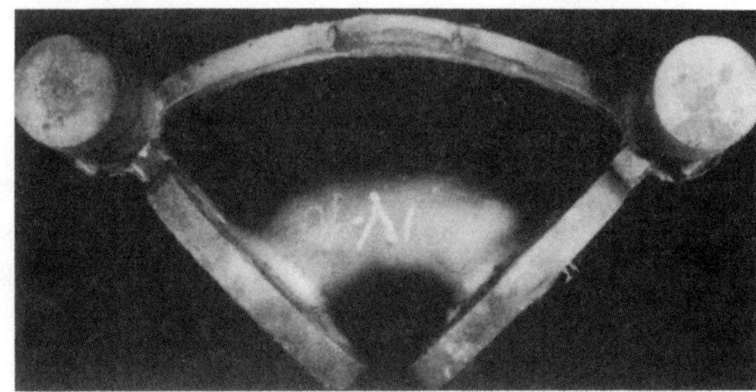

FIG. 3. Elbow casting used in vacuum pulling of regular mould.

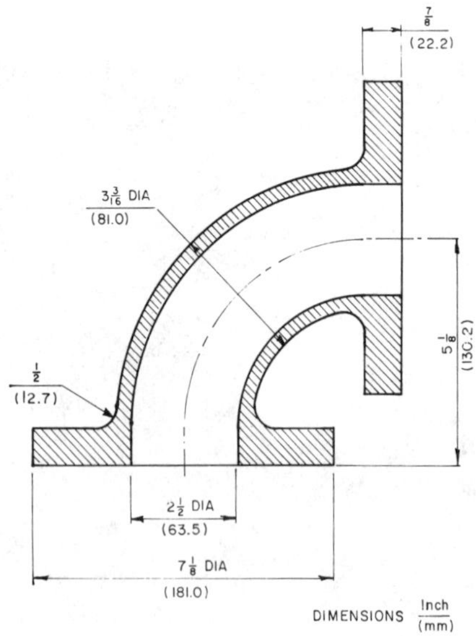

FIG. 4. Dimensions of the elbow casting.

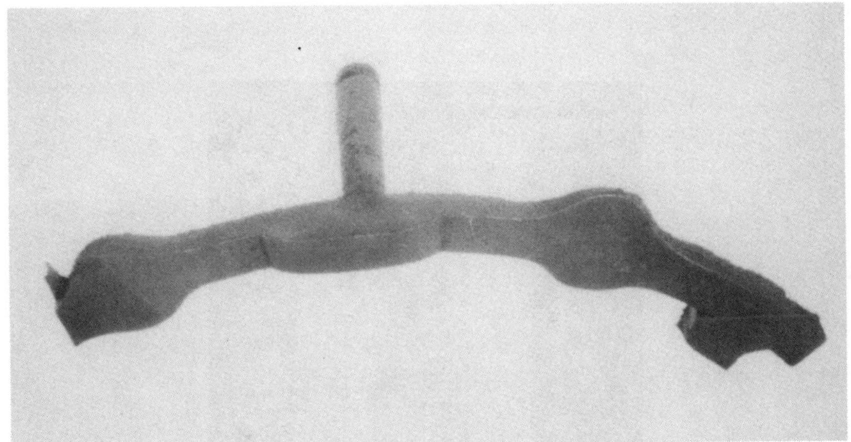

FIG. 5. Runner system for elbow fitting showing a freezing spot (gradual flattened area) on each side of the sprue.

FIG. 6. Steel (left side) and sand (right side) tubes.

FIG. 7. Vacuum tester for sand snorkle on the left, and vacuum tester to study the effect of vacuum on green sand on the right.

FIG. 8. Elbow casting with two types of riser.

FIG. 9. Step block used for tensile test specimens.

FIG. 10. Gating on step block for gravity pouring.

FIG. 11. Examples of gas defect on an aluminum casting.

FIG. 12. Example of shrinkage on the flange of a Cu alloy casting.

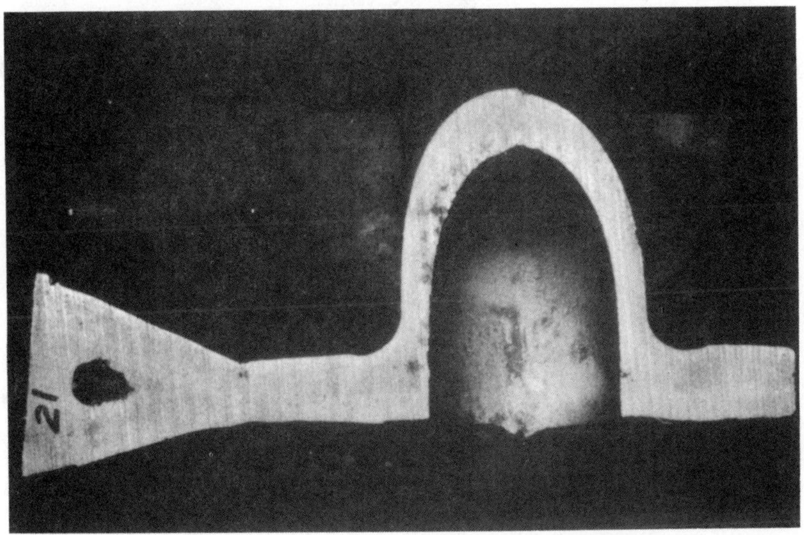

FIG. 13. Steel casting showing swelling on bottom part of flange.

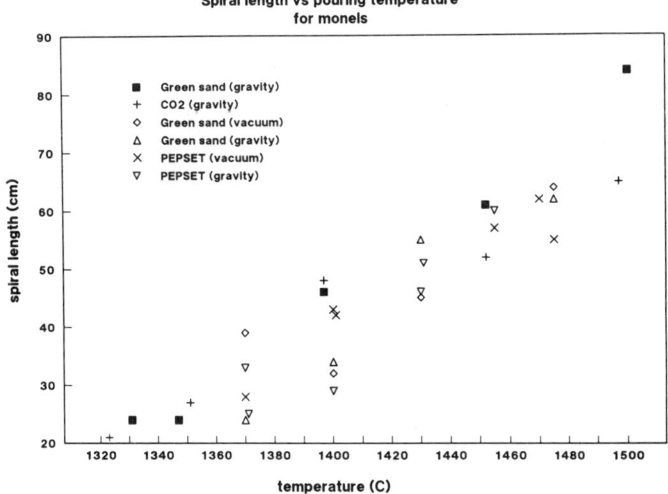

FIG. 14. Fluidity data for gravity and vacuum pouring of monels.

III. Continuous and static casting of steel

Chairman: J.K. Brimacombe
The Centre for Metallurgical Process Engineering,
The University of British Columbia,
Vancouver, British Columbia, Canada

On the solidification of steel ingots and of continuously cast steel castings (Keynote Paper)

H. Fredriksson
Department of Casting of Metals, Royal Institute of Technology S-100 44, Stockholm, Sweden

The macrostructure in ingots as well as continuously cast strands are often divided into three different parts, namely one chill zone at the surface, one zone with columnar crystals and one zone with larger equiaxed crystals in the central part. The lecture will discuss the formation of this three zones and an attempt is done to theoretically describe the formation of the three crystal zones.

Both in normal steel ingots and in continuously cast strands the formation of the three crystal zones influences the macrosegregations. The effect of the crystal formations on the macrosegregations is discussed in the lecture. The effect of the sedimentation of crystals on the macrosegregations are discussed. The effect of electromagnetic stirring on the formation of new crystals is also discussed and its indirect effect on the formation of centerline segregation in continuously cast strands is analysed.

The formation of centerline segregation in slabs as well as in billets are theoretically described. The formation of macrosegregations in ingots is also described. A comparison between the formation of macrosegregations in ingots is made. The different mechanisms behind the formation of macrosegregations in the two casting processes will be delineated.

Solidification shrinkage is the most important parameter for deciding the pipe formation in the ingots. A simple theoretical model of pipe formation in ingots will be used to point out the most important parameters for minimizing the pipe formation. The same model is also used in order to calculate the pipe formation in the last solidified parts of a strand. In the last case it is shown that the cooling shrinkage in the already formed shell is more important than the solidification shrinkage.

Columnar and equiaxed dendrite growth in continuously cast products

A. Etienne
C.R.M., rue Ernest Solvay, 11, B-4000 Liège, Belgium

Introduction

Though many of the macrosegregation problems known in the classical ingots vanished with the application of continuous casting, macrosegregation is still a concerning problem in some applications of high C steel such as steelcords, prestressed wires and ball bearings.

It is clearly recognized that macrosegregation results from fluid flow in the mushy zone but there is not yet any comprehensive model of macrosegregation in cast blooms and billets.

It is a very complex problem which simultaneously involves :

- heat transfer which controls the solidification rate,
- solidification kinetics which control the solidification structure and mushy zone properties,
- fluid flow which controls mass transfer.

Highly sophisticated means are now available which can be used to build up a solution :

- Heat transfer models applied to solidification are now classical tools used to control the casting process.
- Fundamental description of solidification has progressed dramatically during the recent years.
- Powerful computing methods and software packages are now available to cope with the difficult problems of fluid mechanics.

Aiming at a better understanding and control of the macrosegregation, this work describes the solidification kinetics coupled with the heat transfer processes and proposes a model of the solidification structure.

Since the aim of the model is to solve practical, industrial problems, a careful comparison of the model predictions with experimental observations has been made at every step of the work.

---

* This research was carried out with financial support from the ECSC.

Experimental Data

Primary solidification structures have been assessed in a large number of samples taken from slices of continuously cast blooms.

The samples have been cut off in square blooms (220 x 220 mm) made of hard steel grade (C ≃ 0.8 %) used to produce prestressed bars. Most blooms were electromagnetically stirred in the mould and at the end of the liquid pool. These samples covered a wide range of casting conditions : superheats measured in the tundish were between 5 and 35°C, the casting rate was varied from 0.8 to 1.5 m/min as the secondary cooling conditions were changed from soft to very hard.

The sample surface which has been inspected was perpendicular to the casting direction. The treatment used to reveal the primary solidification structure was a heat treatment, austenitization followed by quenching in salt water, and etching with Oberhoffer's reagent (1).

The metallographic observations have consisted in noting the presence of :

- needlelike columns, oriented or branched, forming the columnar structure,
- more or less rounded globules forming the equiaxed structure.

Two parameters have been selected to characterize these structures :

- the thickness of the columnar zone,
- the thickness of the globular zone.

When going from the surface to the centre, there is a progressive change of structure from oriented columnar, to branched columnar and eventually to equiaxed. As there are no sharp boundaries between these structures, it is known that these metallographic observations are subjective and approximate.

The structure assessment has been completed by measuring the average grain size in the equiaxed zone when it was present in the centre of the sample.

Model

Heat transfer

The basic model deals with the heat transfer phenomena associated with solidification. It is a classical solution of the enthalpy field using explicite finite difference equations (2,3).

The location of the beginning of solidification has been identified at every time step in order to describe the convective heat flux at the solidification front (eqt. 1) :

$$Q = h \cdot (TL - TSC) + Qerosion \qquad (1)$$

h is a heat transfer coefficient which depends on the convection velocity of the liquid and which can be evaluated with classical formulas taken in the literature (4,5).

Qerosion is the heat flux associated with the erosion of the solidification front as discussed in the next chapter.

One dimension solution has been worked out in cylindrical coordinates.

The liquid or mushy zone laying beyond the solidification front is considered to be homogeneous. Its enthalpy results from a heat balance. The convection velocity is estimated and introduced as a datum depending on the location in the machine (5).

Columnar growth

The dendrites grow from the surface towards the centre, in the direction opposite to the heat flux. The advancement of the isotherms constrains the dendrite growth velocity which determines the tip undercooling.

The dendrite tip defines the location of the solidification front and the dendrite tip temperature defines the temperature of the beginning of solidification.

Various simplified models have been proposed to describe the dendrite tip undercooling (6,7,8). They mainly differ by :

- the dendrite growth criterium : maximum velocity or stability limit,
- the dendrite tip shape : half sphere or paraboloid of revolution.

These various models lead to quite different undercooling values when applied to steel continuous casting (Fig. 1). The solution proposed by Kurz and Fisher for the paraboloid of revolution which appears to be the most correct has been adopted and may be represented by equation (2) which is valid in a range of solidification rates related to continuous casting, $1.E-5 < V < 1.E-3$ m/s :

$$\Delta T = a\, Co^{(1-b)}\, V^{b} \tag{2}$$

	Ferrite	Austenite
a	91.46	72.13
b	0.3609	0.3727

TABLE 1

Undercooling of the Columnar Front in Fe-C Calculated for a Solidification Velocity of 2.E-4 m/s

C, %		0.05	0.09	0.3	0.4	0.53	0.8
ΔT, °C	Ferrite	0.6	0.9	1.9	2.3		
	Austenite			1.4	1.7	2.0	2.6

Table 1 shows the effect of the C content on the dendrite tip undercooling for a solidification rate of 2.E-4 m/s. The dendrite tip undercooling increases with the carbon content. For the same carbon content, it is larger in the ferrite than in the austenite.

Formation of free cristallites

It is well established that free cristallites can be formed in the liquid either by heterogeneous nucleation or by multiplication due to the erosion of the solidification front (9) but existing data to describe these two mechanisms remain largely insufficient.

Thevoz, Desbiolles and Rappaz have recently proposed a new formulation of the heterogeneous nucleation applied to solidification processes which was used in this work (10,11). Different nucleation sites are available which become active at different undercoolings. The nucleus density is calculated by integration of a distribution curve which is assumed to be of the Gauss type. It is characterized by 3 parameters : the average undercooling, ΔTN, the standard deviation, ΔTσ and the maximum density of sites, Nmax, which are to be determined experimentally.

There is no quantitative description of multiplication of free cristallites by erosion of the solidification front in the literature though there are many experimental evidences supporting this mechanism (9,12).

Since a description of multiplication is compelling in this study, we propose the following relationship assessed from experimental results. The erosion of the solidification front is equivalent to an additional heat flux, Qerosion related to the erosion velocity :

$$Q_{erosion} = V_{erosion} \cdot \rho \cdot (L + c_p (T_L - T_C)) \qquad (3)$$

A first set of information comes from the work of Lesoult et al. who made an estimate of the erosion velocity in the case of trials with electromagnetically stirred slabs (13). At a flow velocity of 40 cm/s, 8 cristallites cm-2 s-1 with 0.03 cm radius were eroded.

A second set of information is the trials made by C.R.M. to cast with low superheat : steel flows as a high speed film through a water cooled Cu nozzle (14). This device has a stable operation and removes a high heat flux. From the results, the following estimation was made. A steady state is achieved and the solidified shell thickness in nozzle depends on the superheat. For instance, a heat flux of 1.36 MW/m2 was measured for an initial superheat of 35°C and an average superheat of 25°C. The shell thickness was estimated to be 12.5 mm thick. The convective heat flux was estimated as 0.048 ΔT MW/m2 for a 1.8 m/s flow velocity. The calculated heat flux difference was attributed to the erosion effect, Qerosion = 0.52 MW/m2.

Table 2 summarizes the available data describing the erosion.

These data can be fitted by equation (4) :

$$V_{erosion} = 6.8 \cdot 10^{-5}\, W^{2.2} \qquad (4)$$

TABLE 2

Evaluation of the Erosion of the Solidification Front

Erosion velocity m/s	Flow velocity m/s	Reference
9.10^{-6}	0.4	Lesoult (13)
$2.5.10^{-4}$	1.8	Present work

The equations (4), (3) and (1) are used to describe the erosion of the solidification front taking place in continuously cast strands.

Both the heterogeneous nucleation model and the multiplication model have been applied to four casting conditions summarized in Table 3.

TABLE 3

Experimental References

	With moderate EMS		Without EMS	
Superheat, in the tundish, °C	25	20	20	10
Estimated, max flow velocity, m/s	0.25		0.15	
Observed structure in cast bloom	Columnar	Globular	Columnar	Globular
Average globule radius, µm		160		150

Figure 2 shows how the calculation results compare with the experimental facts.

The calculation results give the evolution of the total number of free cristallites per unit volume, N, as a function of the superheat. When the structure is equiaxed in the centre of the bloom, there is a direct relationship between the average grain diameter and the number of free cristallites : An average grain radius of 150 µm corresponds to approximately 5.E10 free cristallites per m 3. This is the minimum number of free cristallites required to have a well developed equiaxed structure.

When the number of free cristallites is much smaller they cannot grow sufficiently to modify the structure. A columnar structure prevails when the number of free cristallites is below 1.E8 m-3.

The experimental results are represented by four points defining the limiting conditions to have either an equiaxed structure or a columnar structure in the bloom centre.

The model gives an adequate description of the experimental facts when the four points are on the two calculated curves.

When heterogeneous nucleation alone is considered, the adjusted parameters are : $\Delta TN = 6.5°C$, $\Delta T\sigma = 0.8°C$, Nmax = 1.5 E 16 m-3, where Nmax appears exceedingly high (11) and the model fails to describe the system behaviour when EMS is applied.

Figure 2 also shows that multiplication associated with heterogeneous nucleation can account for the structures observed both with and without EMS. In this evaluation, the initial radius of the eroded fragments was adjusted to 5 µm and multiplication was found to be the dominant mechanism in the cristallite formation.

Equiaxed growth

The free cristallites grow from the melt without any preferred orientation. The heat has to flow from the cristallites towards the liquid. It is the free dendrite growth which can only proceed in an undercooled melt. In this case, the tip undercooling determines the rate of grain growth. Lipton, Glicksman and Kurz made a detailed analysis of this problem (15). Their model yields the undercooling- growth rate relationship which is drawn in Figure 3 for 0.75 % C steel. In the range of growth rates pertinent to continuous casting, $E-8 < V < 5.E-6$ m/s, this relationship can be represented by equation (5) :

$$V = a\,\Delta T + b\,Co^{-1}\,(\Delta T)^2 + c\,Co^{-2}\,(\Delta T)^3 \qquad (5)$$

	Ferrite	Austenite
a	$6.61\ 10^{-7}$	-
b	$4.38\ 10^{-6}$	$4.58\ 10^{-6}$
c	$1.33\ 10^{-6}$	$5.38\ 10^{-6}$

It was checked that, in the case of small globules ($d < 400$ µm) which is effectively observed, the heat transfer between globules and liquid is achieved with a negligible temperature difference. Both liquid and globules are considered to be at the same temperature.

Equiaxed fraction

If the population of cristallites is approximated to spheres of average radius r, it is straitghtforward to describe the equiaxed fraction by equation (6) :

$$FE = \frac{4}{3}\pi r^3 . N \qquad (6)$$

However this equation does not take into account interferences existing when the equiaxed fraction gets close to 1. Equation (7) is often used to translate this effect (13,16,17) :

$$FE = 1 - \exp\left(-\frac{4}{3}\pi r^3 N\right) \quad (7)$$

This equation which associates an infinite radius to an equiaxed fraction of 1 is not appropriate to describe solidification problems. We have chosen to use the impingement model based on a compact arrangement of spheres, proposed by Rappaz (18). Impingement between grains begins when the spheres located in the maximum density plane touch each other. This occurs when :

$$FE = 0.74048 \qquad \text{with } rc = \left(\frac{4}{N}\right)^{1/3} \frac{\sqrt{2}}{4}$$

At this stage, the melt consists in globules and residual liquid which is entrapped between the globules. The growth is then restricted to the grain surface which is in contact with the liquid. When the final melt disappears, the globule radius is :

$$rf = rc\,\sqrt{2}$$

Equiaxed grain composition

The equiaxed grains are considered to be a mixture of solid and liquid in chemical and thermal equilibrium. The grains grow from the residual liquid.

Initially, the liquid has the nominal steel composition, Co. At the liquid grain interface, the concentration is the liquidus concentration at the liquid temperature. As the solidification proceeds, the solute content of the liquid progressively increases and reaches the liquidus value when the remaining volume of liquid gets small. The exact evolution of the liquid composition can only be obtained by the complete solution of the mass transfer problem. It has been approximated in the following way :

$$CL = Cliquidus \qquad \text{when } FE = 0.77$$

$$CL = Co + \frac{FE}{0.77}(Cliquidus - Co) \quad (8)$$

The equiaxed grain composition is calculated according to the mass balance

$$Cequi \,.\, FE + CL\,(1 - FE) = Co \quad (9)$$

Computation sequence

The model solves the system of equations presented in this chapter following the computation sequence summarized in the figure 4. The calculation starts with the computation of the enthalpy field in the solidified shell. Then, it proceeds with the determination of the dendrite tip undercooling, the position of the solidification front and the enthalpy of the melt.

Finally, it ends with the computation of the number of free cristallites, their growth rate, the equiaxed fraction and the fraction of solid in the melt. Several iterations may be needed to satisfy the overall heat balance.

Results

The model predictions have been compared to the metallographic observations made on blooms cast at ARBED-Esch-Schifflange. The Figures 5 and 6 summarize the theoretical and the experimental results.

Solidification structure (Figure 5)

The calculated values are given as curves representing the equiaxed fraction in the melt as a function of the position of the solidification front (dendrite tip) and of the initial superheat. The boundary of the strictly columnar zone is marked by the 1 % equiaxed fraction but columns still appear as the dominant feature of the structure up to some 33 % equiaxed fraction. The strictly equiaxed zone is present for equiaxed fractions larger than 74 % but globules are already predominating the structure since some 66 % equiaxed fraction (16). As it has previously been mentioned there are no sharp boundaries to mark the extent of both columnar and equiaxed zones.

The metallographic observations are summarized as black circles to represent the thickness of the columnar zone and as black triangles to represent the thickness of the equiaxed zone.

The extent of the globular zone observed on macrographs corresponds to a calculated equiaxed fraction comprised between 66 and 74 %.

Some scatter appears in the visual evaluation of the extent of the columnar zone. It approximately fits the 33 % calculated fraction.

The conclusion of this comparison is that the computed values are in close agreement with the experimental observations : The model makes a correct prediction of the solidified structures observed in high C blooms with moderate E.M.S.

Tundish superheat greater than 25°C produces columnar structures but non oriented branched dendrites are present in the bloom centre even at 35°C superheat. The free cristallites in the liquid are too few to grow and change the solidification structure.

Tundish superheat below 22°C allows the formation of a centre equiaxed zone. The extent of this zone becomes sizable for superheats below 17°C.

In the intermediate superheat range, 22-25°C, the growth of free cristallites disturbs the columnar growth but the transition from columnar to equiaxed is not reached.

Average globule diameter (Figure 6)

The Figure 6 shows the effect of the superheat on the average globule diameter measured and calculated in the equiaxed zone when it was present in the centre of the blooms.

The measurements show that the average grain diameter remains constant without bearing any effect of the superheat, the electromagnetic stirring or the cooling. The values range from 300 to 360 μm with a standard deviation of ± 140 μm.

The calculation results are given as curves which show that the globule diameter increases with superheat and is very weakly affected by cooling.

The experimental and the calculated results are obviously in contradiction.

The model predictions are self consistent since both the extent of the equiaxed fraction and the final grain diameter depend on the total number of free cristallites which increases when the melt superheat decreases.

The surprising experimental results can only be explained by some phenomenon which is not taken into account in the model : for instance, the aggregation of cristallites in the liquid which would be limited by some external factor, such as heat transfer.

This experimental result induces to reconsider some aspects of the macrosegregation since the observed decrease of macrosegregation cannot be explained by a finer primary structure.

Conclusions

A model able to predict solidification structures in continuously cast products has been proposed. This model provides a description of the solidification kinetics coupled with heat transfer. It deals with both macroscopic and microscopic aspects of solidification. In a first approach, the fluid flow description was simplified and roughly evaluated.

The columnar-equiaxed transition predicted by the model was found to agree with the observations made on square blooms cast in high C steel with moderate electromagnetic stirring.

The modelling showed that heterogeneous nucleation alone cannot account for the known effect of electromagnetic stirring on the solidification structure which can only be explained by the multiplication of cristallites by erosion of the solidification front. The free cristallite formation is the only part of the model where the lack of fundamental data made necessary to adjust the theoretical parameters to the experimental results.

The description of the constrained and unconstrained dendrite growth was made according to the relationships proposed by Kurz and Fisher and Lipton, Glicksman and Kurz respectively. The description of the dendrite tip as a paraboloid of revolution rather than a half sphere, has a major effect on the system behaviour description.

The work has underlined an important experimental fact. The average primary grain diameter in the bloom centre remains constant with reduced superheat or increased cooling or enforced stirring.

TABLE 4

List of Symbols

C, Solute concentration, weight %

Co, CL, Cliquidus, Cequi refer to initial, liquid, liquidus and equiaxed grain concentration

FE, Equiaxed fraction

FSL, Solid fraction in the melt

H, Enthalpy, J kg^{-1}

L, Heat of fusion, J kg^{-1}

N, Number of free cristallites, m^{-3}

T, Temperature, °K

TL, TC, TSC refer to liquid, and cristallite and beginning of solidification

V, Solidification velocity, m s^{-1}

W, Flow velocity, m s^{-1}

Q, Heat flux, W m^{-2}

cp, Specific heat, J kg^{-1} $°C^{-1}$

d, Diameter, m

h, Heat transfer coefficient, W $°C^{-1}$ m^{-2}

r, Radius, m

t, Time, s

ρ, Specific weight, kg m^{-3}

ΔT, Temperature difference, °C

References

(1) H. Jacobi, K. Schwerdtfeger, Metallurgical Transactions 7A, 811, (1976)

(2) A. Etienne, B. Mairy, C.R.M. Reports, 55, 3, (1979)

(3) A. Etienne, Régime thermique du brin coulé en continu, EUR 6307, C.C.E., Luxembourg (1979)

(4) W.H. McAdams, La transmission de la chaleur, Dunod, Paris (1961)

(5) A. Etienne, 4th International Conference Continuous, Casting, p. 597, Stahleisen, Düsseldorf (1988)

(6) W. Kurz, D.J. Fisher, Fundamentals of Solidification, Trans. Tech. Publications, Aedermannsdorf (1989)

(7) H. Esaka, W. Kurz, Z. Metallkde, 76 (2), 127 (1985)

(8) M.H. Burden, J.D. Hunt, J. Cryst. Growth 22, 99, 109 (1974)

(9) M.C. Flemings, Solidification Processing, McGraw-Hill Inc., New-York (1974)

(10) Ph. Thevoz, J.L. Desbiolles, M. Rappaz, Metallurgical Transactions 20A, Feb., 311 (1989)

(11) J.L. Desbiolles, M. Rappaz, Ph. Thevoz, Zou Jic, J.P. Gabathuler, H. Lindscheid, Micro-macroscopic modelling of solidification, Cost 504 - Project CH1, Jan. (1988)

(12) A. Vogel, R.D. Doherty, B. Cantor, Solidification and Casting of Metals, p. 518, The Metals Society, London, (1979)

(13) G. Lesoult, P. Neu, J.P. Birat, Metallurgical Applications of Magneto-hydrodynamics, p. 164, The Metals Society, London (1984)

(14) S. Wilmotte, P. Naveau, F. Knaff, 4th International Conference Continuous Casting, p. 2351, Stahleisen, Düsseldorf (1988)

(15) J. Lipton, M.E. Glicksman, W. Kurz, Metallurgical Transactions, 18A (2), 341, (1987)

(16) J.D. Hunt, Mater. Sci. Engineer., 65, 75 (1984)

(17) C. Wert, C. Zener, Journal of Applied Physics, 21(1), 5, (1950)

(18) M. Rappaz, International Materials Reviews, 34(3), 93, (1989)

* * * * *

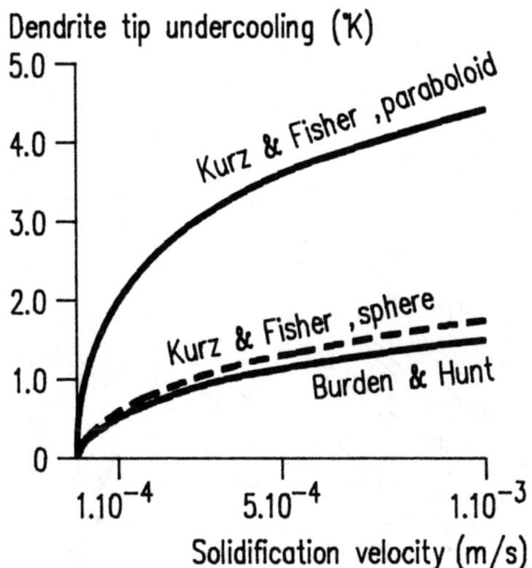

FIG.1_ Dendrite Tip Undercooling in Directional Growth.
(Fe – 0.75% C)

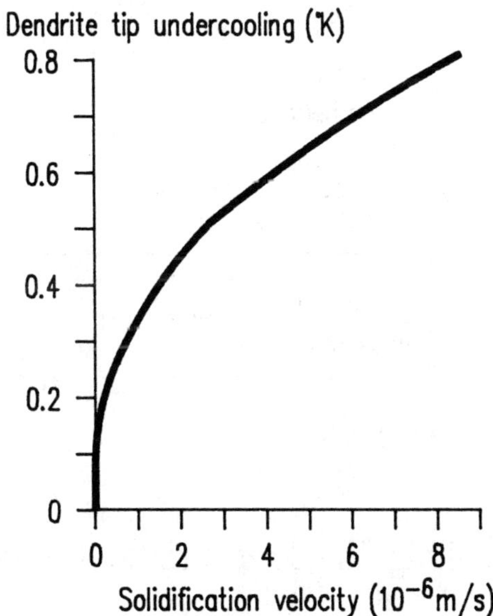

FIG.3_ Dendrite Tip Undercooling in Free Growth.
(Fe – 0.75% C)

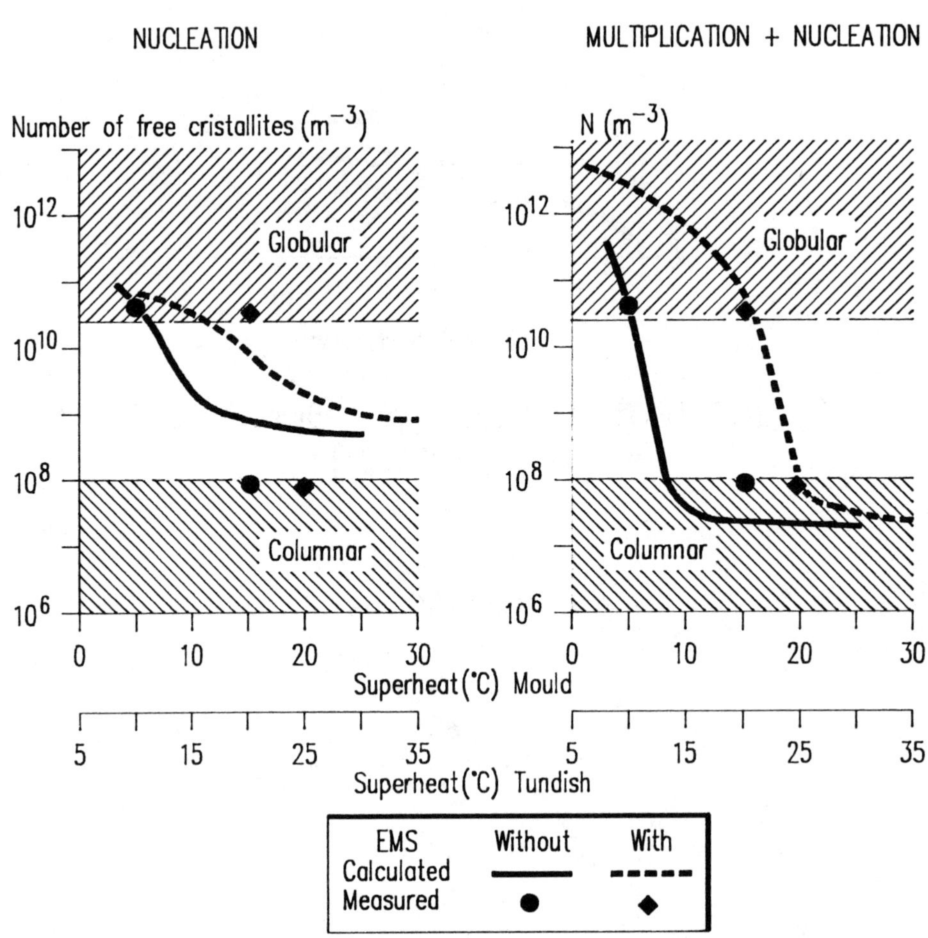

FIG.2_ Modelling the Free Cristallite Formation.

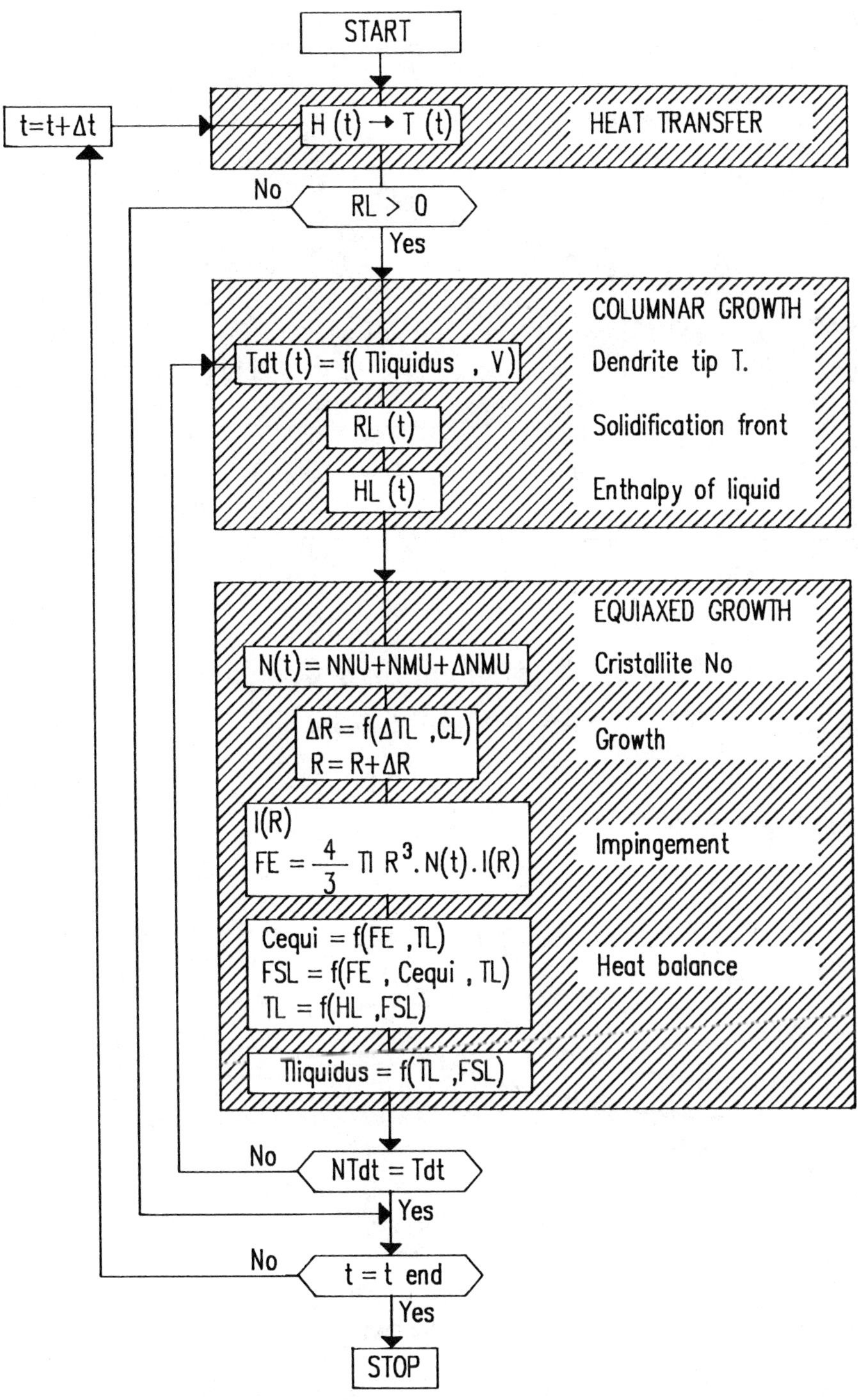

FIG. 4_ Schematic Flow Chart of the Solidification Model.

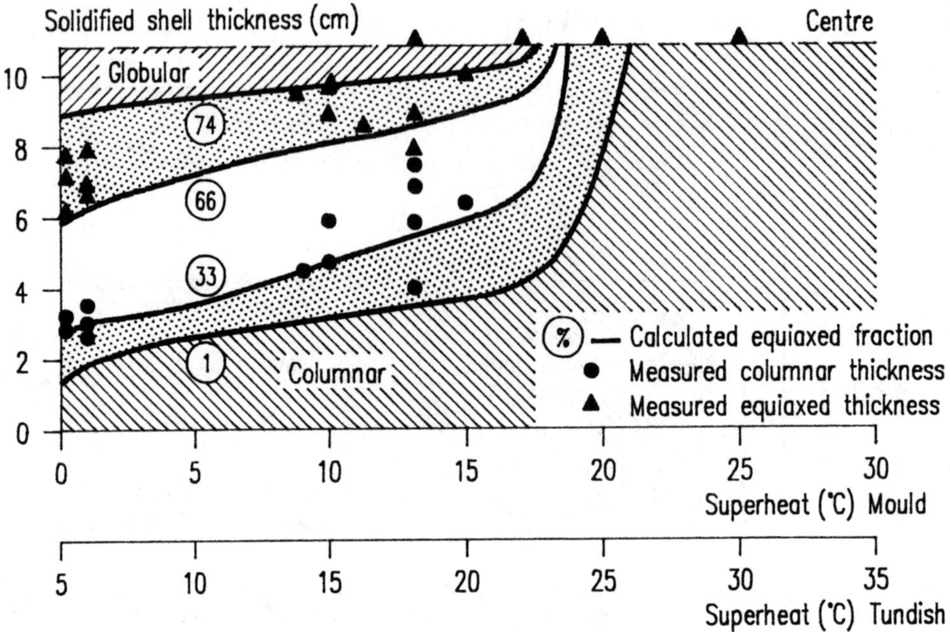

FIG.5_ Solidification in High C Blooms with Moderate EMS.

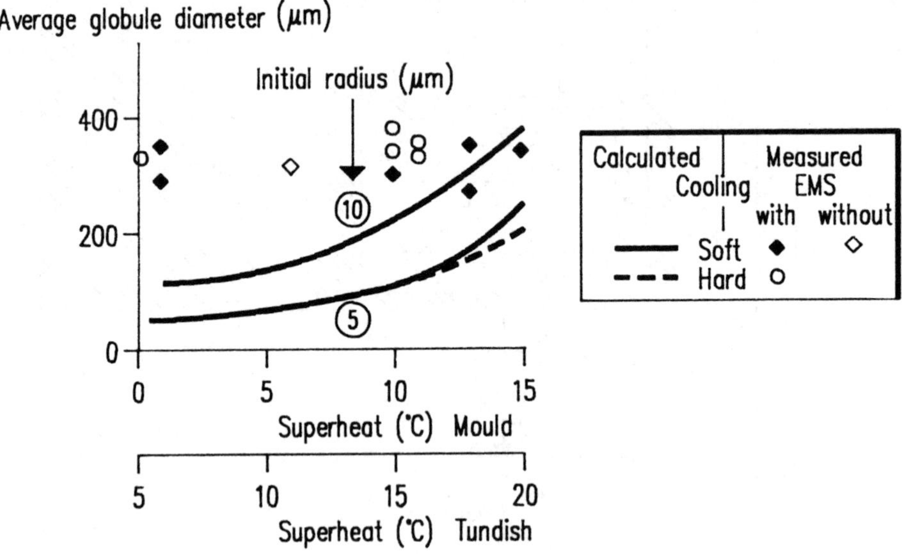

FIG.6_ Average Primary Grain Diameter in the Centre of Continuously Cast Blooms in High C Steel.

Precipitation behaviour of MnS during δ/γ transformation in Fe-Si alloys

Y. Ueshima, Y. Sawada, S. Mizoguchi and H. Kajioka
Nippon Steel Corporation, R&D Laboratories-III, I Kimitsu, Kimitsu-City, Chiba 299-11, Japan

Introduction

During solidification and δ/γ transformation of low carbon steels, solute elements such as C, Si, Mn, S, and P are transported much faster in the δ phase than in the γ phase. Moreover, these elements are redistributed between the two phases, depending on their solubility [1]. Therefore, microsegregation is strongly affected by alloying elements according to their tendencies to stabilize the δ or γ phase [2].

Since the precipitation of MnS depends on the microsegregation of Mn and S, it should also be affected by the δ/γ transformation. As an example, the crystallization and precipitation of MnS have been analyzed quantitatively in the case of low carbon/high sulfur, free cutting steel [3].
On the other hand, there have been some publications concerning the morphology of MnS precipitates during solidification and transformation [4-9]. However, these studies have been very limited with respect to rates of precipitation during the δ/γ transformation.

It is, therefore, the objective of the present paper to add some information on the precipitation of MnS during the δ/γ phase transformation in low carbon Fe-Si alloys with a wide temperature range for the δ/γ two-phase region.

Experiments

Unidirectional solidification experiments have been done with the apparatus shown in Figure 1 [1]. Samples were 15mm in diameter and 150mmin length, which were forged and machined from ingots casted in a vacuum induction furnace. Their chemical compositions were shown in Table 1. The Si content of the samples was changed to study the effect of the degree of the δ/γ

Table 1. Chemical Composition of Specimens(Weight Percent)

Alloys	C	Si	Mn	P	S	N	Al
High Si	0.074	3.17	0.082	0.008	0.0197	0.0079	0.0121
Low Si	0.080	2.34	0.084	0.009	0.0227	0.0066	0.0176

transformation. The sample was heated in this apparatus, by the radiation from the cylindrical graphite susceptor heated by a high frequency induction field. The lower end of the sample was held by a water-cooled stainless steel support. After melting its center, the sample was withdrawn downward at a constant speed so that the dendrites grew upward. During withdrawal, argon gas was admitted into the apparatus to maintain an inert atmosphere. The temperatures of the melted zone were 1550℃ for the 3.17pct Si alloy and 1560℃ for the 2.34pct Si one. The gradient of temperature in the furnace was about 54℃/cm, and the rates of withdrawal were 1, 5, and 10mm/min. Therefore, the cooling rates were 5.4 ℃, 27℃, and 54℃/min, respectively. After the withdrawal by 15cm, the sample was dropped into a water tank for quenching in such a manner that the solute distribution obtained during unidirectional solidification was maintained. The advantage of this method is that it is easy to investigate, in a single experiment, all the phenomena from solidification at high temperatures to precipitation at low temperatures, since these will correspond to different cross-sections of the quenched specimen.

On the horizontal cross-section of a specimen, the distributions of solute elements (Si, Mn, and S) and MnS precipitates were analyzed with a computeraided X-ray microanalyzer (to be called CMA[1]). The analysis conditions are shown in Table 2. The function of this CMA includes the

Table 2. Condition of the Measurement with a Computer-Aided X-Ray Microanalyzer (CMA)

Analysis spot size (μm)	2 × 2
Analysis area (mm)	1 × 1
Integration time (msec)	30
Elements	Si, Mn, S

measurement of compositions, the detection of precipitates, and the analyses of results. The location where the Mn content exceeds 1.32 × $[\text{pct Mn}]_{mean}$ and also X-ray intensity of S exceeds 5 × σ, standard deviation of the X-ray intensity, was recognized as MnS precipitates [10]. The number, the areal ratio, and the mean diameter of MnS precipitates were also analysed with the CMA.

Mathematical analysis

The same mathematical model[1, 3] used in our previous work was used to interpret the results of redistribution of solute elements during δ/γ transformation and precipitation of MnS. With this model, changes in both the fractions of the δ and γ phases and the distributions of solute elements and MnS precipitates during cooling can be calculated simultaneously. The assumptions are summarized as follows:

(1) The γ phase grows as a plate with a constant spacing in the δ phase of uniform concentration. The spacing of the γ phase was given as 500 μm based on the present experimental results.
(2) At the δ/γ interface, the two phases are in thermal equilibrium without undercoolings, and the deviation of the distribution coefficient of solute elements from equilibrium is neglected.
(3) One-dimensional diffusion of five elements (C, Si, Mn, P and S) is calculated by using different diffusion coefficients in each phase.
(4) The precipitation of MnS starts immediately after the product of Mn and S contents in solution reaches the equilibrium solubility limit in each phase. At the interface of MnS and iron, these two phases are in equilibrium.
(5) MnS precipitates are spherical, and their growth is controlled by the diffusion of Mn and S in solid iron.
(6) Precipitation sites are distributed uniformly with a constant spacing (25μm) in both phases based on the present experimental results.

The calculation was carried out by the direct finite difference method[16]. Half of the γ spacing was divided into 20 nodes, as shown in Figure 2.

The δ/γ interface is moved to the next node when A4 or A3 temperature calculated by Eq. [1] or [2] becomes equal to the actual temperature given by a cooling rate.

$$T_{A4} = 1392+1040[pctC]-2000[pctC]^2+\{16.5-481[pctC]+26.9[pctC]^2\}[pct\ Si] + \{-63.1+1490[pctC]-9170[pctC]^2\}[pct\ Si]^2+12[pctMn]-140[pctP] \quad (\text{above } 1100^\circ C)\ (^\circ C) \quad [1]$$

$$T_{A3} = 911-5300[pctC]+35,200[pctC]^2+\{-19.1-54.7[pctC]+5940[pctC]^2\}[pct\ Si] + \{68.6-1390[pctC]+7860[pctC]^2\}[pct\ Si]^2 \quad (\text{below } 1100^\circ C)\ (^\circ C) \quad [2]$$

Both equations were obtained by multiple regression based on the Fe-C, Fe-X, and Fe-C-Si equilibrium phase diagrams[11, 12, 13].

In order to calculate the distribution of MnS precipitates between the δ phase and the adjacent γ phase. 20 spheres are supposed, in which average concentrations of dissolved Mn and S are put equal to those of each node between the δ phase and the adjacent γ phases. When the solubility product of Mn and S of the n-th node meets the equilibrium one, concentrations of Mn and S at the n-th node increase for 1/500 of the stoichiometric concentrations of MnS(0.126 pct Mn, 0.074 pct S), as shown in Figure 3. The equilibrium solubility products in the δ and γ phases are determined by Eqs. [3] and [4], based on the results of Wriedt and Hu[14] and Turkdogan et al[15].

$$\log[pctMn][pctS](\delta) = -10,590/T+4.2489-0.07[pctSi] \quad [3]$$

$$\log[pctMn][pctS](\gamma) = -9,090/T+2.929-(-215/T+0.097)\cdot[pctMn]-0.07[pctSi] \quad [4]$$

The coefficients of equilibrium distribution and diffusion of solute elements used for the present calculation are the same as given previously[1, 2].

Results

Figure 4 shows the results of analysis of the distributions of Si and MnS precipitates during the δ/γ transformation at various temperatures for the 3.17 pct Si alloy with the cooling rate of 27℃/min. The lighter tone gray color corresponds to a higher content of Si. In Figure 4, the lighter region is recognized as the δ phase and the darker region as the γ phase, because Si content should be greater in the δ phase than in the γ phase. Thus, the segregation of Si during the δ/γ transformation is clearly seen in Figure 4. MnS precipitates are shown as white spots.

The composition changes with temperature shown in Figure 4 may be explained by the equilibrium phase diagram, as shown in Figure 5. At 1370℃, the matrix is completely the δ phase, and no MnS precipitates are found. At 1300℃, the γ phase appears like leaves in the matrix of the δ phase. The number of MnS precipitates is small at this temperature but increase drastically below 1200℃. At 1100℃ and 1000℃ MnS precipitates are not mostly included in the γ phase. Some of them may be found particularly near the δ/γ interface. At 800℃, the γ phase becomes unstable, according to Figure 5; nevertheless, the dark region of low-Si content still remains. As will be discussed later, this is is due to the fact that diffusion of Si has been suppressed because of the rather rapid cooling of 27℃/min, so the distribution of Si is not uniform at low temperature. Here, again, it is clear that MnS precipitates segregate in the lighter region of high-Si content, which corresponds to the δ phase at high temperatures.

The behavior of the precipitation of MnS should be related to the diffusion and the redistribution of Mn and S and their solubility product limit. Therefore, it may be interesting to investigate the effects of the cooling rate and Si content of the alloy.

Figure 6 shows the effect of Si content on the distribution of MnS precipitates for a cooling rate of 27 ℃/min. Particularly for the 2.34 pct Si alloy, the amount of the δ phase becomes smaller, and almost all of the MnS precipitates segregate within the narrow region of the δ phase. From this observation, it can be qualitatively concluded that the precipitation of MnS is much more favorable in the δ phase than in the γ phase, particularly for alloys of higher Si content with slower cooling rates.

Next, the quantitative results of the concentration analysis of Mn, Si and S during the δ/γ transformation will be described. Figure 7 is the result of linear analyses across the phase boundary at 1300℃ and clearly shows that amounts of dissolved Si and S are lower in the γ phase than in the δ phase, and these elements are enriched near the boundary in the δ phase. Thus, those elements are redistributed from the γ to the δ phase during the δ/γ transformation.

Figure 8 shows the change in the areal ratio of MnS precipitates with temperature for the specimens of the 3.17 pct Si alloy at different cooling rates. Precipitation started at about 1300℃ and increased rapidly between 1200℃ and 1100℃ but only slightly below 1000℃ at any cooling rate.

Figures 9 and 10 show the change in the number and the mean diameter of MnS precipitates for the 3.17 pct Si alloy. It is interesting to note that the number of MnS precipitates started to increase at about 1200℃ continued to increase below 1000℃, while the diameter became small below 1000℃. This may be explained by considering the change in the nucleation frequency of MnS and the diffusivity of Mn and S with temperatures. At high temperatures the excess amount of Mn and S in solid solution is small, but they can diffuse very fast over a large distance to help the growth of precipitates. However, in order to absorb the excess Mn and S at lower temperatures, increased nucleation will take place with shorter interparticle spacings because of slower diffusion.

A broken line in Figure 11 shows the comparison of the change in areal ratio for tha specimens of different Si contents being cooled at 27℃/min. The amount of MnS precipitates for the low-Si alloy is much smaller than that for the high-Si alloy.

These results confirm that a slower cooling rate and a higher Si content will increase the precipitation of MnS, particularly in the δ phase.

Discussion

The distribution of MnS precipitates during the δ/γ transformation is now explained quantitatively, based on the mathematical model previously presented [1, 3]. First, the changes in the volume fractions of the δ and γ phases with temperature are calculated depending on Si conent of the alloy as shown in Figure 11. In the 3.17 pct Si alloy, the γ phase appears at 1367℃, grows to reach its maximum volume fraction of 69 pct at 1078℃, and then disappears at 701℃. On the other hand, in the 2.34 pct Si alloy, the γ phase appears at 1400℃ and reaches the maximum fraction of 100 pct at 984 ℃. Thus, the γ phase exists over an appreciably wider range of temperature in the 2.34 pct Si alloy. These changes are expected from Figure 5.

Second, the change in the S content as MnS precipitates during the δ/γ transformation has also been calculated by the model. Figure 12 shows the result for the 3.17 pct Si alloy. The precipitation of MnS starts at 1312℃ in the γ phase having the lower limit of equilibrium solubility product. At 1211℃, however, the amount of MnS precipitates increases drastically in the δ phase. An increase of MnS precipitates is clearly seen at the δ/γ interface in particular. On further cooling, the amount of MnS precipitates

still increases in the δ phase and levels off at about 850℃.
On the other hand, the amount of MnS-precipitates in the δ phase levels off below 1200℃. This calculation agrees well with the observation in Figure 5.

Figure 13 shows a similar calculation for the 2.34 pct Si alloy. In this alloy, the proportion of the δ phase is smaller than that in the 3.17 pct Si alloy. Even in this case, precipitation of MnS takes place mostly in the δ phase. That is, MnS precipitates are segregated to the small fraction of the δ phase remaining at the end of the δ/γ transformation. This calculation explains the obseravation in Figure 7.

Figure 14 shows the calculation for the changes in the amount of MnS precipitates with temperature. The precipitation of MnS starts at about 1300℃ and increases drastically at about 1200 ℃. The reaction levels off at about 1000℃. The amount of MnS decreases for the alloy of lower Si content. This calculation is also consistent with the observation in Figure 8.

The distribution of Si has also been calculated at several temperatures, as shown in Figure 15. During the δ/γ transformation, Si is redistributed from the γ to the δ phase, causing sharp spikes near the interface in the δ phase. Although only the δ phase remains at 701℃, the spike and segregation of Si still clearly remain. Thus, the distribution of Si in Figures 4, 6, and 7 may be interpreted quantitatively by this calculation.

The reason for the enhanced precipitation in the δ phase will be discussed in more detail. The distribution of the Mn and S contents around a MnS precipitate in the γ iron at 1091℃ has been calculated, as shown in Figure 16. During the growth of MnS precipitates, excess Mn and S in solution are consumed. The distribution of S content is almost uniform, whereas a significant depletion of Mn is present around MnS. This means that the growth rate of MnS is controlled by the diffusion of Mn, because the diffusion coefficient of S is much larger than that of Mn shown in Figure 17. The depleted zone of Mn was not detected in the present work because of the lack of precision in the measurement. However, in low carbon/high sulfur free-cutting steels, clear evidence of the Mn-depleted zone around large MnS has been obtained[3].

The diffusion coefficient of Mn in the δ phase is about 100 times larger than that in the δ phase at the same temperature, as shown in Figure 17. Therefore, the supply of Mn to MnS precipitates is much faster in the δ phase Moreover, enrichment of S near the δ/γ interface in the δ phase helps precipitaion of MnS.

Conclusion

The precipitation of MnS during the δ/γ transformation has been investigated both experimentally and theoretically, using a low-carbon Fe-Si alloy. The results obtained may be summarized as follows:

1. The precipitation of MnS started above 1300℃, and the amount of the precipitates increased drastically between 1300℃ and 1100℃ but very slowly below 1000 ℃ for both alloys of 2.34 pct Si and 3.17 pct Si content. The amount increases with the decreases in cooling rate and with the increases in the Si content of the alloy.
2. Theoretically, the precipitation of MnS will start in the γ phase because of the lower solubility limit ot MnS in the γ phase. The growth rate of MnS is very small due to the slow supply of Mn ans S in the γ phase.
3. The precipitation of MnS is enhanced in the δ phase because of the higher diffusion rate of Mn and the enrichment of S in the δ phase.

References

1. Y. Ueshima, S. Mizoguchi. T. Matsumiya, and H. Kajioka:Metall. Trans. B, (1986), 17B, 845.
2. Y. Ueshima, N. Komatsu, S. Mizoguchi, and H. Kajioka:Tetsu-to-Hagane, (1987), 73, 1551.
3. Y. Ueshima, K. Isobe, S. Mizoguchi, H. Maede, and H. Kajioka:Tetsu-to-Hagane, (1988), 74, 465.
4. H. Tanaka, I. Bessho, and T. Itoh:Tetsu-to-Hagane, (1976), 62, 1319.
5. Y. Itoh, N. Yonezawa, and K. Matsubara:Tetsu-to-Hagane, (1979), 65, 391.
6. Y. Itoh, N. Yonezawa, and K. Matsubara:Tetsu-to-Hagane, (1979), 65, 1149.
7. Y. Itoh, N. Masumitsu, and K. Matsubara:Tetsu-to-Hagane, (1980), 66, 647.
8. Y. Itoh, N. Narita, and K. Matsubara:Tetsu-to-Hagane, (1981), 67, 755.
9. Y. Itoh, N. Yonezawa, and K. Matsubara:Tetsu-to-Hagane, (1982), 68, 1569.
10. Y. Fukuda, S. Mizoguchi, H. Kajioka, and K. Arihara:Tetsu-to-Hagane, (1985), 71, S218.
11. Metals Handbook, 8th ed., T. Lyman, H. E. Boyer, W. J. Carnes, and M. W. Chevalier, eds., ASM, Metals Park, OH, 1973, vol. 8
12. T. Sato:Technology Reports of the Tohoku University, (1963), vol. 9, p. 515.
13. O. Kubaschewski:Iron Binary phase Diagrams, Springer-Verlag, (1982)
14. H. A. Wriedt and Hsun Hu:Metall. Trans. A, (1976), 7A, 711.
15. E. T. Turkdogan, S. Ignatowicz, and J. Pearson:JISI, (1955), 180, 349.
16. I. Ohnaka:Tetsu-to-Hagane, (1979), 65, 1737.

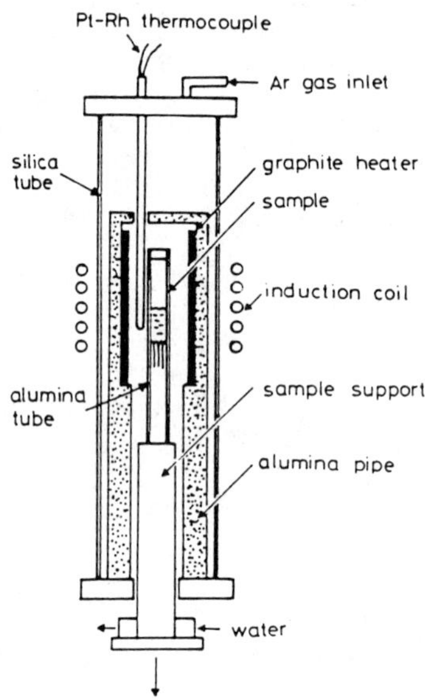

Fig.1 Unidirectional solidification apparatus.

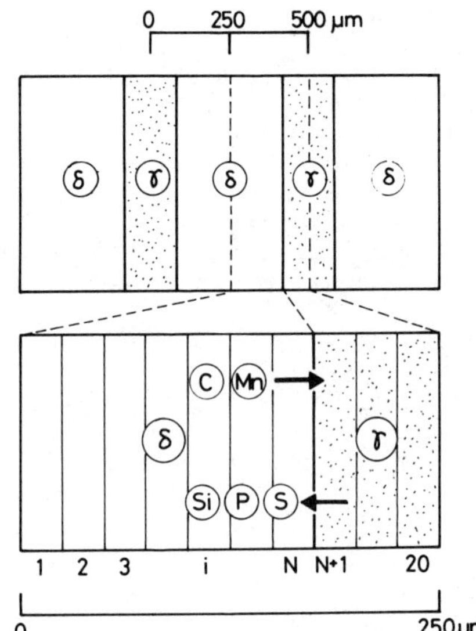

Fig.2 Redistribution of solute elements between the δ and the γ phases.

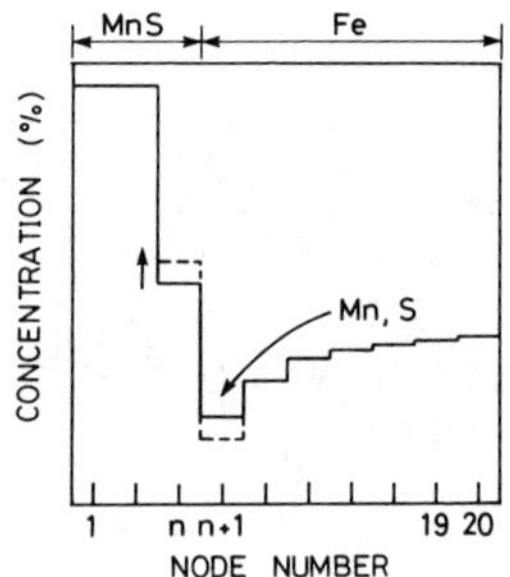

Fig.3 Growth of MnS and diffusion of Mn and S.

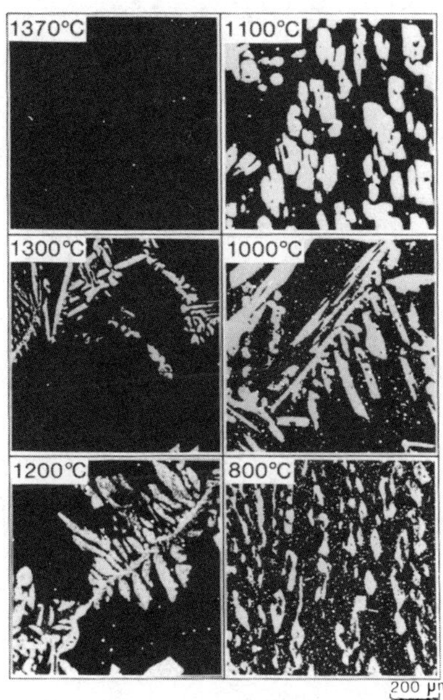

Fig.4 Distribution of Si and MnS. Black regions are δ, gray regions are γ and white spots are MnS. 3.17pctSi.

Fig.6 Effect of Si content on the distribution of MnS. 1000°C.

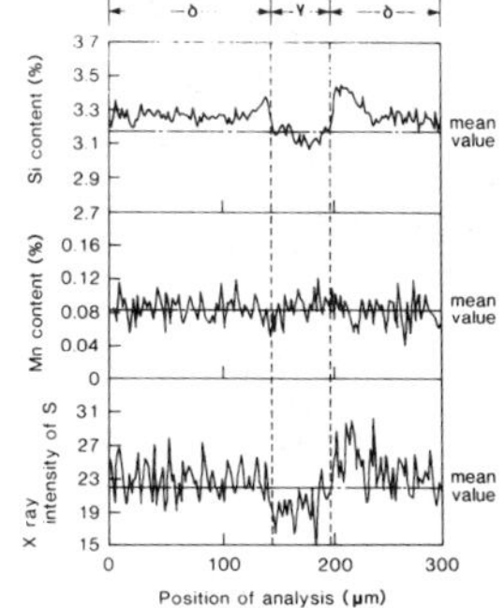

Fig.7 Line analyses of Si, Mn and S.

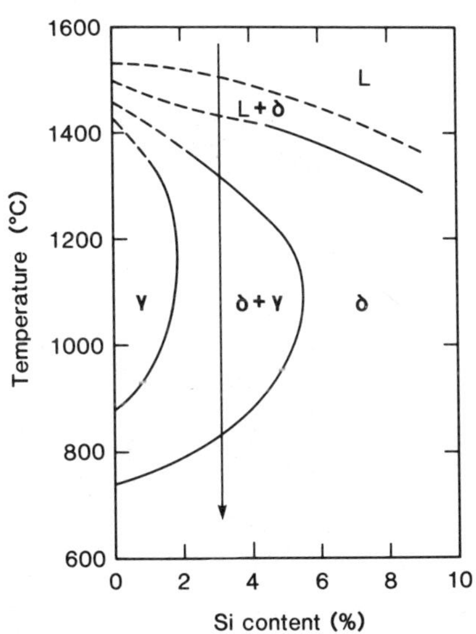

Fig.5 Phase diagram at the section of 0.05pctC of Fe-C-Si alloy[12].

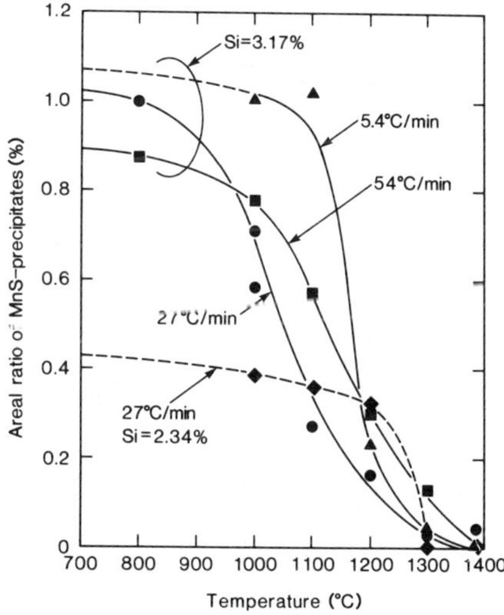

Fig.8 Change in the areal ratio of MnS with temperature.

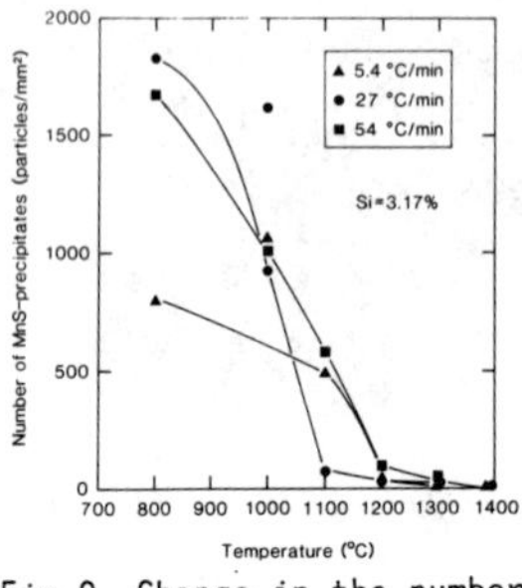

Fig.9 Change in the number of MnS with temperature.

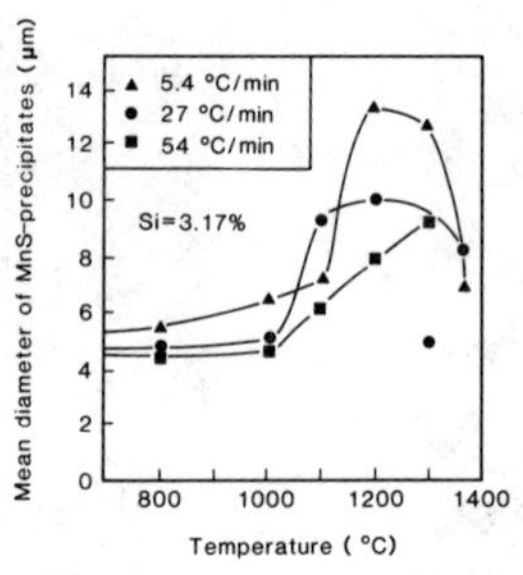

Fig.10 Change in the diameter of MnS with temperature.

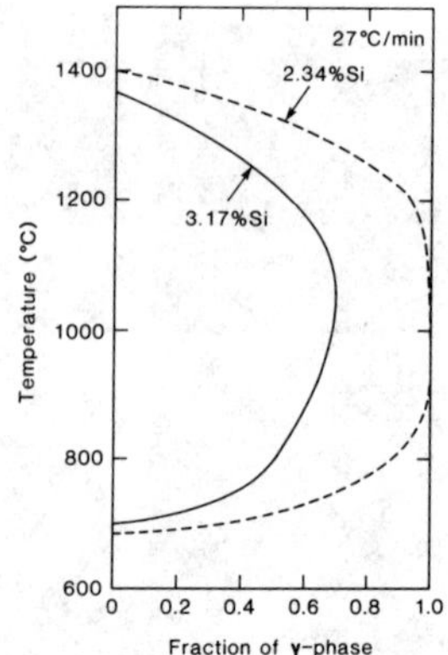

Fig.11 Change in the fraction of the γ phase.

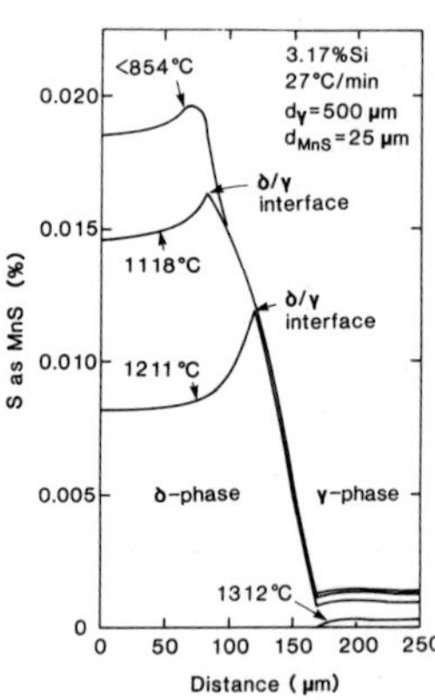

Fig.12 Calculated distribution of MnS. 3.17pctSi.

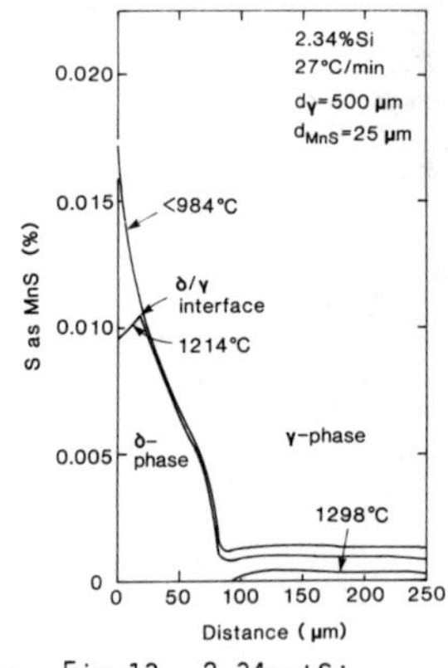

Fig.13 2.34pctSi.

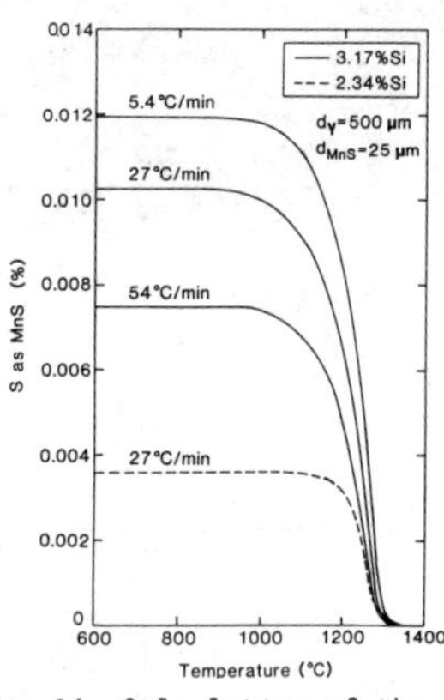

Fig.14 Calculation of the change in the amount of MnS.

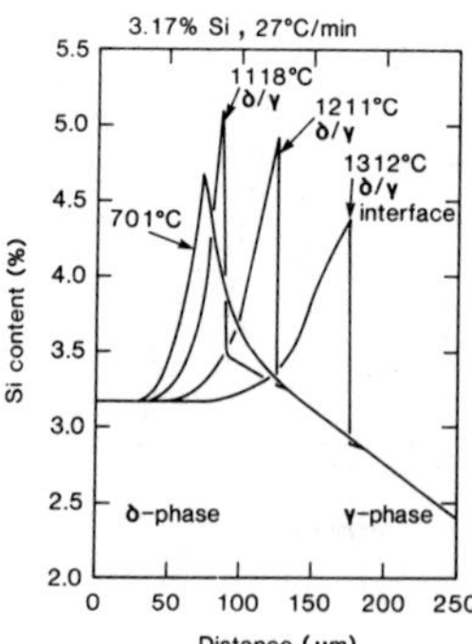

Fig.15 Calculated distribution of Si

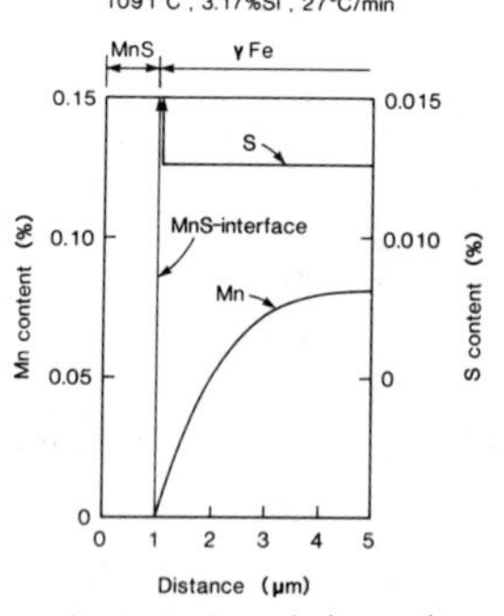

Fig.16 Calculated distribution of Mn and S around MnS.

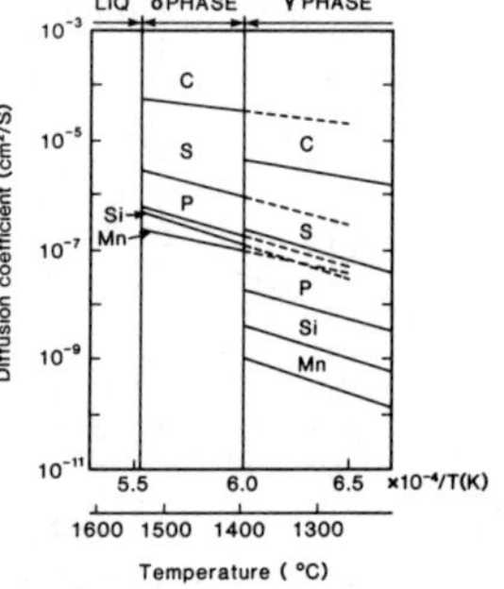

Fig.17 Diffusion coefficients.

The removal of superheat from continuous casting molds

B.G. Thomas, F.M. Najjar and L.J. Mika
Department of Mechanical and Industrial Engineering, University of Illinois at Urbana-Champaign, 1206 West Green Street, Urbana, Illinois 61801, U.S.A.

Abstract

The superheat contained in liquid steel must be removed before it can solidify. This heat represents a significant fraction of the total heat removed by the mold. Its dissipation has a great influence on growth of the solidifying shell and later on development of the microstructure. To investigate this, a finite-element model has been applied to compute the fluid flow and temperature distribution within the liquid pool and heat transfer to the inside of the solidifying shell of a continuously-cast steel slab. It includes separate models of the nozzle and mold cavity using the K-ε turbulence model in FIDAP. Calculated velocities have been compared with experiments using a plexiglass water model. Heat transfer predictions are compared with available liquid temperature measurements. Results indicate that the maximum in heat input to the shell occurs in the vicinity of the impingement point on narrow face, and confirms that most superheat is dissipated in or just below the mold. The model is then used to predict the effect of casting variables, superheat temperature difference, casting speed, nozzle jet angle, mold width, and submergence depth on flow pattern and heat flux to the narrow face shell. Superheat temperature difference and casting speed have the most important effects on heat flux. Calculated heat flux profiles are then input to a 1D solidification model to show their effect on growth of the narrow face shell.

Introduction

Steel is poured into the mold at a temperature above the liquidus. The sensible heat contained in the liquid steel represented by this temperature difference is known as the *superheat*. The average rate of removal of superheat, Q_{sh}, (J/s) can be calculated from:

$$Q_{sh} = (T_o - T_{liq})\, \rho\, C_p\, v\, W\, N \qquad (1)$$

The superheat is directly related to the temperature difference, ΔT_s, in equation 1, (see Table I) referred to as the *superheat temperature* and is sometimes confused with the superheat itself. Superheat is important because:

1, it must be removed before the steel can solidify;

2, it has a great effect on the solidified microstructure; and

3, it affects the formation of defects, such as breakouts, oscillation marks and cracks, through its influence on the formation of the growing shell.

The superheat can be convected to the solidifying steel shell while in the mold, and conducted through the shell to the copper mold walls, or it can be swept out below the mold region, to be dissipated lower in the caster. Assuming 60% of the superheat in Eq. 1 is removed by the mold, rough calculations show that this heat (.48 MW) represents about 20% of the total heat extracted by the mold (2.48 MW). These calculations are given in Appendix I for typical slab casting conditions in Table I.(1) However, uneven dissipation of superheat to the shell will produce a local maximum heat input near the point of jet impingement, so a disproportionately large amount of superheat is delivered to the narrow face. Moreover, the rate of heat extraction by the mold is relatively less from the narrow faces.(2) Thus, this fraction is likely to at least double to 40% for the narrow faces. The total superheat, (.79 MW), is more than twice the heat sustained in a uniform narrow face shell, (.34 MW). Thus, the manner of superheat delivery to the narrow face has a significant effect on slowing shell growth, or even reversing its growth locally. This increases the likelihood of breakouts near the narrow faces and could have an important effect on other defects as well.

Table I Simulation Conditions

Symbol		Standard	Nakato *et al* (1)
W	Slab mold width	1.32 m (52" wide mold)	1.05 m
L_m	Working mold length	0.6 m	
	Total mold length	0.7 m	
N	Slab mold thickness (across NF)	0.22 m	
	Caster length simulated	3.00 m	
°	Nominal nozzle angle (at inlet)	15° down	12° down
	Nozzle port dimensions	90 high x 25.5 mm thick	thicker
L_n	Nozzle submergence depth	0.265 m	0.100 m
v	Casting speed	0.0167 m/s (39"/min.)	0.0267 m/s
T_o	Casting temperature (at inlet)	1550 °C (2822 °F)	1558 °C
T_{liq}	Liquidus temperature (at wall)	1525 °C (2777 °F)	1531 °C
ΔT_s	Superheat temperature (T_o - T_{liq})	25 °C (45 °F)	27 °C
h	Surface heat transfer coefficient	40 W/(m^2K)	
	Ambient temperature (above top surface of powder layer)	27 °C	
ρ	Density	7020 kg/m^3	6968 kg/m^3
k_o	Laminar thermal conductivity	26 W/(mK)	
C_p	Specific heat	680 J/(kgK)	720 J/(kgK)
ΔH	Latent heat of fusion	280,000 J/kg	
μ_o	Laminar (molecular) viscosity	0.00385 kg/(ms)	
K	Inlet and initial turbulent kinetic energy	0.0502 m^2/s^2	
ε	Inlet and initial dissipation	0.457 m^2/s^3	
v_x, v_y	Inlet peak velocity, v_x and v_y	1.062 m/s and 0.471 m/s	
Pr_o	Laminar Prandtl Number, $C_p \mu / k_o$	0.1	
Pr_t	Turbulent Prandtl Number	0.9	

Superheat also has an important effect on the microstructure.(3) Crystals that nucleate in the mold can survive in the liquid without remelting if their melting temperature is higher than that of the alloy in the liquid pool. This is more likely at low superheat, so more solid crystals survive to sink or be carried down through the liquid pool. This provides nucleating crystals ahead of the growing columnar grains lower in the caster, resulting in a larger equiaxed zone for lower superheats.

At high superheat, nucleation is more difficult since steeper temperature gradients exist ahead of the interface. In addition, nuclei convected into the hotter liquid away from the walls cannot survive long before remelting. Thus, few nuclei remain to initiate the central equiaxed zone, and a large columnar zone results. The success of in-mold electromagnetic stirring to improve microstructure has been attributed, in large part, to its effect on superheat removal and movement of these nuclei.(4)

Finally, superheat has an important influence on defect formation. By increasing heat input locally to the inside of the solidifying shell, it can slow down or even stop shell growth, and produce local hot spots on the outer surface of the shell.(5) Upon exiting the mold, this hotter, thinner, weaker shell is more susceptible to deformation, bulging, and crack formation. Equally important is the temperature of the steel near the meniscus. If the superheat is completely removed before reaching the meniscus, the steel liquid temperature may be too low during the critical solidification stage. This would allow both freezing of the meniscus, and excessive solidification of a thick slag rim. This could lead to quality problems such as deeper oscillation marks, which later initiate transverse cracks.

In light of its importance, this project was undertaken to investigate how superheat is dissipated in the continuous slab casting process under various conditions and how this affects shell growth. This was done by developing mathematical models to calculate the temperature field within the molten steel, based on previous calculations of the fluid velocities, and then extracting the heat flux profile down the mold wall. This heat flux is then used to predict the growth and temperature distribution within the steel shell as it moves through the mold.

The effect of important controllable casting variables on superheat removal are investigated, including superheat temperature, nozzle submergence depth, and nozzle port geometry, through its effect on jet angle leaving the nozzle. The effect of other important variables not available for process control are also examined: steel grade, mold width, and thickness, and casting speed.

Previous work

The importance of superheat removal to the continuous casting process is well-known. However, most previous mathematical heat flow models of the process have been simple conduction models that ignore the important effects of fluid flow on superheat distribution. Often, the heat transferred by turbulent fluid convection is roughly accounted for by artificially raising the thermal conductivity of the liquid phase by a constant factor, such as 7. (eg. 6,7)

Mathematical models have recently been applied to calculate fluid flow in the continuous slab casting mold.(5, 8-12). However, only a few models have been applied to better understand superheat dissipation in continuous casting molds. Nakato *et al* (1) extended heat convection measurements for water jets on plates to predict heat flux to the narrow face and compared these with experimental measurements. Recently, Flint (5) used a finite difference model (PHOENICS) to solve the coupled 3D turbulent fluid and energy transport equations, including the effects of latent heat. This model calculated both the fluid flow pattern and temperature contours, including the solidification and withdrawal of the strand, and predicted significant thinning of the solid steel shell on the narrow face and adjacent edges of the wide face.

Heat flow model in liquid

To further understand superheat removal in continuous slab casting, a finite element model has been developed to calculate the temperature distribution within the molten steel pool inside the shell in the mold region of a continuous slab casting machine, fed by a bifurcated, submerged entry nozzle.

The model is based on fluid velocities and turbulence parameters calculated by a separate model of fluid flow in the mold.(9-10) Figure 1 presents an example of three-dimensional velocity predictions within the nozzle, for standard conditions. Calculated velocities such as these and turbulence parameters at the nozzle exit are used as inlet boundary conditions to the mold cavity.

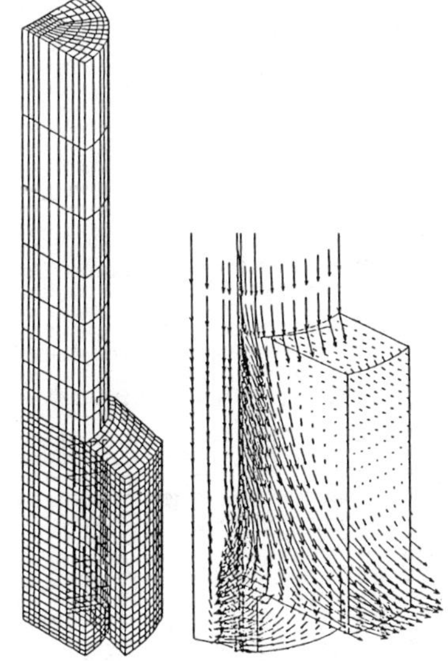

FIG. 1. Finite element mesh and calculated velocities in bifurcated entry nozzle

Figure 2 presents a typical 3D flow pattern in the mold. Since the bifurcated nozzle generates a flow pattern that has relatively little recirculation through the thickness of the mold, the 3D results can be reasonably approximated with a 2D model of a vertical section through the center of the mold, parallel to the wide face, as seen in Figure 3. The major difference in these two flow patterns is the increased curvature of the jet in the bulk of the mold in 3D, which results in slightly higher impingement points on the narrow face wall for the 3D calculations. This curvature is caused by the upward lifting force on the jet resulting from the reduced pressure in upper recirculation zone. Since the

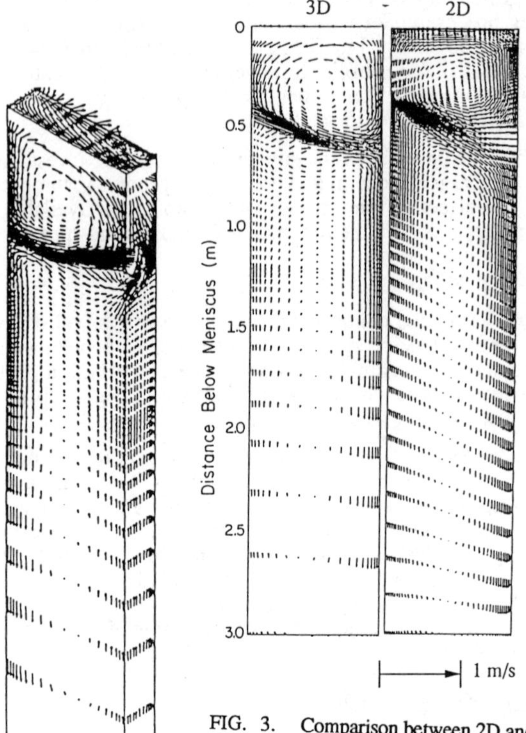

FIG. 2. Flow in the mold cavity calculated with 3D fluid flow model

FIG. 3. Comparison between 2D and 3D velocity predictions in centerline

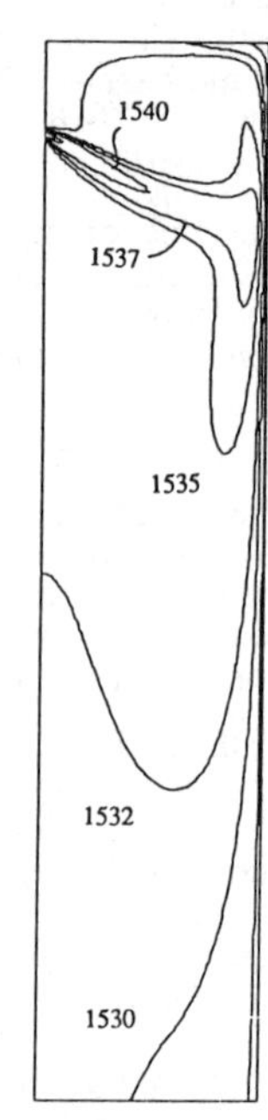

FIG. 5. Calculated temperature contours (Standard conditions Table 1)

2D jet broadens less, it has more momentum to resist this upward bending. It should be noted that slightly different submergence depths and mesh refinements were used in these two simulations.

Figure 4 shows the 40 x 75 node mesh of two-dimensional 4-node finite elements used to model the liquid pool in the present work. The heat transfer model calculates temperatures in this domain by solving the 2D energy equation:

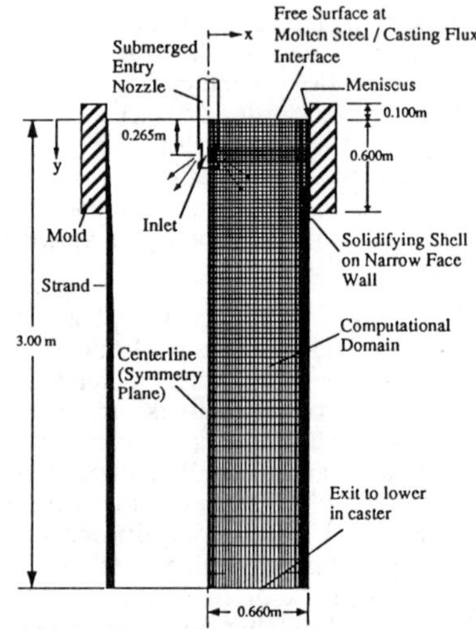

FIG. 4. Mesh and domain used in the 2D fluid flow and heat transfer model

$$\rho C_p \left(v_x \frac{\partial T}{\partial x} + v_y \frac{\partial T}{\partial y} \right) = \frac{\partial}{\partial x}\left(k_{eff} \frac{\partial T}{\partial x} \right) + \frac{\partial}{\partial y}\left(k_{eff} \frac{\partial T}{\partial y} \right) + (\mu_o + \mu_t)\, \Phi \qquad (2)$$

$$\text{where } \Phi = 2\left(\frac{\partial v_x}{\partial x}\right)^2 + 2\left(\frac{\partial v_y}{\partial y}\right)^2 + \left(\frac{\partial v_y}{\partial x} + \frac{\partial v_x}{\partial y}\right)^2$$

The thermal convection terms in this equation depend upon previously calculated velocity fields, which are discussed and compared with water model observations elsewhere.(9) Flow in the continuous casting mold is highly turbulent, as indicated by the minimum Reynolds number of 12,000 found far below the mold, so the standard K-ε turbulence model was used in calculating velocities.(13) In addition, conductive heat transfer is enhanced greatly by turbulent eddy motion, so the effective thermal conductivity, k_{eff}, consists of two components:

$$k_{eff} = k_o + \frac{C_p\, \mu_t}{Pr_t} \qquad (3)$$

The turbulent component can be seen to depend on the previously calculated turbulence parameters via the turbulent viscosity, $\mu_t = 0.09\ \rho\ K^2/\varepsilon$, and the turbulent Prandtl number, Pr_t, set to 0.9.

The effect of buoyancy forces (due to temperature-induced density differences) on the velocity fields were neglected. This allowed the uncoupling of the energy equation from the other equations. This approximation is reasonable for the purposes of calculating the overall flow pattern, jet impingement point, and accompanying heat transfer to the narrow face, since the governing dimensionless parameter, Gr / Re^2, is 0.06 in the high-velocity regions of the upper mold. In the stagnant regions both in and below the mold, however, natural convection becomes an important force driving flow up the walls. Single phase flow was also assumed, so effects such as those from argon gas bubble injection are not considered.

Boundary conditions

Inlet and outlet.

Velocity and turbulence inlet conditions are known to be important, so are calculated in the separate model of the nozzle.(9) Temperature across the inlet plane (nozzle exit) was simply fixed to the casting temperature, T_o, corresponding to a tundish temperature. Calculations revealed that heat losses through the nozzle are very small, which agrees with experimental findings shown in Figure 6. (14) Temperature gradients across the symmetry plane (the centerline) and the bottom outlet plane were set to zero.

Top surface.

Calculations to account for heat losses conducting through the molten flux and powder layers and radiating to ambient, determined an average heat transfer coefficient of 40 W/m^2 (15) which was applied to the top surface.

Mold wall.

To avoid the computational difficulties associated with modelling latent heat evolution at the solidification front, fluid flow was modelled up to, but not including, the mushy zone. Consequently, a fixed temperature, nominally equal to the liquidus, is imposed along the mold wall (narrow face) side of the model. Empirical "wall law" functions were employed for tangential velocity, K, and ε, along the mold wall. These had a great effect on heat transfer, although there was very little effect on fluid flow. Standard wall laws where velocity and heat flux are proportional to tangential velocity vastly under-predicted heat flux near the impingement point, so a function proportional to K was used instead.(10)

This approach differs from other recent models, which couple the fluid flow and solidification calculations. The latter models use a function (based on flow through porous media) to radically reduce velocity and turbulence levels within the mushy zone.(5) By separating the fluid flow and solidification calculations, the present approach reduces the complexity of the heat conduction and solidification model of the solid shell. Results from this model are then more easily coupled with other thermal stress and shrinkage models that incorporate heat flow across the growing gap.(16)

Solution method

Obtaining a reasonably-converged velocity field for this problem is difficult and is discussed in depth elsewhere.(10) Solution of the energy equation is relatively easy, requiring only a single iteration, using the FIDAP program.(13) However, the temperature solution is very sensitive to the convergence of the velocity and turbulence fields, so more stringent convergence requirements must be satisfied when velocity solutions are used in subsequent heat transfer calculations. The turbulent Prandtl number is also very important, since it controls k_t, but the standard value of 0.9 has been found to produce reasonable results.(10)

The heat flux predictions described here are input to other models of the solidifying steel shell, described later. The results of this work will be used to increase our understanding of the early stages of shell solidification and structure development and to provide insights into the prevention of quality problems, such as breakouts and longitudinal cracks.

Model results and discussion

The fluid flow and heat flow models for the liquid were run under standard conditions in Table I. Model calculations of temperature and heat flux are now presented and compared with available measurements.

Temperature

The isotherms shown in Figure 5 clearly outline the path of the hot steel making up the jet and show how the fluid carries heat to the narrow face wall. If the jet impinges the narrow face before the top surface, as in this case, it then splits to flow both upwards and downwards. These flows cool against the solidifying steel shell along the narrow face wall as they travel. The steel flowing downward is almost at the liquidus temperature and retains little of its original superheat by the time it leaves the computational domain, 3m below the meniscus. The steel flowing upward also loses most of its superheat by the time it reaches the meniscus. Heat is actually lost from the jet even faster than this figure suggests, since the wide face extracts heat also. The upper recirculation zone delivers the fastest-moving, hottest liquid close to the center of the molten powder layer (between the nozzle and narrow face). Thus, the meniscus and the region adjacent to the nozzle at the top corners of the domain are the coldest locations in the liquid pool.

The model temperature predictions are compared in Figure 6 with thermocouple measurements in the molten steel from Offerman (14) To acheive the comparison, the standard model conditions were altered to set $T_o = 1540$ °C and $T_{liq} = 1510$ °C. Overall agreement is reasonable, as the same general trends outlined above can be seen in the measurements.

However, temperatures are somewhat colder in the central regions of the liquid pool than calculated by the 2D model. In particular, the measured temperature of the jet loses 2/3 of its superheat to drop from 1540 to 1519 °C, in travelling only half way across the mold. The model predicts the jet to retain 2/3 of its superheat until very near its impingement point on the narrow face wall. This discrepancy is most likely due to the two dimensional nature of the calculation, which ignores the continuous loss of heat from the jet to the wide faces, as it moves. It might also be caused by melting dendrites broken off the inside of the shell (14), or problems locating the jet when making the experimental measurements. Finally, the assumed Pr_t of 0.9 might be too high.

Heat flux

The model is also used to calculate the heat flux across the liquid/solid shell boundary calculated as a function of distance down the narrow face wall. The results in Figure 7 show that heat input to the shell is a maximum near the bottom of the mold at the region where the jet impinges upon the wall for these standard conditions.

Model heat flux predictions were compared against experimental correlations for water jet impingement on a heated flat plate by Kumada and Mabuchi.(17) These correlations were adapted by Nakato *et al* (1) and shown to reasonably predict heat transfer due to steel jet impingement within a continuous casting mold. The same method was used in the present study to find an independent measure of heat flux as a function of position down the narrow face shell wall. In this procedure, heat flux is first calculated at the point of jet impingement, based on the total superheat. The heat lost to the wall is used to decrease the superheat remaining, and the heat flux in the two

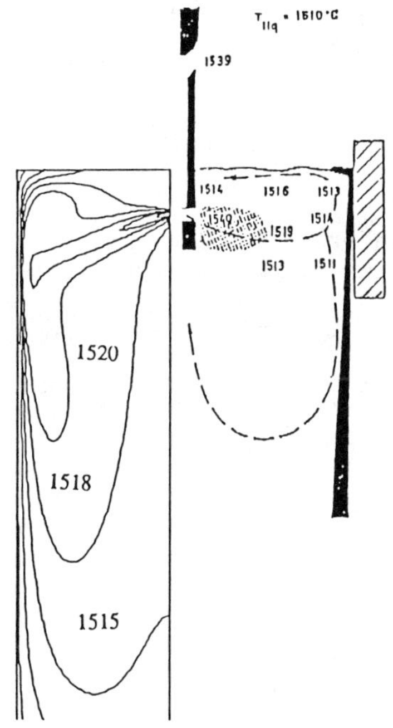

FIG. 6. Comparison between calculated and measured (14) molten steel temperatures

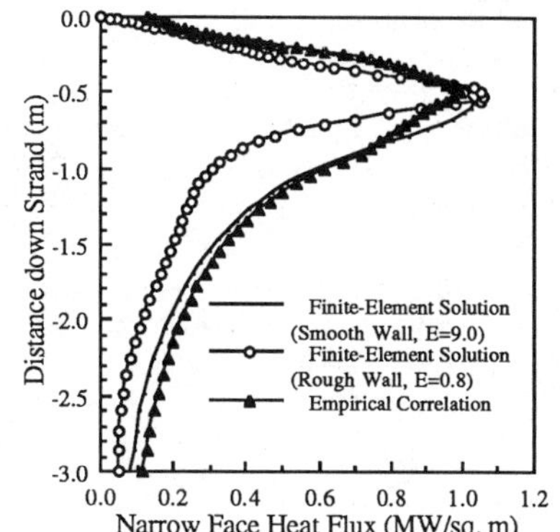

FIG. 7. Comparison between model heat flux predictions and experimental correlations

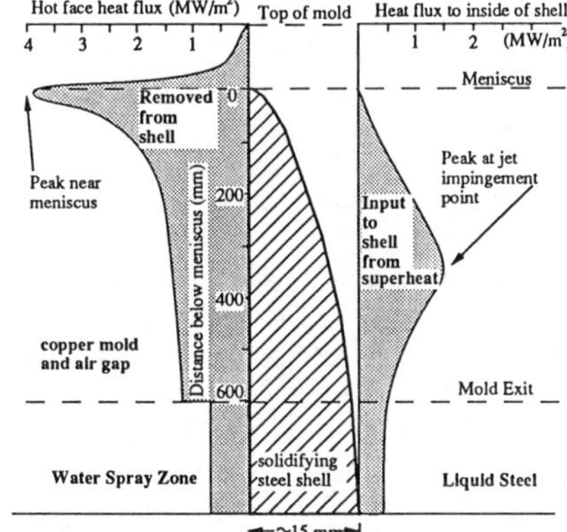

FIG. 8. Heat transfer through solidifying steel shell

adjacent regions are calculated. This recursive procedure continues in both directions, until the entire profile is produced.(15)

The heat flux predicted by this correlation was compared with the results of the finite element simulation, based on the standard conditions in Table I, and assuming a smooth wall. Figure 7 shows a surprizingly close match all the way down the mold. The maximum heat flux at the jet impingement point on the narrow face predicted by both methods is about 1.0 MW/m^2. This heat transfer rate is comparable with convection rates measured by Argyropoulos *et al* for solid steel melting in a stirred liquid steel bath.(18)

Finally, Figure 7 also shows that the roughness of the wall, used in the turbulent wall law, has an important effect on the heat flux profile. The inner surface of the dendrites in the continuous caster has larger crevices than the flat plate used in the correlations. This promotes faster local heat dissipation and creates a sharper heat flux peak at the impingement point.

<u>Superheat distribution</u>

The relationship between the heat flux calculated from superheat dissipation and the heat extracted to the mold wall is illustrated schematically in Figure 8 for the narrow face. Growth of the shell naturally depends upon the combination of these two boundary conditions. The superheat distribution is important since this figure shows that the two curves are of the same magnitude.

Table II presents a calculated distribution of superheat dissipation down the caster. This involved integrating the area between the heat flux curve and the shell in Figure 8 for the conditions in Table I reported by Nakato *et al.* (1). This table shows that almost 60% of the superheat is dissipated through the mold and 83% within 1.6m from the meniscus. This agrees with previous estimates that the majority of the superheat is removed in or just below the mold. (3) It also confirms the great importance of superheat found in the calculations in Appendix I.

It is interesting to note that the model predicts the rate of superheat removal to drop off substantially below this distance so that over 10% of the superheat is convected to very deep in the caster. With deeper nozzle submergence and / or steeper nozzle jet angles, other results show that considerably more (even > 50%) of the superheat can be removed below the mold.

Table II Predicted Superheat Distribution

Superheat lost to:	Heat (MW)	%
Conduction through top surface flux powder layer	0.02	2
Convection to shell inside mold (.6m)	0.44	58
Convection to shell below mold (.6 -1.6m) (first meter below mold exit)	0.19	25
Convection to shell below mold (1.6 - 3.0m) (next 1.4 m)	0.03	4
Dissipation very low in caster (beyond 3m below meniscus)	0.09	11
Numerical convergence errors	0.03	
Total	0.79	100

The model was then used to investigate the effect of important casting variables on heat extraction to the narrow face, based on the standard conditions given in Table I. Steel grade was assumed to affect only the liquid properties, and these effects were found to be almost negligible.

Effect of superheat temperature

Superheat temperature has the most direct influence on superheat dissipation of all casting variables. Since natural convection in the mold is negligible, this temperature difference has no effect on the flow pattern in the mold. Thus, superheat delivered to the shell should increase in direct proportion to the difference between the liquidus and pouring temperatures. This trend is illustrated in Figure 9, which shows the expected linear increase in heat flux to the narrow face with increasing superheat temperature, calculated by the model. Since heat extraction to the mold is only barely influenced by superheat, higher superheats will result in thinner shells leaving the mold. Thus, accommodations should be made when casting with a high superheat, to avoid breakouts. Higher superheat also increases the temperature gradient at the solidifying interface, which promotes a columnar grain structure.

Effect of casting speed

The importance of casting speed on superheat dissipation can be inferred from its definition in Eq. 1 and Appendix I. This equation states that the superheat that must be dissipated increases in direct proportion with the casting speed. Since heat extracted by the mold increases only slightly with increasing casting speed, this implies an increased importance of superheat at high casting speeds.

Figure 10 shows the effect of casting speed on heat flux to the inside of the narrow face shell calculated by the model for conditions in Table I used by Nakato *et al* .(1) This figure confirms that increasing casting speed greatly increases superheat to the narrow face. It also shows that the peak heat flux is sharpened as a slightly higher fraction of the superheat is removed near the impingement point at higher speeds. This finding is consistent with the increase in heat

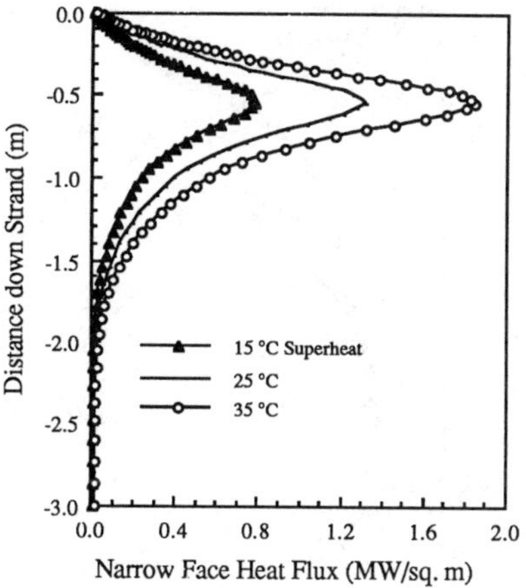

FIG. 9. Effect of superheat temperature difference on superheat dissipation

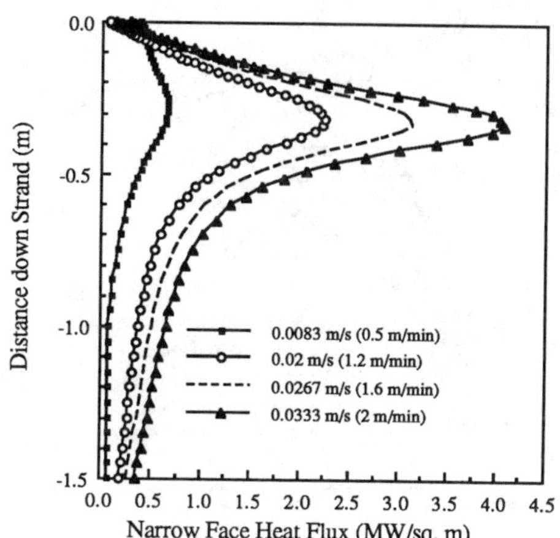

FIG. 10. Effect of casting speed on superheat dissipation (Table I conditions (1))

transfer rate encountered when fluid velocity across the solidifying interface is increased, that has been observed experimentally in stirring of liquid metal baths.(18)

Increasing casting speed has an additional effect on the calculated results not shown in this figure. It tends to straighten out the jet leaving the nozzle, by giving it more momentum to overcome its tendency to bend toward the surface. This makes the jet impinge lower on the narrow face at higher casting speeds. It should be cautioned, however, that higher casting speed also makes the edges of the nozzle ports more effective (9), which would tend to make the original jet angle less steep, for a typical downward nozzle. The two effects partially cancel, so the flow pattern is not affected much by casting speed unless near critical conditions, as discussed later.

Impingement of the jet below mold exit has important implications when casting at high speeds. The efficiency of spray water cooling on the narrow face and off-corner region of the wide face then becomes critical. The thinnest regions of the shell, accompanying the internal jet impingement, could coincide with the hottest outer surface temperature, if spray cooling heat extraction was nonuniform or less than in the mold. The combination would increase the likelihood of breakouts, bulging, or distortion of the unsupported shell just below mold exit, where the shell is its thinnest, hottest, and weakest. These phenomena likely contribute to the increased quality problems found at higher casting speeds.(1)

These problems could be a concern even at moderate casting speeds, if a nozzle port becomes severely blocked by alumina build-up, and increases flow through the other port. This would result in jet impingement on one narrow face equivalent to that experienced at much higher casting speeds.

Since casting speed and superheat temperature both increase the heat flux directly, it is important to slow down the casting speed when high superheats are encountered, in order to prevent a breakout. The exact relationship should depend on the mold width, nozzle angle, and submergence depth.

Effect of mold width

Figure 11 compares the predicted superheat flux to the narrow face for a range of mold widths at different flow rates and casting speeds. Increasing the mold width for a constant flow rate through the mold (same inlet velocity conditions) decreases the peak heat flux to the narrow face significantly, since the jet travels farther and spreads more.

However, to keep the casting speed constant, an increased flow rate is required. This greatly increases the heat flux, since there is more heat to extract through the same narrow face area. Figure 11 shows that the combined effects on the peak heat flux almost cancel. In fact, peak heat flux to the narrow face actually decreases slightly as width increases for constant casting

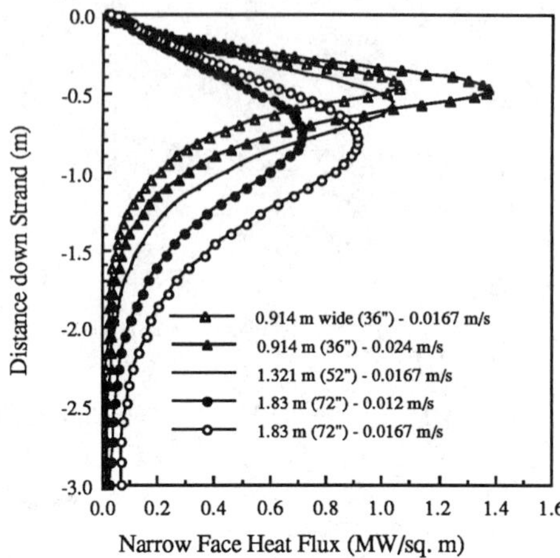

FIG. 11. Effect of mold width on superheat dissipation

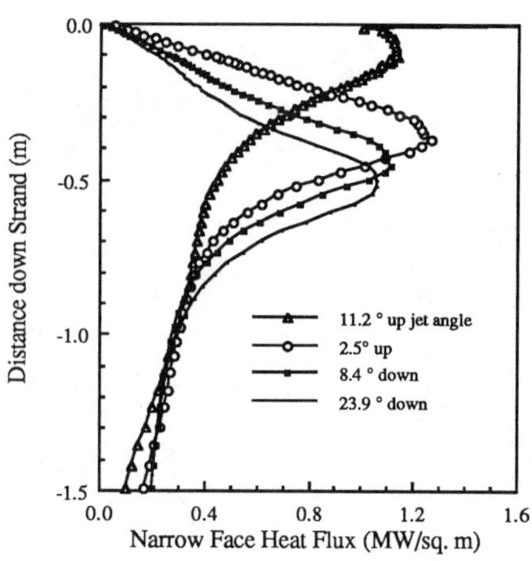

FIG. 12. Effect of nozzle angle on superheat dissipation

speed. The increased heat extraction required for increasing width at constant casting speed is indicated in Figure 11 by a larger area under each curve. The high heat of the jet impingement region is broadened and extended to lower down the narrow face. The peak heat flux even moves below the mold for the two widest molds under the standard conditions simulated. This makes narrow face spray cooling just below the mold more critical for wider slabs.

This analysis assumes that the narrow faces extract all of the superheat. Since increasing mold width will significantly decrease the fraction of the superheat extracted by the narrow faces, problems with wider molds will be less than these figures imply. Note that increasing mold thickness has the same effect as reducing mold width, since both actions reduce the aspect ratio.

Effect of jet angle

The angle of the steel jet is controlled by the nominal angle, size and thickness of the nozzle ports. When oversized, thin-walled, downward ports are used, the impingement point is always low on the narrow face, near the mold exit. In these cases, the effect of changing nozzle angle on the heat flux to the narrow-face wall is relatively slight, as shown in Figure 12. Steeper nozzle angles move the impingement point point lower on the narrow face wall and slightly decrease the magnitude of the peak, so a greater fraction of the superheat is removed below the mold.

However, by adjusting the size and angle of the nozzle ports, the impingement point of maximum heat input can be moved to any desired position along the narrow face wall. Using smaller, thicker, upward angled ports moves the impingement point higher up the mold. Jet angle was increased to 2.5° up by modifying the standard nozzle ports to 15° up and 45mm high. It was further increased to 11.2° upward by increasing the thickness of these new nozzle ports to 38 mm.

Between these latter two conditions, a radical change in the flow pattern occurs, which is shown in Figure 13. The upper recirculation region shrinks, which forces the fluid jet to bend upward to impinge first on the liquid surface beneath the molten powder layer. This flow pattern is similar to that observed in a water model by Soboleski *et al*, for similar casting conditions.(19) It is interesting to note that this tendency of the jet to bend upward and impinge the top surface has been calculated without consideration of buoyancy from argon gas injection.

This new flow pattern produces a maximum superheat dissipation heat flux at the meniscus. Almost all of the superheat is removed in the mold for this case, and heat flux at mold exit is small. Delivering high heat flux to the top of the mold will prevent meniscus freezing and melt back the solid flux rim, in addition to avoiding shell thinning lower in the mold. This might be beneficial. However, the increased flow across the top surface towards the meniscus, that accompanies impingement on the surface, also increases surface turbulence, which could promote entrainment of molten mold flux. It also increases heat input to the powder layer, which would increase its melting rate. To avoid this impingement on the top surface, the nozzle should be designed to produce a sufficiently steep jet angle for the given conditions (casting speed, mold width, and submergence depth).

Effect of submergence depth

When the liquid jet leaving the nozzle points downward, then changing the nozzle submergence depth changes the point of impingement on the narrow face wall in direct proportion. Figure 14 shows the expected trend that the heat flux profile was shifted downward by about the same distance as the nozzle submergence was lowered from .15 to .365m. This deeper submergence brings colder fluid to the top surface of the liquid pool, particularly at the meniscus, with the accompanying potential quality problems discussed earlier.

Reducing the submergence depth shifts the heat flux profile up the wall. Eventually, a small decrease in submergence depth substantially changes the flow pattern. At too shallow a submergence, the jet bends upward to impinge first on the top surface for a combination of critical submergence depth and nozzle angle. This produces a maximum heat input to the mold at the meniscus of the narrow face, as seen in Figure 14. The effect is very similar to the results obtained when the jet angle leaving the nozzle is too high, as shown in Figure 13. This might help to explain the increase in quality problems, such as longitudinal cracks, observed at shallow submergence depths.(20)

A critical submergence depth appears to govern the substantial change in the flow pattern, for a given jet angle, casting speed, and mold width. In the present example of 1.6 m/min, 1.05 m wide, and 12° down nozzle angle, (1) decreasing submergence from 0.15 m to only 0.1 m completely changes the flow pattern. The jet impingement point moves from 0.35 m down the narrow face to the midway across the top surface. This critical submergence is highly dependent upon the nozzle angle, and also changes with casting speed and mold width. For example, with 0.1 m submergence, the normal flow pattern was restored by steepening jet angle from 12° to 20° downward. Higher casting speeds allow shallower submergence or shallower angles before the critical point is reached. To avoid impingement with the top surface, the nozzle submergence should be deeper than the critical depth for the given set of conditions.

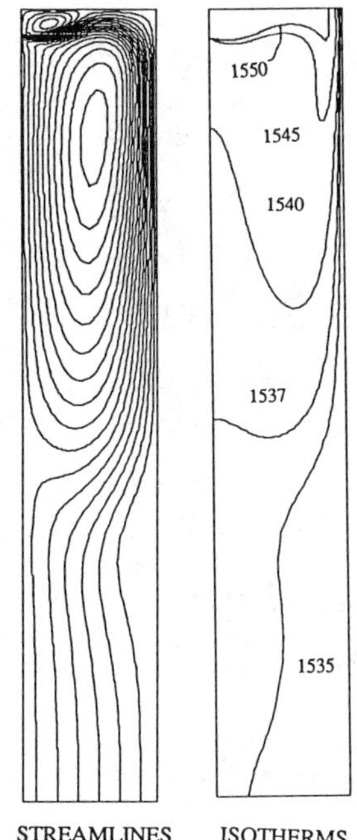

FIG. 13. Velocity and temperature results showing jet impingement on the top surface (Standard conditions with 11.2° upward jet angle)

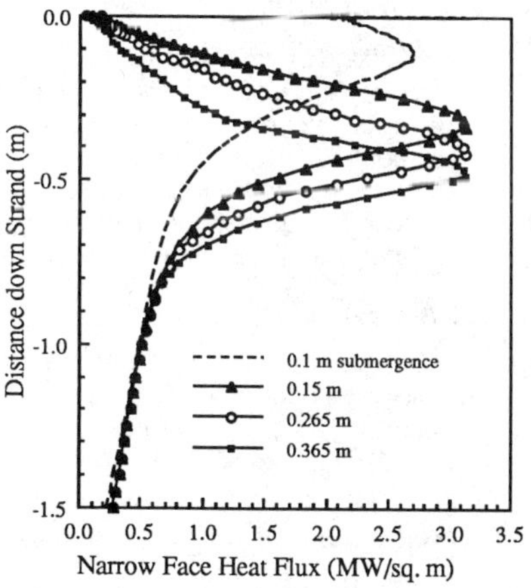

FIG. 14. Effect of nozzle submergence depth on superheat dissipation (conditions from Table I (1))

Solidification model

To determine the effect of the heat flux due to superheat on temperature development and growth of the solidifying steel shell, a simple solidification model was developed. This model solves the one-dimensional transient heat-conduction equation using a finite difference discretization:

$$\text{Interior node:}\quad T_i^{new} = T_i^{old} + \frac{\Delta t \quad k}{\Delta x^2 \rho C_p}(T_{i-1} - 2T_i + T_{i+1}) + q_n \frac{\Delta t}{\Delta x \rho C_p} \tag{4}$$

$$\text{Surface node:}\quad T_1^{new} = T_1^{old} + \frac{\Delta t \quad k}{\Delta x^2 \rho C_p}(2T_2 - 2T_1) - 2q \frac{\Delta t}{\Delta x \rho C_p} \tag{5}$$

These nodal equations were made explicit by evaluating all RHS terms at the old time level. Solidification was accounted for by simply enhancing C_p between the solidus and liquidus temperatures:

$$C_{peff} = C_p + \frac{\Delta H}{T_{liq} - T_{sol}} \tag{6}$$

To ensure latent heat was not missed, a post-iterative correction was performed after each time step. Whenever a solidifying node fell below the solidus, or a liquid node fell below the liquidus, its temperature was adjusted to account for the incorrect change in enthalpy that occurred during that time step.

Boundary conditions for the model are illustrated schematically in Figure 8. They can be adjusted to simulate lateral positions on either the wide or narrow faces. Heat flux q was extracted from the surface node according to a prescribed function of time and distance down the mold. Below the mold, heat flux is applied as a function of spray cooling practice (based on water flux) and natural convection (based on a coefficient of 10 W/m^2K).

To mathematically describe how superheat is delivered to the inside of the shell, the initial temperature of every node is set to the liquidus temperature. When modelling the narrow face shell, the heat flux curves calculated in the previous section are used to define q_n down the mold. This function is applied at each time step to the location corresponding to the inside of the shell. This location is presently taken to be the hottest node in the mushy zone (below T_{liq}). At all other nodes, q_n is set to zero.

The heat flux curves in the previous section assume all of the superheat is transported through the narrow faces. The solidification model assumes only a fraction of this total to apply to the narrow face. Currently, the narrow faces are given 40% of the total superheat, which represents about 2.3 times more than a uniform share, for a 1.05 x .22 m mold section. Once heat flux distributions are extracted from the 3D model simulation, this fraction can be predicted more precisely. Results will then be input to a two-dimensional transient thermal-stress model, to enable a complete analysis of the shell behavior in 3 dimensions.(16)

Results and discussion

To demonstrate the ability of this model to predict the effect of superheat on shell growth, simulations were conducted for the typical slab casting conditions (Table I), where measurements had also been made by Nakato *et al.*(1) These conditions include casting at .0267 m/s with a superheat of 27 °C from a 12° downward nozzle into a 1.05 x .22 m mold.

Boundary conditions for heat extraction by the mold, q, were taken from Nakato *et al.*(1) The heat flux profile from superheat dissipation, shown in Figure 8, was found from the fluid flow model results for these conditions, discussed in the previous sections and given in Figure 10. Figure 15 compares the solidification model predictions of shell thickness down the narrow face with measurements.(1) The latter were obtained by injecting FeS tracer into the jet stream and taking sulfur prints.

Figure 15 shows that the thick, regular growth of the shell on the wide face, calculated assuming simple conduction in the liquid, follows the usual parabolic relationship. It is similar and

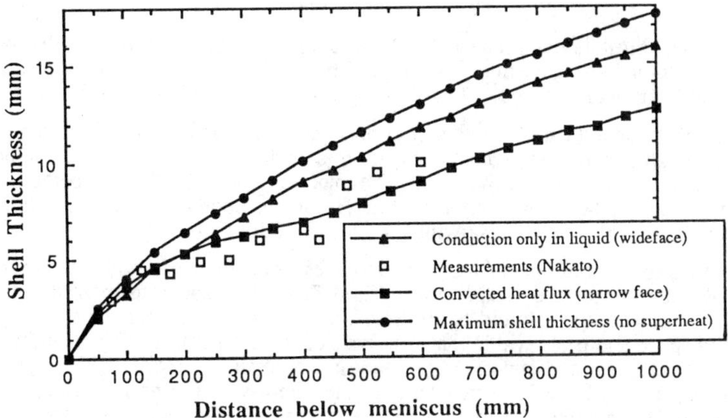

FIG. 15. Calculated shell thickness profiles compared with measurements (1)

close to the maximum shell thickness possible without superheat. This figure also shows that growth of the narrow face shell is significantly less than the wide face, which agrees with the findings of Flint.(5) In addition, its growth is non-uniform, with a sharp reduction in growth rate between about 150 and 450 mm below the meniscus. This coincides with the vicinity of the impingement point of the jet on the shell. Remelting of the shell was not measured or predicted for these conditions. However, a higher casting speed or superheat could produce actual shell remelting. This would increase the danger of a breakout, particularly if the shell had an irregular growth pattern with local thin spots, such as due to casting a medium carbon grade or deep oscillation marks.

Slight misalignment of the nozzle could direct the jet away from the narrow face centers towards the slab corners. This would move the jet impingement points to the off-corner regions of the wide faces. These critical locations would then experience the thinning calculated in Figure 15. Defects found near opposite corners of the slab would be indicators of this potential effect.

The results from this model agree approximately with the measurements and predictions of a simpler model used by Nakato *et al.*(1) However, the shell growth predicted in the present study is smoother, over-predicting shell thickness in the impingement region, and under-predicting it lower in the mold. The differences might be due to the two-dimensional nature of the present liquid model, which introduces uncertainty in both the fraction of heat flux delivered to the narrow face and its exact distribution. A sharper heat flux profile might be found at the impingement point in a 3D calculation. The results also depend on the assumed surface roughness parameter, and other aspects of the wall law used in the computations.

A significant difference can be seen between shell thickness predictions obtained from conduction only in the liquid and both the measurements and calculations obtained considering superheat delivery at the impingement point lower in the mold. This crude approximation has been made in many previous models that have been successful in predicting average growth of the shell.(6,7) However, it produces a continual removal of the superheat down the caster. The result is a slight under-prediction of solidification at the meniscus, where the start of solidification is often delayed. This is accompanied by an over-prediction of shell growth near mold exit on the narrow face and the adjacent off-corner regions of the wide face. These areas remain much thinner even far below the mold, since the thermal resistance of the shell dominates heat transfer below the mold. This discrepency persists until very deep in the caster after all superheat has been extracted in both cases.

These slight errors in predicting shell thickness could be important because simple heat conduction models are commonly used to predict the initiation point of defects in the cast slab. For example, an over-prediction of shell thickness near mold exit could imply the cause of a defect was *inside* the mold, when it actually occurred just *below* mold exit. The differences are also important for subsequent thermal stress calculations.

Conclusions

A mathematical model has been developed to calculate how superheat is dissipated from the liquid in the continuous slab casting process. The results are readily input into further models that calculate solidification of the shell. The following conclusions are based on these calculations.

Most of the superheat is removed in or just below the mold. The heat flux delivered to the inside of the narrow face as superheat dissipation is of the same magnitude as the heat extracted from the cold side of the shell by the mold. The maximum heat flux from superheat dissipation is always centered about the point where the jet first impinges upon the steel shell solidifying against of the narrow face. This results in slowing the growth of the shell.

Nozzle angle, submergence depth, and to a lesser extent, mold width and casting speed all combine together to produce the pattern of fluid flow and heat transfer in the mold. Nozzle geometry and submergence should be designed and controlled to produce a flow pattern, for a given casting speed and mold width, that delivers steel to the meniscus that is neither too cold nor too hot. Casting speed and superheat temperature both increase heat flux to the narrow face greatly. Thus, to avoid breakout problems in the vicinity of the narrow face, these variables should be controlled together. A maximum safe casting speed exists for each superheat. Lower casting speed should be used when superheat is very high. This relationship should be altered to take into account nozzle angle, mold width, and submergence depth.

The models described in this paper are part of a larger system of models being developed to simulate the behavior of the solidifying steel, including the effects of fluid flow, heat transfer, solidification, shrinkage, stress generation, mold distortion, and crack formation. They are being applied to predict and understand the effects of such diverse variables as nozzle design, mold distortion and taper on the generation of defects in the solidifying shell. Together, these models should provide a comprehensive analysis tool for the continuous casting process. The goal is to provide insights that will increase understanding of quality problems such as breakouts, depressions, and cracks and aid in optimizing nozzle and taper design and caster operation.

References

1. H. Nakato, M. Ozawa, K. Kinoshita, Y. Habu, and T. Emi, Trans. Iron Steel Inst. Japan, 24, #11, pp. 957-965, (1984).
2. K. Saito and M. Tate, OH-Proc. AIME, Cleveland, p. 238, (1973).
3. A. Etienne, Proc. of 4th Int. Conf. Continuous Casting, Brussels, May 17, 1988, 2, Stahl Eisen, pp. 597-608, (1988).
4. J. P. Birat and J. Chone, Ironmaking and Steelmaking, 10, # 6, p.273, (1983).
5. Paul Flint, ISS 73rd Steelmaking Conference, Detroit, March 25, (1990).
6. J. Lait, J.K. Brimacombe, and F. Weinberg, Ironmaking and Steelmaking (Quarterly), #2, pp. 90-98, (1974).
7. A. Grill, J.K. Brimacombe and F. Weinberg, Ironmaking and Steelmaking, 1, pp. 38-47, (1976).
8. B.G. Thomas, Iron and Steel Society Transactions, 12, #16, 53-66, (1989).
9. B.G. Thomas, Mika, L. M., and F.M. Najjar, Met. Trans. B, 21B, pp. 387-400, (1990).
10. F. M. Najjar and B.G. Thomas, Third FIDAP Users Conference, Evanston, IL, Sept. 25, (1989)
11. M. Yao, M. Ichimiya, K. Syozo, K. Suzuki, K. Sugiyama, and R. Mesaki, Steelmaking Proceedings, 68, Iron and Steel Society, Inc., pp. 27-34, (1985).
12. T. Robertson, P. Moore, and J.F. Hawkins, Ironmaking and Steelmaking, 13, #4, pp. 195-203, (1986).
13. M.S. Engelman, FIDAP Theoretical Manual - Revision 4.0, Fluid Dynamics International, Inc., 1600 Orrington Ave., Suite 400, Evanston, IL, (1986).
14. C. Offerman, Scand. J. Metallurgy, 10, (1981).
15. F. Najjar, Masters Thesis, University of Illinois, Mechanical Engineering, (1990).
16. B.G. Thomas, A. Moitra, and W. R. Storkman, to be presented at 6th International Iron and Steel Congress, Nagoya, Japan, Oct. (1990).
17. M. Kumada, and I. Mabuchi, Trans. Japan. Soc. Mech. Eng., 35, pp. 1053-1061, (1969).
18. P. Sismani, and S. Argyropoulos, Met. Trans. B, V. 19B, Dec., pp. 859-870, (1988).

19. R. Sobolewski and D.J. Hurtuk, Second Process Technology Conf. Proc., 2, pp. 160-165, (1982).
20. E. Hofken, H. Lax, and G. Pietzko, 4th Int. Conf. Continuous Casting, Brussels, Stahl Eisen, 2, p461, May 17, (1988).
21. J. Savage and W.H. Pritchard, J. Iron Steel Inst., 178, pp. 269-77, (1954).

Appendix I Rough heat balance on typical slab casting mold

For conditions in Table I (Nakato *et al*, (1)), casting at .0267 m/s in a 1.05 x .22 x .6 m mold:

Heat extracted by mold

Time-average (total) mold heat flux (based on Savage & Pritchard) (21):

$$q_t = 2.68 - 0.222\sqrt{\frac{L_m\ (m)}{v\ (m/s)}} = 1.63\ MW/m^2$$

Total heat extracted: $Q_{total} = q_t\, L_m\, (2W+2N) =$ 2.48 MW

This heat comes from the following sources:

Dissipation of most of the 27° C Superheat:

$$q_{sh} = \Delta T_s\, \rho\, C_p\, v = (27\ °C)\ (7020\ kg/m^3)\ (680\ J/kgK)\ (.0267\ m/s) = 3.44\ MW/m^2$$

60% of total superheat: $Q_{sh} = 60\% * q_{sh}\, W\, N = 60\% * 0.795\ MW =$ .48 MW

Latent heat (needed to solidify a shell thickness, X):

$$q_{\Delta H} = \Delta H\, \rho\, v\, X = 52.5\ MW/m^2\, X$$

Sensible heat, (needed to cool the shell surface to 1050 °C assuming a linear temperature profile in the shell):

$$q_c = \left(\frac{T_{liq} - 1125\ °C}{2}\right) \rho\, C_p\, v\, X = 25.5\ MW/m^2\, X$$

The heat required to solidify a uniform shell of X=10 mm thick around the entire mold is:

Q_{wide}	$= (q_{\Delta H}+q_c)\ (2W)\ X$	= 1.64 MW
$Q_{narrow\ faces}$	$= (q_{\Delta H}+q_c)\ (2N)\ X$	= 0.34 MW
Total		2.46 MW

Acknowledgements

The authors wish to thank the steel companies: Armco Inc., (Middletown, OH), Inland Steel Corp., (East Chicago, IN), and BHP Co. Ltd., (Wallsend, Australia), for grants which made this research possible and for the provision of data. This work is also supported by the National Science Foundation under grant #MSS-8957195. Valuable discussions with Vahe Haroutunian, Richard Sussman, Ismael Saucedo, Alan Cramb, Paul Flint, Ed Szekeres, and Vaughn Voller are gratefully acknowledged. Thanks are also due to Fluid Dynamics Inc., (Evanston, IL) for help with the FIDAP program and to the National Center for Supercomputer Applications at the University of Illinois for time on the Cray-XMP/48 and Cray 2 supercomputers.

Measurement of the mechanical interaction between strand and mould in the continuous casting of steel billets

I.A. Bakshi, I.V. Samarasekera, J.K. Brimacombe
The Centre for Metallurgical Processing Engineering, The University of British Columbia, Vancouver, British Columbia, Canada, V6T 1W5

J.L. Brendzy
Alcan International Limited, Arvida Research and Development Centre, Jonquière, Québec, Canada, G7S 4K8

Abstract

The mechanical interaction, or friction, between the mould and strand was measured on an operating billet casting machine with a load cell system. It was found that there were two distinct modes of mould-strand interaction present during each oscillation cycle. One mode occurred during the positive-strip portion of the oscillation cycle where the compressive load was a maximum and influenced by oil lubricant type, lubricant flow rate and steel carbon content. The other mode of interaction was observed during the mould negative-strip time when the measured load varied as a function of the mould lead (the distance the mould travels past the strand during negative strip) which in turn is directly affected by the billet casting speed. The load cell system has demonstrated promise as a tool in the evaluation of mould tapers, mould oil types and mould oil flow rates with respect to their influence on mould/strand friction, and hence, billet quality.

Introduction

In the continuous casting of billets, oil lubricant is fed continuously into the mould to prevent the newly formed steel shell from sticking to the copper mould wall and rupturing. In addition, the mould is oscillated in a sinusoidal manner to provide a stripping action, again to prevent sticking. This movement, combined with the thermal distortion of the copper mould at the meniscus which creates a local negative taper, results in the formation of oscillation marks (1). If non-uniform in depth about the billet periphery, the oscillation marks may lead to defects such as rhomboidity and off-corner cracks (2). Thus understanding the mould-strand interaction that gives rise to these defects is crucial to the optimization of the steel casting process and production of high quality billets.

In order to study the influence of mould design, oscillation characteristics and operating parameters on billet quality, a means of measuring the interaction between the mould and solidifying steel shell was sought. The technique chosen was to monitor the changes in compressive loading on the mould during the casting operation by installing four load cells in the mould housing of a production casting machine. Linear variable displacement transducers (LVDT's) also were employed in order to monitor the precise location of the mould during its oscillation cycle so as to correlate measured mould load and mould position. While similar techniques have been adopted to determine frictional forces on slab casters (3,4,5) as well as on stationary pilot-plant scale casting machines (6), this is believed to be the first attempt to do so on a production billet caster.

During the industrial test campaign, the influence of the following variables on measured mould loading was examined:

- position of mould during the oscillation cycle
- casting speed
- oil lubricant type
- oil lubricant flow rate
- carbon content of the steel

In the course of the trials, three different types of oil lubricants were fed at flow rates of between 20 and 54 ml/min to the mould producing the 120-mm square billet section under study. The load cell system proved itself capable of monitoring the changes in compressive load on the mould as a function of the different oil types and flow rates. In addition, the formation of transverse depressions on the billet surface was detected.

The measurements of mould friction were part of a larger study involving the University of British Columbia and seven Canadian companies focussing on changes to the design of the mould (eg. taper, wall thickness) and its operation (eg. oscillation characteristics, lubrication) in order to improve billet quality in the production of a broad range of steel grades.

Experimental Procedure

Instrumentation

A schematic diagram of one of the four load cells positioned between the mould housing and the oscillator frame is shown in Fig. 1. Each load cell was recessed into the mould housing such that the measuring tip of the load cell protruded against the oscillator frame. The four load cells were located at the corner positions in order to support the full weight of the housing on the oscillator frame and ensure that all changes in load on the mould would be detected. However, as will be discussed later, a problem arose in maintaining the mould water seal with the commonly used rubber 'O' ring gaskets, located at the front of the housing, which interfered with the front load cell measurements. In order to track the movement of the mould precisely during its oscillation cycle, two LVDT's were mounted against the mould housing.

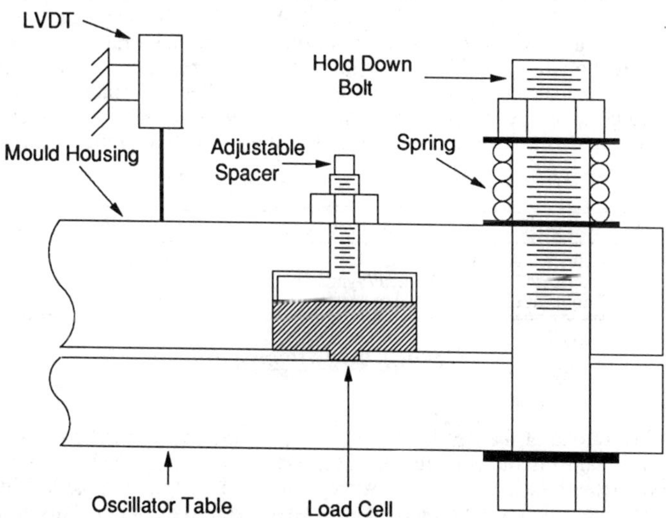

FIG. 1. Schematic of load cells and LVDT installation on billet mould housing.

The mould housing was secured to the oscillator table and the load cells were pre-loaded using four hold-down bolts. A pre-load of approximately 9000 N was applied to each bolt to ensure the mould housing was securely fastened. However, owing to the coarse threads on the one-inch hold-down bolts, a modification incorporating a spring was made to control the application of pre-load. This pre-loading of the system also was necessary as the load cells could only measure compressive, not tensile, loads. But the use of the hold-down bolts meant that the load cells were not measuring the absolute compressive load on the housing. Any external load applied to the assembly is partitioned between the hold-down bolt assembly and the load cell, with the load seen by each component being a function of its stiffness constant. Thus, the measured load changes provided only a relative indication of how the total compressive load on the mould housing varied. It is planned that this limitation will be overcome in future work through an on-site calibration procedure.

The load cells employed a bonded foil strain gauge design which provided a 0 to 20 mv output proportional to the compressive load. Both the load cells and LVDT's were monitored with a PC-based data acquisition system. The resolution of this digital system with the load cells was 0.012 mv or 25 N. Additional caster operating variables also recorded with this system included an array of mould wall thermocouples, the billet casting speed and mould cooling water temperatures.

Plant Trial

A plant trial, using the load cell system to monitor mould friction, was undertaken in 1989 on an operating production billet casting machine. Details of the casting machine involved in the trial are given in Table 1.

TABLE 1. Features of Billet Casting Machine Involved in Plant Trial

Casting Machine Details	
Machine type	Curved mould
Machine radius	7.9 m
Mould taper	Parabolic
Mould material	Cr-Zr copper
Mould thickness	12.7 mm
Cooling water velocity	14 m/s
Oscillation type	Sinusoidal
Stroke length	9.5 mm
Oscillation frequency	2 Hz
Nominal casting speed	2-2.5 m/min
Oil distribution system	UBC design

A total of 22 heats were cast with the load cell monitoring system in place. The steel grades cast during the trial included 1008, 1010, 1012, 1015, 1018 and 1039 having a carbon range of 0.04 to 0.40%. During the course of the trial, the mould oil type and flow rate were varied in order to determine their effect on the mould load. The lubricants and flow rates tested are listed in Table 2. A mould oil distribution system, designed at UBC, was installed to ensure uniform delivery of oil around the mould periphery, even at reduced oil flow rates. Billet samples, for each combination of oil type and flow rate tested, were collected for subsequent metallographic examination at UBC.

TABLE 2. Mould Oil Types and Flow Rates Tested

Oil Code	Oil Type	Flow Rates Tested
Lubricant A	Mineral	54, 44, 34 ml/min
Lubricant B	Mineral/Synthetic	54, 44, 34, 24 ml/min
Lubricant C	Soybean	40, 30, 20 ml/min

Results and Discussion

Typical Load Cell Response

A typical signal from one of the two load cells located at the rear of the housing is shown in Fig. 2. The measured signal from the other rear load cell was virtually identical at all times, whereas the response from the front two load cells was somewhat dampened, in comparison, due to the close proximity of the rubber inlet/outlet mould water 'O' ring seal to the latter. This gasket acted like a soft spring and absorbed a portion of the load, reducing the sensitivity of the front load cells. Thus, the ensuing analysis focussed only on data obtained from the two rear load cells.

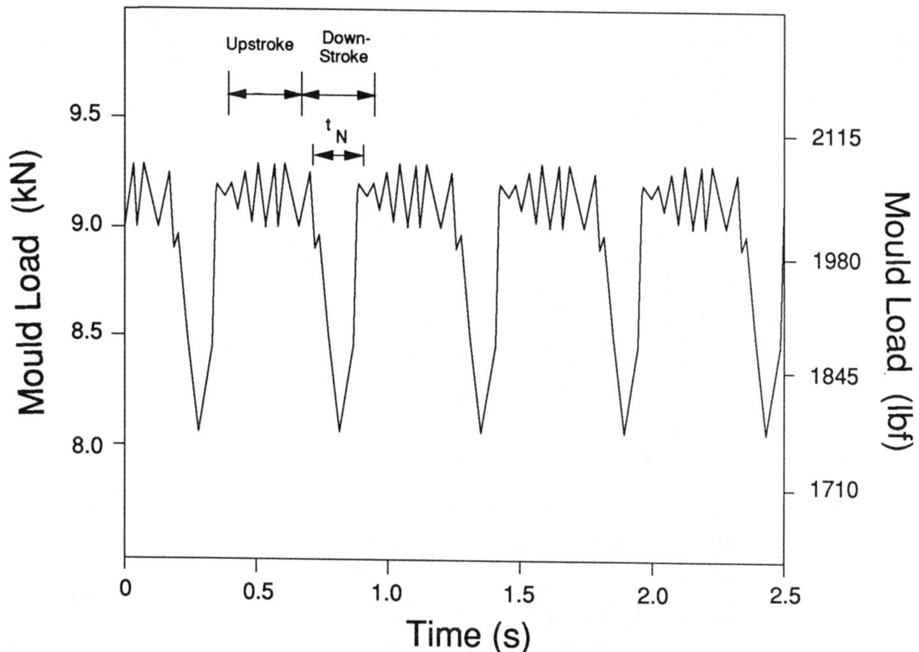

FIG. 2. Typical load cell signal.

The load cell signal in Fig. 2 is seen to be periodic with a frequency of 2 Hz, the same as the oscillator, but it is not sinusoidal as is the mould movement. Instead, the load cell response during an oscillation cycle is characterized by two distinctly different regions related to the mould movement. In general, the maximum load corresponds to the mould upstroke, while the mould downstroke movement results in the minimum load. The direction of mould movement, based on the response of the LVDT's, is indicated relative to the mould load for one oscillation stroke.

The upstroke period is dominated by the presence of numerous small peaks along the maximum load plateau. It will be shown that these peaks can vary from being quite small to large and are sensitive to the effects of different mould lubricating oils and their flow rate as well as steel composition. The casting of lower carbon grades produced variable peaks, which will be shown later to be the result of binding in the mould, while the higher carbon grades exhibited peaks which were invariant over time.

The mould downstroke period is characterized by a smooth decompression of the load centred about the negative-strip period. The negative-strip time, t_N (the time period during which the mould travels downward faster than the strand), is also indicated in Fig. 2. This minimum load value varied regardless of lubricant type, flow rate or carbon content. Instead it was found that the minimum load was highly dependent on the billet casting speed which, as will be shown later, is due to the mechanical interaction between the mould and the solidifying shell at the meniscus. The compressive load decreases smoothly until the maximum downward velocity of the mould is reached (or shortly thereafter), at which point the load begins to increase monotonically until the mould upstroke begins.

The observed compressive mould loading signal can be explained by the mechanical interaction of the oscillating mould with the strand. One need only consider the direction of the reaction force on the mould for the two cases where the mould is travelling downward faster than the strand (negative strip) and vice versa (positive strip), as shown in Fig. 3. In the first case, during negative strip (V_M>V_S), the strand pushes upward on the mould, in reaction to the jamming action of the latter which assumes a negative taper at the meniscus and squeezes downward on the newly forming shell (1). The reaction force of the strand decompresses the load cells and reduces the measured load.

Conversely, during the positive-strip period ($V_M<V_S$), any sticking, or friction, between the mould and strand results in a downward force on the mould and therefore an increase in the measured compressive load.

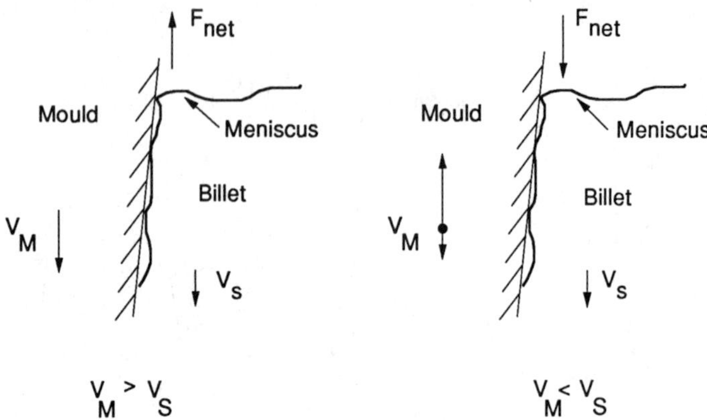

FIG. 3. Direction of the net force acting on the mould for the two cases of the mould travelling downward faster than the strand (negative strip) and the strand travelling downward faster than the mould (positive strip).

The squeezing action of the mould on the shell at the meniscus during negative strip causes the solid to buckle and form an oscillation mark as shown in Fig. 4. In this schematic diagram the mould displacement and load cell response are shown in relation to the formation of an oscillation mark on the billet. The mould is shown to have acquired a negative taper 90 to 100 mm below the meniscus due to differential thermal expansion (1). The mould does not begin to bear down on the steel meniscus until the start of negative strip which is reflected in the start of load cell decompression (Fig. 4a). At the midpoint of the negative-strip time the maximum mould lead causes the greatest deformation of the steel shell and maximum decompression of the load cells (Fig. 4b). Finally, as the end of negative strip is approached, the mould lead distance decreases and the strand gradually disengages itself from the mould (Fig. 4c), having completed the formation of the oscillation mark. The measured load then increases as the mould lead decreases until the upstroke period begins.

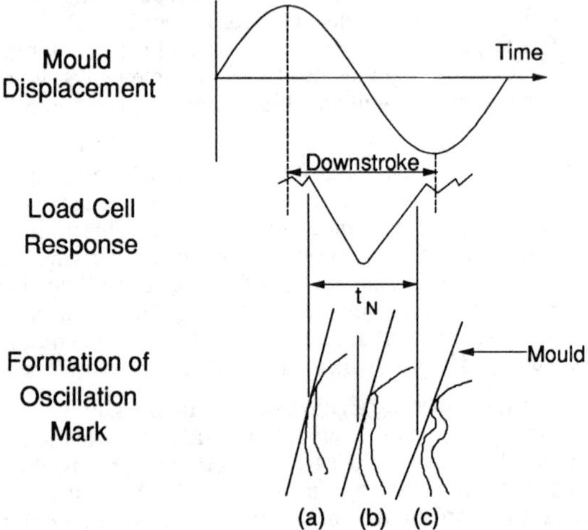

FIG. 4. The formation of an oscillation mark due to the mechanical interaction between mould and strand during negative strip.

Effect of Casting Speed on Minimum Load

The measured minimum load was found to have a strong correlation with casting speed as shown in Fig. 5 where the measured load from one of the rear load cells is plotted together with casting speed against time. It can be seen that as the billet casting speed increases, so does the minimum load value. Conversely, as the billet casting speed drops, the minimum load decreases.

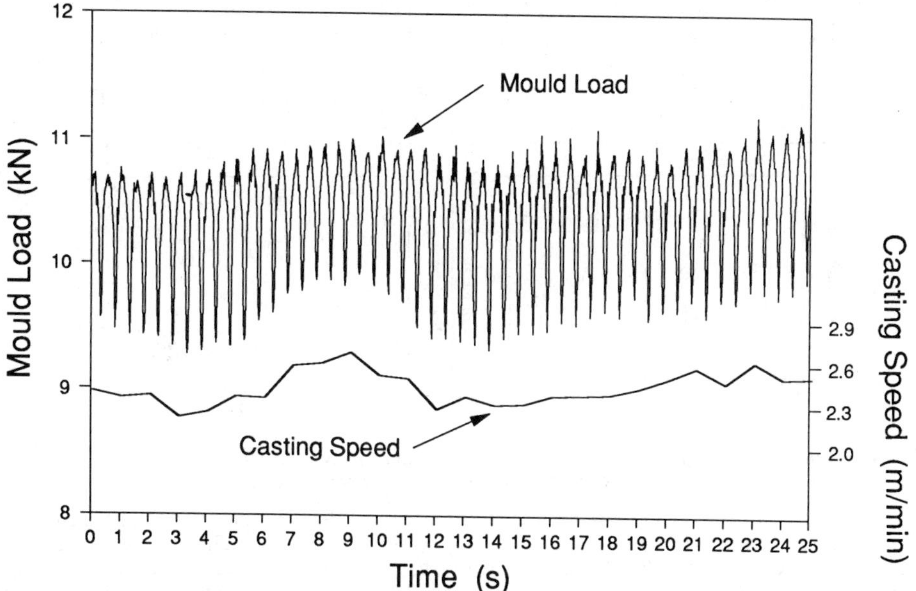

FIG. 5. Load cell response showing change of minimum compressive load with casting speed.

This result can be explained by considering the forces generated during the negative-strip period as discussed earlier. When the casting speed decreases, the negative-strip time and mould lead (the distance that the mould travels past the strand during negative strip) increases, so that the overall interaction between the shell and mould is greater, causing increased decompression which is sensed by the load cells.

Effect of Mould Oil Lubricant Type and Flow Rate

The maximum load peaks could not be correlated to the billet casting speed; instead they were found to be affected by the mould oil lubricant type and flow rate as well as steel grade. Figure 6 presents an example of this influence; the mould load response is shown for Lubricant B fed at flow rates of 54, 44, 34 and 24 ml/min. It can be seen that at the higher flow rates, especially at 54 ml/min, the peaks form a plateau consisting of short jagged spikes. However, as the oil flow rate was gradually reduced, the plateau disappeared and the peak load continually increased until the end of mould upstroke. In Fig. 6, the maximum loads increase in value from a series of multiple peaks on a plateau of 9.5 kN at 54 ml/min to steeply rising peaks reaching a maximum value of approximately 10.5 kN at 24 ml/min of lubricating oil.

From these observations, it is clear that the oil lubrication is most crucial during the upstroke of the mould while during the downstroke, mechanical interaction between the mould and solidifying shell at the meniscus dominates. With hindsight, this result is not surprising since it is during the upstroke that the oil must continue to find its way into the steel/mould gap and provide lubrication while the mould and strand are moving in opposite directions. If, due to inadequate lubrication, the steel begins to stick to the copper mould, the friction coefficient between the mould and strand will increase and transfer more of the strand weight to the mould, consequently increasing the measured compressive load. Indeed it was observed that as the oil rate was reduced, small patches of the steel shell moved upward above the meniscus during mould upstroke, a certain indication of enhanced shell/mould sticking.

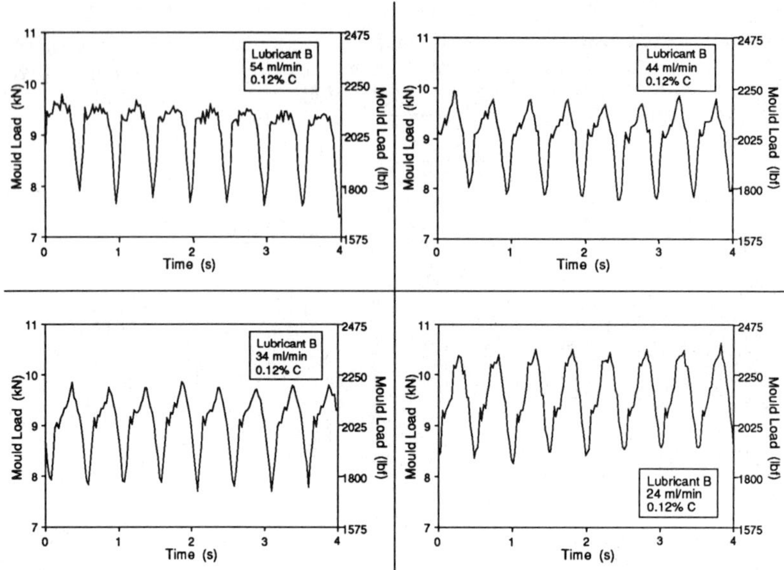

FIG. 6. Load cell response while feeding Lubricant B at flow rates of 24, 34, 44 and 54 ml/min during the casting of a 0.12% carbon steel.

It is more difficult to make direct comparisons among the three oil lubricant types since it was impossible to change the mould lubricant quickly within a single cast. This resulted in data from heats in which different mould lubricants were used to cast different grades of steel having different casting conditions (eg. superheat). Still, some general observations could be made. As in the case with varying lubricant flow rate, the differences seen among the mould lubricants were reflected solely in the load cell peak response occurring during the mould upstroke. While there was a perceptible difference in the maximum load curves among the different oil lubricants, it is uncertain whether this difference was entirely due to the lubricants themselves or was also influenced by the different steel compositions and casting conditions. As an example, Fig. 7 shows mould load data for Lubricant A when casting a 0.09% carbon steel, Lubricant B when casting a 0.12% carbon steel and Lubricant C when casting a 0.15% carbon steel. In all three cases the mould oil flow rate was 44 ml/min. It can be seen that the peak load characteristics of Lubricants A and C are significantly different from those of Lubricant B. The load peaks of Lubricant B consisted of a series of spikes steadily increasing in value, in sharp contrast to the much flatter multiple peaks present with the other two lubricants. There was also a difference in the load peaks between Lubricants A and C in that the peaks of Lubricant C are much more pronounced than those of Lubricant A. It can be speculated that Lubricant B, with the load increasing during the mould upstroke (similar to that seen when reducing the mould oil flow rate) was not as effective as the other two lubricants, which appear to provide better release between the mould and strand during upstroke.

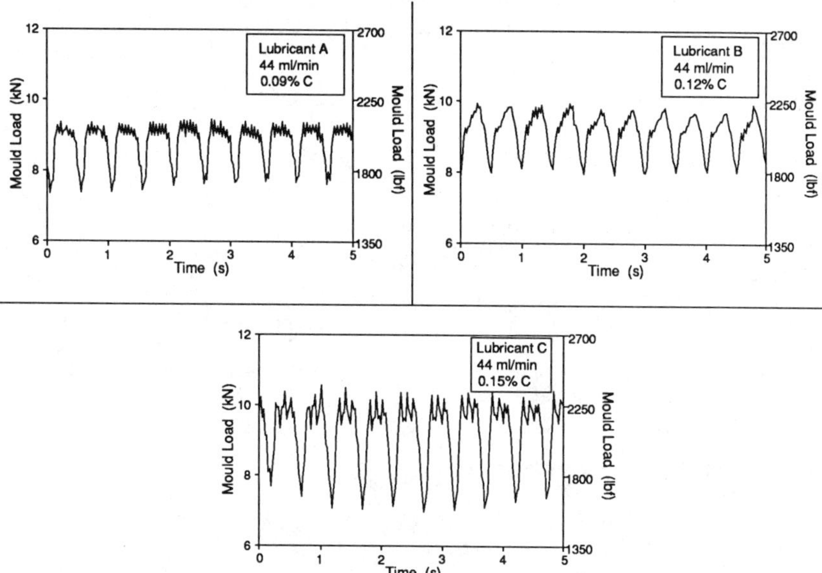

FIG. 7. Load cell response while feeding Lubricants A, B and C at a flow rate of 44 ml/min.

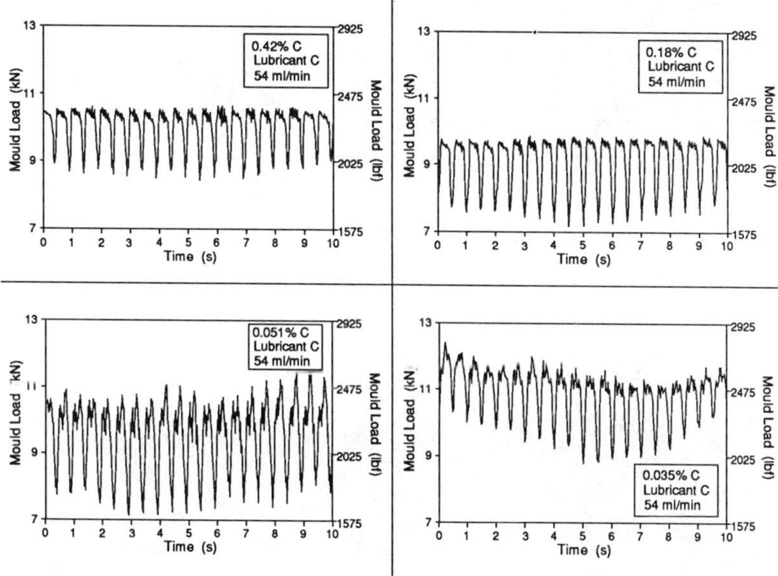

FIG. 8. Load cell response while casting 0.42, 0.18, 0.051 and 0.035% C billets with Lubricant C fed at 54 ml/min.

Effect of Steel Carbon Content

Figure 8 presents load cell data for a series of heats having carbon contents of 0.42, 0.18, 0.051 and 0.035% respectively. Lubricant C was fed to the mould at 54 ml/min in all of these tests. Figure 8 clearly shows the change in the load cell response as the carbon content is decreased. The low carbon data, shown in the lower two quadrants of Fig. 8, can be seen to exhibit significant variations in the magnitude of the maximum load peaks. Fluctuations within and between the peaks are present, in sharp contrast to the load response of the medium (0.18% C) and high (0.42% C) carbon grades which are shown in the upper two quadrants of Fig. 8. The peaks of the latter are seen to be constant in magnitude over time and exhibit a reduced severity of load spikes. Also seen in the low carbon data is considerable variation in the minimum compressive loads, caused by rapid casting speed changes.

Transverse Surface Depressions

An examination of the low-carbon billets revealed the reason for the erratic load cell signals. Severe transverse depressions were observed on the surface of these billets as can be seen in Fig. 9. Sections cut longitudinally through the depressions revealed the presence of subsurface cracks as shown in Fig. 10. Because the cracks form close to the solidification front, comparison of the measured crack depth of 3 to 4 mm to the model-predicted shell profile yields a distance of 410 to 480 mm below the meniscus at which they formed. Thus it could be established that the cracks, and presumably the depressions, were generated in the lower half of the mould. Another observation of note is that during the casting of the low-carbon billets, intermittent withdrawal, or jerking, of the strand was evident as if the steel was binding in the mould.

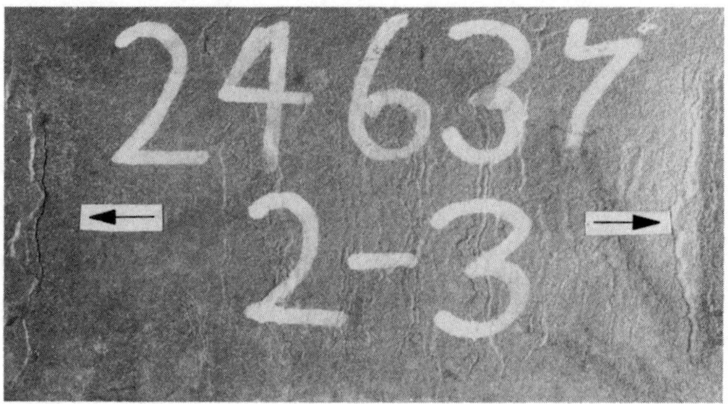

FIG. 9. Photograph of transverse depressions on a 0.05% carbon billet.

FIG. 10. Photograph of longitudinal section cut through transverse depressions of billet in Fig. 9 revealing subsurface cracks at the base of the depressions.

Binding between the mould and descending strand was most likely due to excessive mould taper relative to the shrinkage of the low-carbon billets which transfer less heat to the mould than higher carbon grades (1). The transverse depressions then form in response to the resulting axial tensile stress imposed on the solidifying shell. The cooler surface region of the shell, which is ductile, necks under tension much as a ductile specimen in an Instron machine; the zone of low ductility adjacent to the solidification front, however, cracks under tension. This is illustrated schematically in Fig. 11. From the position of the cracks, the binding likely occurred in the lower half of the mould although it could have been higher.

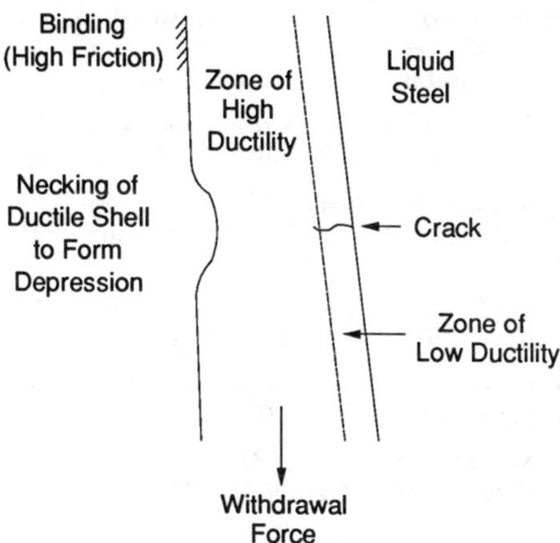

Fig. 11. Depression formation caused by binding in the mould.

Evidence linking the binding and transverse depressions can be found in the variation of the load peaks, which are a measure of the binding, seen in Fig. 8 for the low-carbon billets and the spacing between depressions on the billet surface, Fig. 9. Typically the time between maximum upper peaks (binding) was 2 to 5 s which, with a casting speed of approximately 2 m/min, indicates that the depression spacing should be 60 to 160 mm. This is in fairly good agreement with evaluation of the billet surfaçes which revealed depressions occurring every 50 to 150 mm.

Summary

The mechanical interaction, or friction, between a continuous casting mould and steel strand was measured on an operating billet casting machine using a load cell system. The following conclusions can be drawn:

(1) A system employing load cells has proven effective in measuring the mould-strand interaction in an operating production billet caster.

(2) The measured load cell response was found to be sensitive to: mould position, billet casting speed, mould oil lubricant type, mould oil lubricant flow rate and the carbon content of the steel.

(3) The effects of mould lubrication are measurable during the mould upstroke but are masked during the mould downstroke period due to the mechanical interaction between mould and strand.

(4) The load cell system was able to detect binding and the formation of transverse depressions on the surface of billets when casting low carbon steel grades.

Work is continuing to gather more data from additional plant trials. The goal is to develop the load cell system into an on-line quality tool which, for the first time, will allow operators to monitor the effect of mould tapers, mould oil types and flow rates on mould/strand friction and therefore, billet quality.

Acknowledgements

The authors wish to thank Neil Walker and Nick Hemingway for their assistance in carrying out the plant trials. The financial support of the Natural Sciences and Engineering Research Council of Canada and the following companies - Courtice Steel, Hatch Associates, Ivaco Rolling Mills, Sidbec-Dosco, Slater Steels, Stelco Steel and Western Canada Steel - is also greatly appreciated.

References

1. I.V. Samarasekera, J.K. Brimacombe and R. Bommaraju, *Mould Behaviour and Solidification in the Continuous Casting of Steel Billets. Part 2. Mould Heat Extraction, Mould-Shell Interaction and Oscillation Mark Formation*, ISS Trans., Vol. 5, p. 79-94, (1984).
2. I.V. Samarasekera and J.K. Brimacombe, *Thermal and Mechanical Behaviour of Continuous Casting Billet Moulds*, Ironmaking and Steelmaking, No. 1, p. 1-15, (1982).
3. B. Mairy and M. Wolf, *On the Importance of Mould Friction Control in Continuous Casting of Steel*, Fachberichte Huttenpraxis Metallweiterverarbeitung, Vol. 20, No. 4, p. 222-227, (1982).
4. C. Schacht, *Kinematic and Dynamic Behavior of a Continuous Caster During Start-up*, Steelmaking Proceedings (Atlanta), Vol. 66, p. 269-275, (1983).
5. C. Schacht, *An Investigation of the Cyclic Behavior at the Caster Mold-Strand Interface*, Steelmaking Proceedings (Pittsburgh), Vol. 70, p. 443-448, (1987).
6. M. Komatsu, T. Kitagawa and K. Kawakami, *Measurement of Mold Friction Force in Experimental Continuous Casting Machine*, Transactions ISIJ, Vol. 23, No. 10, p. F-1, (1983).

The effects of solid and liquid-phase diffusion on micro-segregation in castings

T.P. Battle
Centre for Numerical Modelling and Process Analysis, Thames Polytechnic, Woolwich, London SE18 6PF, England

ABSTRACT

Diffusion in castings can have a significant effect on microsegregation. Most models of segregation treat only diffusion in one phase directly; diffusion in the other phase is assumed to be zero or infinite. In this study these assumptions are examined for several alloy systems and solidification processes, using the dimensionless Peclet and Fourier numbers, and a combined solid and liquid-phase diffusion model. The results verify the diffusion-dependent assumptions made in most casting models, but caution is required in use of the Peclet and Fourier numbers, which are generally time-dependent. The results of the combined diffusion model are more dependable, and, with modern-day computing power, easily utilized.

Introduction

Microsegregation is a phenomenon common to all types of castings, and can lead to poor physical and/or mechanical properties. It has been shown by model and experiment that many factors affect segregation in typical metallic alloys. These include the equilibrium partition coefficient of the solute(s) of interest, the local cooling rate, dendrite-arm coarsening, geometric effects, convection, and solid-state phase transformations. Two of the most important parameters affecting segregation are solute diffusion in the solid and liquid phases. Traditionally two different types of mathematical models have been developed to account for diffusion phenomena. In the casting industry unlimited liquid-phase diffusion is assumed, with most models attempting to accurately include solid-state diffusion effects. In crystal growth processes, on the other hand, the assumption of no solid-state diffusion is generally made, with finite liquid-phase diffusion of paramount interest. All too often these assumptions are automatically made in more recently created models, with little or no justification for the specific application. In this work various calculations are made to show the effects of diffusion on segregation resulting from several traditional solidification processes. A literature review will be followed by calculations based on two different approaches: the use of dimensionless numbers, and results from a model of combined solid- and liquid-phase diffusion during solidification.

This subject is of particular significance to this symposium because of Professor Weinberg's own research in the field of alloy microsegregation. In a series of papers published over the past twenty years (1-4), he and his co-workers have measured microsegregation in both ferrous (1,2,4) and non-ferrous (3) alloys, and related these results to process models. Unlike most metallurgists doing similar work, he compared experiment to the predictions not only of the 'traditional' casting models of microsegregation, but the crystal growth models as well. His microsegregation research has inspired the present work.

Literature Review

It is impossible in the limited space to review the many works relating to diffusion effects on microsegregation. This topic, along with others relating to mathematical modelling of microsegregation, will be considered fully in a forthcoming review (5). Here brief consideration is given to those few models which consider both finite solid- and liquid-phase diffusion.

There has been limited interest in considering solid-phase diffusion effects in crystal growth systems. The analysis of Tiller and Sekerka (6) solved the diffusion equations analytically, using Laplace transforms. A solution was readily obtainable only for the condition $D_S = D_L$. They found significant differences between the Scheil equation and their relation for $k > 1$. Fischer (7) considered the effect of D_S on the zone-refining process. He found significant effects only when D_S values approached typical metal D_L values along dislocations and subgrain boundaries. Calculations for the melting/solidification problem indicated differences only when D_S exceeded 0.1 D_L (8). Deviations from the Tiller/Sekerka model were found at high values of D_S.

Two recent models of casting segregation have considered the combined effects of solid- and liquid-state diffusion. Roosz, et al. (9) used dimensionless parameters along with their numerical model to show the rapidity of liquid diffusion in typical castings. The model of Battle and Pehlke (10) treated microsegregation as a combined heat and mass transfer problem. They found that for typical liquid-phase diffusivities in the Al-Cu (10) and Fe-Ni (11) systems, only a very small concentration gradient is present in the liquid phase during solidification.

Test Cases

In order to examine the effects of diffusion on segregation, five different binary alloys have been chosen for examination by this work. The systems chosen and the pertinent physical properties are listed in Tables I and II (where the equilibrium solid/liquid intervals are defined by equation (1); see Appendix for symbols). Note that for simplicity, all parameters are assumed constant. This may be reasonable for order-of-magnitude calculations, but will not suffice for precise measurements (12,13). In addition, process parameters have been collected for various solidification techniques (Table III). Here also the parameters chosen are only representative of the process, and may differ between different applications of a process or different locations in a sample.

$$\Delta T_o = - m_L C_o \frac{1 - k}{k} \qquad (1)$$

Use of Dimensionless Parameters

The utilization of dimensionless numbers in engineering analyses is not new. Use of dimensional analysis is particularly important in heat transfer and fluid flow problems, and are used by chemical engineers in many situations involving mass transfer. These parameters can be of great use, since they allow users to quickly determine the relative importance of various physical processes, and may lead to simpler models and analyses than otherwise. However, use of these parameters can be dangerous, since it is not always clear which dimensionless number to use in a given situation. In addition, dimensionless quantities are often treated as if they are constant throughout the course of a given process, which is rarely true. The parameters that go into a dimensionless number may have to be estimated, which is another source of error. A brief discussion of dimensionless parameters which have been used in the modelling of diffusion during solidification is given below.

To estimate the effects of diffusion on mass transfer, two different dimensionless parameters have been used to date. The first is a mass-transfer form of the Peclet number. Equation (2) has been used to define the relative importance of convection by relating flow velocity V to the diffusion velocity D/L, where L is a length parameter. This has been adapted in solidifying systems to form the relation of system size to the diffusion length D/V, where V is now the velocity of the solidification front. This has been used in analyses of plane-front crystal growth (26), but it has been shown (27) that the same equation may not apply under typical dendritic growth conditions. The most commonly

TABLE I:

Important Physical Properties for Segregation Calculations, Selected Systems (References in parentheses)

System Solvent-Solute	Solid Phase	k (mole%/ mole%)	D_L (10^{-9} m^2/sec)	D_S (10^{-9} m^2/sec)	m_L (K/mole %)
Fe-C	δ	0.24 (13)	7.9 (15)	5.8 (18)	-18.12 (13)
Fe-Ni	γ	0.90 (13)	4.8 (16)	0.0002 (19)	-4.05 (13)
Fe-S	γ	0.014 (13)	4.5 (16)	0.16 (19)	-61.04 (13)
Al-Cu	α	0.145 (10)	4.9 (17)	0.005 (20)	-6.57 (21)
Cu-Ni	α	2.7 (14)	3.5 (3)	0.002 (14)	+4.70 (14)

TABLE II:

Typical Alloy Concentrations and Equilibrium Solid/Liquid Intervals for Systems of Interest

	Fe-C	Fe-Ni	Fe-S	Al-Cu	Cu-Ni
C_o (mole %)	0.03	25	0.03	2.14	3.24
ΔT_o (K)	17.2	11.3	129	82	11

TABLE III:

Typical System Parameters for Solidification Processes

Process* (and abbreviation)	Cooling Rate (K/sec)	Interface velocity (mm/sec)	Length Scale (mm)	Ref
Bridgman growth (BR)	10^{-4}	$2\ 10^{-5}$	100	(22,23)
Al DC Casting (DC)	0.5-2	2	0.03-0.07	(12,24)
DS Turbine Blade (DS)	4-0.2	0.1-0.001	0.03-0.04	(25)
CC Carbon Steel (surface) (CCs)	70	0.5	0.08	(25)
CC Carbon Steel (center) (CCc)	5	0.03	0.20	(25)

*DC = Direct Chill DS = Directionally Solidified CC = Continuously Cast

used dimensionless parameter in diffusion modelling during solidification is that represented by equation (3), which is the solute Fourier number. It can be derived from heat transfer principles, although the metallurgical application arises from the microsegregation model of Brody and Flemings (BF) (28). Low values of α cause the BF relation to reduce to the Scheil equation, while larger values lead toward equilibrium freezing and the lever rule. Allen and Hunt (27) later showed that this equation is similar to that derived from a consideration of diffusion from a moving source, where the diffusion distance is proportional to the square root of the diffusivity times the elapsed time (29).

$$Pe = LV/D \tag{2}$$

$$\alpha = D_S t_f / L^2 \tag{3}$$

The results of the Peclet number calculation are given in Figures 1 and 2, with those utilizing Fourier numbers in Figure 3. Since the length scale L and the interface velocity V are assumed to be functions only of the solidification process itself, a graphical depiction of Pe number effects is to plot the product LV against the solute diffusivity D. In the Fourier number plot as well, process-dependent parameters can be plotted against alloy properties. The product εL^2 is plotted against $D\ \Delta T_0$, where it is assumed that the solidification time is given by the equilibrium solidification range ΔT_0 divided by the cooling rate ε. Figure 1 shows the Peclet number plot for all alloy systems and processes of interest, while Figure 2 focuses on those processes which have low values of LV. Figure 3 includes Fourier number calculations over the entire range of parameters. In all figures, the solidification processes are indicated by their abbreviations (see Table III) across the top, with the alloy systems of interest on the right.

The alloy effects in Figures 1-3 are in broad agreement. The Fe-C system, for example, has comparatively low calculated Peclet numbers and high Fourier numbers for both solid and liquid. This verifies the general assumption of the lever rule (which assumes infinite diffusion in both phases) for calculations in the Fe-C. The other solid phases have comparatively low values of Fourier and high Peclet numbers. This justifies the Scheil equation (no diffusion in solid, infinite diffusion in liquid), if the opposite condition holds for the liquid phase. The Fourier number approach better separates the alloy systems. For example, Fe-S and Al-Cu are differentiated from Fe-Ni and Cu-Ni as a result of the large solidification ranges of the former alloys, leading to higher Fourier numbers. This implies that liquid diffusion is more nearly complete in these two systems. The problem of deciding the critical values of Pe and α remain. What is the minimum value of Fourier number before diffusion is insignificant? What is the maximum before infinite diffusion can be assumed?

Figures 1-3 can also be examined with relation to particular processes, where the results are not entirely self-consistent. For example, calculations using the parameters of DC Casting result in high Peclet numbers, due to high interface velocities, and high Fourier numbers, due to low cooling rates. Similarly, Bridgman Growth results in low Fourier numbers due to the large system size, and low Peclet numbers due to low growth velocities. Directional solidification and continuous casting, under these conditions, seem more consistent. It should also be noted that, as expected, the center of the continuous casting possesses a higher value of α and lower Pe than the surface (thus, equilibrium freezing conditions are more closely approached in the center). In spite of the disagreements, however, a plot of Pe against α for each system (Figure 4) indicates broad agreement between the parameters: high values of one correspond to low values of the other.

It is clear that the dimensionless numbers discussed above can be useful to indicate the relative importance of solid- and liquid-phase diffusion. However, α and Pe are not constants but may change during the solidification process. In a recent publication, Roosz, et al. (9) considered the importance of liquid- and solid-phase diffusion in castings. They discussed the disadvantages of the 'average' Fourier number of equation (3), and defined a more general parameter by equation (4). This general Fourier number takes into consideration that diffusivity may be temperature-dependent (and, therefore, a function of time), and that the system size may be a function of time (this can be due to coarsening or the general change in phase size due to solidification). Roosz, et al. (9) made calculations based on equation (4) and two simpler approximations. It was found that the general value of α only differed by a factor of two or three from the approximate relations.

$$\alpha = \int_0^{t_f} \frac{D(T)}{x^2(t)} dt \qquad (4)$$

It might be argued, however, that it is meaningless to attempt to characterize segregation by a single parameter, be it equation (2), (3), or (4). Solidification is a continuous process, and α and Pe change continuously. In Figure 5 the result of calculations for Al-Cu are presented, which show the changes in α_L and α_S using the model of Battle and Pehlke (10). During the course of solidification, α_S decreases from near infinity to 0.01, while α_L

increases rapidly to approach infinity at the end of solidification. The importance of finite liquid diffusion in this system will apparently be restricted to short times. The variability of α_L and α_S in this instance further indicates that for many systems and processes, dimensionless numbers can only be the roughest of guides. More detailed calculations are required to determine whether diffusion has significant effects on the segregation occurring during a particular solidification process.

Combined Solid and Liquid Phase Diffusion

Most microsegregation algorithms consider finite solid or liquid-phase diffusion, not both. It is of interest to examine the predictions of a complete diffusion model, such as the BP model mentioned earlier (10). In this algorithm, the combined heat and mass transfer problem in one dimension has been solved using the method of lines with invariant imbedding. For this study, calculations were made in the Al-Cu and Cu-Ni systems.

Figures 6 and 7 show concentration profiles in the Cu-Ni system as a function of time during and after solidification, using thermal parameters from the work of Weinberg and Teghtsoonian (3). The predictions of Figure 6 are based on the currently-accepted physical property values of Cu-Ni alloys (Table IV), while Figure 7 changes only the value for the liquid-phase diffusivity, decreasing D_L by a factor of 1000. The position of the solid/liquid interface is apparent after 2 and 4 seconds in both plots, as is the solid-phase concentration gradient due to the low diffusivity of nickel in solid copper (14). The liquid-phase concentration varies little with position in Figure 6, thus verifying the assumption of infinite liquid-phase diffusion made by most casting microsegregation models. From Figure 7, it is apparent that reducing D_L has led to a situation where liquid-phase diffusivity can no longer be ignored. A concentration gradient has built up in the liquid ahead of the interface, slowing down the solidification process and changing the microsegregation characteristics of the alloy.

TABLE IV:

Physical Property Values for the Cu-Ni System

Parameter	Value	Units	Ref
Liquidus Equation	1356 + 5.37*mole % Ni	K	14
Latent Heat	2970*mole% Ni + 2050*mole% Cu	J/kg	30
Solid-State Diffusivity	$1.1*10^{-4}$ exp (-27076/T)	m^2/sec	14
Liquid-State Diffusivity	3.5 10^{-9}	m^2/sec	3
Partition Coefficient	2.7 + 4.70*mole% Ni	molar units	14

The effects of liquid-phase diffusion on segregation in this system can be seen more clearly by referring to Figures 8 and 9. Here the concentration profile after 4 seconds is reproduced for D_L varying from 10^{-9} to 10^{-15} m^2/sec. A concentration gradient in the liquid is first apparent when D_L is about 0.01 of its original value, and increases in magnitude significantly as D_L decreases further.

Figure 10 shows the effect of increasing the experimentally-measured value of D_S by a factor of 100. As expected, the concentration profile in the solid becomes almost flat, and the maximum concentration is significantly reduced.

Discussion

The results for the combined model have tended to follow the results of the analysis using dimensionless numbers. Under ordinary casting conditions, in typical alloys, liquid-phase diffusion is quite rapid; further calculations would be required to determine if this is true in extreme casting conditions (such as near a chill) as well. It is of interest to note the calculated values of liquid-phase Peclet number for the tests depicted in Figures 8 and 9. For the experimentally-determined value of D_L, 3.5 10^{-9} m^2/sec, the liquid-phase Peclet number varies from 0.002 to 0.004 during solidification, while α increases from 0.16 to 137. When D_L has been decreased by a factor of 100, and appreciable concentration gradients appear in the liquid, Pe has increased to 0.02-0.6, with α now 0.002-1.2. This indicates that the critical values for both Pe and α in this system are perhaps somewhat less than one. More important would be the effect of changing Pe and α on the calculated microsegregation. Of particular interest would be the reproducibility of the critical values of the dimensionless parameters across different systems and processes. This would require a further series of computations.

The use of dimensionless numbers is still a significant technique in segregation calculations. However, computing hardware and software are becoming more and more powerful and (relatively) less expensive. A complete numerical analysis, which can be run in a manner of minutes on a workstation, is now a viable technique for determining the significance of solid-and/or liquid-phase diffusion in castings.

Conclusions

1. Solid and liquid-phase diffusion are two critical parameters which affect alloy microsegregation.

2. The majority of current models treat diffusion in one phase only; diffusion in the other phase is held to be infinite or non-existent.

3. Dimensionless parameters can be used to give an indication of the importance of diffusion in segregation, but should only be used as guidelines.

4. A combined heat and mass transfer model, which allows finite solid- and liquid-diffusion, has been utilized to show the effects of varying liquid- and solid-phase diffusion in the Cu-Ni system.

5. Under ordinary casting conditions, and for typical binary alloys, liquid-phase diffusion can be assumed infinite. However, this may not pertain for conditions where particularly large dendrite-arm spacings exist, or cooling rates are rapid. More detailed calculations should certainly be made in these situations.

Acknowledgments

The assistance of Andrew Chan in improving the BP model software made this work easier than it would otherwise have been. The continuing support and encouragement of Professor Mark Cross, Head of School, Department of Mathematics, Statistics, and Computing, Thames Polytechnic, and the financial support of BNF Metals Technology Centre, Wantage, Oxfordshire, United Kingdom are greatly appreciated.

REFERENCES

1. Weinberg, F.and R. K. Buhr. in *The Solidification of Metals.* The Iron and Steel Institute, London 295 (1968).
2. Thresh, H., M. Bergeron, F. Weinberg and R. K. Buhr. *Trans. Met. Soc AIME.* Vol 242, 853-858 (1968).
3. Weinberg, F. and E. Teghtsoonian. *Metall. Trans.* Vol 3, 193-211 (1972).
4. Boeri, R. and F. Weinberg. *AFS Trans.* 179-184 (1989).
5. Battle, T. P., Thames Polytechnic, work in progress (1990).
6. Tiller, W. A. and R. F. Sekerka. *J. Appl. Phys.* Vol 35, 2726-2729 (1964).
7. Fischer, D. *J. Appl. Phys.* Vol 44, 1977-1982 (1973).
8. Verhoeven, J. D. and K. A. Heimes. *J. Crys. Growth..* Vol 10, 179-184 (1971).
9. Roosz, A., E. Halder and H. E. Exner. *Mat. Sci. Techn.* Vol 1, 1157-1162 (1985).
10. Battle, T. P. and R. D. Pehlke. *Metall. Trans. B.* Vol 21B (1990).
11. Battle, T. P. and R. D. Pehlke. in *Solidification Processing 1987.* Institute of Metals, London 71-74 (1988).
12. Kirkwood, D. H. *Mat. Sci. Eng.* Vol 65, 101-109 (1984).
13. Battle, T. P. and R. D. Pehlke. *Metall. Trans. B.* Vol 20B, 149-160 (1989).
14. Brandes, Eric A. (Ed.) *Smithells Metals Reference Book.*, 6th edition. Butterworths, London (1983).
15. Turkdogan, E. T. *Physical Chemistry of Oxygen Steelmaking, Thermochemistry and Thermodynamics.* United States Steel, Monroeville, Pa.
16. Kubicek, P. and T. Peprica. *Intl. Metals Rev.* Vol 28, 131-157 (1983).
17. Ejima, T., T. Yamamura, N. Uchida, Y. Matsuzaki and M. Nikaido. *J. Jpn. Inst. Met..* Vol 44, 316-323 (1980).
18. Bester, H. and K. Lange. *Arch. Eisenh..* Vol 43, 207-213 (1972) (in German).
19. Oikawa, H. *Techn. Rep., Tohoku Univ.* Vol 47, 215-224 (1982).
20. Murphy, J. B. *Acta Met..* Vol 9, 563-569 (1961).
21. Murray, J. L. *Intl. Metals Rev..* Vol 30, 211-233 (1985).
22. Brice, J. C. *Crystal Growth Processes.* Blackie, London (1986).
23. Sloan, G. J., and A. R. McGhie. *Techniques of Melt Crystallization.* John Wiley and Sons, New York (1988).

24. DeRoss, A.B. and L. F. Mondolfo, in *Aluminum Transformation Technology and Applications*. C. A. Pampillo, H. Biloni, and D. E. Embury (Editors). American Society for Metals, Metals Park, Oh. 81-140 (1980)
25. Professor Brian Thomas, University of Illinois, private communication, 1990.
26. Verhoeven, J. D., W. N. Gill, J. A. Puszynski and R.M. Ginde *J. Crys. Growth..* Vol 89, 189-201 (1989).
27. Allen, D. J. and J. D. Hunt. *Metall. Trans. A..* Vol 10A, 1389-1397 (1979).
28. Brody, H. D. and M. C. Flemings. *Trans. Met. Soc. AIME.* Vol 236, 615-624 (1966).
29. Crank J.. *The Mathematics of Diffusion.* Clarendon Press, Oxford (1975).
30. Pehlke, R. D., A. Jeyarajan, and H. Wada. *Summary of Thermal Properties for Casting Alloys and Mold Materials.* National Science Foundation, Washington (1982).

APPENDIX: TABLE OF SYMBOLS

C_o Initial concentration (mole percent)
D_i Solute diffusion coefficient in phase i (m^2/sec)
k Equilibrium partition coefficient (dimensionless molar units)
L Length scale of system (m)
m_L Liquidus slope (K/mole percent)
Pe Dimensionless Peclet number
T Temperature (K)
t Time (sec)
t_f Solidification time (sec)
V Velocity of interface (m/sec)
x Position (m)
α Dimensionless solute Fourier number
ΔT_o Equilibrium freezing range (K)
ε Cooling rate (K/sec)

Subscripts:

L Liquid s Solid

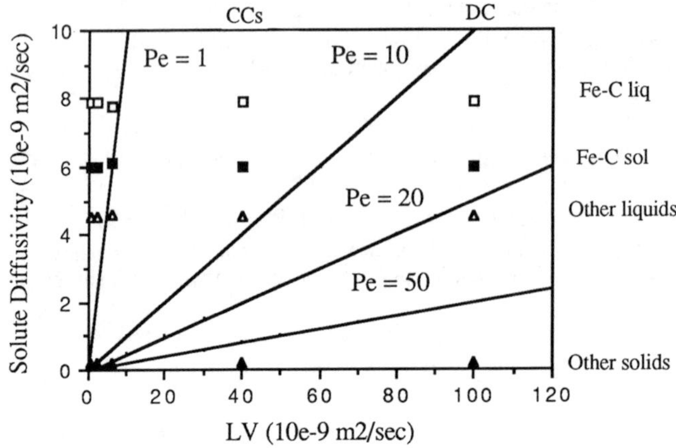

Fig. 1 Peclet Number Plot for All Systems

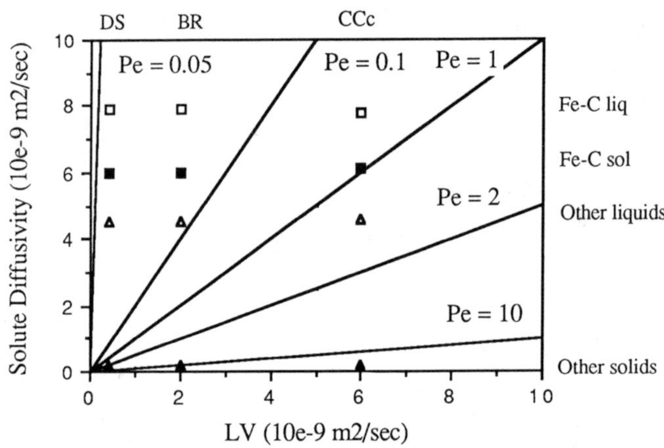

Fig. 2 Peclet Number Plot for Systems with Low Values of LV

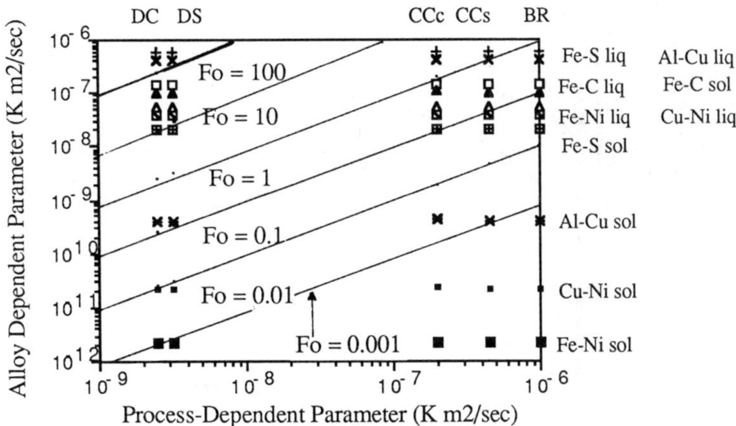

Fig. 3 Fourier Number Plot for All Systems

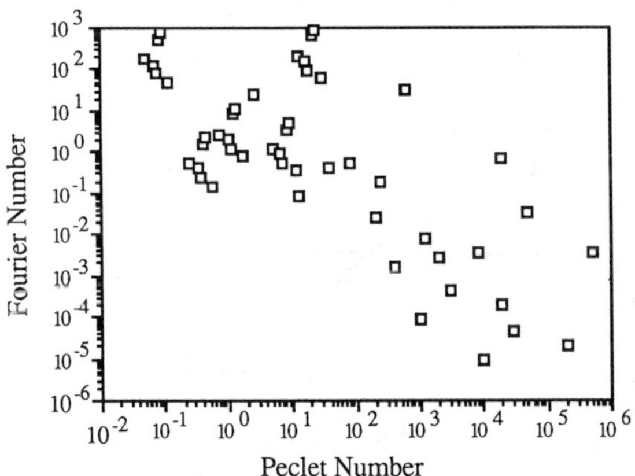

Fig. 4 Comparison of Fourier Numbers with Peclet Numbers for Same Systems

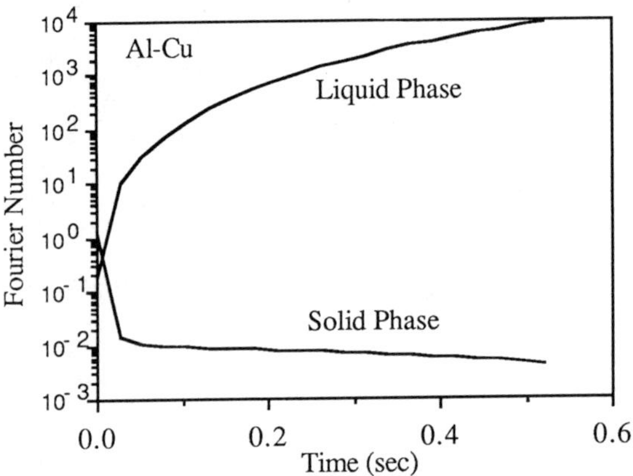

Fig. 5 Solid and Liquid-Phase Fourier Numbers as a Function of Time, Al-Cu Alloy

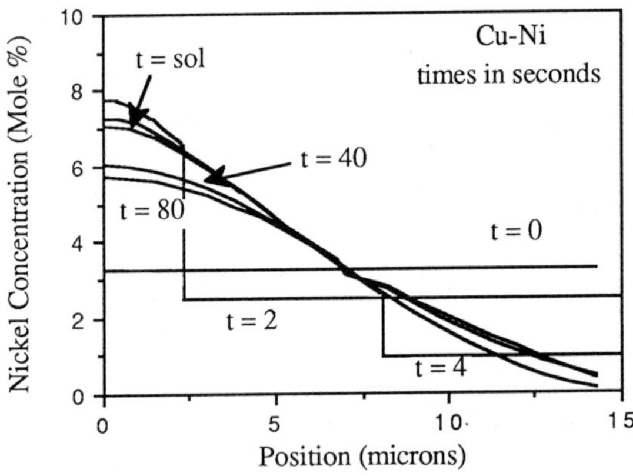

Fig. 6 Predicted Concentration Profiles During and After Solidification, Cu-Ni

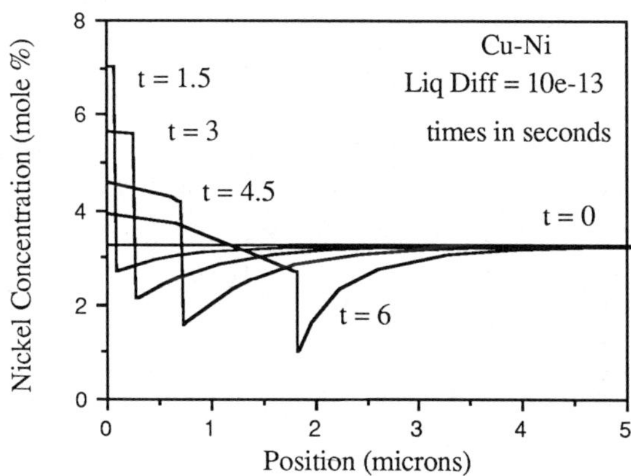

Fig. 7 Predicted Concentration Profile for Cu-Ni Alloy, with Artificially Reduced Liquid Diffusivity

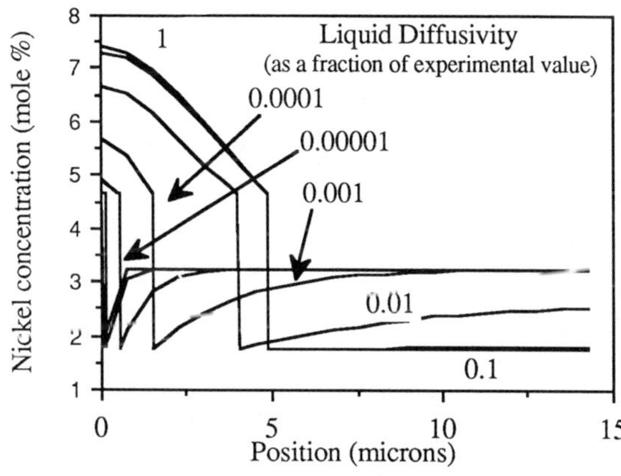

Fig 8 Predicted Concentration Profiles for Cu-Ni System After 4 Seconds of Solidification, as a Function of Liquid-Phase Diffusivity

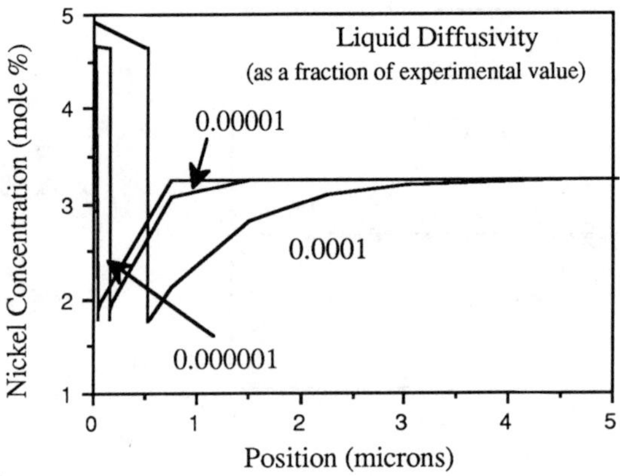

Fig 9 Predicted Concentration Profiles for Cu-Ni System After 4 Seconds of Solidification, as a Function of Liquid-Phase Diffusivity: Low Diffusivities

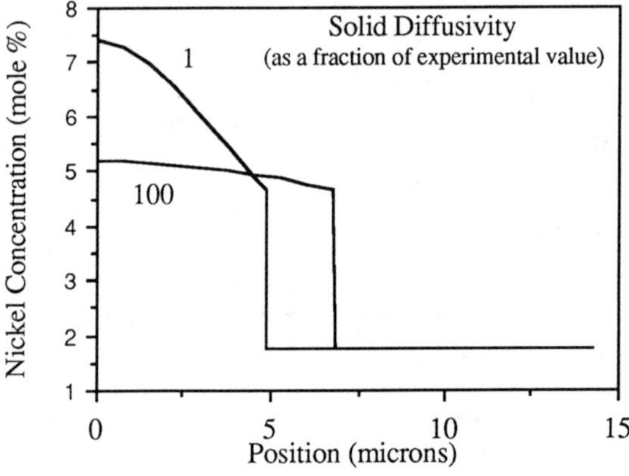

Fig 10 Predicted Concentration Profiles for Cu-Ni System After 4 Seconds of Solidification, as a Function of Solid-Phase Diffusivity

IV. Novel solidification studies

Chairman: D. Apelian
LeBow Engineering Centre,
Drexel University,
Philadelphia, Pennsylvania, U.S.A.

Solidification processing at near-rapid and rapid rates (Keynote Paper)

M.C. Flemings
Head, Department of Materials Science and Engineering, Massachusetts Institute of Technology, Cambridge, Massachusetts 02139, U.S.A.

Abstract

In recent years a number of novel casting processes have been developed which involve solidification at near-rapid or rapid rate. The processes include powder production, strip casting, wire casting, surface melting, and pressure die casting. Other processes are spray forming, semi-solid forming, and metal matrix composite casting. These processes are discussed in this paper, with special reference to dendritic growth and microsegregation. Processes which take place at cooling rates in the range of 1 to 10^3 Ks^{-1} are referred to as "near-rapid" rate processes, and those at greater cooling rates "rapid rate" processes. For both rapid and near-rapid rate processes, it is useful to distinguish between two types of solidification: columnar and equiaxed. It is also useful to distinguish between processes where heat flow is through the growing liquid-solid region, and those where heat flow is into undercooled bulk liquid.

Solidification Processes

Solidification processes involving dendritic solidification are carried out commercially over a range of cooling rates of more than 12 orders of magnitude — from 10^{-4} Ks^{-1} for large ingots to nearly 10^9 Ks^{-1} for surface treatments. Isotherm velocities range from as little as 10^{-5} ms^{-1} to more than 10 ms^{-1}. Table 1 lists examples of the various solidification processes and their respective ranges of cooling rates.

TABLE 1
Range of Cooling Rates in Solidification Processes

Range of Cooling Rate		Production Processes	Dendrite Arm Spacing (μm)
Ks^{-1}	Designation		
10^{-4} to 10^{-2}	Slow	Large castings and ingots	5000 to 200
10^{-2} to 1	Medium	Small sand castings and ingots; billet and bar continuous castings	200 to 50
1 to 10^3	Near-rapid	Strip casting; die casting; coarse powder atomization; "Premium Quality" casting; spray casting; semi-solid forming; metal matrix composite casting	50 to 5
10^3 to 10^9	Rapid	Melt spinning; fine powder atomization; thin strip casting; electron beam or laser surface melting	5 to 0.05

A regime of growing industrial significance is that of "near-rapid" solidification, the range of about 1 to 10^3 Ks^{-1} cooling rate, where dendrite arm spacings in the range of 5 to 50 μm are obtained. Here, casting structure is usually fine-grained equiaxed, and segregate spacings are sufficiently fine that a high degree of homogeneity can be obtained through subsequent thermal processing (or thermomechanical processing in the case of ingots and continuous castings). Processes carried out in this region of cooling rate include strip casting, die casting, coarse powder atomization, "premium quality" casting, spray casting, semi-solid forming, and metal matrix composite casting. At more rapid rates, processes of fine powder atomization, thin strip casting, laser surface melting, and others are becoming of increasing importance.

In considering solidification at these near-rapid and rapid rates, we need first of all to distinguish between columnar and equiaxed solidification, and secondly to distinguish between cases in which heat flow during solidification is into the bulk liquid, and those in which it is through the growing semi-solid.

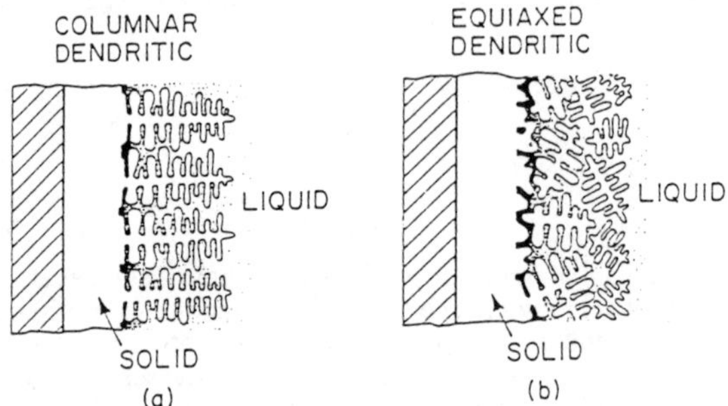

Figure 1. Solidification of an alloy against a cold chill wall. (a) Columnar solidification; (b) equiaxed solidification.

Figure 1a and b illustrates schematically the growth of equiaxed versus columnar grains. From the point of view of solute redistribution, the major difference between these two cases is that, for the case of equiaxed grains, the velocity of the isotherm where solidification begins, R_T, can be much higher than velocity of any given dendrite tip. It is:

$$R_T = \delta \dot{n} \tag{1}$$

where δ is linear distance in the growth direction between nucleation events, and $\dot{n}$ is rate of nucleus formation in the growth direction. Hence, even at high rates of movement of the solidification front, R_T, dendrite tip velocity can be quite low. The same is not true in columnar or cellular growth. Here, isotherm velocity and dendrite tip velocity must be very nearly equal.

"Near Rapid" Soldification; Heat Flow Through the Growing Semi-Solid

Here we assume the dendrite tip velocity is sufficiently low that the tips are growing at a temperature quite close to the liquidus. This may either be because effective nuclei are present, resulting in an equiaxed structure, or because, with columnar growth, isotherm velocity is sufficiently low. Tip undercooling in columnar growth is discussed in a later section.

Figure 2 shows the model schematically. Liquid composition at any temperature within the liquid-solid zone, Figure 2c, is given by the equilibrium liquidus line, Figure 2b. Fraction liquid (or solid) at any point within the liquid-solid zone, Figure 2d, is calculated by the "Scheil equation" or one of its derivative analyses to be described below. In addition to the assumptions of interfacial equilibrium and negligible undercooling implicit in Figure 2, the Scheil equation assumes no macroscopic mass transport, and no effect of ripening or other remelting process on growth. In differential form it is written (1):

$$\frac{\partial C_L}{\partial f_L} = -(1-k)\frac{C_L}{f_L} \tag{2}$$

where C_L is liquid composition, f_L is liquid weight fraction, and k is the equilibrium partition ratio. For constant partition ratio, this yields the familiar equation

$$C_S^* = kC_0 (1-f_s)^{k-1} \tag{3}$$

where C_S* is the composition of solid forming when fraction solid equals f_S, and initial composition is C_0 (2). Equation (3) describes the microsegregation within fully solidified castings or ingots. It is written for a binary alloy with constant partition ratio, k. Calculated results for a simple binary alloy are shown schematically in Figure 3.

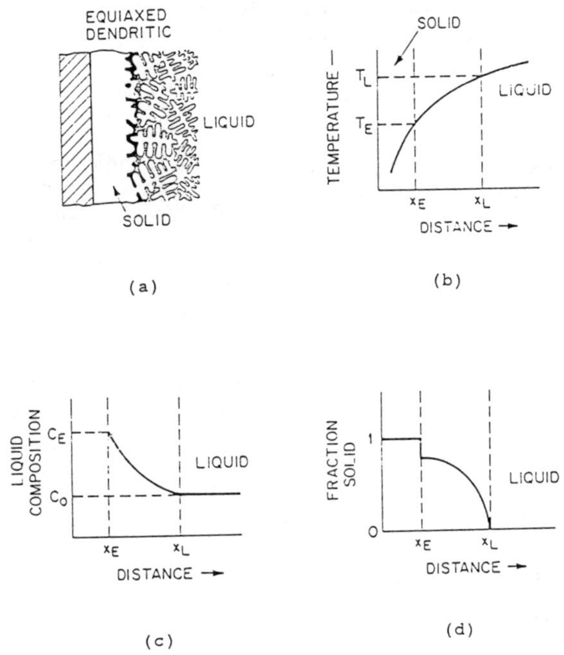

Figure 2. Model of equiaxed solidification.

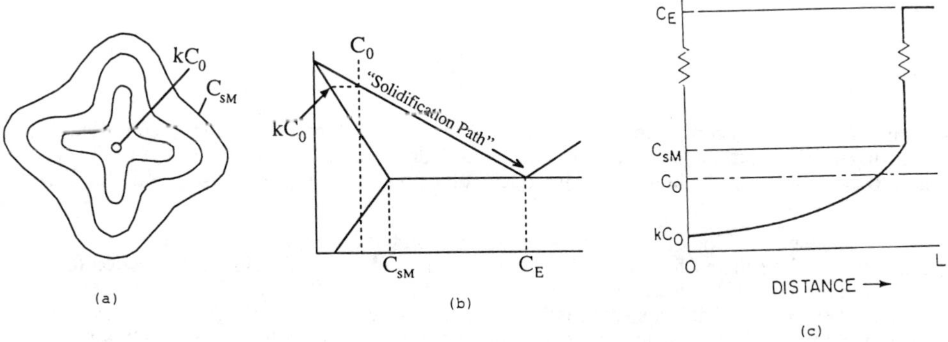

Figure 3. Formation of microsegregation. (a) Isoconcentrates as seen in a cross-section of a primary columnar dendrite arm after solidification. (b) Phase diagram showing "solidification path." (c) Microsegregation, measured along a radial path from the center of the dendrite.

It has long been understood that diffusion in the solid during dendritic solidification alters microsegregation in most real systems. This was first quantitatively analyzed by Brody and Flemings (3), who derived a simple approximate analytical expression (modified Scheil analysis) to account for solid diffusion:

$$C_S^* = kC_0\left[1 - \frac{f_s}{1+\alpha k}\right]^{k-1} \tag{4}$$

and

$$\alpha = \frac{4\,D_s t_f}{d^2} \tag{5}$$

where D_s is the diffusion coefficient of solute in the solid, t_f is solidification time, and d is dendrite arm spacing. This analysis is quite approximate, valid only for small values of α. Calculated results from the same work using a more exact numerical solution are shown in Figure 4 for Al-4.5%Cu alloy.

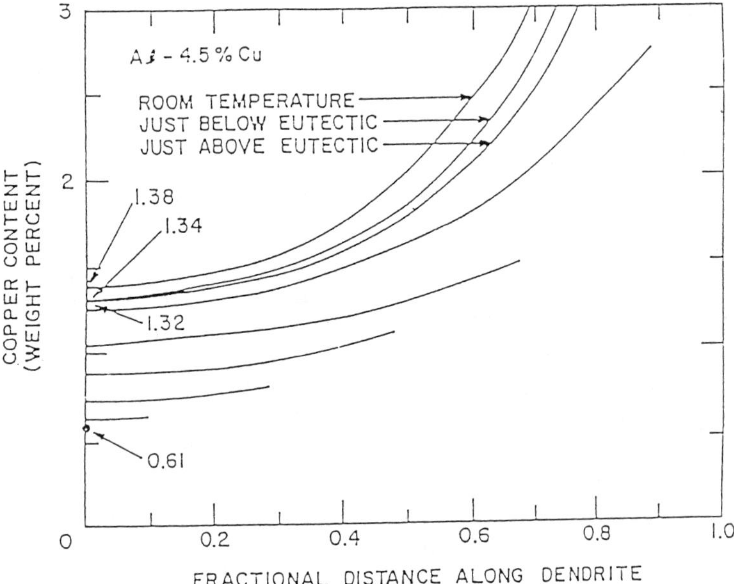

Figure 4. Calculated composition across a dendrite arm at different times during solidification, showing effect of diffusion in the solid. Each curve drawn extends from the center of the dendrite arm to the position of the liquid-solid interface at the given time.) (From Brody and Flemings[3])

Since publication of the above work and its companion experimental study (3,4), many refinements of the theory have been developed. Ohnaka (5) derived a simple expression for an improved estimate of liquid composition during solidification. Clyne and Kurz (6) presented an analytical expression that is also approximate but has the correct limit at high values of α. Kobayashi (7) has recently developed an improved analytic solution, although it is more difficult to use. Yeum, Laxmanan, and Poirier (8) presented a computationally efficient finite element method to calculate the microsegregation. Kirkwood (9) and Ogilvy and Kirkwood (10) first incorporated the phenomenon of dendrite arm coarsening into the microsegregation model, and Mortensen later derived an analytical expression for this effect (11). Allen and Hunt (12,13) first showed that dendrite arms tend to "climb" up the temperature gradient as a result of concentration-driven solute diffusion. Others, most recently Riedl and Fischmeister (14), have amply demonstrated the effect of dendrite arm climb and its influence on microsegregation. No attempt has yet been made, however, to incorporate this effect into the more general microsegregation models described above.

In equiaxed structures, with heterogeneous nucleation at low undercoolings, the main factor causing microsegregation to deviate from that predicted by the Scheil equation appears to be diffusion in the solid during solidification. Figure 5, from Ohnaka (5), compares calculated results on segregation of manganese in an Fe-1.52wt%Mn alloy, obtained using several of the different analyses referred to above. The analyses differ only in the analytic or computational method used to account for solid diffusion (including assumed dendrite geometry).

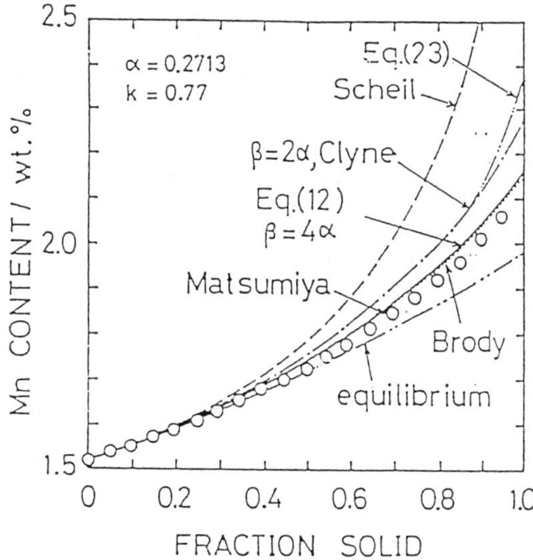

Figure 5. Comparison of calculated microsegregation for manganese in a Fe-1.52%Mn alloy. (From Ohnaka[5])

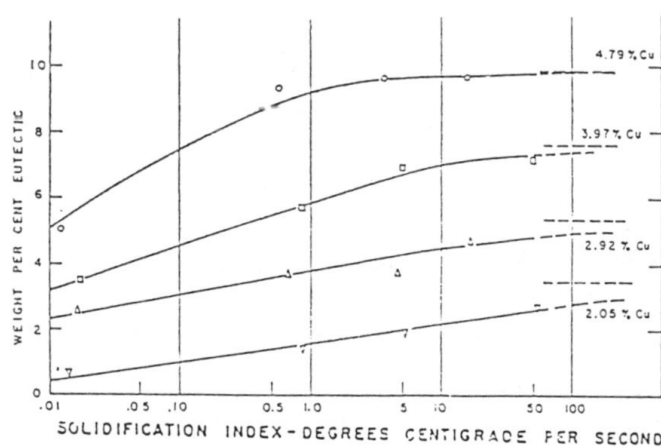

Figure 6. Weight percent eutectic vs. "solidification index" (cooling rate) for aluminum-copper alloys. (From Michael and Bever[15])

The extent of diffusion in the solid occurring during solidification depends on solidification time and inversely on the square of the diffusion distance (and hence, interdendrite arm spacing). Since dendrite arm spacing, as discussed below, depends on the cube root of solidification time, the result is a weak dependence of final microsegregation on solidification time (*i.e.*, for a given alloy or cooling rate). Figure 6 is an example from work of Michael and Bever (15), in which severity of microsegregation is measured by amount of non-equilibrium eutectic in aluminum-copper alloys. For the alloys studied, the amount of interdendritic eutectic increased by a factor of two or more over the range of cooling rates from 0.01 Ks^{-1} to 10^2 Ks^{-1}. At the highest cooling rate employed, about 10^2 Ks^{-1}, the amount of non-equilibrium eutectic measured is about that predicted by the Scheil equation. More recently, these and similar experimental results have been quantitatively interpreted on the basis of the solidification model discussed above, considering the effect of solidification and coarsening (3,4,16-20).

While the degree of final segregation is not much influenced by cooling rate in these equiaxed structures, even at the maximum cooling rates studied, the segregate spacing (*i.e.*, the dendrite arm spacing) is very much affected. Figure 7 is a summary of data from a wide range of investigations for the binary alloy Al-4.5%Cu, from Jones (20). The line through the points is

$$d = 50\varepsilon^{-1/3} \tag{6}$$

where d is dendrite arm spacing (μm) and ε is cooling rate (Ks^{-1}). The factor determining dendrite arm spacing is now well understood to be ripening, which occurs during solidification (11,21). It is seen that this relation, and hence the importance of ripening, extends to well above the range of cooling rates of near-rapid solidification.

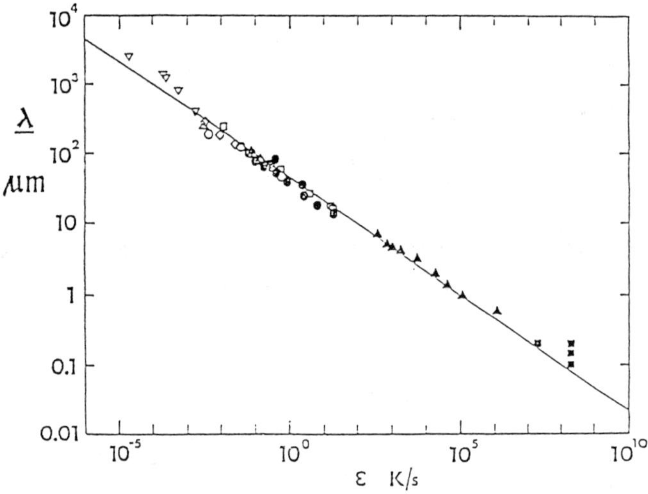

Figure 7. Dendrite arm spacing versus cooling rate for alloys of approximately Al-4.5wt%Cu. Experimental data are from a number of different investigations. (From Jones[20])

Dendrite arm spacing is of engineering importance primarily because it determines the distance over which solute elements must diffuse to obtain fully homogeneous material. For dendrite arm spacings above those obtained in near-rapid solidification (above about 50 μm), times required for full homogenization in most commercial alloys become prohibitively long. Moreover, experience indicates that even extensive hot working of coarse dendritic structures does not fully eliminate the compositional heterogeneities.

It should be added that, in the solidification processes that occur at rates up to those of near-rapid rate, solidification behavior (including microsegregation and secondary dendrite arm spacing) is not influenced by grain size or by whether the grain structure is equiaxed or columnar.

Heat Flow in Near Rapid Solidification; Mechanical Properties

Modern mathematical models of heat flow in solidification of alloys are generally based on physical models similar to that of Figure 2 (22,23). Figure 8 shows some calculated results from work of Campagna (24,25), in which his physical model was exactly that described by Figure 2 and for which the model alloy chosen was Al-4.5%Cu. The results plotted are for solidification against a water-cooled chill, with a Biot number of 0.6. Movement of the dendrite tips and the liquidus isotherm versus the square root of time are shown at the upper left. Average cooling rate during solidification is shown in the center plot, and dendrite tip velocity at the bottom. We may use this and Figure 7 (or Equation 6) to illustrate some engineering aspects of several important commercial processes.

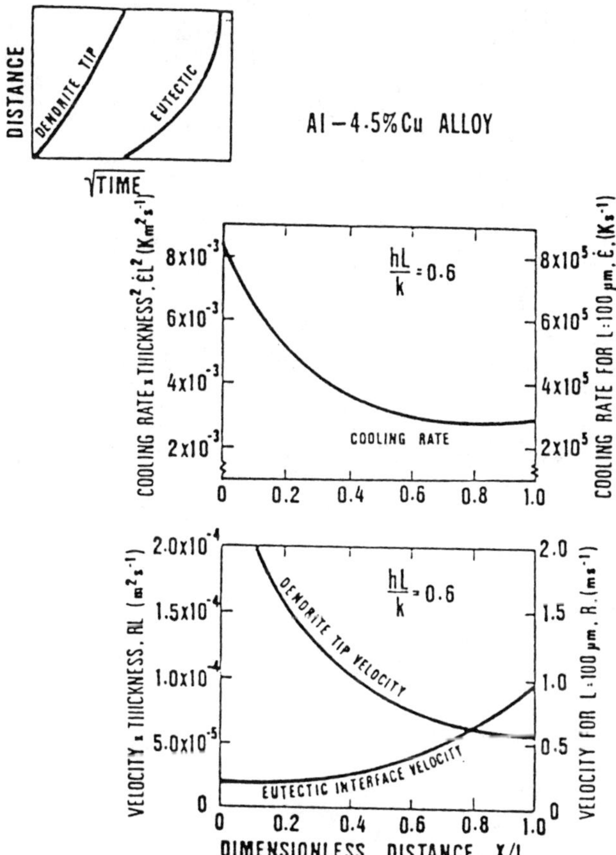

Figure 8. Solidification of Al-4.5%Cu alloy against a water-cooled chill. Biot Number is 0.6.[25]

Consider first a 100 mm thick plate of Al-4.5%Cu alloy, chilled on both sides. Mold-metal interface resistance is typical of that encountered in casting processes (about 0.04 cal $cm^{-2}s^{-1}K^{-1}$) so that the Biot Number is about 0.6. From Figure 8, cooling rate varies from about 3.0 Ks^{-1} at the surface to 1.0 Ks^{-1} at the centerline. Corresponding dendrite arm spacings vary from about 35 μm at the surface to 50 μm at the center — a suitable range for good mechanical properties, as seen by the experimental data in Figure 9. Figure 10 plots this expected variation in dendrite arm spacing and mechanical properties across the plate thickness.

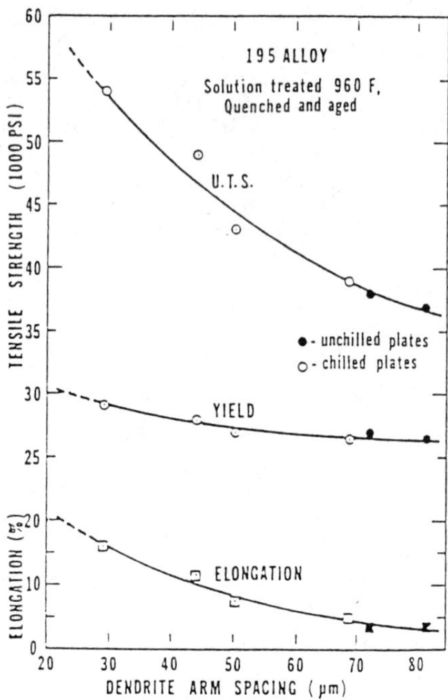

Figure 9. Mechanical properties of Al-4.5%Cu alloy versus dendrite arm spacing.[25]

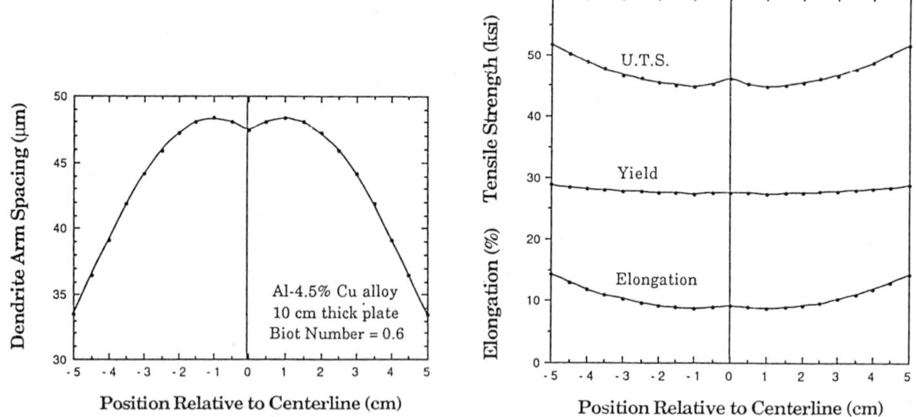

Figure 10. Calculated dendrite arm spacings and mechanical properties across the thickness of a 100 mm chilled plate, Al-4.5%Cu alloy.

A casting of the same thickness cast in a sand mold would solidify, of course, at a much slower rate. An aluminum casting this size would solidify throughout at a cooling rate of about 0.02 Ks^{-1}, or 50 times more slowly than the center of the above chilled plate. The dendrite arm spacing throughout this sand casting would be about 200 μm. Full homogenization would be impossible (2), and, as can be inferred from Figure 9, very poor properties would result.

The foregoing relationships are used by foundrymen in producing "Premium Quality" castings (2), in which high mechanical properties such as those of Figure 9 are reliably produced throughout castings. In sand castings, this is done by placing metal "chills" in the mold in appropriate locations and at appropriately spaced intervals so that the sand casting solidifies at rates comparable to those of the chilled casting described above — and final dendrite arm spacings are under about 50 μm. Dendrite arm spacing measurements from actual castings are sometimes requested by customers as a measure of process control.

We may also consider solidification behavior of a die casting. Die castings typically solidify with much higher mold-metal heat transfer coefficients than do chilled sand castings, and we may estimate that a die casting of about 3 mm thickness would solidify with a Biot Number of about 0.6, so that we may again use Figure 8 to estimate cooling rate during solidification. Cooling rate at the surface would be over 3 x 10^3 Ks^{-1} and cooling rate at the center about a third of that. Dendrite arm spacing would be about a tenth that of the chilled casting described above: about 3 μm at the surface and 5 μm at the center.

Such high cooling rates would normally be expected to result in excellent properties, but, for a number of reasons, high mechanical properties are not obtained in most of today's die castings. Reasons for this include entrapment of air and other gases during mold filling, and difficulties in fully feeding shrinkage. In addition, alloys used for most die castings are not amenable to development of high properties. The alloys used are chosen for low cost and ease of casting.

It is clear, however, that the very high cooling rates obtainable in die casting could result in castings of high mechanical properties and reliability. Perhaps new processes such as semi-solid forming ("Rheocasting") will permit die casting to reach its potential for high performance products.

These same analyses can be extended for calculation of solidification at "rapid" rate although it must be recognized that assumptions made (such as no undercooling at the dendrite tip) become increasingly poorer approximations. As an example of solidification at the very high rates we may consider the example of a "splat cooled" sample of Al-4.5%Cu solidifying from both sides. The high heat transfer coefficient here will be quite high and a Biot Number of 0.6 is reasonable for a sample 100 μm thick; thus, we can again use Figure 8 to calculate cooling rate. Now cooling rate is about 3 x 10^6 Ks^{-1} at the surface and a third of that at the center. Dendrite arm spacing, from Figure 7, varies from 0.35 μm at the surface to 0.5 μm at the center. Table 2 summarizes the cooling rate and dendrite arm spacing calculations outlined in the previous paragraphs.

TABLE 2
Cooling Rates and Dendrite Arm Spacings for Some Practical Examples.
Al-4.5%Cu Alloy. Equiaxed Solidification.

	Surface		Center	
Example	Cooling Rate (Ks^{-1})	DAS (μm)	Cooling Rate (Ks^{-1})	DAS (μm)
100 mm thick sand casting	0.02	184	0.02	184
100 mm thick chilled casting	3.2	35	1.2	47
3 mm thick die casting	3.6 x 10^3	3.3	1.3 x 10^3	5
100 μm thick splat-cooled droplet	3.2 x 10^6	0.35	1.2 x 10^6	0.5

Novel "Near-Rapid" Solidification Processes

It is of interest to consider several processes which are often practiced in the near-rapid range of cooling rates, but in which cooling behavior is greatly different from that in the more conventional processing described above. One of these is "spray casting," shown schematically in Figure 11, in which metal is spray atomized, partially solidified in flight, and then deposited in plate form or other shape. The first part of this solidification may involve some undercooling, and certainly involves rapid rate; the last part of solidification is slower, depending on heat extraction to the substrate or surroundings. Final segregate spacing is found to be somewhat less than would be expected from a direct linear correlation of dendrite arm spacing with the logarithm of solidification time (26).

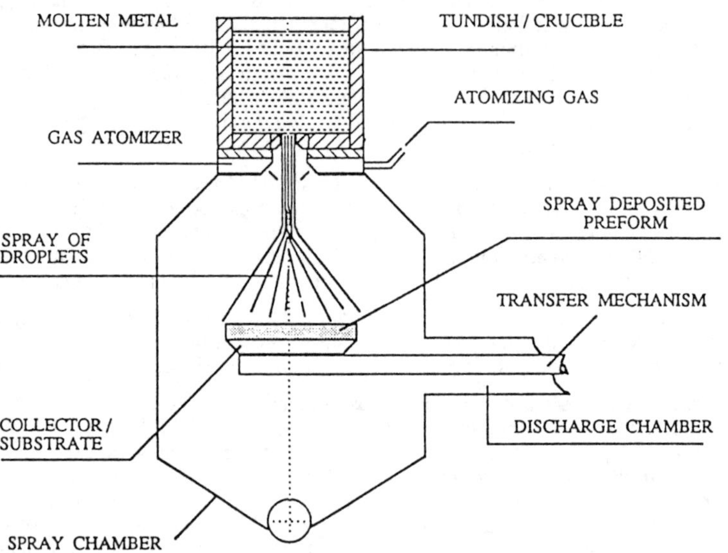

Figure 11. Schematic of the Osprey^{TM} process.[26]

In semi-solid forming (Rheocasting) (2,27), Figure 12, solid particles initially present in the metal may be relatively large (50 µm or more), but solidification in the last stages, (*e.g.*, in a metal mold) may be very rapid. Microsegregation, and homogenization of that microsegregation, are only now being studied in these alloys (28).

A third process in this category is casting of metal matrix composites. In one type of such casting, metal is infiltrated into a ceramic preform. If the preform is cooler than the liquidus temperature of the alloy, initial solidification of the advancing front is very rapid; subsequent solidification is then determined by heat extraction to the mold (29,30). The amount of dendrite coarsening that can occur in these composites is, however, limited, since (at least in the alloys studied to date) coarsening ceases when dendrite size becomes about that of the interstices of the preform. Final microsegregation in these cases can therefore be much less than in conventional solidification (31). Figure 13 shows the coarsening process in a fibrous composite schematically. Figure 14 illustrates that, after a critical time during solidification (which depends on the size of the interstices in the composite), the dendritic structure has disappeared as a result of solid state diffusion into arms which are limited in their ability to coarsen by the ceramic fibers.

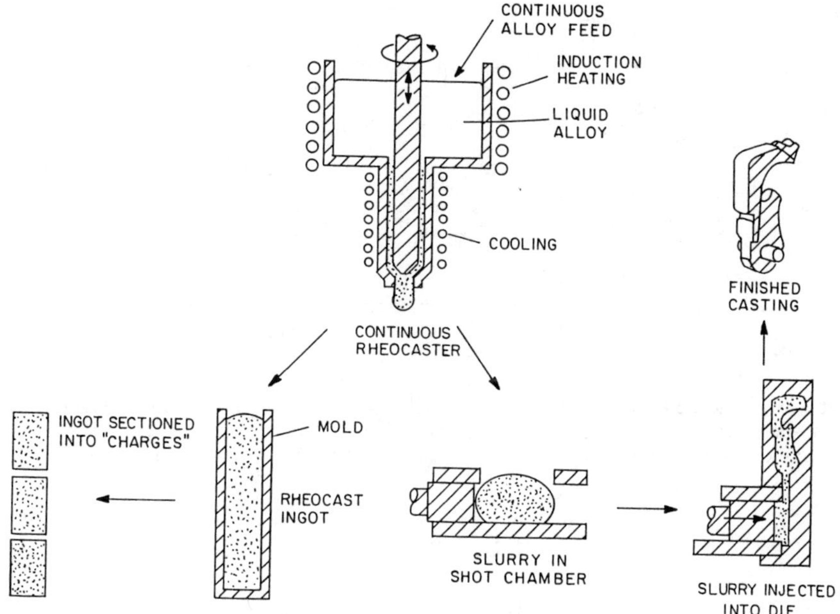

Figure 12. Rheocast process.[27]

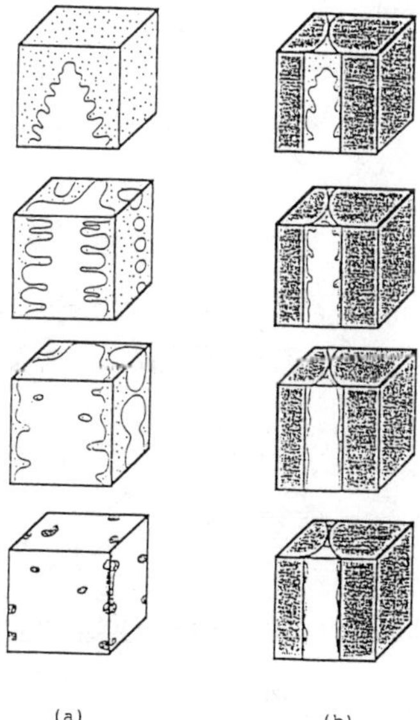

Figure 13. Schematic rendition of coarsening in (a) a usual casting or ingot; (b) the interstices between fibers in a metal matrix composite.) (From Mortensen et al.[31])

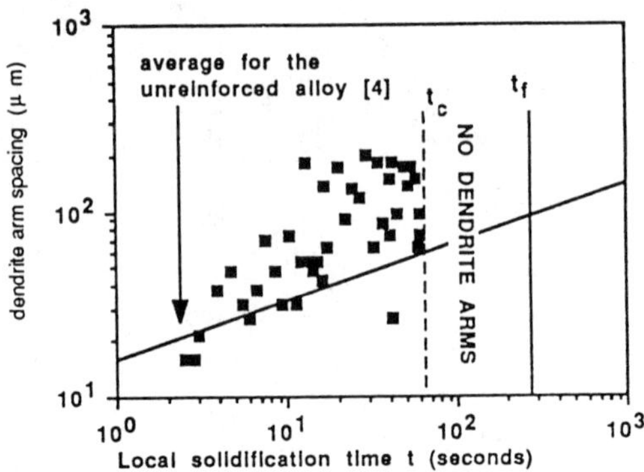

Figure 14. Dendrite arm spacing versus solidification time in an aligned fiber metal matrix composite. Dendrite structure (and microsegregation) are eliminated when dendrite arms coarsen to the maximum size permitted by the interstices between fibers. (From Mortensen et al.[31])

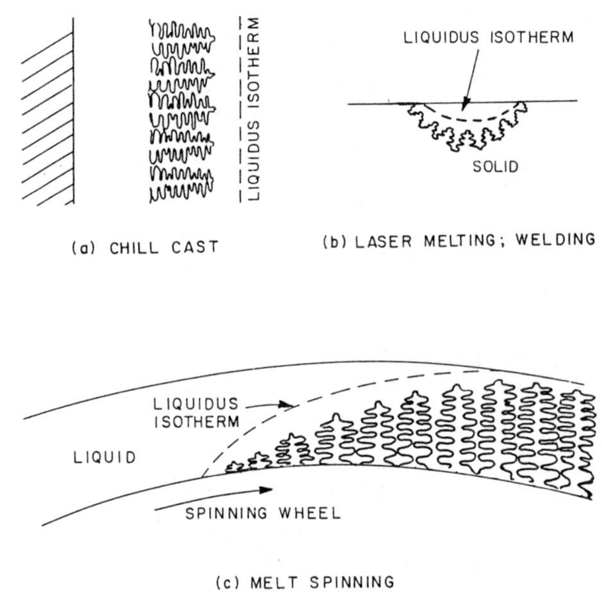

Figure 15. Examples of rapid solidification processing with columnar dendritic growth; heat flow through the growing solid.

"Rapid" Solidification; Heat Flow Through the Growing Solid

Dendrite tip temperature for this type of solidification, especially in the case of columnar growth, can fall significantly below the liquidus isotherm, as illustrated in the practical examples of Figure 15. Calculation of the extent of tip undercooling has occupied the attention of solidification scientists for a number of years.

It has long been understood that dendrite tip undercooling can arise from three sources: the effect of interface curvature on equilibrium melting point; diffusion of heat and/or solute from the growing tip; and interface kinetics. A recent model, that of Kurz, Giovanola, and Trivedi (the "KGT model") incorporates all three effects using the Ivantsov solution for the diffusion field, the shortest marginally unstable wavelength for determining the tip radius, and a growth rate dependent diffusion coefficient, k (32).

Figure 16 shows results of calculations of dendrite tip temperature versus growth velocity for Al-4.5%Cu alloy using the KGT model. The calculations assume growth into an isothermal melt at the tip temperature, so are drawn for G/R=0. Equilibrium interface kinetics are also assumed. Note that significant tip undercooling is present at tip velocities as low as 10^{-2} to 10^{-3} ms^{-1}. The undercooling increases with increasing tip velocity, reaching a maximum at about 6.3 ms^{-1}, the "limit of absolute stability" first described by Mullins and Sekerka (33,34). At this velocity, solidification is with a plane front of uniform solid composition equal to C_0.

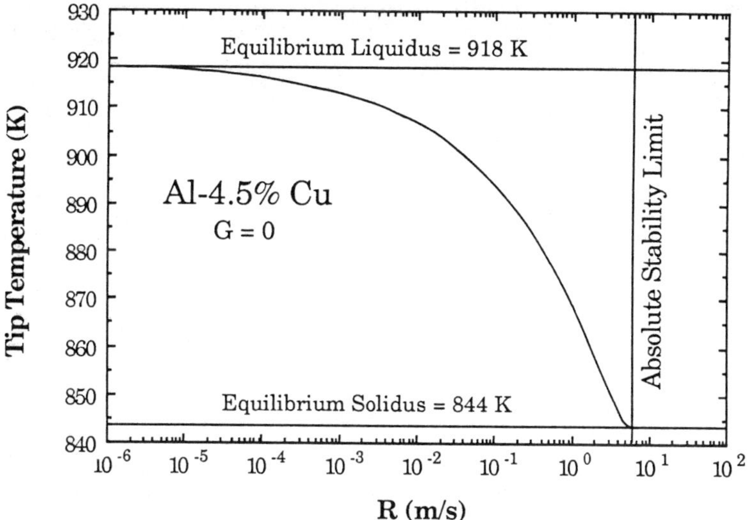

Figure 16. Dendrite tip temperature versus tip velocity. Columnar growth of Al-4.5%Cu calculated using KGT model 32 with equilibrium partition ratio.

From a simple kinetic argument we expect that, at sufficiently high growth velocities, the partition ratio will differ from the equilibrium ratio as a result of "solute trapping." A functional relationship proposed by Aziz (35) and incorporated in the KGT model for dendritic growth is:

$$k = \frac{k_0 + \frac{a_0 R}{D}}{1 + \frac{a_0 R}{D}} \quad (7)$$

where k and k_0 are the kinetic and the equilibrium partition ratios, respectively, D is the solute diffusion coefficient at the interface, and a_0 is a length scale related to the interatomic spacing. Equation (7) predicts a rise in k from k_0 to unity over a range of velocities. For the Al-4.5% Cu alloy used as example here, it is over the range of about 1 to 100 ms^{-1}, assuming a_0=0.5 nm.

For a sufficiently dilute alloy, we expect the limit of absolute stability to be at a lower velocity than that where solute trapping occurs, but for higher solute contents solute trapping can be important. For Al-4.5%Cu alloy, it is possible, as indicated above, that some significant trapping can occur at as low as about 1 ms^{-1}.

Figure 17a shows how dendrite tip composition increases with increasing tip velocity for Al-4.5%Cu alloy. Tip composition varies from k_0C_0 at the lowest velocities to C_0 when absolute stability is reached. Figure 17b shows the calculated final microsegregation, using the recent empirical formulation of Giovanola and Kurz (36) to "patch" the KGT solution to the Scheil relation. These calculations may underestimate the tip undercooling and tip composition at the higher tip velocities, because of the assumption of no solute trapping ($k=k_0$ at the dendrite tip). Experimental work of Boettinger et al. (37) has apparently verified solute trapping for rapidly solidified Ag-15wt%Cu alloy. The KGT analysis with $k=k_0$ predicted a tip composition substantially less than that observed (Figure 18a) while assuming a higher value of k resulted in close agreement of the KGT theory with experiment (38). The resulting solute distribution across the dendrite arm is shown in Figure 18b.

For columnar growth, the relation between dendrite tip velocity and casting cooling rate can readily be determined from heat flow calculations. Results of some calculations are shown in Table 3, using the heat flow model of Campagna described earlier. Note that tip undercooling has already become significant (25 K) in the 3 mm die casting which solidifies with a cooling rate at the upper end of the "near rapid" range. Of course, tip undercooling would be low to negligible, even at these high cooling rates, for equiaxed solidification.

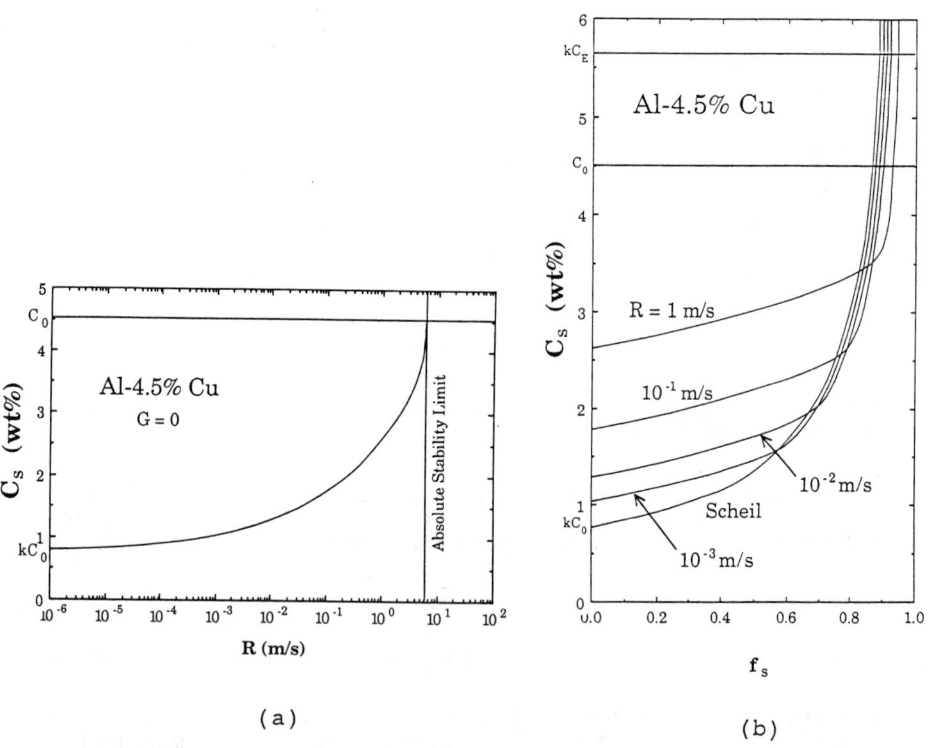

Figure 17. Microsegregation in rapidly solidified Al-4.5%Cu, calculated using the KGT model 32 with equilibrium partition ratio. (a) Dendrite tip composition. (b) Solute redistribution using the empirical patchy relation of Giovanola and Kurz.[36]

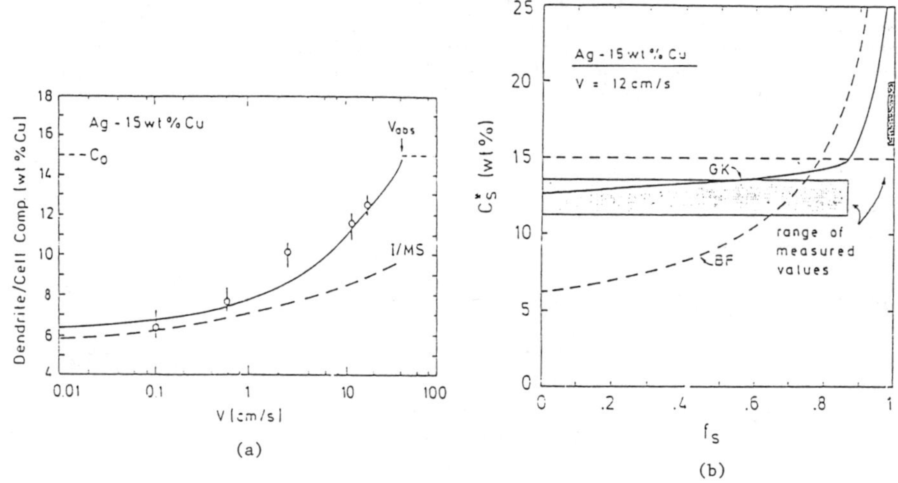

Figure 18. Microsegregation in rapidly solidified Ag-15wt%Cu. (a) Dendrite tip composition. Data from Boettinger et al.[38] Solid curve, KGT model 32 with $k=k_0$. Dashed curve, other models assuming $k=k_0$. (b) Solute redistribution. Solid curve calculated from KGT model using experimental patching.[37] Dashed curve, Brody and Flemings.[3]

TABLE 3
Calculated Tip Undercoolings for Some Practical Examples.
Columnar Growth of Al-4.5%Cu Alloy.

Example*	Tip Velocity (ms^{-1})	Cooling Rate (Ks^{-1})	Tip Undercooling (K)
100 mm thick chilled casting	2.5×10^{-3}	1.5	7
3 mm thick die casting	0.09	2×10^{3}	25
100 μm thick splat-cooled droplet	2.5	1.5×10^{6}	67

* Calculation in each case is for a location halfway between the surface and center of the casting; columnar growth assumed with Biot Number equal to 0.6.

Heat Flow into the Bulk Liquid

When heat flows from the bulk liquid directly to the surroundings before or during solidification, conditions prevail that alter solidification behavior in an important way. Heat then flows from the growing dendrite tips into the liquid, and recalescence occurs (at least locally). We sometimes see a small amount of this recalescence at the beginning of solidification in conventional castings and ingots. We can obtain much larger undercoolings in small droplets, bulk specimens, or continuous processes, by employing clean molten metal and solidifying it without contact with materials which catalyze crystallization.

In Figure 19, for example, a liquid droplet is originally undercooled to a temperature T_i below the equilibrium liquidus, T_L. Nucleation occurs at the left of the droplet and dendrites grow from left to right. Temperature acrôss the droplet at this time is as shown in Figure 19b with heat flowing predominantly into the liquid. Movement of the dendrite front across the specimen is quite rapid at high undercoolings, and so the rate of local recalescence can be very rapid indeed.

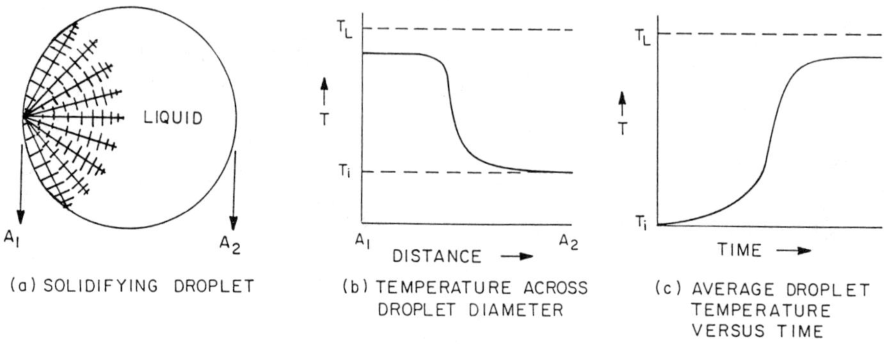

Figure 19. Solidification of an undercooled alloy droplet.

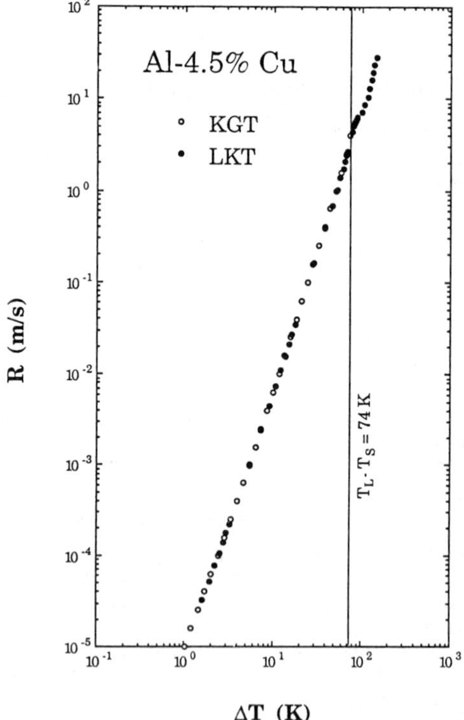

Figure 20. Comparison of predictions of growth velocity versus undercooling for the models of Kurz et al.[32] and Lipton et al.[40] for constrained and free dendritic growth, respectively.

The growth velocity of dendrites into an undercooled melt is treated in a manner similar to the constrained growth discussed in the previous section. Moreover, numerical results of dendrite tip velocity versus undercooling are very close to those calculated earlier for constrained growth. For Al-4.5%Cu alloy, using the model by Lipton, Kurz, and Trivedi (LKT model) (39), the results are as shown in Figure 20. Superimposed on this model is the curve of Figure 16 (KGT model), which is for growth into a non-undercooled melt (constrained growth). The KGT and LKT models yield essentially identical results up to an undercooling of T_L-T_S. The KGT model, however, does not extend below the solidus temperature, T_S, since at this temperature "absolute stability" is attained.

Many workers are now studying dendrite growth in undercooled alloys, and comparing experimental results with these and other analyses. Results of one such study by Wu *et al.* (40) are summarized in Figure 21.

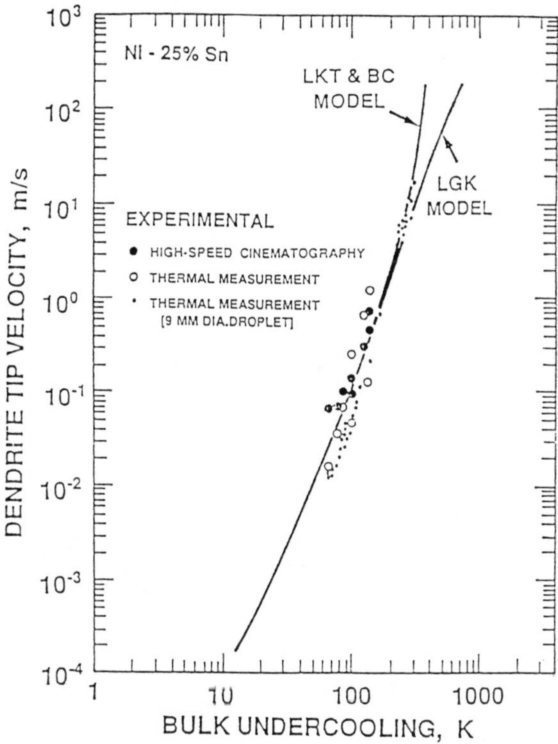

Figure 21. Dendrite tip velocity vs. undercooling in Ni-25wt% Sn alloy. (From Wu et al.[45])

Solidification of this type is obtained in continuous processes when heat flow from the liquid is rapid and nucleation is hindered. Figure 22, for example, shows equipment for direct casting of steel wire developed at Michelin several decades ago (41). The molten steel is ejected from a small orifice at speeds up to about 15 ms^{-1} and solidifies before Rayleigh breakup occurs. The wire is only about 200 μm in diameter, so radial temperature differences, even at these high velocities, are small. Solidification, in the usual case, occurs by axial dendritic growth, Figure 23, and so, at steady state, dendrite growth velocity is just equal to the ejection speed. Figure 24 shows results of actual experimental temperature measurements in wire spinning for a range of ejection velocities. Dendrite tip undercooling is seen to increase with increasing tip velocity and to be significant at these tip velocities of some meters per second. The undercoolings, in fact, are quite close to those of the Ni-25%Sn alloy of Figure 21.

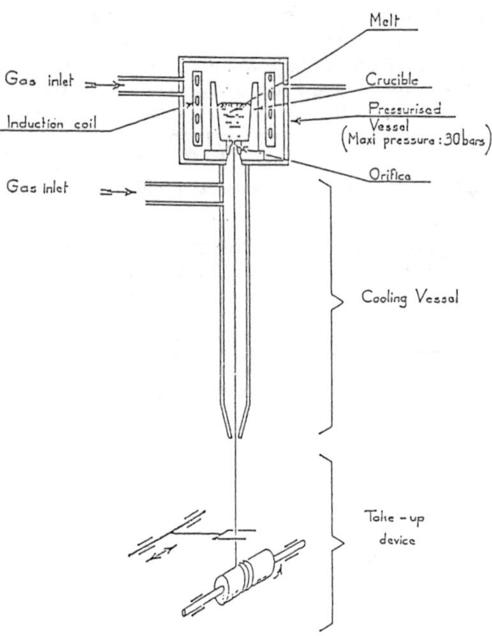

Figure 22. Apparatus for spinning steel wire from the melt. (From Massoubre and Pflieger[41])

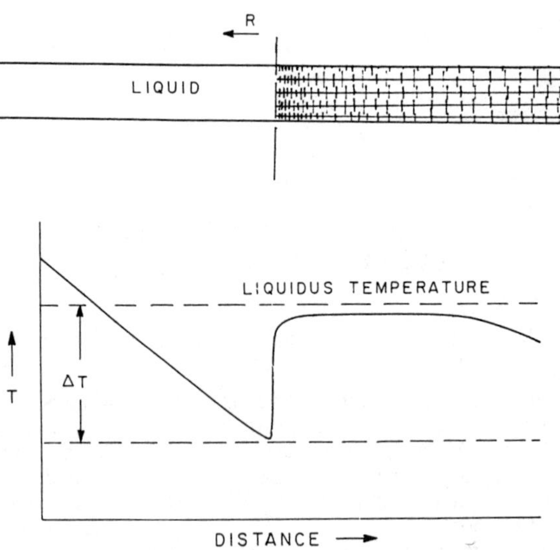

Figure 23. Solidification of spun cast wire. At steady state, wire moves to right with velocity R and dendrites grow at constant velocity R into liquid metal undercooled an amount ΔT.

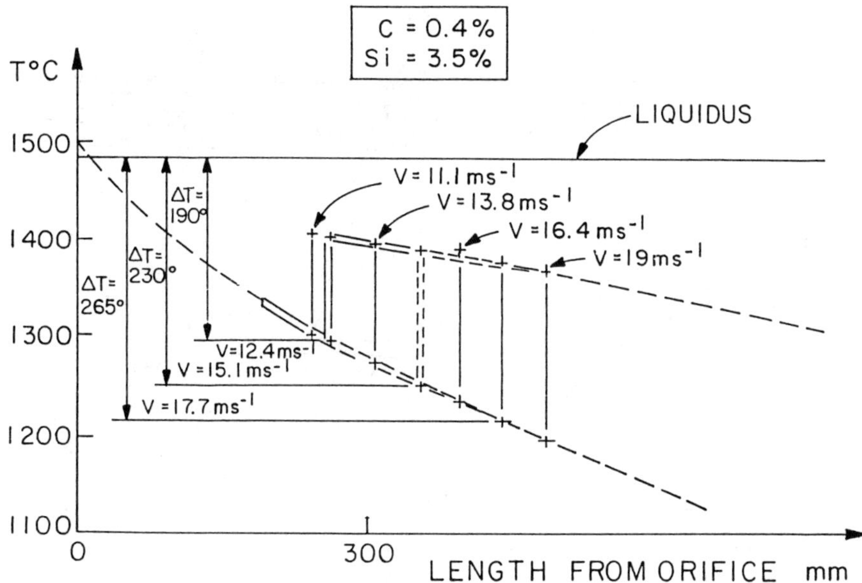

Figure 24. Measured temperature distributions along spun wire during steady state solidification. (From Massoubre and Pflieger[41])

Mixed Cases of "Rapid" Solidification

The classes of rapid solidification cited above are not necessarily mutually exclusive with respect to any given casting operation. For example, consider the wire spinning operation shown schematically in Figure 23 at steady state, and suppose that at a time, t, a large number of effective heterogeneous nuclei are continuously introduced into the molten stream. The "upstream front" then moves quickly up to the liquidus and subsequent solidification is by heat flow through the liquid-solid zone, comparable to the schematic examples of Figure 15 In the unsteady state, there remain two additional solidification fronts which grow together as shown schematically in Figure 25. This phenomenon has been observed in metal wire spinning (41).

Mixed types of solidification can also occur when a thermal boundary layer reaches the surface of a casting, as in solidification of atomized powders or in melt spinning. Figure 26 illustrates this for melt spinning. Growth is assumed to occur only by columnar growth of existing grains, with no nucleation at the growing interface. Hence, substantial undercooling must occur in the melt to propagate growth upstream, and growth is rapid. But if the strip thickness is small compared with the thermal boundary layer, then this type of solidification can occur only in the bottom portion of the strip. In later stages of solidification, the undercooling in the bulk liquid will be depleted, tip temperature will rise, and tip velocity will slow to that which results from external heat extraction. The result is a strip of two distinct regions — an outer fine structure and an inner, much coarser structure, as observed by a number of workers and described and explained by Chu *et al.* (42,43).

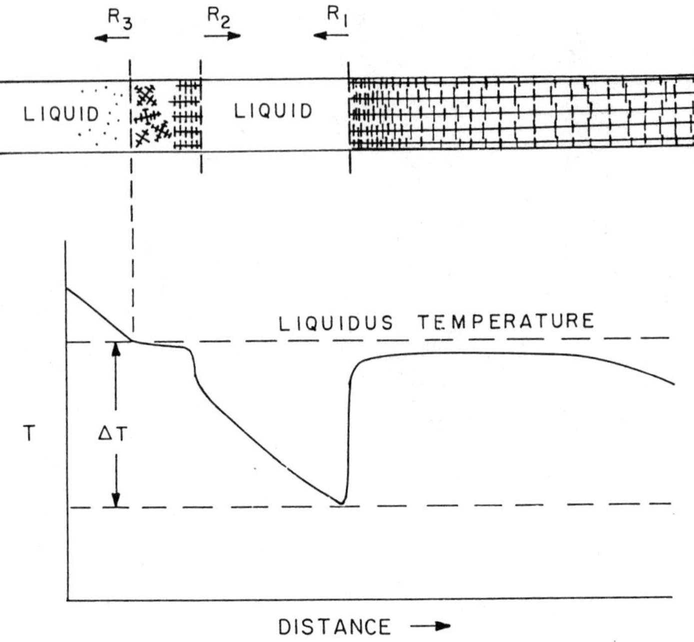

Figure 25. Schematic illustration of transient solidification resulting when nuclei are introduced to the melt stream in wire casting. Three solidifiication fronts result.

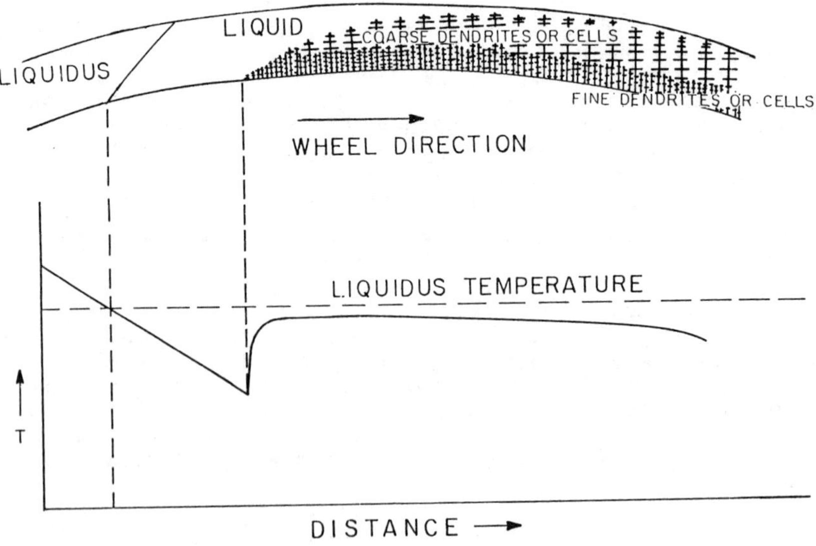

Figure 26. Mixed solidification in melt spinning.

Figures 27a and 27b are from that work, and employ the theory of Lipton *et al.* (39) to interpret the two distinct regions observed in Al-Fe alloy melt-spun strip of 100 μm thickness. Figure 27a shows calculated growth velocity, assuming 75 K undercooling at the start of solidification, and Figure 27b shows calculated dendrite (or cell) tip radius. Note that this radius increases by nearly an order of magnitude over the first 20 μm of growth (as undercooling is dissipated) and stays relatively constant as further growth is limited by heat transfer to the surroundings. The points in Figure 27b indicate experimental estimates of tip radius and agree quite well with calculations. In this case the very narrow freezing range of the Al-Fe alloy, combined with a Biot number that was not too low, apparently permitted the fine structure to be retained. In other instances, if ripening takes place more uniformly throughout a specimen, it may not be possible to distinguish as readily between the two regions of solidification.

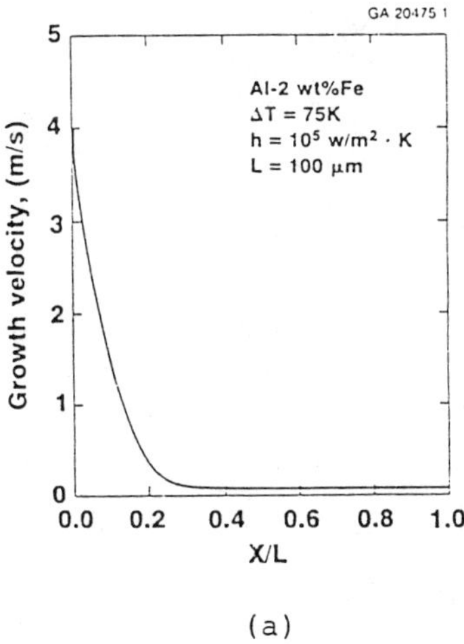

(a)

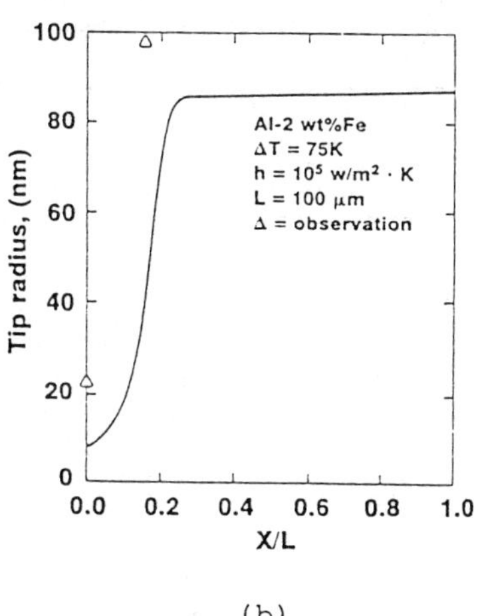

(b)

Figure 27. Dendrite tip velocity and tip radius from melt-spun Al-2wt%Fe alloy. Curves are calculated from Lipton, Kurz, and Trivedi.[40] Triangular points on Figure 27b are experimental. From Chu, Granger, and Ludwiczak.[42]

References

1. E. Scheil, Z. Metallkunde 34, 70 (1942).
2. M.C. Flemings, Solidification Processing, McGraw-Hill Book Company, New York (1974).
3. H.D. Brody and M.C. Flemings, Trans. AIME 236, 615 (1966).
4. T.F. Bower, H.D. Brody, and M.C. Flemings, Trans. AIME 236, 624 (1966).
5. I. Ohnaka, Trans. ISIJ 26, 1045 (1986).
6. T.W. Clyne and W. Kurz, Met. Trans. 12A, 965 (1981).
7. S. Kobayashi, J. Crystal Growth 88, 87 (1988).
8. K.S. Yeum, V. Laxmanan, and D.R. Poirier, Met. Trans. 20A, 2847 (1989).
9. D.H. Kirkwood, Mat. Sci. Eng. 65, 101 (1984).
10. A.J.W. Ogilvy and D.H. Kirkwood, Appl. Sci. Res. 44, 43 (1987).
11. A. Mortensen, Met. Trans. 20A, 247 (1989).
12. D.J. Allen and J.D. Hunt, Met. Trans. 7A, 767 (1976).
13. D.J. Allen and J.D. Hunt, Solidification and Casting of Metals, p. 39, The Metals Society, Book 192, London, (1979).
14. R. Riedl and H.F. Fischmeister, Met. Trans. 21A, 264 (1990).
15. A.B. Michael and M.B. Bever, Trans. AIME 200, 47 (1954).
16. M. Basaran, Met. Trans. 12A, 1235 (1981).
17. A. Roosz, Z. Gacsi, and E.G. Fuchs, Acta Met. 32, 1745 (1984).

18. J.A. Sarreal and G.J. Abbaschian, Met. Trans. 17A, 2063 (1986).
19. E. Halder, A. Roosz, H.E. Exner, and H.F. Fischmeister, Mat. Sci. Forum 13/14, 547 (1987).
20. H. Jones, J. Mat. Sci. 19, 1043 (1984).
21. T.Z. Kattamis, J.C. Coughlin, and M.C. Flemings, Trans. AIME 239, 1504 (1967).
22. T.W. Clyne, Mat. Sci. Eng. 65, 111 (1984).
23. M. Rappaz, Int. Materials Rev. 34, 93 (1989).
24. A.J. Campagna, PhD Thesis, Department of Materials Science and Engineering, Massachusetts Institute of Technology (1970).
25. M.C. Flemings and Y. Shiohara, Tetsu-to-Hagane 71, A204 (1985).
26. P. Mathur, D. Apelian, and A. Lawley, Acta Met. 37, 429 (1989).
27. M.C. Flemings and K.P. Young, Yearbook of Science and Technology, p. 49, McGraw-Hill Book Company, New York (1978).
28. J.M.M. Molenar and W.H. Kool, J. Mat. Sci. 24, 1782 (1989).
29. A. Mortensen, L.J. Masur, J.A. Cornie, and M.C. Flemings, Met. Trans. 20A, 2535 (1989).
30. L.J. Masur, A. Mortensen, J.A. Cornie, and M.C. Flemings, Met. Trans. 20A, 2549 (1989).
31. A. Mortensen, J.A. Cornie, and M.C. Flemings, Met. Trans. 19A, 709 (1988).
32. W. Kurz, B. Giovanola, and R. Trivedi, Acta Met. 34, 823 (1986).
33. W.W. Mullins and R.F. Sekerka, J. Appl. Phys. 34, 323 (1963).
34. W.W. Mullins and R.F. Sekerka, J. Appl. Phys. 35, 444 (1964).
35. M.J. Aziz, J. Appl. Phys. 53, 1158 (1982).
36. B. Giovanola and W. Kurz, Met. Trans. 21A, 260 (1990).
37. W.J. Boettinger, L.A. Bendersky, S.R. Coriell, R.J. Schaefer, and F.S. Biancaniello, J. Crystal Growth 80, 17 (1987).
38. W. Kurz and B. Giovanola, J. Crystal Growth 91, 123 (1988).
39. J. Lipton, W. Kurz, and R. Trivedi, Acta Met. 35, 957 (1987).
40. Y. Wu, T.J. Piccone, Y. Shiohara, and M.C. Flemings, Met. Trans. 18A, 915 (1987).
41. J.M. Massoubre and B.F. Pflieger, AIChE Symposium Series 74, 48 (1978).
42. M.G. Chu, D.A. Granger, and E.A. Ludwiczak, Solidification Processing 1987, p. 271, Institute of Metals, London (1988).
43. M.G. Chu and D.A. Granger, Met. Trans. 21A, 205 (1990).
44. L.A. Bendersky and W.J. Boettinger, Rapidly Quenched Metals, p. 887, S. Steeb and H. Warlimont (eds.), North Holland Physics Publishing (1985).

Influence of matrix solidification during infiltration on the structure of a cast fibre reinforced alloy

Ph. Jarry, A. Dubus, G. Regazzoni
Centre de recherches de Voreppe SA (CRV), A Pechiney Group Company, BP27, 38340 Voreppe, France, Paris

V. Michaud, A. Mortensen
Department of Materials Science and Engineering, Massachusetts Institute of Technology, Cambridge, Massachusetts 02139, U.S.A.

Abstract:

Results from experiments on infiltration of alumina short-fiber preforms by an aluminum-copper alloy are presented. Composite samples were cast using industrial apparati at CRV. Variations in macrosegregation, matrix grain size and fiber volume fraction as a function of initial preform temperature are characterized and compared with predictions based on theory developed at MIT.

Introduction

Pressure casting offers considerable potential for producing fiber reinforced metals: adverse wetting is overcome and shrinkage pores are eliminated in a process that allows for high production rates. As in most casting processes, casting parameters significantly influence the final macro- and microstructure of the resulting material in a way that must be understood before the process can be used in a reliable and optimized fashion.
We present recent results from ongoing research on the pressure casting of aluminum-based alloys into chopped fiber preforms. It is known that when the initial fiber preform temperature is below the metal liquidus, substantial macrosegregation can result in the final composite [1], a phenomenon analyzed by two of the authors in a forthcoming publication [2]. In this collaborative effort, we report results from pressure infiltration experiments using apparati and procedures that replicate actual industrial processing conditions. We vary the initial fiber temperature, analyze microstructure and macrosegregation in the final composite, and investigate the applicability of theoretical analysis for the present casting conditions.

Experimental Procedure:

Fiber preforms were Saffil™ Alumina preforms, 30mm-thick and 120 mm in diameter. The initial fiber volume fraction was 20 %, and cohesion of the preforms was ensured by the use of a few wt% silica binder. The matrix was aluminum alloy AU5GT (4.4 wt% Cu, 0.27 wt% Mg, 0.18 wt% Ti, < 0.02 wt% impurities, balance Al). The pressure casting apparatus was a horizontal shot-chamber die casting machine where the fiber preform was held vertically against the die wall opposite the shot chamber. Fiber preforms were preheated to a controlled temperature (T_f) in a separate furnace, and inserted into the die chamber. Metal was then poured at 1070 K into the shot chamber, and forced by a piston into the die. The time elapsed between retrieval of the preform from the preheating furnace and subsequent infiltration was on the order of 20 s. The initial mold temperature was between 470 and 570 K depending on location within the mold. Separate investigations at CRV showed that by the time the metal penetrates the preform, it has cooled to a temperature of about 970 K. Three samples were produced, in which the fiber preforms were preheated to T_f = 973, 873 and 773 K (Samples 1, 2 and 3, respectively).

Samples were cut off from the center of the castings, where infiltration was unidirectional toward the die wall. These were solutionized for 8 hours at 798 K and quenched in water in order to eliminate microsegregation within the matrix. Samples were then polished, and examined under the optical microscope and using an Analytical Scanning Electron Microscope (ASEM) at CRV.

The ASEM was composed of a scanning electron microscope equipped with a LaB_6 filament operated in the Back Scattered Electron (BSE) mode. The ASEM was interfaced with an energy dispersive microanalyser (EDX) and with an automated image analyzer. The latter drove both the SEM functions (stage, beam, magnification) and the EDX acquisition routines.

The local surface fraction of fibers was measured with the image analyzer using a high sensitivity BSE detector to obtain good contrast between the (darker) fibers and the matrix. A relatively low acceleration voltage (15kV) was used to render the fiber-matrix interface as sharp as possible. A mask, comprising the fibers enlarged by a shadow 2 µm wide, was then created in the image analyzer. This was used to guide the electron beam away from the fibers, so as to scan only the remainder of the sampled area during chemical analysis by EDX. This procedure ensured that signal was only collected from the matrix during analysis. Data were collected in this manner from square areas of side 200 µm, once every millimeter along the infiltration direction. The acceleration voltage was 15 kV, the primary beam current was $0.15 \cdot 10^{-9}$A and data acquisition times were 100 to 200 s.

Experimental results

Composites were well infiltrated, except in a thin zone along the die wall opposite the shot chamber. A typical as-cast microstructure is included in Fig. 1, showing the matrix and small quantities of θ phase (Al_2Cu) near the fibers.

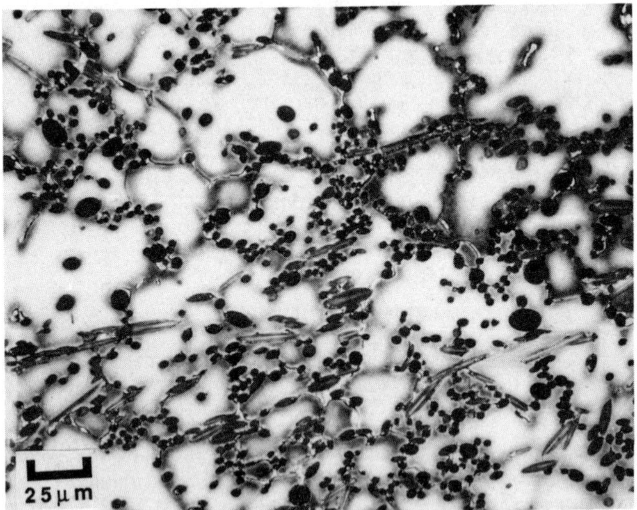

FIG.1 Typical microstructure of infiltrated composites in the as-cast condition.
Sample 2, 15 millimeters from the die wall.
Keller's etch reveals darker aeras, close to the fibers, with higher Cu concentration than away from them.

Small quantities of silicon were also detected in the matrix, generally in the vicinity of the fibers, Fig. 2.
Examination of the macrostructure of the samples showed a region rich in copper in the composite along the die wall opposite the shot chamber. Significant compression of the fiber preforms was also evident, Table I, which resulted in an increase in the final average fiber volume fraction in the composite. Compression of the preforms increased as fiber preform temperature decreased.

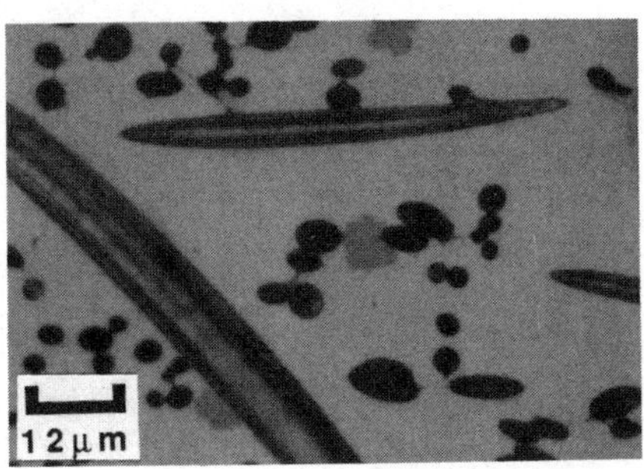

FIG. 2 Silicon (light grey) in the matrix near the fibers. Sample 2, 2 millimeters from the die wall.

Table I - Initial and final preform thickness

Sample #	Initial preform height (mm)	Final preform height (mm)	% deformation
1	30.2	24	20
2	30.2	22	27
3	30.4	20.5	33

Results from quantitative analysis are given in Figs. 3 (a) - (c). General features of macrosegregation and fiber volume fraction distribution were similar in all three samples, although values differed. One can see from the figures that a region of nominal copper composition extends from the preform side facing the gate. In this region the fiber volume fraction is lower than average and close to the initial preform volume fraction (20 %). The size of this region decreases with decreasing initial preform temperature T_f. Beyond this region, the copper concentration first drops, then rises toward the preform side facing the die wall. Concomitantly, the fiber volume fraction abruptly increases (Fig. 4) and stays roughly constant over the remainder of the composite length, extending to the die wall.

Along with fiber volume fraction and copper concentration, the grain size, revealed by Keller's etch, varies along the sample. In Sample 1, a 1 mm wide zone of equiaxed grains is present near the die wall. Thereafter, a columnar zone, of grains about 100 μm wide and orthogonal to the die wall, extends 9 mm into the composite, followed by an equiaxed zone filling the remainder of the composite and the unreinforced matrix, where the grain size is about 250 μm, Fig. 3 (a). In a sample of the same matrix, die-cast under identical conditions but without fibers, there is no equiaxed zone near the die wall, the columnar region is about 9 mm wide, is composed of large grains 1000 μm in diameter, and is followed by an equiaxed region of grain size decreasing gradually from 1000 μm to 250 μm. In Samples 2 and 3, fine equiaxed grains 20 μm in diameter extend throughout and a few millimeters beyond the copper-depleted region. The remainder of the matrix, both in the composite and beyond, is composed of equiaxed grains 250 μm in diameter, Fig. 3 (b) and (c).

Theory:

When metal is injected into a fiber preform of initial temperature below the metal liquidus or melting point, local matrix solidification takes place between the fibers during infiltration. Because the solid metal is retained by the fibers, while the liquid metal is flowing at a relatively high velocity (typically 1 to 10 $cm \cdot s^{-1}$), redistribution of solute occurs within the casting. The origin of macrosegregation in pressure cast composites is thus the same as in conventional unreinforced castings, namely fluid flow within a region where solid and liquid metal coexist [3]. Quantitatively, however, the phenomenon is quite different, since convection is forced and solidification is induced by the fibers. In a theoretical analysis of the infiltration of fiber preforms by a binary alloy (under publication [2]), equations are derived to calculate the local temperature, matrix composition and fraction solid within the composite, during unidirectional adiabatic infiltration of a rigid fiber preform under constant applied pressure. Other assumptions of this

FIG. 3a - Macrostructure, volume fraction fibers, and experimental and computed copper distribution, as a function of distance in mm from die wall for Sample 1. The dotted line on the plot of volume fraction fiber versus distance from die wall is the two-step volume fraction fiber distribution used in theoretical analysis.

The scale of the macrograph is given by the horizontal axis of the graphs.

$T_f = 700°C$

The strong temperature gradient between the mold wall and the hot preform (700°C) results in the formation of columnar grains.

The fibers's volume fraction and copper profiles are flat, along with a theoretical zero fraction solid metal. A deviation is only observed close to the die wall, where the preform must have been locally cooled down by the mold.

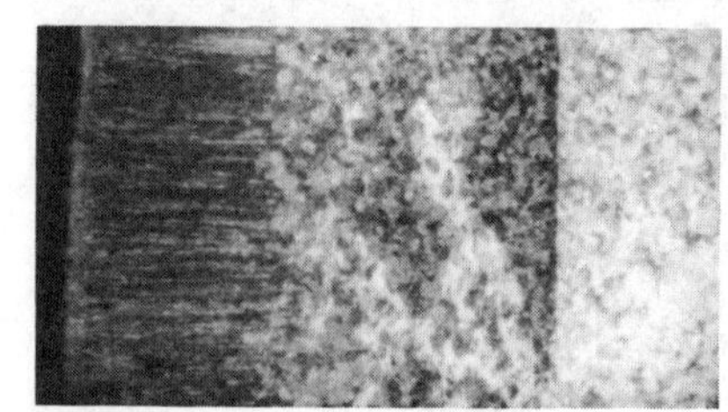

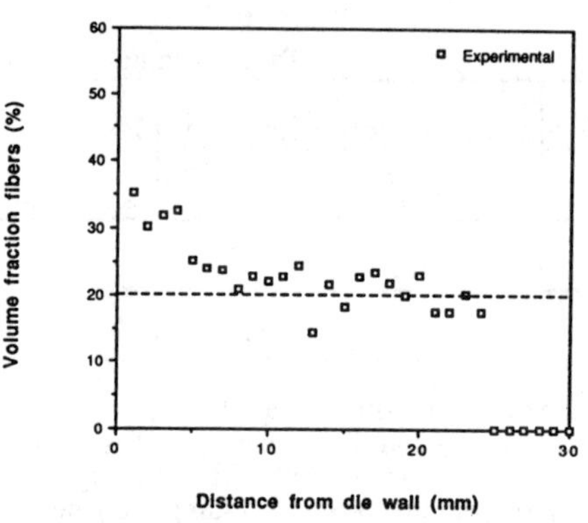

theoretical fraction solid = 0

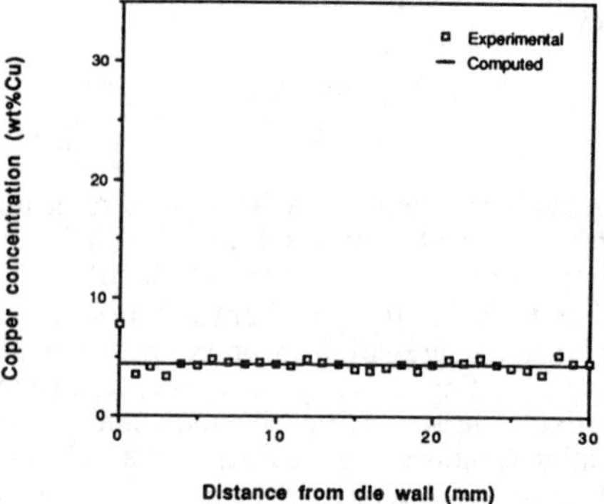

FIG. 3b - Macrostructure, volume fraction fibers, predicted volume fraction solid metal, and experimental and computed copper distribution, as a function of distance in mm from die wall for Sample 2. The dotted line on the plot of volume fraction fiber versus distance from die wall is the two-step volume fraction fiber distribution used in theoretical analysis.

The scale of the macrograph is given by thehorizontal axis of the graphs.

$T_f = 600°C$

Note the correspondence between
- the darker zone on the macrograph within the composite region,
- the higher fiber density zone,
- the computed zone with non zero fraction solid metal during infiltration,
- and the Cu depleted zone (where the Cu concentration can drop down to 3.5%).

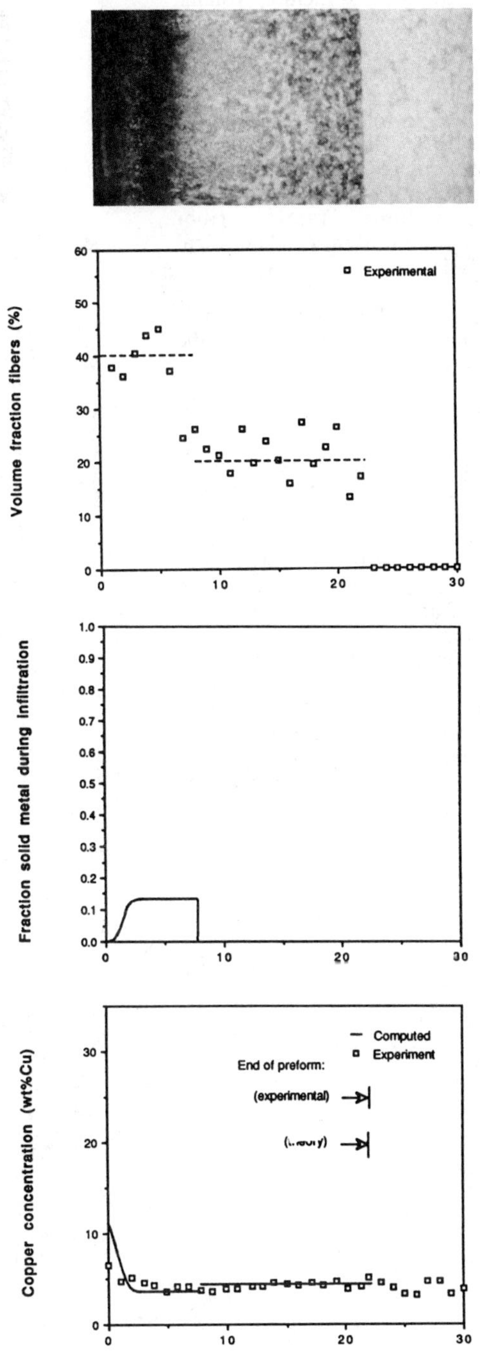

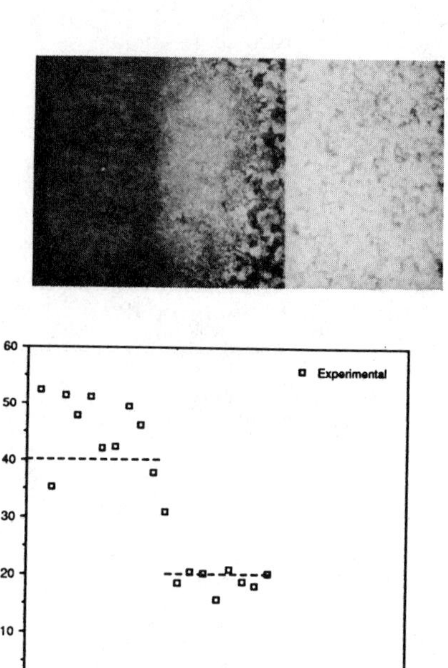

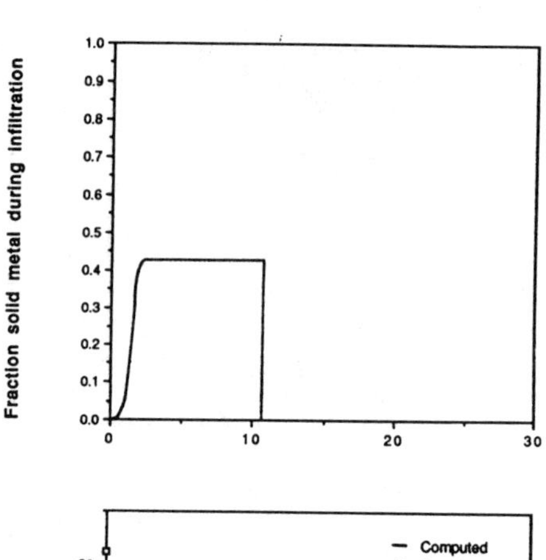

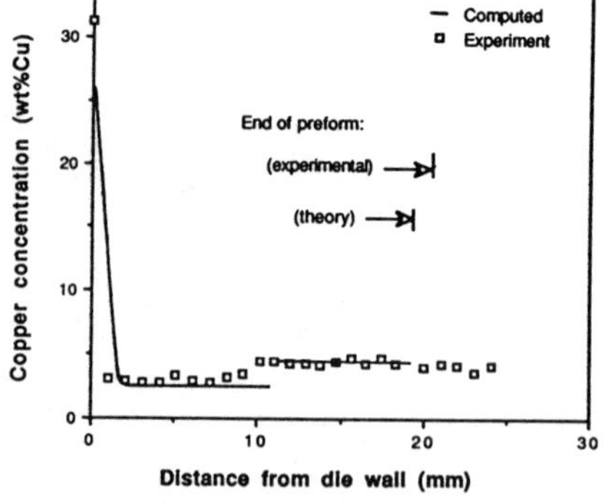

FIG. 3c - Macrostructure, volume fraction fibers, predicted volume fraction solid metal, and experimental and computed copper distribution, as a function of distance in mm from die wall for Sample 3. The dotted line on the plot of volume fraction fiber versus distance from die wall is the two-step volume fraction fiber distribution used in theoretical analysis.

The scale of the macrograph is given by the horizontal axis of the graphs.

$T_f = 500°C$

Note the correspondence between
- the darker zone on the macrograph within the composite region,
- the higher fiber density zone,
- the computed zone with non zero fraction solid metal during infiltration,
- and the Cu depleted zone.

The average Copper concentration in the depleted zone is 3%, whereas it is 4,4% (nominal concentration) in the region on the gate side.

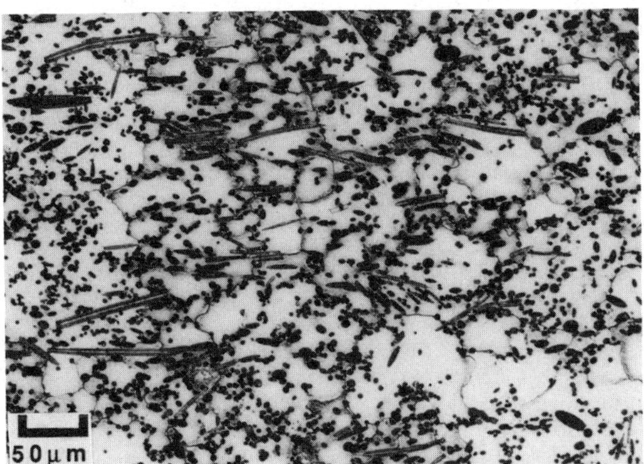

FIG.4 Microstructure of two different zones in Sample 2. The first one belongs to the region of nominal composition and the second one to the copper depleted zone, showing the transition in volume fraction fibers and in grain size (revealed by Keller's etch).

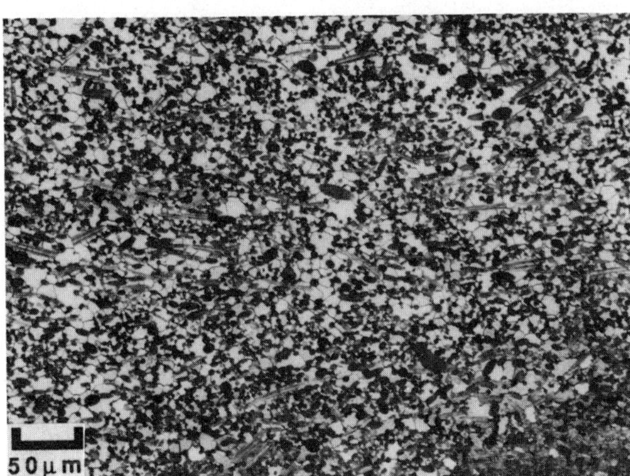

analysis are listed in Reference [2], and include equilibrium at the fiber-matrix interface, negligible diffusion of solute along the length of the composite, and negligible temperature and composition gradients on a length scale equal to one fiber diameter. A similarity method is used to reduce the mathematical complexity of the problem. Temperature, matrix composition and fraction solid profiles in the infiltrated composite are derived as functions of infiltration parameters and a single variable χ which combines position along the preform, x, and time:

$\chi = x/L$.

L is the infiltrated length at time t, and is proportional to $\sqrt{t}$ with the assumptions above. Process parameters include thermal characteristics of the fiber preform and of the alloy, the matrix phase diagram, as well as:

- T_f, temperature of the fibers before infiltration
- T_m, temperature of the metal before infiltration
- V_f, fiber volume fraction
- $\Psi = L/\sqrt{t}$, which depends on applied pressure, and is constant during unidirectional adiabatic infiltration with constant applied pressure. Experiments are in progress at MIT in which the assumptions of the model (adiabaticity, constant applied pressure) are replicated and process variables are independently controlled. Preliminary data from these experiments show good agreement with theory [4].

In the present work, neither pressure nor velocity during infiltration could be measured with sufficient accuracy because of the very short duration of the actual infiltration event compared to the entire casting cycle. It is likely that neither was constant during infiltration. In using the model, a simplifying assumption is therefore made, whereby Ψ is taken constant during infiltration, at 0.022 $m \cdot s^{-1/2}$. This estimated value is based on the average piston displacement velocity. Calculations were also conducted with values of Ψ at both extremes of its range of plausible variation (0.014 to 0.022 $m \cdot s^{-1/2}$), and results showed little variation with this parameter. The fiber volume fraction is taken as 40 %, which is the average fiber volume fraction in the compressed region of the sample (measured values seem somewhat higher in Sample 3, however, this must be due to experimental error since the total mass of fibers initially in the preforms must be conserved). Cooling of the fiber preforms by the die is neglected, and T_f is taken as the initial fiber preform temperatures given above. T_m is 973 K, as mentioned above. The matrix is approximated as a binary alloy of Al containing 4.4 wt% Cu, i.e., the small amounts of magnesium and silicon present in the matrix are ignored.
With these parameters, a uniform composition and no solid metal are predicted within the composite for Sample 1, in which T_f was above the metal liquidus (917 K). In Samples 2 and 3, matrix solidification and concomitant macrosegregation are predicted. Because T_m is above the metal liquidus, a region of remelted matrix is predicted, extending from the preform close to the gate roughly to half of its infiltrated length.

Discussion

It is clear from the experimental data that the fibers were compressed by the metal during infiltration. This occurs because viscous drag exerted by the flowing metal on the surface of the fibers transmits the applied pressure to fibers ahead of the infiltration front [5]. Similar observations have been reported by other investigators [6,7]. When the preform is fully infiltrated, metal flow stops, and the preform relaxes. Focussing our attention on Samples 2 and 3, we see that full relaxation is observed only in the region of nominal copper concentration, whereas the remainder of the sample features a fiber volume fraction far above the nominal value. Theoretical analysis predicts that there is no solid metal in the region of nominal copper content, whereas in the rest of the sample, the matrix is partly solid. This provides an explanation for the stepped fiber volume fraction distribution within these two composites: solid metal present in the segregated portion of the composite prevents relaxation of the fibers, leaving the local fiber volume fraction at roughly the value it has during infiltration when metal solidifies.
This justifies the use of the higher fiber volume fraction in the calculations. We now make the additional assumption that after infiltration with a uniformly high fiber volume fraction, the preform relaxes only in that portion of the composite where no solid metal is present (this is an approximation because the fiber volume fraction varies along the length of the preform during infiltration [5]). As the preform relaxes, more metal of nominal composition flows between the fibers to produce a composite of lower fiber volume fraction than where solid metal is present. Resulting solute profiles are superimposed on experimental data in Fig. 3. Agreement is very satisfactory, with respect both to copper content and position of the remelting front.

With Sample 1, theory agrees with experiment, except for variations in copper content and fiber volume fraction observed experimentally near the die wall, downstream of the infiltration path. We interpret these localized variations in composition and fiber volume fraction as resulting from cooling of the preform against the die wall, around 500 K, which in turn results in local solidification of the metal in the affected zone of the preform. Estimating the preform thermal diffusivity α as $4 \cdot 10^{-7}$ $m^2 \cdot s^{-1}$, and using the usual error function solution for the temperature profile in the preform before infiltration, motion of the liquidus isotherm T = 923 K is given by $x = 1.4 \sqrt{t}$, with x in millimeters and t in seconds. Since the preform is cooled for about 20 s before infiltration, a zone about 6 mm deep in the preform is expected, wherein the fiber temperature is below the matrix liquidus when infiltration starts. This is of the same order of magnitude as the observed segregated zone length in Sample 1, 5 mm, Fig.3 (a). Cooling of the preform by the die can therefore account for segregation in the composite when T_f is above the matrix liquidus.

The fine grain sizes observed in the composites are explained by the solidification that took place during infiltration. The fine-grained equiaxed regions extend farther than the solute-depleted and fiber-enriched zones. This is not observed in samples which were produced at MIT with much lower infiltration pressures and no preform compression. This is an indication that the high rate of infiltration and solidification in die casting must have left small nuclei of primary metal, initially formed by cooling from the fibers, a few millimeters into the remelted region. The macrostructure in Sample 1 is similar to that with no preform in the die, save for the equiaxed zone near the die wall. This zone formed in Sample 1, as in Samples 2 and 3, due to cooling by the fibers, and is not the chill zone encountered in ordinary castings. The similarity of the unreinforced sample with Sample 1 is an indication that the central equiaxed zone largely results from a high nucleation rate in the die, aided by the Ti grain refiner, rather than grain multiplication by convection as in most castings. Trends observed here are similar to those found by Clyne and Mason [8] and in earlier studies at MIT [9].

The presence of silicon in the matrix is an indication that the fibers (which contain 4 % SiO_2) or the silica binder are reduced by the matrix. The small magnesium addition probably accelerates this reaction, because composites produced by infiltration of similar preforms with binary Al-4.5 wt%Cu at MIT showed no silicon in the matrix under an optical microscope.

Conclusion

The theoretical analysis presented in Reference [2] successfully predicts macrosegregation and microstructural features in the infiltrated composites produced in this work, despite the use of casting equipment and procedures typical of industrial practice, which are less well controlled than laboratory experiments.
Composite pressure casting procedures are in general dictated by a compromise of (i) low initial fiber, mold and metal temperatures to minimize chemical interaction between the metal and the fibers or the mold, and (ii) sufficiently high temperatures to ensure full and rapid infiltration of the preform and minimize segregation in the composite. The interplay between these parameters and the choice of optimal casting conditions are complex, but can be modeled to optimize the structure of the resulting composite casting.

Acknowledgement:

We want to thank M. Tirard-Collet for all the metallographical work, M. Chastagner for the ASEM analysis, and MM Pluchon and Burtin for fabricating the composite materials at CRV workshops.
The present work was funded by the Centre de Recherches de Voreppe, S.A..

References

1. L.J. Masur, experimental work cited by J.A. Cornie, A. Mortensen and M.C. Flemings, Proceedings of the Sixth International Conference on Composite Materials ICCM 6, F.L.Matthews, N.C.R.Buskell, J.M.Hodgkinson and J.Morton Eds., London, Elsevier, 1987, pp.2.297-2.319.

2. A. Mortensen and V.J. Michaud, "Infiltration of Fiber Preforms by a Binary Alloy Part 1: Theory", *Metall. Trans. A*, in print.

3. M.C. Flemings, *Solidification Processing*, McGraw-Hill, Inc., New York, NY, 1974, pp. 244 - 252.

4. V. Michaud, experimental work cited by A. Mortensen, "Infiltration of Metal Matrix Composites", Proc. of the Japan-US Cooperative Science Program Seminar on Solidification Processing of Advanced Materials, Proc. Conf. May 29-June 1 1989, Oiso, Kanagawa, Japan, Japan Soc. for Promotion of Science/N.S.F., pp. 203-212.

5. A. Mortensen and J. L. Sommer, ongoing research, M.I.T..

6. H. Fukunaga, in *Cast Reinforced Composites*, Proc. Int. Symp. on Advances in Cast Metal Composites, S.G. Fishman and A.K. Dhingra Eds., Chicago, Sept. 1988, ASM International, pp. 101-107.

7. Y. Nishida, H. Matsubara, M. Yamada and T. Imai, Proc. of the Fourth Japan-U.S. Conf. on Composite Materials, June 27-29, 1988, Washington D.C., Technomic Publishing Co., Lancaster, pp. 429-438.

8. T.W. Clyne and J.F. Mason, *Metall. Trans. A*, 1987, vol. 18A, pp. 1519-1530.

9. L.J. Masur, A. Mortensen, J.A. Cornie and M.C. Flemings, *Metall. Trans. A,* 1989, vol. 20A, pp. 2549-2557.

Microstructure development during spray casting

S. Annavarapu, D. Apelian and R.D. Doherty
Department of Materials Engineering, Drexel University, Philadelphia, Pennsylvania 19104, U.S.A.

ABSTRACT

Spray casting is emerging as an attractive technology for the production of net or near-net-shaped components of a variety of materials. The process (e.g. the Osprey™ process) involves sequential atomization and droplet consolidation at deposition rates in excess of 0.25kg/s. It is possible to fabricate axisymmetric shapes such as disks, billets, tubes and sheet or strip, by suitably manipulating substrate configuration and motion. A major advantage of spray casting is that the product has a uniform distribution of equiaxed grains (20μm-200μm), no macroscopic segregation of alloying elements, a uniform distribution of second phases, and low oxide content. In this paper, the scientific aspects of the development of microstructure during spray casting are reviewed. Successful utilization of spray casting mandates control of preform shape with accompanying metallurgical integrity. Consequently, it is a critical to quantify the contribution of the mechanisms involved in the formation of preform shape, microstructure and overall yield.

INTRODUCTION

In response to increasing global competition, the traditional strategy of utilizing high plant capacity and mass production is now being replaced or complemented by flexible manufacturing methods. This new climate has resulted in a strong interest in net- or near-net shape manufacturing (NNSM) processes which are materials and energy efficient. Experts predict that NNSM practices will increase yields, improve efficiency, enhance product quality and generate higher profits [**1-3**]. Spray casting is an emerging NNSM technology which utilizes sequential atomization and droplet consolidation. Currently there are two approaches to spray casting, namely the Osprey™ Process [**4-7**] and Liquid Dynamic Compaction (LDC) [**8,9**]. The primary difference between the two approaches is in the mode of atomization. LDC utilizes high velocity pulsed gas jets from an ultrasonic gas atomizer (USGA), while the Osprey process uses a patented gas atomizer of conventional design.

Since the pioneering work of Singer in the early '70s on spray casting [**10-11**], the first (and only) viable process for bulk fabrication of spray cast preforms was developed by Osprey Metals Ltd. [**4-7**]. Initial efforts were targeted towards the production of preforms suitable for subsequent hot forging into finished shapes. Today, the Osprey™ process is being evaluated for the commercial fabrication of disks, billets, tubes and strip/sheet. The process is shown schematically in Fig. 1. The alloy charge is melted in a crucible located on top of the spray chamber with an inert gas cover to prevent oxidation of the melt. The molten alloy exits through a re-fractory nozzle in the bottom of the crucible and is atomized by nitrogen or argon into a spray of droplets (30-300μm in dia.). Liquid droplets are subsequently cooled by the gas during flight and accelerated to the substrate (collector) which is positioned 0.3-1.4m below the atomization zone. The droplets impinge and consolidate on the substrate to form a thick deposit devoid of porosity. Various preform shapes can be produced by appropriate maneuvering of the substrate beneath the spray, as shown schematically in Fig. 1.

The primary advantage of the process is that a near-net shape product can be fabricated in a single operation directly from the melt at rates in the range 0.25 - 2.5 kg/s. The microstructure is characterized by a uniform distribution of fine, equiaxed grains (20μm-200μm), no macroscopic segregation of alloying elements, a uniform distribution of second phases, low oxide content and the absence of particle boundaries **[5-7,12-16]**. Consequently, preforms produced via the Osprey process are ideally suited to subsequent hot consolidation (i.e. rolling, forging, extrusion...) operations to the final shape yielding isotropic mechanical properties which are at least comparable to those of conventional products.

Despite the apparent potential of spray casting as a NNSM process, the materials processing industry has been slow to implement this technology on a commercial scale (see Table 1). The challenge for manufacturers using the spray casting process is to consistently meet shape and property requirements. This entails a clear understanding of the mechanism of the formation of the deposit and the preform structure. In this paper, the current understanding of the spray casting process is briefly described, and recent results obtained at our laboratory, which highlight the mechanisms controlling preform microstructure, are discussed.

PROCESS FUNDAMENTALS

The spray casting process consists of the three broad stages of *spray formation*, *spray transport* and *preform consolidation*. The process incorporates a multitude of dependent and independent process parameters which govern the outcomes of each stage and the process; a comprehensive review of these critical process parameters is available in Ref **[19]**. While the preferred approach for commercial application has been empirical optimization, the analytical approach has been to correlate the influence of these parameters on the process outcome by the means of mathematical modeling based on fundamental scientific theories; the process outcome is defined in terms of the yield, shape and microstructure of the deposit.

Spray formation involves the delivery of the metal and gas to the atomization zone, and atomization of stream of metal into droplets. The primary independent parameters in this stage are the melt superheat, the metal flow rate, gas flow rate, the gas and the alloy composition. The primary output of this stage is the formation of a spray of droplets with a characteristic size and spatial distribution.

Spray transport consists of the transfer of the droplets towards the substrate with concurrent solidification via gas cooling at rates of 10^2-10^3°C/s **[20]**. Since experimental analysis of the microscopic events of droplet flight is difficult and imprecise, mathematical modeling has been extensively utilized to analyze the thermal history of the material (see Fig. 2) **[9,19,20,21]**. The primary output of this stage can be termed as the state of the spray at impact characterized by the velocity and extent of solidification of droplets prior to impact. The state of the spray at impact is determined by the velocity field of the gas jet entraining the droplets and the stand-off distance. In order to represent the state of the spray at a macroscopic level, the weighted mean of the fraction of liquid in individual droplets is utilized i.e the average fraction of liquid in the spray. The predicted variation of $\%L_{spray}$ with X is plotted in Fig.3 for the five alloys under a fixed set of parameters illustrating the effect of material properties **[19]**. If the spray (and/or the top surface of the deposit) contains a high volume fraction of solid (f_l =0), a majority of the droplets will be solidified and no coherent deposit will be formed; the process will then resemble powder production. At the other extreme, spray casting will be analogous to conventional casting if the fraction of liquid equals unity and the "ideal" fraction of liquid lies between the two extremes cited above.

Preform consolidation comprises of the consolidation of the droplets in the spray to yield a preform (i.e. deposit formation) and its subsequent solidification. Consolidation of the droplets into a preform (i.e process yield) is governed by the product of the target *efficiency* , Π_t (i.e. the fraction of the droplets reach the target) and the *sticking efficiency*, Π_s (i.e. the fraction of the droplets arriving at the substrate that "stick" to the deposit) i.e. **Yield** % = $100.(\Pi_t \, . \, \Pi_s)$. Π_t is largely a function of substrate parameters (size, motion,..) and the spatial distribution of droplet mass in the spray. Π_s is governed by the thermal and physical states of the spray and the deposit top surface at impact, and by the substrate configuration/motion.

Preform cooling and solidification at the substrate, occurs by the removal of the remaining enthalpy of the material (H_{rem}), via conduction through the deposit and convection at the top surface (see Fig. 2). The problem has been analyzed utilizing macro- and microscopic energy balances between the heat influx from the spray H_{rem}, heat extracted by the substrate Q_S, and heat extracted by the atomizing gas Q_g, to calculate the solidification time, t_f, required to remove the heat H_{rem} from the preform **[9,19-23]**. Model predictions indicate that except in the initial zone at the bottom which freezes rapidly, a preform (~15cm high) solidifies over a period of 70-100s and experiences a cooling rate < 20°C/s i.e. the small fraction of the liquid (10-40%) carried into the deposit undergoes relatively slow cooling. A partially liquid layer (f_S<1.0) forms on the surface of the preform during deposition due to the imbalance between the heat influx and the heat extraction rate.

The presence of a thick partially liquid layer allows fluid flow into interstices between presolidified droplets yielding a uniform structure throughout the thickness of the deposit. The fraction of solid in the liquid layer is estimated to be ≥0.8 **[19,20,24]**. Grain size and segregate spacing in the preform are larger than those in atomized powders but *smaller* than the predicted values (Table 2) based on empirical correlations of dendrite arm spacing and cooling rates **[20, 22,23,25]**. This overestimation of the grain size and/or segregate spacing is attributed to (i) the two-stage solidification process in spray casting, (ii) retarded coarsening of the dendrite arms at a high volume fraction of solid **[20]**, and (iii) nucleation by presolidified droplets from the spray. The parameters which determine the final microstructure are (a) the condition of the spray at impact, (b) the spatial distribution of solid particles after impact, and (c) the time required for complete solidification of the preform **[19,20]**.

EVOLUTION OF PREFORM MICROSTRUCTURE

The microstructure of spray cast deposits forms in two stages: solidification in flight and preform consolidation. Solidification in flight produces a droplet microstructure consisting of very fine dendrites as shown in Fig. 4. Subsequent to preform consolidation, the dendritic structure is transformed into one containing equiaxed grains as shown in Fig. 5. The proposed mechanisms for this transition are described and assessed. In addition, key assumptions of existing models are analyzed in the light of recent experimental results, and their implications for microstructure evolution are discussed.

Solidification in the Flight

Solidification during flight is determined by the extent of convective cooling of the droplet mass in the spray by the gas. Consequently, the critical parameters in this stage are the droplet size and spatial distribution in the spray, and the velocity flow field of the gas jet entraining the droplets. Therefore, atomization plays a pivotal role in terms of the overall solidification of droplets in flight, and is briefly discussed.

Atomization: The process of atomization is poorly understood and accurate models predicting the droplet size distribution are unavailable. Consequently, empirical correlations **[26]** between the mean droplet size and the process parameters have been utilized by fitting them to experimental data. However, in all the work to date, the size and spatial distribution of the droplets in the spray have been quantified by the measurement of powders collected downstream via invasive experimental procedures **[19]**. Droplet size distributions at conditions suitable for deposit formation in the Osprey process have been reported to have a mean droplet diameter d_m of 50-100μm with a log-normal standard deviations (d_{84}/d_{50}) of 1.8-2.0 for a range of ferrous **[20,22]** and non-ferrous alloys **[6,16]**. In contrast, the droplet sizes reported for the LDC process are smaller with a d_m of 30-50μm **[8,9]**. However, recent atomization studies using Cu-6Ti have shown that in the absence of secondary atomization (characteristic of invasive techniques), d_m is 150-200μm and σ is 1.5-1.7. Furthermore, the spatial distribution of the droplet mass and sizes in the spray is relatively invariant.

Cooling in Flight:: In most of the mathematical models describing cooling in flight, the gas flow field is assumed to approximate a decaying sonic jet of diameter equivalent to the total outlet area of the gas nozzles. The effects of turbulence of atomization, mixing of the individual jets, recirculation zones inside the chamber, collisions between the droplets and interaction of the droplets with the medium subsequent to atomization, are neglected. Some of these issues have been addressed in a recent analysis of the flow field **[27]** based on a multi-phase Navier-Stokes equation using a PHOENICS code, to predict the particle trajectories and the solidification of the droplet mass in the spray. However, there has been limited confirmation of the predictions of these models via experimental measurements.

Droplets of diameter less than a critical value d* (typically 30-125µm) are completely solidified prior to impact, and droplets greater than about 300µm dia. for Al and 900µm dia. for the other metals arrive at the deposition surface in a completely liquid state **[19]**. Droplets with intermediate diameters impact the deposition surface in a "mushy" condition; these droplets comprise a solidified dendritic skeleton, as observed from glass slides [**20**]. Therefore the initial solidification structure is dendritic with a high fraction of liquid filling the interstices. The average fraction of liquid in the spray using the new data for the size distribution of droplets in the spray is about 0.40, which is significantly higher than those predicted previously for similar conditions.

Preform Consolidation

Preform solidification is determined by a balance between the heat and mass influx from the spray, and the heat extraction by the substrate and the gas. Consequently, the critical parameters in this stage are the average fraction of liquid in the spray, the fraction of droplets in the spray participating in the deposit formation, the state of the deposit surface, and the external rate of heat extraction. Therefore, the issues affecting deposit formation are pivotal in terms of the preform solidification, and are briefly discussed. Preform solidification may occur by one or more of several competing mechanisms and, their contributions are assessed in light of new results.

Deposit Formation: Currently, only qualitative knowledge of the dependence of sticking efficiency of different droplets exists. Similarly, the functionality of target efficiency on the substrate parameters such as the tilt angle, rotational speed etc. is only qualitatively known; current predictions of preform shape and solidification account for Π_t and Π_s by the means of factors which have to be confirmed empirically **[28]**. Recent atomization studies using Cu-6Ti have shown when a substrate is allowed to intercept the core of the spray, the mass of overspray powder increases by about 10%; this increase is concentrated in the -250 sieve fraction (-75µm). Since the starting size distribution of the spray consisted of a significant proportion of large fully liquid particles and relatively few fines, it is inferred that this increase fines is due to secondary atomization of the large droplets (>300µm) on impact with substrate. Introduction of substrate motion (i.e rotation and reciprocation) exacerbated secondary atomization and further shifted the droplet size distribution towards the fines. From this evidence, it appears that the sticking efficiency of droplets is highest for droplets in the size range of 100-300 µm. Therefore, the enthalpy and average fraction of solid in the spray supplied to the deposit is given by deducting the contributions of the droplets that did not consolidate. If we assume that the increase in the mass of droplets in the overspray is primarily due to secondary atomization of large liquid droplets (200-400µm), then the spray enthalpy decreases by 12% i.e. starting average fraction of solid in the deposit is 0.28.

Preform Solidification: Upon consolidation, the material added to the top surface of the deposit consists of about 70% solid in the form of fine dendrites wetted by a film of liquid. Therefore, the evolution of the solidification microstructure is driven by the reduction of surface energy of the dendritic nuclei via coarsening, under an imposed rate of heat extraction. This can occur by one or more of the several competing mechanisms which are described in Fig 6 and discussed in chronological order in the following section.

Thermal Equalization: On consolidation, droplets that have undergone solidification independent of one another are instantaneously unified. Fine droplets are fully solid and arrive at the deposit at an enthalpy lower than the average spray enthalpy. At the other extreme, large droplets are fully liquid and land on the deposit with an enthalpy higher than the spray enthalpy. Mushy droplets contain liquid in an undercooled state and have an intermediate enthalpy. Therefore, the consolidated material is initially in a non-equilibrium thermal state and will move towards equilibrium by heat flow over ~100-200μm from the liquid to the center of the dendritic nuclei. The time scale for thermal homogenization was calculated to be in the range 10 to 100μs **[25]**. This transition to equilibrium may occur by diffusion-solidification with the dissolution of fine particles and growth of coarse particles decreasing the fraction of liquid. However, the extent of the change in fraction of liquid has not been demonstrated and will only become significant when the fraction of liquid in the spray is less than 0.2.

Dendritic Coarsening at High Fraction of Solid: The physical state of the consolidated material can be described as a mixture of fragments of dendrites from the solid and mushy droplets in the spray, and liquid metal. The dendritic fragments have a fine structure (secondary dendrite arm spacing = 0.5-2μm) **[6,19]** i.e. a large surface area and energy. Therefore, on consolidation the dendritic structure will have a strong driving force to coarsen into spheroids, in order to reduce the surface area and, hence, the surface energy. The time scale of coarsening of the dendrite arms to form spheroids assuming $t^{1/3}$ coarsening **[29]** is about 10-100ms. As the preform continues to cool, the spheroids will continue to coarsen dendritically as long as a continuous film of liquid wets the dendritic cells. Recent experiments of isothermal coarsening of Al-7Cu indicate that when the fraction of liquid decreases below ~0.15, this liquid film breaks down forming separate pools of liquid. Subsequently, the coarsening rate i.e. increase in segregate spacing occurs by Ostwald ripening of these liquid pools via diffusion of the solute through the solid separating them. Since $D_S << D_L$, coarsening rate will decrease dramatically arresting any further change in cell spacing. Additional experiments are currently being conducted to confirm these preliminary results.

SUMMARY

- Microstructure development during spray casting occurs in two stages: droplet solidification in flight and preform consolidation. The state of the droplets/spray at impact is primarily determined by the size distribution of droplets produced by atomization, and material parameters such as density, melting point and heat contained within the freezing range of the alloy. Measurements via non-invasive methods indicate that the size distribution of the droplets in the spray is relatively coarse so that the average fraction of liquid in the spray at a flight distance of 400mm is ~0.4.

- Preform yield is given by the product of the target and sticking efficiencies. Sticking efficiency is highest for droplets in the size range of 100-300 μm. Droplets of diameter larger than 300μm contain a high proportion of liquid and underego secondary atomization on impact with substrate. The average fraction of liquid in the consolidated material is therefore lowered to ~0.28.

- Preform solidification involves thermal equalization of the consolidated material and, the coarsening of the dendrites from the solid and mushy droplets in the spray. Dendritic coarsening is dominant until the fraction of liquid decreases below ~0.15 after which coarsening is arrested yielding the typical equiaxed microstructure.

REFERENCES

1. P. W. Wright: *Materials and Design*, Vol. 8, 3, (1987)

2. Net Shape Technology in Aerospace Structures, I-IV, National Academy Press, Washington, D.C., (1986)

3. D. Apelian: "Innovations in Materials Processing", G. A. Bruggeman and V. Weiss, eds., Plenum Press, New York, NY. (1983)

4. Osprey Metals Ltd., U.K. Pat.No.147239

5. A.G. Leatham, W. Reichelt and O. H. Metelmann: "Near Net Shape Manufacturing Processes", eds. P.W. Lee and B.L. Ferguson, ASM International, Metals Park, OH, p. 259 (1988)

6. R.W.Evans, A. G. Leatham and R. G. Brooks: *Powder Metallurgy*, Vol. 28, 1, p.13 (1985)

7. D. Apelian, G. Gillen and A. G. Leatham:

"Processing of Structural Metals by Rapid Solidification", eds. F.H. Froes and S.J. Savage, ASM International, Metals Park, OH, USA, p.107 (1987)

8. E. J. Lavernia and N. J. Grant: *Metal Powder Rep.*, Vol. 4, p.255 (1986)

9. E. Gutierrez-Miravete, E.J. Lavernia, G.M. Trapaga, J. Szekely and N. J. Grant: *Metall. Trans.*, Vol. 20A, 1, p.71 (1989)

10. A.R.E.Singer: *J. Inst. Metals*, Vol. 100, p.185 (1972)

11. A.R.E. Singer: "Casting of Near Net Shape Products", eds. Y. Sahai, J. Battles, R.S. Carbonara, and C.E. Mobley, TMS, Warrendale, PA 15086, USA, p.245 (1988)

12. A.G. Leatham, A.J.W. Ogilvy, P. F. Chesney and O. H. Metelmann: *Modern Developments in Powder Metallurgy*, Compiled by : P. U. Gummeson and D. A. Gustafson, Metal Powder Industries Federation, Princeton, NJ, Vol. 19, p.475 (1988)

13. R. H. Bricknell: *Metall. Trans.*, Vol. 17A, 4, p.583 (1986)

14. H.C. Fiedler, T.F. Sawyer and T.F. Kopp: "Spray Forming-An Evaluation Using IN718", General Electric Technical Information Series, 86CRD113, May (1986)

15. A. Moran and W.A. Palko :"Progress in Powder Metallurgy", eds. C. L. Freeby and H. Hjort, Metal Powder Industries Federation, Princeton, NJ, 43 (1987)

16. J. Duszczyk et al.: op. cit. 12, p.489

17. J.F. Faure and L. Ackermann: op. cit. 12, p.425

18. Olin Corp.: private communication, H. Cheskis (1989)

19. P. Mathur, S. Annavarapu, D. Apelian and A. Lawley: *JOM*, 41, 10, p.23 (1989)

20. P. Mathur, D. Apelian and A. Lawley: *Acta Metall.*, Vol. 37, 2, p.429, (1989)

21. L.H. Kallien, P.N. Hansen, and P.R. Sahm: "Modeling and Control of Casting and Welding Processes IV", eds. A. F. Giamei and G. G. Abbaschian, The Minerals, Metals and Materials Soc., Warrendale, PA, p.543 (1988)

22. S. Annavarapu, A. Lawley and D. Apelian: *Metall. Trans*, Vol. 19A, 12, p.3077 (1988)

23. E. Gutierrez-Miravete, G.M. Trapaga and J. Szekely: "Casting of Near Net Shape Products", eds. Y. Sahai, J.E. Battles, R.S. Carbonara and C.E. Mobley, TMS, Warrendale, PA, p.133, (1988)

24. P.S. Grant, W.T. Kim, B.P. Bewlay and B. Cantor: *Scripta Metall.*, 23, p.1651 (1989)

25. M.C. Flemings: Solidification Processing, McGraw Hill, N.Y. (1974)

26. R.K. Upadhyay and H.S. Spacil: *Surface & Coatings Tech.*, 37, p.379 (1989)

27. P.S. Grant, S. Rogers, B. Cantor and L. Katgerman: presented at the TMS Annual Meeting at Anaheim, Feb. 18-22 (1990)

28. P. Mathur, S. Annavarapu, D. Apelian and A. Lawley: *J Mat. Sci.& Eng.*, In Press.

29. T.Z. Kattamis, J.C. Coughlin and M.C. Flemings: *Trans. AIME*, 239, 1504, 10 (1967)

ACKNOWLEDGMENTS

This study was supported by the Office of Naval Research. The authors wish to thank Dr. Pravin Mathur for many valuable discussions and to extend their appreciation to Robert McCormick and Ronesh Dalal from Howmet Corporation for assistance with experiments.

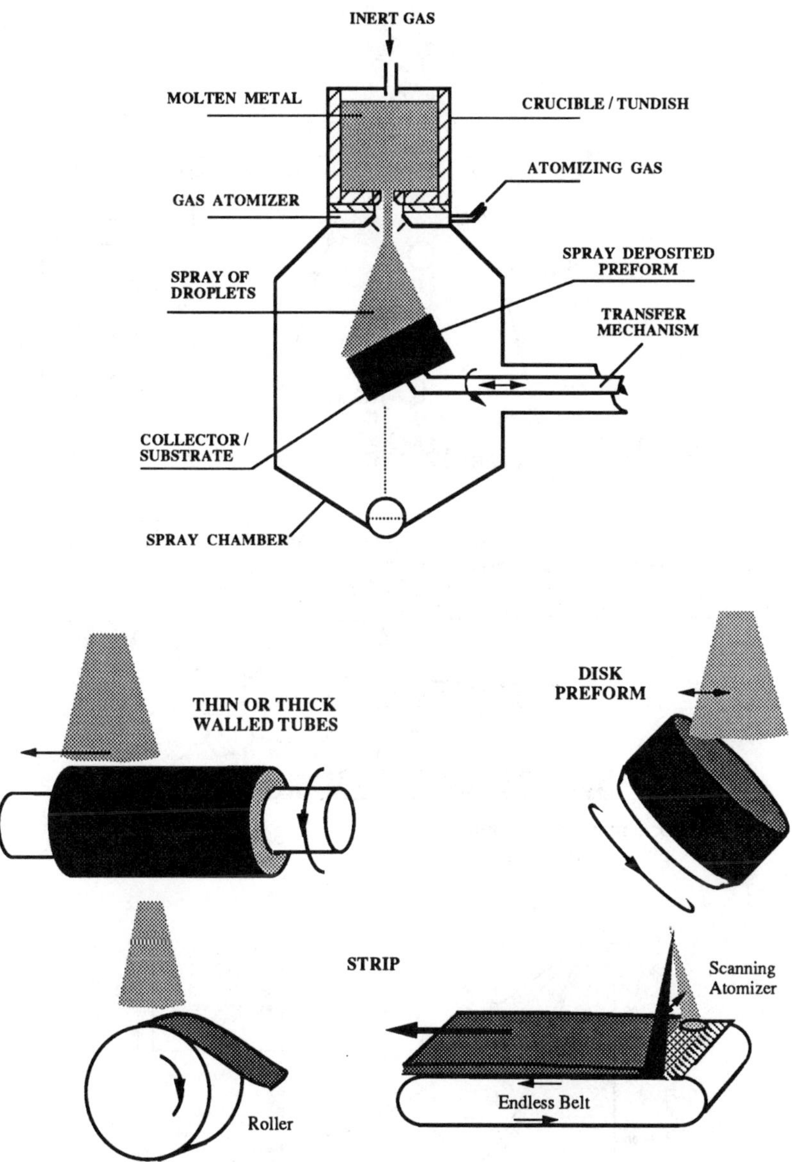

Fig.1 : Schematic of the Osprey™ process and possible preform shapes obtained by manipulation of the substrate configuration/motion

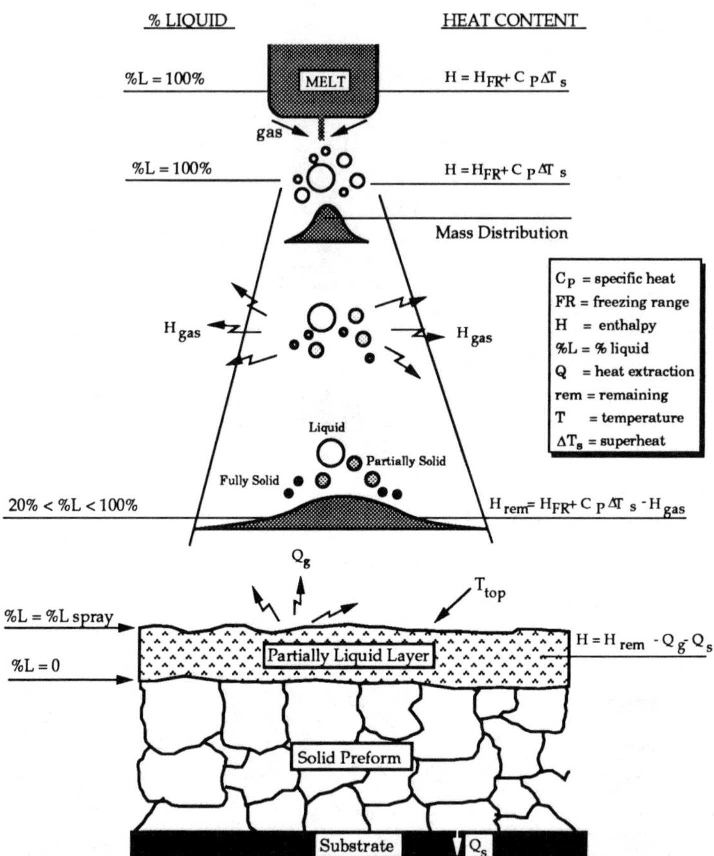

Fig. 2 Schematic representation of spray casting showing the physical and thermal states of the spray and of the deposit

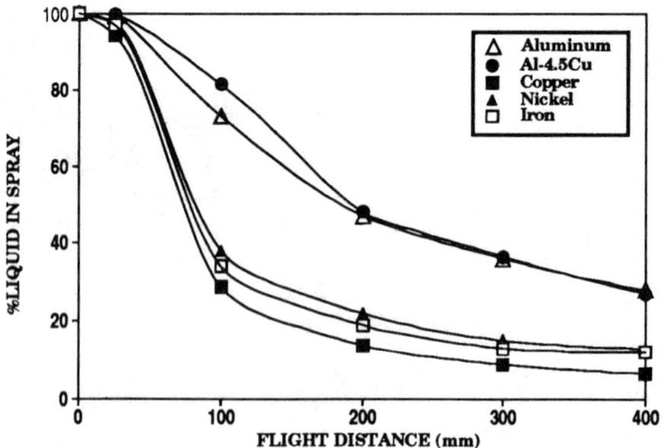

Fig. 3 Predicted variation of % liquid in the spray with flight distance showing the effect of material properties; nitrogen gas [19]

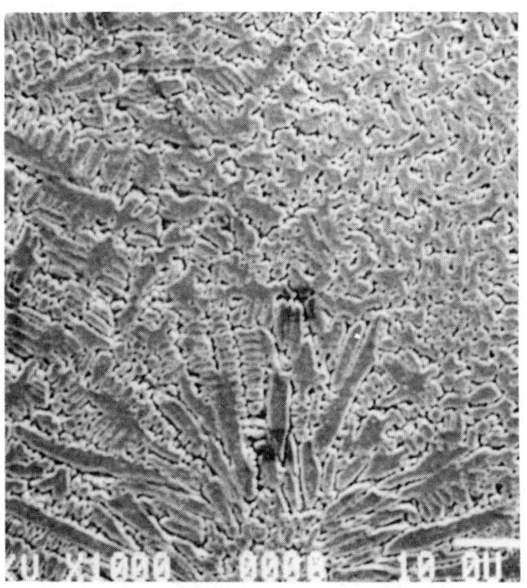

Fig. 4. Micrograph depicting the typical fine dendritic microstructure of a 100μm droplet of Cu-3Ti.

Fig. 5. Micrograph depicting the typical final microstructure of equiaxed grains in a Cu-3Ti preform.

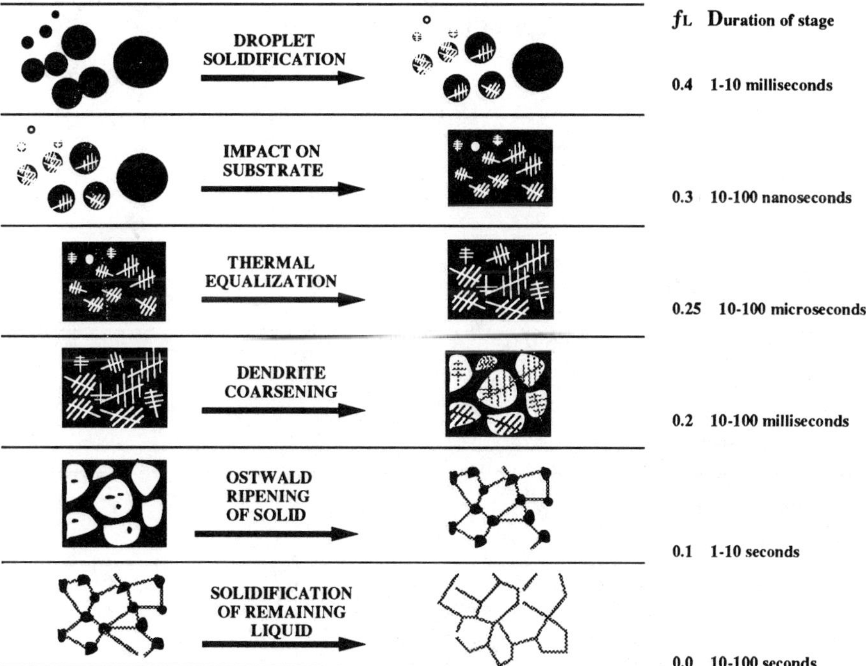

Fig. 6 Schematic illustrating the various proposed mechanisms involved in the evolution of the final microstructure during sprray casting

TABLE 1. List of Osprey Licensees for Commercial Application

MANUFACTURER	MATERIAL	APPLICATION	STAGE OF DEVELOPMENT	REF.
Billets/Disks:				
Osprey Metals Ltd.	Aluminum matrix composites w/ SiC	Automotive	Process & property evaluation	5,12
Alcan	Aluminum matrix composites w/ SiC	Automotive	Process & property evaluation	5,12
Howmet Corp.	Nickel based super alloys	Aerospace	Process & property evaluation	7
General Electric	Nickel based super alloys	Aerospace	Process evaluation	14
Alusuisse	Aluminum alloys	Automotive	Process evalaution	16
Pechiney	Aluminum alloys	Automotive	Process & property evaluation	17
Sheet/Strip:				
Mannesmann-Demag	Ferrous alloys	Structural	Process evaluation	5,12
Olin Corp.	Non-ferrous alloys	Electrical	Process & property evaluation	18
Tubes/Rolls:				
Sandvik	Stainless Steel	Structural	Semi-commercial production	5,12
Sumitomo	Tool steel	Structural	Process evaluation	5,12

TABLE 2. Comparison Between the Predicted and Measured Cell Sizes for Fe-20Mn. [20]

Deposit thickness, z (mm)	tf (s)	Predicted Cell Size (μm)	Measured Cell Size
20	50	115	40
40	110	140	57
80	270	175	79
140	20	91	13

Development and application of OCC technology

A. Ohno
Department of Metallurgical Engineering, Chiba Institute of Technology, Chiba-ken, Japan
H. Soda
Department of Metallurgy and Materials Science, University of Toronto, Toronto, Ontario, Canada

Introduction

A study undertaken in 1978 to create a casting with no equiaxed crystals was the starting point for the development of a new continuous casting system. Since the patent application was granted in late 1980, OCC process technology has advanced rapidly and the advantages of the process have been quickly recognized by the metal process industries in Japan. At the present time, over thirty industrial members including seven licensees have a strong interest in the OCC technology. Research and development work is being carried out from various standpoints:

- new product development
- product - process integration
- process route improvement
- process step elimination
- evaluation and characterization of the OCC products

The aim of this paper is to review and introduce some of the ongoing research and development activities as well as commercial developments in Japan and Canada.

Principle and General Properties

As has been described elsewhere (1,2,3,4), the key difference between the OCC and conventional continuous casting process lies in the mold concept. The OCC process involves a heated mold as opposed to the conventional water cooled mold in which solidification takes place. In the conventional system, crystals nucleate on the mold surface and as casting progresses, crystals grow from the mold surface towards the inner part of the casting, and the direction of crystal growth is perpendicular to the casting direction. In the OCC process, there is no nucleation on the mold surface since the mold is heated above the solidification temperature of the metal to be cast. At the start of casting, a retractable dummy bar is placed at the mold exit to solidify the first portion of molten metal with which it comes in contact, when metal is fed into the mold. The dummy bar is then slowly withdrawn taking into consideration such parameters as cooling intensity and mold temperature. The cooling water is directly applied only to the cast product, thus differentiating the direction of heat flow in the OCC process from that in the conventional process. This gives rise to the change in the solidification morphology which influences the metallurgical and mechanical properties of the solidified materials.

During the last several years, attention has been given to the application of the process to the casting of various metals and alloys, and at the same time, an interest has centered on the characterization of OCC products. The major features of the process include the abilities to:

- produce a single crystal or unidirectionally solidified castings
- produce net or near net shape castings
- produce a clean surface with no witness marks
- produce cast products with fewer cavities and porosity defects
- provide good workability
- provide good fatigue and corrosion resistances
- provide a compact melting and casting system

Casting Configurations

The OCC process was first applied to the vertical downward casting system, since many of the conventional continuous casting systems in the metal industry have a vertical configuration and even laboratory experiments of unidirectional solidification use a vertical system which solidifies from bottom to top.

After a number of casting trials using tin, this system was abandoned because of technical difficulties. It required a considerable amount of control to have a continuous casting without breakout. However, it proved that the principle was valid.

The OCC process has been developed stage by stage during the past decade and is continuously developing into new phases. During the first stage of development between 1980 and 1983, much attention was given to the development of casting configurations, and casting trials of rod, tube and sheet were carried out using low melting point metals such as tin, zinc and aluminum. The sequence of development on casting configuration is illustrated in Figure 1.

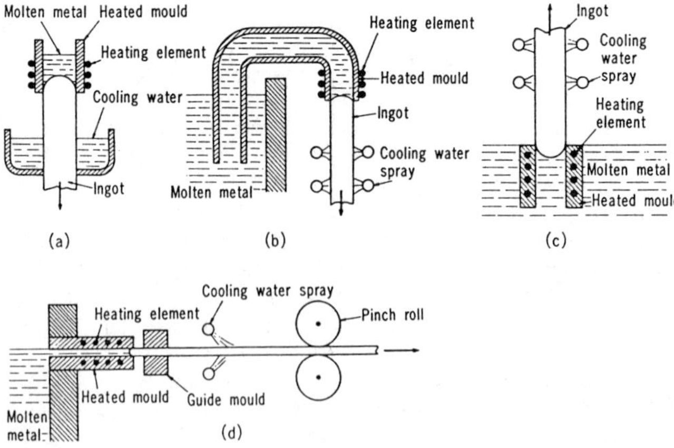

Figure 1. Sequence of development of casting configurations.[5]

It became apparent that the OCC technology provides virtually a new kind of cast product, and it was during this period that the first industrial application was sought. The principle adopted by the majority of companies was the horizontal casting configuration. Mitsui Engineering and Shipbuilding Co. adopted the vertical downward casting system, which was considered to be the most difficult method, and developed the prototype compact casting machine in 1987 to the stage where a net shape cast tube could be produced. Vertical upward casting equipment was also constructed by Osaka Fuji Kogyo Co. for the casting of high melting point materials. The vertical upward and downward OCC equipment for laboratory use is shown Figures 2 and 3.

Figure 2. Vertical upward OCC apparatus (OCC Research Center)

Figure 3. Vertical downward OCC apparatus (Mitsui Engineering and Shipbuilding Co.)

The development of OCC strip casters - heated rotary mold casting and heated stationary mold casting - is underway at the Chiba Institute of Technology. In these processes, the mold is also heated above the solidification temperature of the metal to be cast. A schematic illustration of these casting arrangements is shown in Figures 4 and 5. Investigations are ongoing to establish a basis for application of these processes to the casting of unidirectionally solidified crystal and single crystal strips.

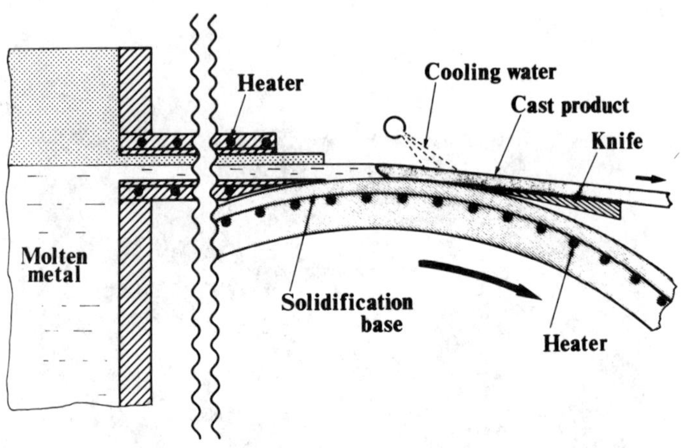

Figure 4. Schematic illustration of the heated rotary mold continuous casting configuration for strip.

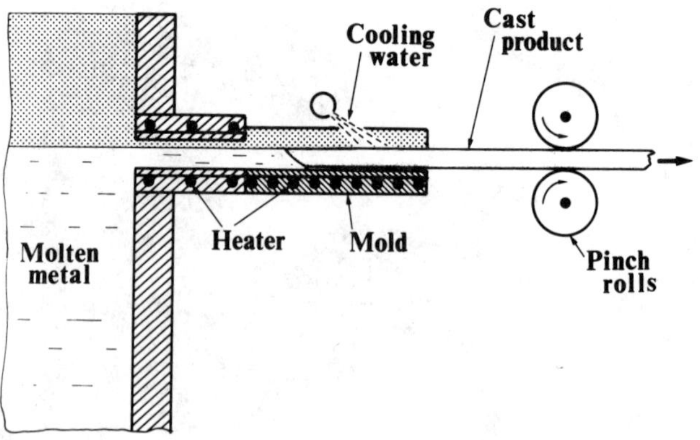

Figure 5. Schematic illustration of the heated stationary mold casting arrangement for strip.

Development of Mold Configurations

With the OCC process, the cross sectional shape of castings is defined by the mold outlet design. In the early stage of development of the OCC process, the casting of rod, tube and sheet were performed using the corresponding mold configurations shown in Figure 6.

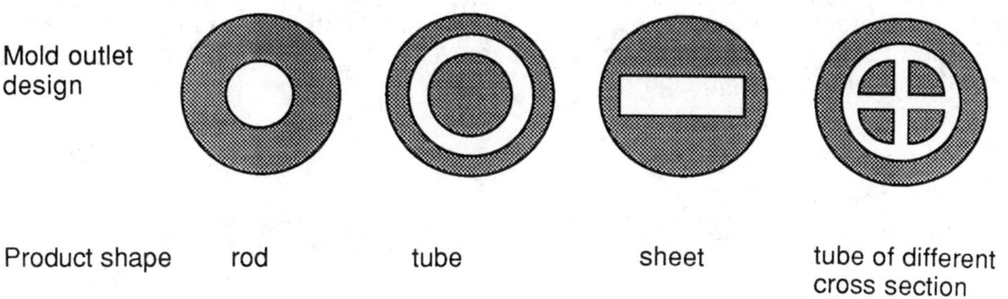

Figure 6. Examples of mold outlet design

The casting trials proved that the process not only provides a unidirectional solidification morphology and extremely clean surface condition but also gives good dimensional stability with respect to the mold outlet profiles. The development of the net shape casting of wires, the diameter of which ranged from 0.5 to 1.5 mm, during 1986 and 1987, and tubes of complex cross sectional design in 1988 were a major advance in the OCC technology. Examples of cast copper tubes having a complex cross sectional design are shown in Figure 7. One example of mold outlet design for these cast tubes is shown in Figure 6.

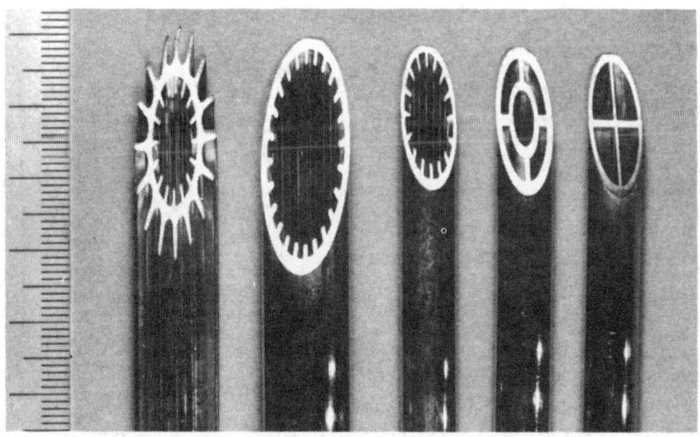

Figure 7. Examples of cast tubes having complex cross sectional shape. (Courtesy of Mitsui Engineering and Shipbuilding Co.)

Commercial Development and Application

First commercial development of the OCC product was an aluminum alloy bonding wire of 25 μm in diameter by Nippon Light Metals Co. This was made possible by the fact that cast aluminum alloy rod containing 1% silicon produced by the OCC process has a high surface quality and contains fewer or no grain boundaries. These two factors greatly contributed to the ability to withstand severe strain imposed by the drawing operation. Following this development, the most successful commercial application of the process was the casting of copper rods and wires of different diameters by Furukawa Electric Co. and Sumitomo Electric Co. in 1985 and 1986. Typical examples of copper wires of different sizes cast by the OCC process are shown in Figures 8 and 9.

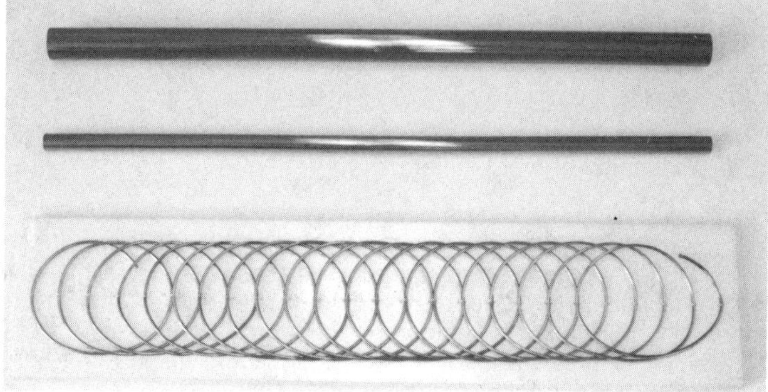

Figure 8. Examples of cast copper wires, 15 mm, 8 mm and 1.5 mm in diameters respectively.

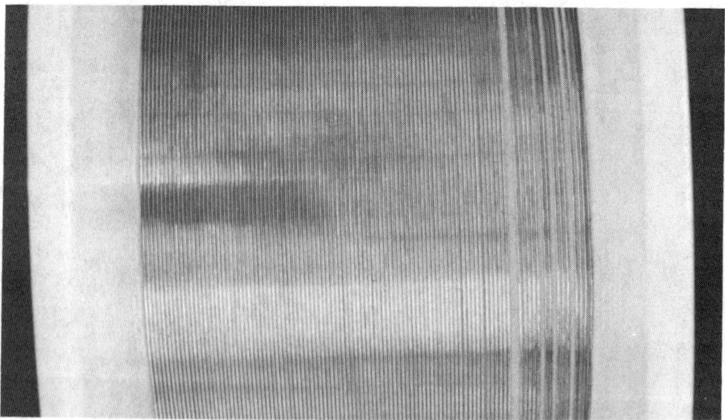

Figure 9. Example of cast copper wire, 0.8 mm in diameter

The figures show the net and near net shape casting capability of the OCC process, and the successful implementation of this casting technique in their production has produced what can be considered as virtually new and high quality products which are widely used in the audio and video cable industry. Some examples of final commercial products are shown in Figure 1

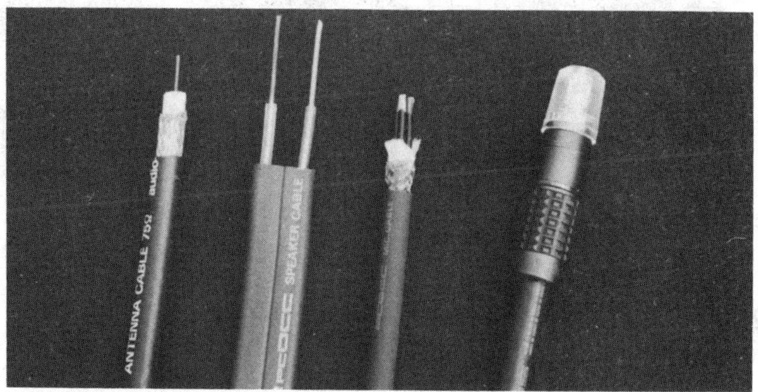

Figure 10. Examples of commercial products of audio and video signal cable.

Recently, a casting of 18 karat gold wire was performed using the OCC technique and evaluations of products were carried out by Ijima Kin-Gin Kogyo Co., one of the precious metals and jewelry suppliers. The cast 18 karat gold alloy wire of 8 mm in diameter was cold rolled followed by the drawing of wire to 0.25 mm in diameter. There was total elimination of breakage of wire arising from the drawing operation, while the conventional polycrystalline material often broke when the diameter reached about 0.7 mm during the operation. When the 0.25 mm wire processed from the OCC cast rod was made into the chains, there was no difficulty observed in welding of chain links, whereas the wire produced from the conventionally cast material exhibits chain linkage defects arising from the immediate opening of the chain links upon application of heat to the chain during the welding operation. Examples of the joint defects which occurred in a material produced from coventionly cast rod and the chain link showing the good bonding which was made from the OCC material are shown in Figure 11.

When these types of defects occur, the automatic welding of chainlinks becomes impossible and the chainlinks have to be welded by hand or the entire lot of chain has to be scrapped leading to considerable loss in production time. The company is now contemplating the implementation of the OCC process in their jewelry making operation at the scheduled new plant location in Tsukuba City on the outskirts of Tokyo.

The 18 karat gold wire of 8 mm in diameter cast by the OCC process and the defect free chain necklace produced using the 0.25 mm diameter wire drawn from the cast 8 mm diameter wire of the OCC material are shown in Figure 12 (courtesy of Ijima Kin-Gin Kogyo).

The OCC technique provides high quality cast products as well as the ability to cast a small quantity of products economically because of the relatively compact size of the integrated melting and casting configuration in addition to its near net and net

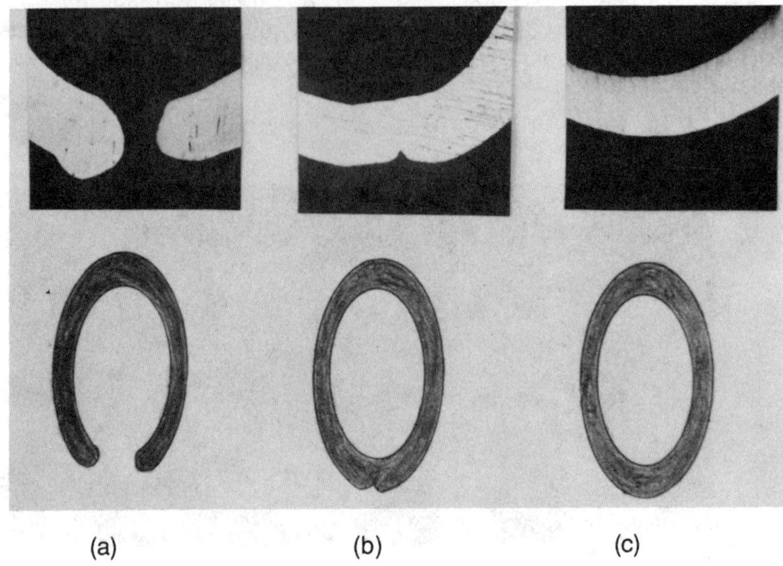

Figure 11. Examples of chainlinks showing defects (a) and (b) in a conventional material and good bonding (c) in the OCC material. (Courtesy of F. Morozumi, Ijima Kin-Gin Kogyo)

Figure 12. (a) 18 karat gold rod of 8 mm in diameter produced by the OCC process
(b) Final commercial product: a gold necklace produced from the 0.25 mm wire.

shape casting capability. These features of the OCC technology are well suited for the casting of aluminum wire with a purity of 99.9996% to be used as vapour deposition material. It is also being considered to cast copper wires of 99.999% or higher purity by the OCC process.

Sn-Bi eutectic solder wire is also cast by the OCC technique at Osaka Fuji Kogyo Co. While in conventional production, the Sn-Bi alloy billet is cast and subsequently extruded slowly into a final wire form of 0.7 mm in diameter with an extrusion rate of about 1.5 kg per day. In this regard, the OCC process offers a good alternative process route.

A series of aluminum alloy wires (Al-Cu, Al-Mg and Al-Mn), for arc welding applications have been cast by the OCC process. These alloys are used for hardfacing and weld depositing as well as repairing the cast products. Efforts are being made to commercialize these products by Osaka Fuji Kogyo Co.

Research and Development

To date, various metals and alloys have been cast by the OCC process. The casting capability of the process has expanded gradually starting with low melting point metals such as tin, zinc and aluminum. At present, the systems in use in industry are capable of casting wires, rods and sheets in a range of aluminum, copper and their alloys. Application of the OCC process is now moving towards high melting temperature materials.

At the OCC Research Center, a division of Osaka Fuji Kogyo Co., casting trials have been carried out for rods and sheets of cobalt base alloys which have a wide spectrum of end uses ranging from welding to dental applications due to their good wear, heat and corrosion resistance properties. Very encouraging results have been obtained. Examples of rod and plate of cobalt alloys cast by the OCC process are shown in Figure 13. Work is ongoing to cast small wires of different diameters which have the potential to provide new products. Casting trials of Ni-Ti, Ni-Al alloys and pure silicon are also being considered.

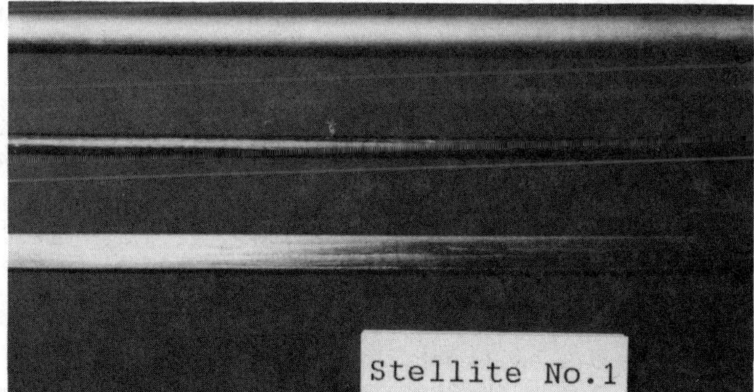

Figure 13. Examples of rod and plate of cobalt alloy cast by the OCC process (courtesy of OCC Research Center Inc.)

The OCC Research Center has succeeded in casting pure magnesium single crystal sheet 75 mm wide x 5 mm thick and a pilot project is currently undeway to cast sheets of 150 mm x 10 mm. The horizontal OCC system for the casting of magnesium sheet is shown in Figure 14. As cast and rolled magnesium products are shown in Figure 15.

Figure 14. Horizontal OCC equipment for casting magnesium sheet

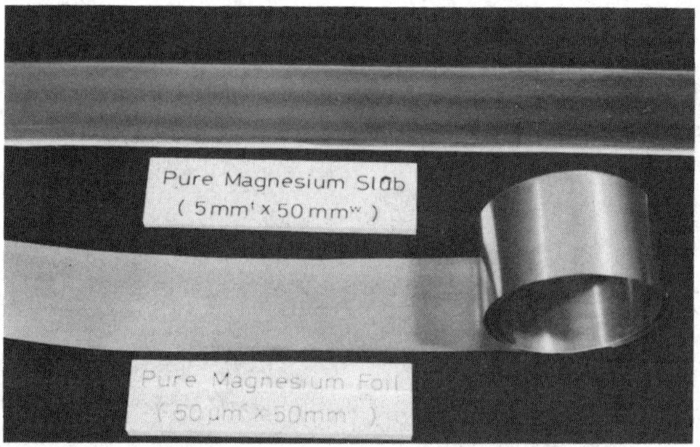

Figure 15. Single crystal magnesium plate and rolled strip.

Feasibility of the OCC process for the casting of various alloys has been examined using alloys of aluminum with Ni, Mn, Mg, Cu, and Si over a wide range of compositions. Some of the alloy wires are being used as hardfacing and weld deposition materials.

Al- 12 % Si die cast alloy containing 1 % Cu, 1 % Mg and 1.8 % Ni has been cast recently. Mechanically superior cast product was obtained and weld deposit applications for die castings are being considered.[6] Since the OCC process does not involve turbulent metal transfer from tundish, melting pot or ladle to the mold as occurs in the conventional casting techniques, there is less chance of metal oxide particles, other foreign inclusions and gas being trapped in the casting produced by the OCC process. The result is mechanically sound cast products.

The casting of Zn-22 % Al containing small amounts of copper, silicon and manganese which has a potential application as an oscillation damping material is underway. Preliminary results show that the sheet cast by the OCC technique exhibits extremely good workability compared to the material cast in an iron mold. The workability comparison between the OCC and conventional cast materials of Zn-Al alloy is shown in Figure 16.

OCC material showing no sign of edge cracking even after 75% reduction

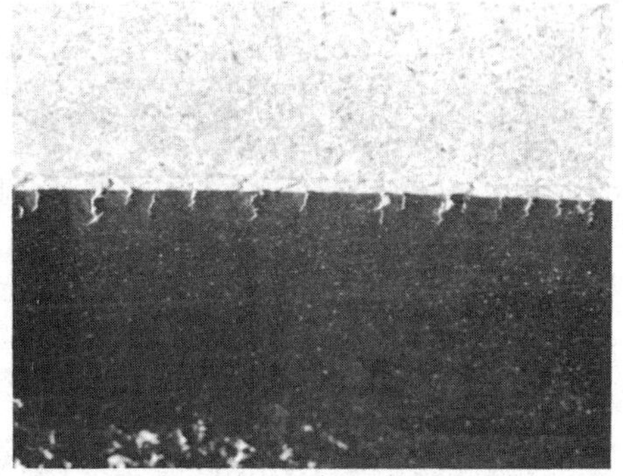

Conventional material showing showing edge cracks after 30 % reduction

Figure 16. Comparison of workability of Zn-22% Al, 0.5 Cu, 1.5 Si, 0.3 Mn and less than 0.0005 Mg alloy sheet, hot rolled.

Other areas of investigation performed at the Research Center include the feasibility of casting Al-Sn alloys, and characterization of cast products. Alloys with an addition of 1% copper or 2-4% silicon are widely used as bearing materials.(7) Binary Al-Sn alloys with tin contents up to 50% have been cast by the OCC process. An interesting phenomenon observed was that the alloys with tin contents above about 10% produce a large amount of hydrogen gas upon contact with water. An alloy containing 15% tin produces about one liter of hydrogen per gram of alloy. The ability to produce hydrogen gas can be attributed to the solidification morphology of material cast by the OCC process. In the binary Al-Sn eutectic system, the solid solubility of tin in aluminum and the solubility of aluminum in tin are virtually zero below the melting points of aluminum and tin. When the tin content exceeds 7%, the formation of tin film occurs around the primary crystals of Al resulting in the coupling of two metals which gives rise to a galvanic effect.(8) The solidification morphology of the OCC materials with fine unidirectional structure may cause the accentuated dissolution of the more reactive metal in the presence of water resulting in a large amount of of hydrogen gas.

At the Chiba Institute of Technology, new research efforts are directed towards the development of the heated rotary mold and stationary mold strip casting processes, Figures 4 and 5. Casting trials have been carried out using pure tin and zinc to obtain a base for application of the OCC principle to the rotary strip caster. The effects of casting parameters such as cooling methods, design of the cooling device, metal delivery, casting speed and mold temperature on the structure of strips have been investigated. Changes and improvements were added during the investigations, and unidirectionally solidified strips of pure tin, zinc and aluminum have been formed with a maximum casting speed of 0.9 m per min for tin, 1.5 m per min for zinc and about 0.5 m per min for aluminum.(9,10)

Recently, trial castings of the stationary mold strip caster have been carried out using pure tin and preliminary results are encouraging. Investigations on both casting systems are continuing.

The group at Chiba Institute of Technology has been carrying out a number of developmental works as well as characterization of the cast products. A feasibility study on the casting of high temperature material by the OCC process was initiated in this laboratory. 4 mm and 6 mm diameter rods of Fe, Co and Ni base magnetic alloys have been successfully cast and various casting conditions were investigated.[11] This study gave the basis for the developmental works on cobalt base alloys which are being carried out at the OCC Research Center.[12]

Characterization of cast products by the OCC process has been conducted. One of the features of OCC cast products is the distinct appearance of a mirror like surface quality. This quality has been studied and expressed in terms of brightness and roughness indexes quantifying the superior quality of the surface.[13]

Workabilities of binary Al-Cu, Sn-Zn and Sn-Pb alloys have been studied extensively revealing excellent rolling, drawing and bending capabilities. [14,15,16]

Recently, using the abilities of the OCC cast products to withstand severe stress imposed by mechanical work such as repeated bending and twisting, work is under way to produce mechanically strong and sound materials without the addition of alloying elements. Because of the nature of solidification morphology and excellent quality of the OCC products, it is possible to workharden materials by twisting or repeated bending without fracturing.[17] Examples of twisted wire of OCC and conventional material are shown in Figure 17.

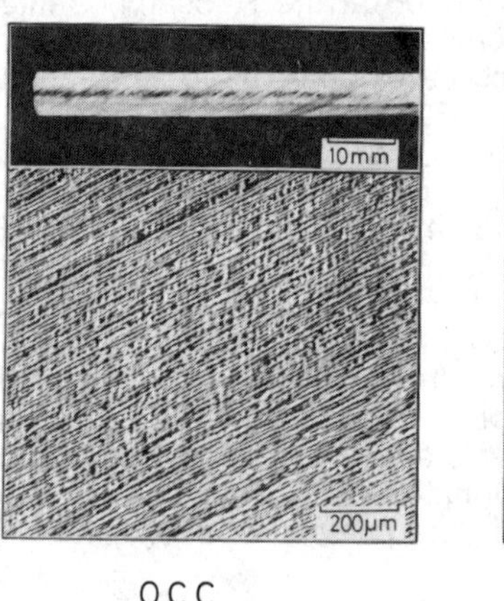

Figure 17. Example of twisted rods and surface appearance after application of twisting (0.5 times per cm).

G. Motoyasu and A. Ohno[18] reported that with the conventional material which was cast into a preheated iron mold, the surface became kinked easily upon application of twisting, and surface cracks were observed with twisting of as little as 0.5 times per cm. With the OCC material, it was possible to apply twisting evenly along the rod and it required 2.5 times per cm of twisting before fracture. This indicates the uniformity of OCC material in that a significant amount of stress can be imposed and substantial workhardening can be carried out before it fractures. With twist hardened OCC material, a considerable improvement of fatigue properties has been observed.

In 1988 an OCC laboratory was established at the University of Toronto. This facility includes integrated melting and casting units for the horizontal casting of low temperature metals such as aluminum, tin, bismuth and their alloys, and the vertical casting of high temperature metals such as nickel, cobalt, stainless steels and super alloys. Novel mold materials containing for example, boron nitride and titanium diboride are being evaluated in conjunction with the Electrofuel Manufacturing Company Ltd. of Toronto. Cooperative work between researchers at the University of Toronto, McMaster University and Chiba Institute of Technology is ongoing with respect to the casting and evaluation of composite materials. Clad rods consisting of an outer tube of tin containing an inner core of tin bismuth eutectic have been produced on the horizontal casting unit and this work will be extended to include aluminum-nickel alloys and metal matrix composite materials. The horizontal system is also being used to develop precise control strategies based on continuous monitoring of process variables and the construction of process models.

Summary

To date, over thirty different metals and alloys have been cast by the OCC process, and it has been found that the concept of the OCC process is viable for a wide range of materials.

The evolution of casting arrangements in the laboratory started with the downward configuration followed by the upward casting method and finally the horizontal casting method.

The casters currently in use in production employ the horizontal casting configuration and are capable of casting wires, rods, tubes and strips of copper, aluminum and their alloys with an excellent surface finish.

Horizontal casters both at Furukawa Electric Co. and Sumitomo Electric Co. are in a multistrand arrangement.

The application of the OCC process for the production of copper wires and rods has created virtually new and high quality products for the acoustics industries.

Prototype downward casting equipment has been developed by Mitsui Engineering and Shipbuilding Co., and is capable of casting wires, rods and tubes of different cross sectional design in a range of copper and brass alloys.

Upward casting equipment was constructed by Osaka Fuji Kogyo Co. and the casting of high temperature materials has been successfully performed at the OCC Research Center. Development work is progressing to cast stellite wire of small dimensions, which may crate new products for the welding industry.

A single crystal of magnesium sheet of 75 mm wide x 5 mm thick has been cast by the OCC Research Center and a developmental pilot project is currently underway to produce a single crystal of magnesium sheet 150 mm x 10 mm at Osaka Fuji Kogyo Co.

New strip casters - heated rotary mold and heated stationary mold casting techniques - have been developed to form strip with unidirectionally solidified crystals or a single crystal at the Chiba Institute of Technology and casting trials are ongoing.

With the application of workhardening, mechanically strong and sound materials can be produced form the OCC products. Preliminary study has indicated a significant improvement in mechanical properties.

Cooperative research studies on metals, alloys and composite materials are ongoing between the group at the University of Toronto, Chiba Institute of Technology and McMaster University. These efforts have been facilitated through the Ontario Centre for Materials Research.

Acknowledgements

The authors would like to express their appreciation to the OCC Research Center Inc., Osaka Fuji Kogyo, Ijima Kin Gin Kogyo and Mitsui Engineering and Shipbuilding Co. for their cooperation and assistance in providing helpful information for this paper. The financial support provided by the Ontario Centre for Materials Research for work on OCC technology at the University of Toronto is gratefully acknowledged.

References

1. A. Ohno, Proceedings, Casting of Near Net Shape Products, The Metallurgical Society, p. 177, (1988).
2. A. Ohno, Proceedings, Metallurgical Processes for the Year 2000 and Beyond, The Mineral, Metals and Materials Society, p. 155, (1989).
3. A. Ohno, H. Soda, A. McLean and H. Yamazaki, Proceedings, Advanced Materials - Application of Mineral and Metallurgical Processing Principles, Society for Mining, Metallurgy and Exploration, p. 161, (1990).
4. A. Ohno, Journal of Metals, Vol. 38, Jan. p. 14, (1986).
5. A. Ohno, Solidification, p. 116, Springer-Verlag, Berlin Heidelberg, (1987).
6. T. Shimizu, H. Yamazaki and A. Ohno, Abstract, Keikinzoku Spring Conference, (1990).
7. I.J. Polmear, Light Alloys, p. 129, Edward Arnold, London, (1989).
8. B.T.K. Barry and C.J. Thwaites, Tin and its Alloys and Compounds, Ellis Wood, West Sussex, (1983).
9. M. Takahashi, M.A. Eng. Thesis, Department of Metallurgical Engineering, Chiba Institute of Technology, (1990).
10. A. Ohno and K. Tada, Abstract, Keikinzoku Spring Conference, (1990).
11. H. Yamazaki, M.A. Eng. Thesis, Department of Metallurgical Engineering, Chiba Institute of Technology, (1986).
12. H. Yamazaki, Casting of Cobalt Base Alloys by the OCC Process, Osaka Fuji Technical Report, 1989.
13. G. Motoyasu, A. Ohno and T. Motegi, Imono, Vol. 59, No. 11, p. 670, (1987).
14. G. Motoyasu and A. Ohno, Keikinzoku, Vol. 39, No. 1, p. 38, (1989).
15. G. Motoyasu, T. Motegi and A. Ohno, Kinzoku gakkai-shi, Vol. 51, No. 10, p. 35, (1987).
16. G. Motoyasu and A. Ohno, Kinzoku gakkai-shi, Vol. 54, No. 3, p. 75, (1990).
17. A. Ohno and G. Motoyasu, Keikinzoku, in print, (1990).
18. A. Ohno and G. Motoyasu, Abstract, Keikinzoku Spring Conference, (1990).

Solidification and metallurgical phenomena in lost foam casting of aluminum alloys

S. Shivkumar, L. Wang and D. Apelian
Aluminum Casting Research Laboratory, Department of Materials Engineering, Drexel University, Philadelphia, Pennsylvania 19104, U.S.A.

ABSTRACT

The lost foam process has generated considerable interest among foundrymen because of the inherent advantages associated with the production of the casting. It has been predicted that the number of foundries utilizing this innovative process will continue to increase rapidly over the next few years. The principal feature of this casting technique is the use of foamed polymer patterns for the production of the component. On contact with the liquid metal, the polymer pattern undergoes thermal degradation through a series of complex transitions to produce liquid and gaseous degradation products. The formation of these degradation products in the mold has a significant effect on the solidification behavior and on the microstructural parameters. Furthermore, the interaction of the degradation products with the solidifying metal may introduce various defects in the casting which are unique to this process. The fundamental aspects associated with the solidification of lost foam castings have been described. The effects of polymer degradation on microstructural characteristics and on defects have been discussed.

KEY WORDS

Casting, Lost Foam Casting, Aluminum Base Alloys, Solidification

INTRODUCTION

In contrast to conventional empty-cavity casting techniques, the primary characteristic of the lost foam casting process is the use of foamed polymer patterns for the production of the component [1-6]. In order to fabricate the metallic part, initially, a polymer pattern of the required shape is produced by injection molding. Expanded polystyrene has been the most common polymer that is currently being used for the production of the casting. In recent years, other polymers such as polymethyl methacrylate and poly alkylene carbonate are being utilized with ferrous castings. Complex polymer shapes may be produced by joining together different sections with a hot-melt adhesive. The pattern is coated with a water based refractory slurry, dried and placed in a box (Fig. 1). Loose sand is then poured around the pattern. The sand is periodically compacted through pneumatic vibration devices. Molten metal is directly poured onto the solid polymer pattern to produce the casting. As the metal fills the mold, the foamed pattern undergoes thermal degradation through a series of complex transitions and the depolymerized products are vented into the sand, leaving an exact metal duplicate in place of the polymer pattern (Fig. 2). This new casting technique offers several benefits to casting manufactures and has therefore emerged as a viable alternative to conventional empty-cavity casting methods for the fabrication of near net shape components. The design flexibility offered by this innovative technique and its cost effectiveness can form a basis for enhancing the overall manufacturing efficiency. Other advantages of this process include features such as the use of unbonded sand, absence of cores, inexpensive flasks, minimum casting cleaning and the ease of automation. Since the invention of the process by Shroyer [7], intensive development work has culminated in a reliable method for the commercial production of a wide variety of ferrous and non-ferrous castings. Typical parts produced by this process include intake manifolds, cylinder blocks and cylinder heads.

POLYMER DEGRADATION

Expanded polystyrene is a linear hydrocarbon with a chemical formula of $(C_8H_8)_n$ and with a molecular weight of about 300000. This polymer initially collapses to 1/10 of its original size at temperatures above 100C. Increasing the temperature above this value leads to gradual thermal depolymerization (or degradation) [8-10]. The depolymerization proceeds by a mechanism referred to as "random scission". The bonds are broken randomly along the polymer chain. This mode of degradation occurs at relatively slow rates and favors the formation of both "liquid" and gaseous products. The liquid degradation products consist predominantly of partially depolymerized products such as dimer, trimer, and tetramer. The gaseous products consist mainly of the monomer. The amounts of liquid and gaseous products formed depend on the casting conditions. There is now a growing body of evidence to suggest that more than 60% of the polymer is converted to liquid degradation products during the formation of the casting. The liquid degradation products are absorbed by the coating and thus may accumulate at the metal/coating interface. Liquid degradation products may remain at this location even after the mold is filled. Liquid products are gradually vaporized at the metal/coating interface and the gases formed escape into the sand.

The gases formed at the metal front escape into the sand very rapidly [11]. Sequential stages during the production of a strip casting as observed through a transparent glass plate are shown in Fig. 3. It can be observed that the gas layer ahead of the metal front is essentially negligible. Thus, at the moving boundary, the metal is almost in direct contact with the polymer at most locations. It should be noted, however, that a large gas layer may be present at the metal front when other polymers (which yield significant quantities of gaseous products) are used for the production of the casting. For example, it has been observed that when PMMA patterns are used for the production of aluminum castings gas layers as large as 2 cm may be present ahead of the metal front [11]. Furthermore, an extensive gas layer may also be present in ferrous castings. At temperatures greater than about 1000C, the monomer molecule (C_8H_8) undergoes additional fragmentation and large quantities of lighter hydrocarbons are produced in the mold. In ferrous castings, the amount of gas produced per gram of polystyrene is greater than about 800 cm^3 (STP). These gases may remain at the metal front for short durations before escaping into the sand.

SOLIDIFICATION OF THE CASTING

The thermal degradation of the polymer during the production of the casting may influence the filling of the mold and the solidification of the component as indicated below:

1) Because of the presence of the polymer pattern in the mold, flow velocities in lost foam castings are significantly lower than in other bonded-sand molds. Typical flow velocities have been measured to be of the order of about 10 cm/s and about 100 cm/s for lost foam and conventional green sand castings respectively [10]. Thus filling patterns in lost foam castings may be significantly different from the filling sequence in other sand casting processes. Furthermore, the filling patterns in lost foam castings may also be altered by placing "holes" within the polymer. When the metal encounters a hollow section, it accelerates rapidly to try and attain the velocity in empty-cavity molds. Hence, the local flow velocity may be altered by positioning the hollow areas at appropriate locations to obtain the desired filling pattern in the mold. Fluidity in lost foam molds is generally less than in conventional empty-cavity sand casting processes.

2) Degradation of expanded polystyrene is a highly endothermic process requiring energies in excess of 1000 J/g [12,13]. Because of the endothermic losses at the metal front, steep thermal gradients are generated within the casting. An example of the thermal gradients in strip castings is presented in Fig. 4. It can be seen that very large thermal gradients are present in a region near the metal front. These gradients in the liquid metal establish

directional solidification of the casting. As a result, local solidification time varies with location in the casting and is inversely proportional to distance from the downsprue (Fig. 5). Typical solidification times in 1.3 cm thick strip castings poured at a temperature of 800C have been measured to be 250 s and 70 s at a distance of 1 cm and 52 cm from the downsprue respectively [14]. For both pure metals and alloys, stoppage of flow occurs because of solidification at the front. This behavior is in contrast to empty-cavity molds, wherein stoppage of flow in pure metals occurs because of a choking phenomenon occurring at the entrance of the channel [12].

3) The interaction of the liquid and gaseous degradation products formed in the mold with the solidifying metal may introduce several unique defects in the casting.

Consequently, microstructures in lost foam castings may be significantly different from those observed in other sand casting processes. Some of the vital aspects pertaining to the microstructure of lost foam castings are reviewed in the following sections.

MICROSTRUCTURES IN CASTINGS

Typical microstructures for alloy 319 containing about 6%Si and 3.5% Cu in a box casting (Fig. 2) at two different locations are shown in Fig. 6. The as cast structure consists of aluminum rich dendrites and other eutectic phases in the interdendritic regions. At locations close to the downsprue, a coarser dendritic structure is observed than at distances far away from the downsprue. A similar refinement of the microstructure is also observed in hypereutectic Al-Si (390) alloys. In this case, the primary Si particle size varies inversely with the length of flow (Fig. 7)

Dendrite morphologies in 319 castings produced by various techniques are shown in Fig. 8. In lost foam castings, the refinement of the dendritic structure with increasing distances from the downsprue is readily evident. Secondary dendrite arm spacing values are of the order of 48 μm at a distance of 1 cm and about 25 μm at 50 cm from downsprue [14]. The dendrite arm spacing in the casting is determined by the local solidification time (θ) according to the relation [15]:

$$DAS = 5.5\,(M\theta)^{1/3} \qquad (1)$$

where M is the dendrite coarsening parameter. M is dependent on the alloy system and for the Al-Si system containing 6% Si is estimated to be 5.8×10^{-18} m^3/s. DAS values calculated from equation (1) compare well with experimentally measured values. For example, the local solidification time at a distance of 1 cm in a casting poured at 750C is of the order of 150 s. The dendrite arm spacing is calculated to be 52 μm while the measured value is about 48 μm. Because of the higher cooling rate, dendrite arm spacings in lost foam castings are generally smaller than in conventional green sand castings (Fig. 8). The microstructures in lost foam castings are, however, coarser than in permanent mold castings.

The grain structure in the box casting at two different locations is shown in Fig. 9. The samples were etched with Boss's reagent to accentuate the grain structure. Large grains are observed near the downsprue while the grain structure is very fine in sections of the casting that were filled towards the end of pouring. Typical grain size values in a section 1.3 cm in thickness are of the order of 3000 μm at a distance of 1 cm from downsprue and about 300 μm at a distance of 50 cm for a casting poured at 750C. Also, in Fig. 9, it can be observed that the grain structure at a distance of 20 cm from downsprue is finer than in well grain refined samples.

The composition and morphology of interdendritic phases depends on the location in the casting. The interdendritic regions essentially contain Si, Cu and Fe-rich phases. In unmodified alloys, Si is present as platelets while in modified samples a fibrous Si eutectic may be detected (Fig. 10). When the castings are not modified, a significant change in the Si particle morphology may be observed with increasing distance from the downsprue. Both the average particle diameter and aspect ratio decrease inversely with the length of flow [14]. The copper-rich phase exhibits two different morphologies: blocky Al_2Cu particles and a eutectic structure (Fig. 10). Both $FeSiAl_5$ needles and $(Fe,Mn)_3Si_2Al_{15}$ chinese script particles may be observed in the microstructure (Fig. 11). Some $(Fe,Mn)_3Si_2Al_{15}$ particles may be present within the dendrites indicating that there was primary precipitation of this phase along with aluminum dendrites. At locations close to the downsprue, clusters of $(Fe,Mn)_3Si_2Al_{15}$ particles may be present in the microstructure. With increasing distances along the length of flow, the clusters gradually disintegrate into isolated particles within the interdendritic region.

At the metal front, large areas over which there was a strong segregation of alloying elements may be detected at several locations. A typical microstructure near a lap type of defect generally observed at the junction of two metal fronts is shown in Fig. 12. The segregation of Si near the metal front is readily evident. Many primary silicon particles (obtained at hypereutectic compositions) are also observed in the microstructure. The concentration of Si particles in the segregated areas has been measured by image analysis techniques to be between 9 to 13%. By comparison, the amount of interdendritic Si in other regions is of the order of 5 %. Large amounts of $FeSiAl_5$ needles with aspect ratios of about 50 to 100 may be present in the segregated region and the proportion of $(Fe,Mn)_3Si_2Al_{15}$ particles is much smaller than at other locations (Fig. 13). The length of the region over which this segregation occurs was measured to be about 0.5 to 1 cm. The segregation of Si at the metal front may contribute to defect formation.

DEFECTS IN CASTINGS

Several defects, some of them unique to lost foam castings, may be present in the casting. These defects can be broadly classified as surface defects, sand compaction defects, lap defects, porosity, inclusions and folds. Surface defects may arise from the pattern material. For example, the bead structure in the polymer pattern is often duplicated in the casting (Fig. 14). Although this defect does not affect the properties significantly, this bead structure may not be suited for some applications. A sand compaction defect is shown in Fig. 15. In this case, because of poor compaction of the sand, the mold is not rigid and the metal may infiltrate into and entrap the sand. Lap type defects shown in Fig. 12 are often observed in complex patterns. This defect is characterized by a partition line which corresponds to the junction of two metal fronts. Thus, the defect forms because of improper fusion of two metal fronts.

The formation of gaseous and liquid degradation products during the production of the casting may affect the amount and distribution of porosity. Each gram of solid polystyrene yields about 205 cm^3 (STP) of gases [9]. These gases may be trapped in the casting if they are not properly eliminated. Liquid degradation products may also be entrapped in the casting during mold filling. The entrapped liquid residue may gradually vaporize to form gases. If these gases cannot escape through the dendritic network, they may contribute to porosity.

Bulk porosity levels in untreated castings (no grain refiner or modifier added to the melt) produced from well degassed melts are comparable to the values measured in green sand castings. Typical porosity values in lost foam, green sand and permanent mold castings produced under identical conditions are of the order of 1.0 to 1.5%, 1.0 to 1.4% and about 0.6% respectively. A significant difference in pore morphology may be observed between lost foam, green sand and permanent mold samples as can be seen from Fig. 16. In lost foam castings, both interdendritic porosity and spherical gas bubbles (up to 5 mm in diameter) may be observed while green sand and permanent mold castings, essentially interdendritic pores are observed. Recent results [11] suggest that a pouring temperature on the order of 730 to 750C and terminal flow velocity of about 9 to 14 cm/s may be necessary to insure uniform filling of the mold and to minimize porosity formation. Uncoated samples generally exhibit a higher porosity than coated specimens. The thickness of the coating has a very small effect on the amount of porosity suggesting that gaseous degradation products are eliminated into the sand very rapidly and that the presence of liquid degradation products in the mold is the principal cause of defect formation. Hot melt adhesives used to join pieces of the polymer may also enhance porosity levels in the casting. The degradation of the adhesive produces enormous quantities of gases, typically about 2500 cm^3 (STP)/g [16]. The production of such large amounts of gases increases the probability of entrapment. Grain refinement with Ti-B and modification with Sr may have a significant effect on the amount and distribution of porosity in the casting. It has been observed that grain refiner leads to an overall reduction in the bulk porosity while Sr modification may increase the porosity levels in the casting [17]. In addition, the average pore size decreases with grain refinement while the opposite trend is observed with Sr modification.

Extensive porosity may be detected near the metal front and at the surface of the casting. Porosity levels of up to 5% have been measured near the metal front [14]. A typical example of porosity near the metal front in a widget pattern is shown in Fig. 17. Porosity values as high as 7% were measured in a region near the surface. The thickness of this region was of the order of 5 mm. The size of the pore in this region was much smaller than in the rest of the cross section. The average equivalent pore diameter in this region was estimated to be about 35 to 40 μm while the pore diameter is of the order of 57 μm in the bulk of the casting. This surface porosity may arise from the accumulation of the liquid degradation products at the metal/coating interface. The vaporized gases may diffuse into the solidifying metal and may be trapped as porosity.

Two defects which strongly impair the quality of lost foam castings may be observed on fracture surfaces: folds and black inclusions. Folds are discontinuities in the casting extending inward from the casting surface and severely affect pressure tightness and mechanical properties. In some respects, folds are analogous to cold shuts and form because of improper fusion of two streams of liquid metal. A typical photograph of a fold is presented in Fig. 18. It can be seen that the area of the fold contains a relatively smooth and striated surface. The size of a fold is typically of the order of a few mm. In extreme cases, folds whose largest dimension is greater than a few cm have also been identified. A scanning electron fractograph of the fold area is shown in Fig. 19. The matrix exhibits a typical cellular structure, while the fold is characterized by longitudinal delaminations. It appears as though the matrix and the fold have been pulled apart during fracture. The presence of folds in the casting severely impairs casting quality and more work is needed to clearly elucidate the mechanism of fold formation.

A typical black inclusion on the fracture surface is shown in Fig. 20. The size of this defect varies from about a few mm to a cm. Clusters of black inclusions may also be observed in some areas. Sections of the casting that were filled towards the end of pouring generally contained a large number of inclusions. EDX analysis indicates that these inclusions originate from the coating material [18]. Depending on the local flow conditions, a small potion of the coating may be fragmented because of impingement from the metal. The fragments of coating may then be engulfed by the metal or may be pushed at the metal front until they are entrapped in the last sections to solidify. Several mechanisms have been proposed to explain the formation of folds including the collapse of a gas bubble and improper fusion of two or more metal fronts.

MECHANICAL PROPERTIES

Typical as cast tensile data for lost foam castings are shown in Table I [17]. Because of a refinement in the dendritic structure and in the interdendritic phases, the mechanical properties generally increase with distance from the downsprue. No significant differences can be detected between the tensile properties of untreated and grain refined (0.2%Ti) and modified alloys (0.01%Sr). In many cases, the beneficial effects of grain refinement and modification may be offset by increased porosity levels originating from the introduction of Sr to the melt. The tensile properties of lost foam castings are better than or comparable to those observed in green sand castings, but are inferior to the properties in permanent mold specimens.

SUMMARY

The thermal degradation of the polymer during mold filling introduces several unique features during solidification, which may affect the microstructure in the casting. Because degradation is highly endothermic, steep thermal gradients are established in the casting. Consequently, for pure metals and alloys, solidification begins at the metal front and proceeds inwards. The thermal gradients promote directional solidification and hence the microstructure is not uniform at all locations in the casting. The dendritic structure and the interdendritic phases are gradually refined along the length of flow. The amount, morphology and distribution of various interdendritic phases may depend on location in the casting. At the metal front, the flow of interdendritic liquid, which is rich in alloying elements, may lead to the formation of a segregated region. Tensile properties are generally better than or comparable to the values in green sand castings.

The gaseous and liquid degradation products formed in the mold may have an influence on the amount and distribution of porosity. While bulk porosity levels in castings are comparable to the values measured in green sand castings, relatively large amounts of porosity may be detected at the metal/coating interface and at the metal front. The interaction of gaseous and liquid degradation products with the metal during solidification may produce several unique defects which may impair casting quality. Additional developmental work is necessary to solve problems pertaining to metal quality and to insure that the process can establish itself as a viable alternative to other conventional casting techniques.

REFERENCES

1. A.J. Clegg, *Foundry Trade Journal Int.*, **9**(30), (1986), 51-69

2. H.J. Heine, *Foundry M & T*, **114**(10), (1986), 36-41

3. G. del Gaudio, G. Serramoglia, G. Caironi and G. Tosi, *Metall. Sci. and Tech.*, **3**(3), (1985), 76-86

4. A.J. Clegg, *Foundry Trade Journal*, **145**(3143), (1978), 393-403

5. A.J. Clegg, *Foundry Trade Journal*, **145**(3144), (1978), 149-160

6. E.J. Sikora, *Trans AFS* , **86**, (1978), 65-68

7. H.F. Shroyer, US Patent No. 2830343, (1958)

8. S.L. Madorsky and S. Straus, *Soc. Chem. Ind.*, Monograph 13, (1961), 60-68

9 S. Shivkumar and B. Gallois, *Trans AFS*, **95**, (1987), 791-800

10. S. Shivkumar and B. Gallois, *Trans AFS*, **95**, (1987), 801-812

11. L. Wang, S. Shivkumar and D. Apelian, *Trans AFS*, **98**, (1990), in press

12. S. Shivkumar, Ph.D. thesis: "Fundamental characteristics of metal flow in the full mold casting of aluminum alloys", Stevens Institute of Technology, Hoboken, NJ, (1987)

13. S. Shivkumar, E. Cesmebasi and B. Gallois, submitted for publication in *Met Trans B*

14. S. Shivkumar, L. Wang and B. Steenhoff, *Trans AFS* , **97**, (1989), 825-836

15. W. Kurz and D.J. Fisher: "Fundamentals of Solidification", Trans Tech Publications, Rockport MA, (1984)

16. K. Vaught, 4th Annual Evaporative Foam Pattern Casting Conference, Rosemont, IL, June 6-7, 1989, 93-106

17. L. Wang, S. Shivkumar and D. Apelian, 2nd International Conference on Molten Aluminum Processing, Orlando, Fl, November 6-7, 1989, Paper #5

18. S. Shivkumar, L. Wang, B. Steenhoff and D. Apelian, 4th Annual Evaporative Foam Pattern Casting Conference, Rosemont, IL, June 6-7, 1989, 123-141

Table I

As cast tensile properties in test castings

Condition	Distance (cm)	YS (MPa)	UTS (MPa)	%Elongation
Lost Foam Untreated	3.8	112.1	137.5	2.3
	11.4	107.2	152.4	3.7
	19.0	134.9	149.8	2.4
	26.6	130.1	159.9	2.7
Lost Foam Grain refined and Modified	3.8	127.1	144.4	2.0
	11.4	125.1	140.0	2.0
	19.0	117.9	137.9	2.2
	26.6	136.8	165.0	2.6
Sand Cast†	-	125.0	148.6	2.3
Permanent Mold†	-	138.0	199.0	3.5

† Grain refined (0.2% Ti) and Modified (0.015% Sr)

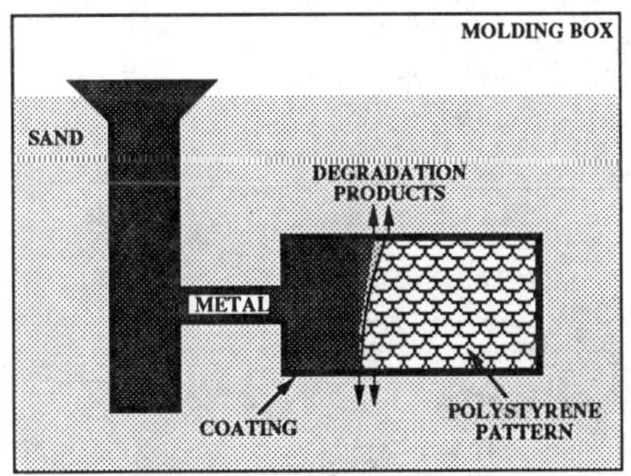

Fig. 1 Schematic of the lost foam casting process.

Fig. 2 Photograph of box pattern and casting.

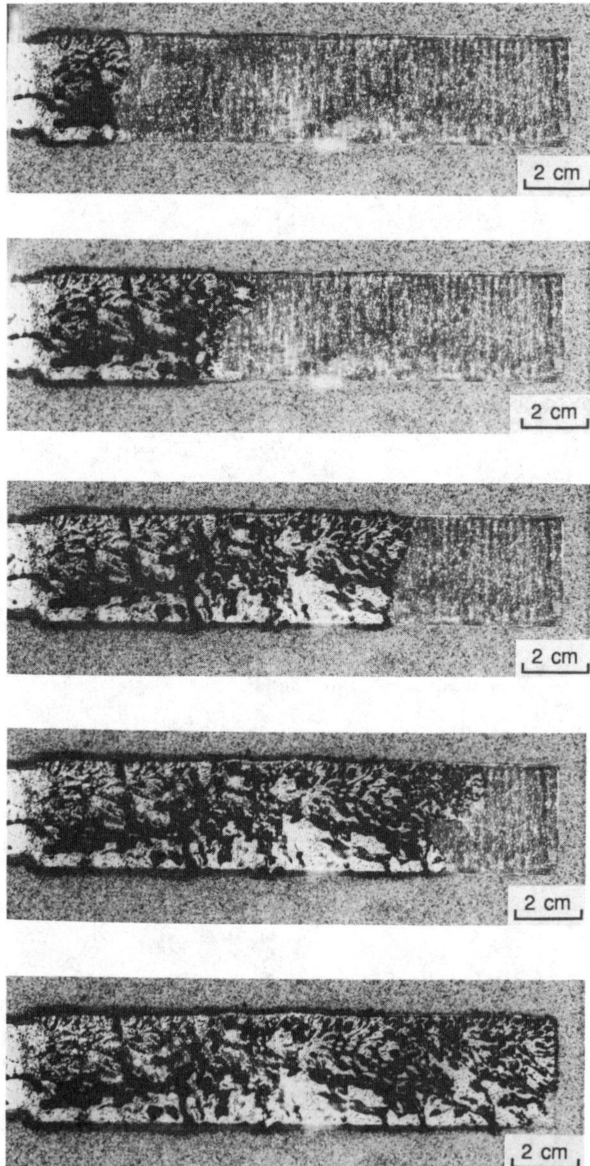

Fig. 3 Sequential stages in the production of the casting as viewed through a transparent glass plate.

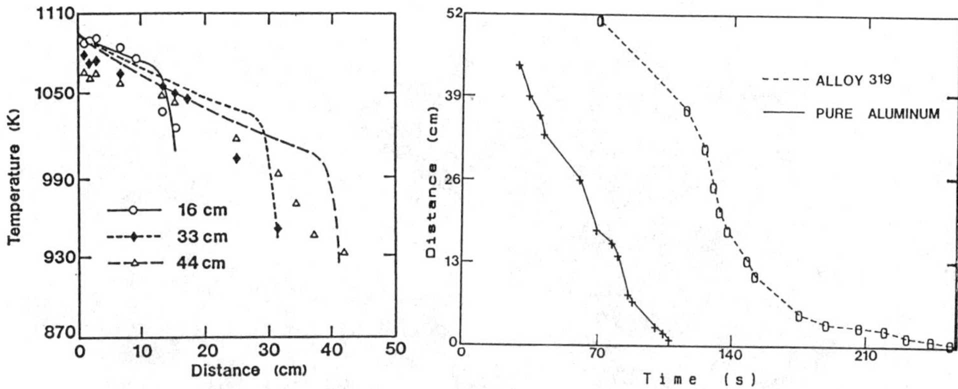

Fig. 4 Thermal gradients in a strip casting at various stages of flow. Symbols correspond to measured values while lines are calculated from a theoretical model [13].

Fig. 5 Solidification times at various locations along the length of a strip casting estimated at the end of the primary arrest (aluminum) and at the end of a eutectic arrest (alloy 319).

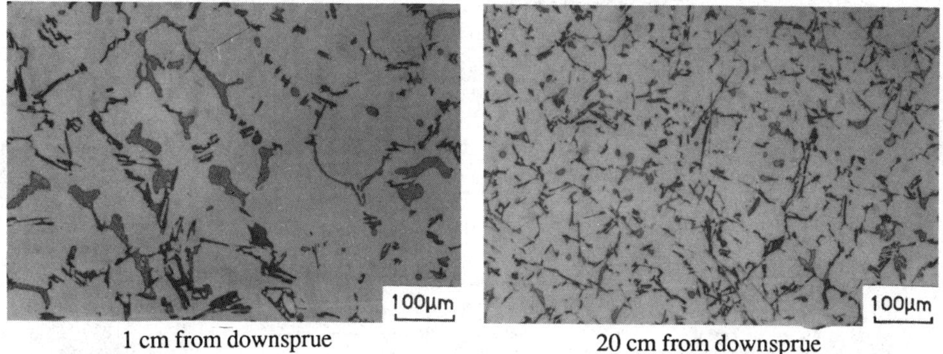

1 cm from downsprue 20 cm from downsprue

Fig. 6 Typical microstructures with alloy 319 at various locations in a box casting.

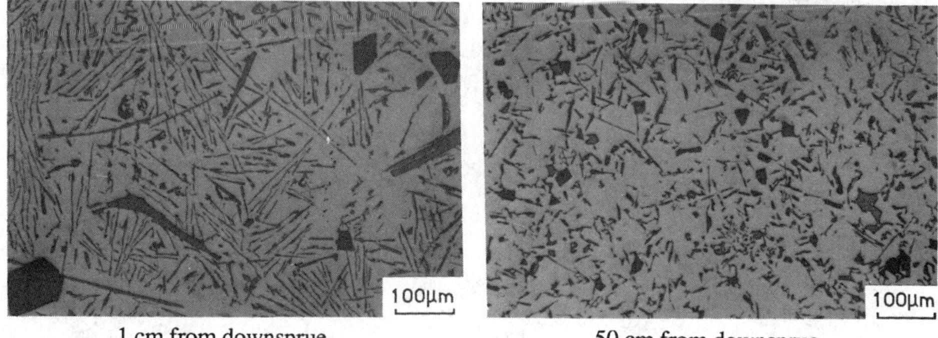

1 cm from downsprue 50 cm from downsprue

Fig. 7 Typical microstructures in a hypereutectic Al-Si alloy (390) at various locations in the casting.

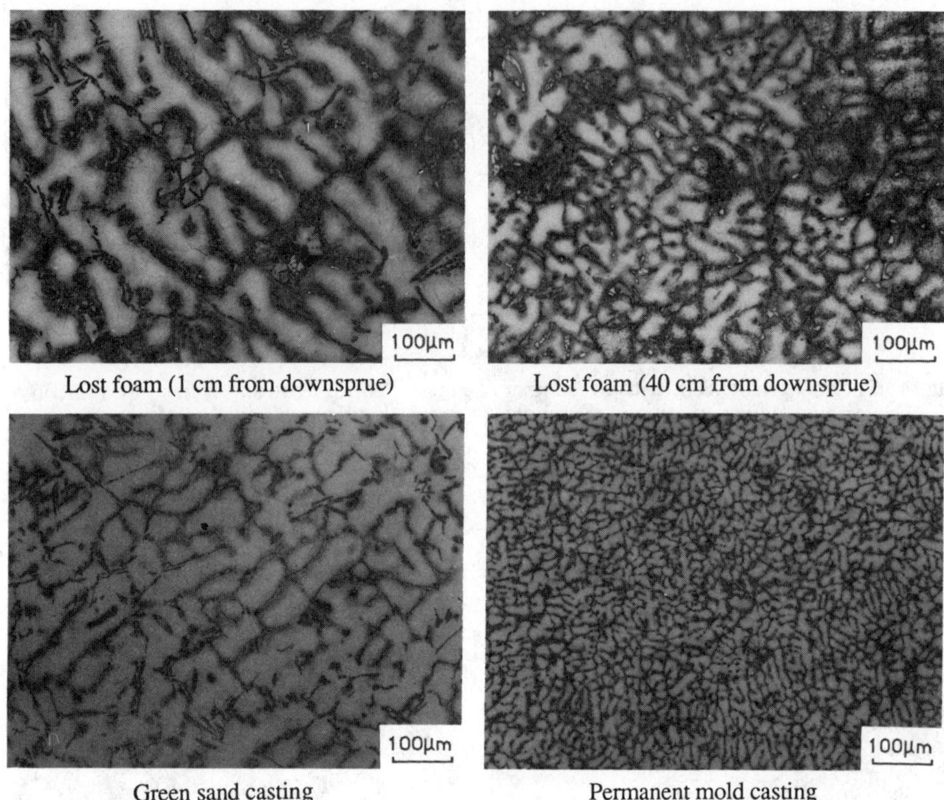

Lost foam (1 cm from downsprue)

Lost foam (40 cm from downsprue)

Green sand casting

Permanent mold casting

Fig. 8 Dendritic morphology in lost foam, green sand and permanent mold castings (alloy 319).

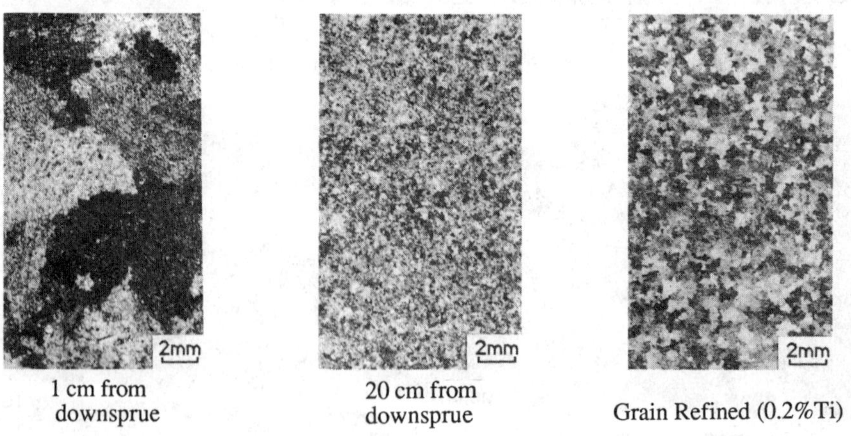

1 cm from downsprue

20 cm from downsprue

Grain Refined (0.2%Ti)

Fig. 9 Grain structures at various locations in a box casting (alloy 319).

Unmodified

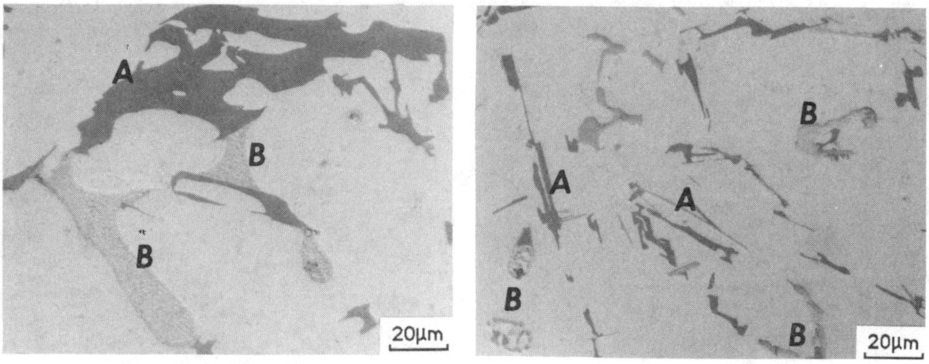

1 cm from downsprue

40 cm from downsprue

Sr modified

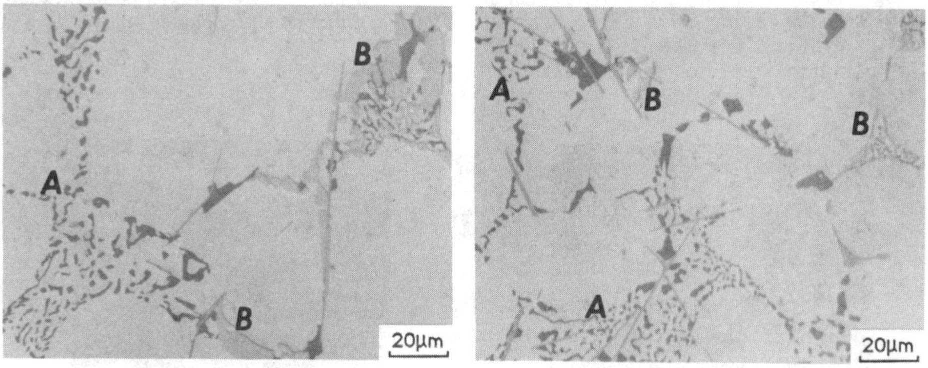

1 cm from downsprue

40 cm from downsprue

Fig. 10 Photographs showing morphology of Si (A) and Cu rich phases (B) in unmodified and Sr modified alloys (alloy 319).

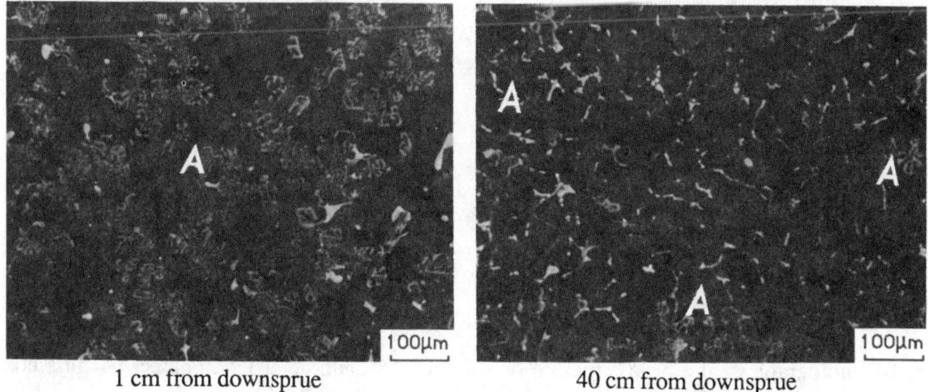

1 cm from downsprue

40 cm from downsprue

Fig. 11 Photographs illustrating morphology of Fe rich phases (A) in a box casting (alloy 319).

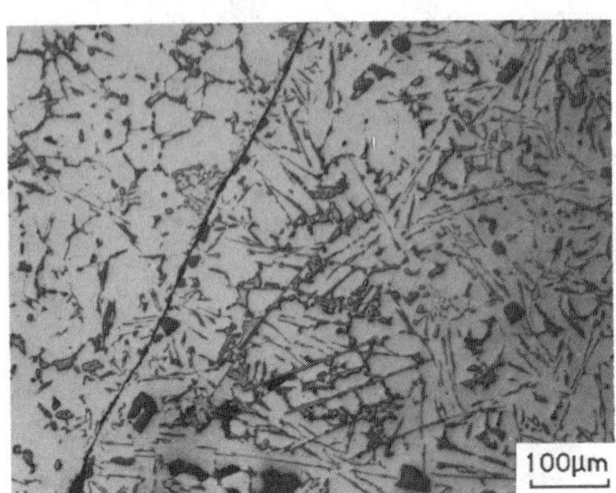

Fig. 12 Photograph of a lap defect at the junction of two metal fronts in a widget casting. Note the segregation of Si and other interdendritic phases near the metal front (alloy 319).

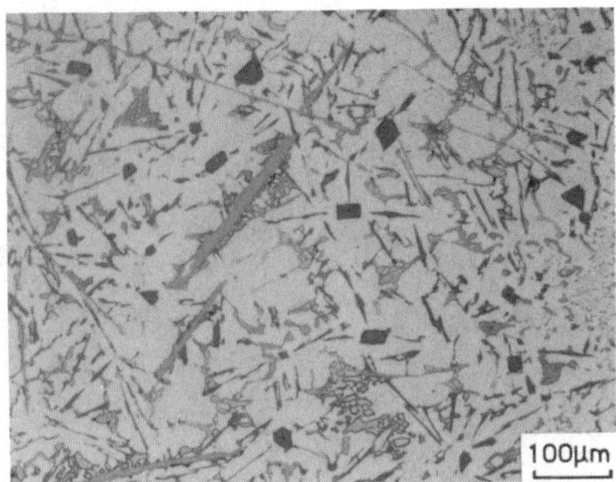

Fig. 13 Photograph at the metal front in a strip casting showing a high concentration of interdendritic phases (alloy 319).

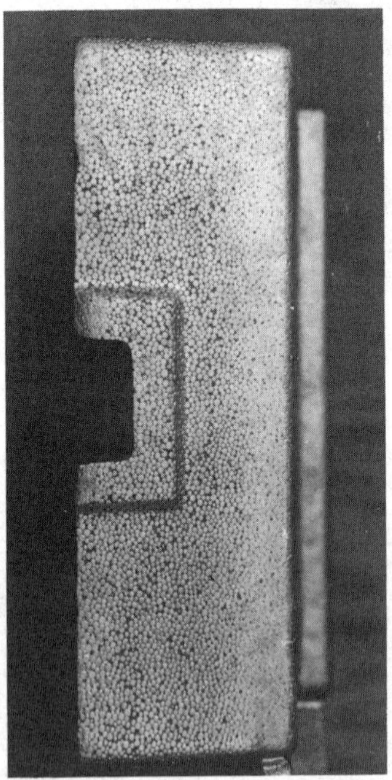

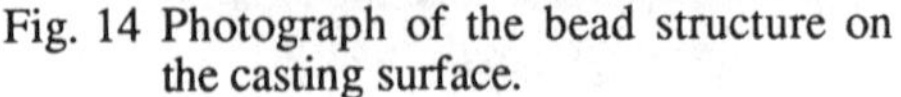

Fig. 14 Photograph of the bead structure on the casting surface.

Fig. 15 Sand compaction defect (A) in a box casting.

Lost Foam

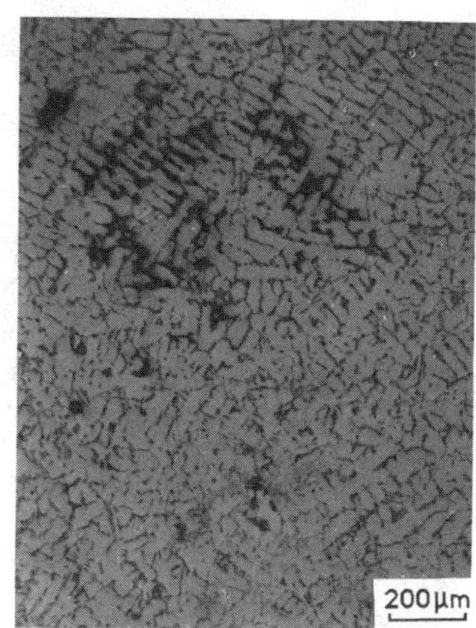

Green Sand

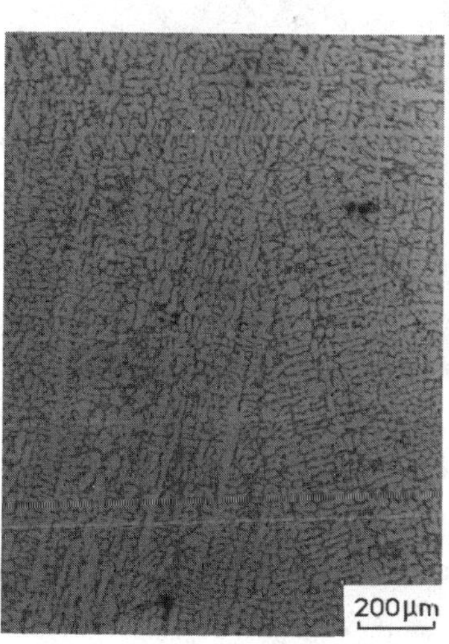

Permanent Mold

Fig. 16 Photograph illustrating pore morphology in lost foam, green sand and permanent mold castings.

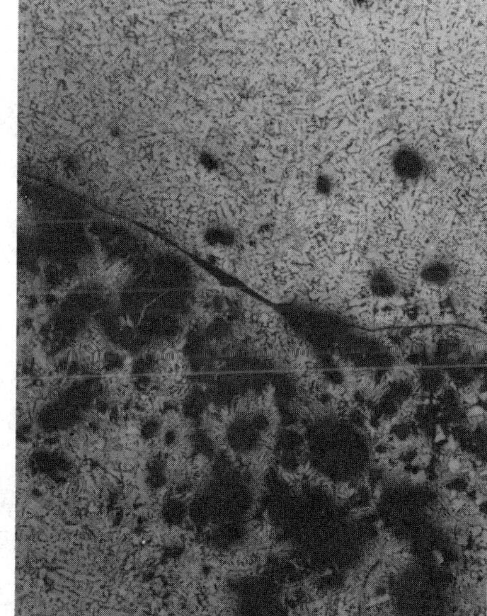

Fig. 17 Photograph of a lap defect in a widget pattern showing large amounts of porosity near the metal front.

Fig. 18 Photograph of fold type of defect observed on fracture surfaces.

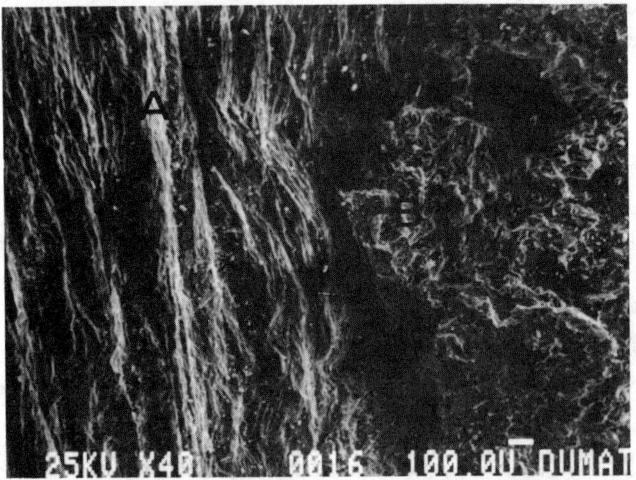

Fig. 19 Photograph of the fracture surface (B) showing the fold area (A).

Fig. 20 Photograph of inclusions observed in the casting.

Interaction of ceramic particles with an advancing solid-liquid interface in aluminum-based composites

Y. Fasoyinu and C.E. Schvezov
Department of Metals and Materials Engineering, The University of British Columbia, Vancouver, British Columbia, Canada, V6T 1W5

Abstract

The interaction of silicon carbide and alumina particles with an advancing solid/liquid interface in aluminum and aluminum alloy has been examined. The results showed that both particles are segregated at cell walls or interdendritic regions. For a planar interface no evidence of particle pushing is observed. Alumina particles have a strong tendency to clump in the melt and the clumps are also segregated at cell walls. It is proposed that the particles or clumps of particles which are segregated at cell walls or interdentritically arrive there by a mechanism which involves fluid flow and collision of the particles or clumps with the advancing interface.

Introduction

The mechanical properties of particle reinforced metal matrix composites (1-3) are directly related to the size and distribution of particles. Since the metal is cast, it is therefore important to establish whether the particles are pushed by an advancing solid-liquid interface during solidification.

The interaction of foreign particles with advancing interfaces has been extensively studied in water and organic materials (4-9). For these materials it has been shown that particles can be rejected by an advancing interface at velocities below a critical value. The critical velocity decreasing with increasing particle radius. Theories have been developed to predict the critical velocity for a given material/particle system (5,7,9). These theories however do not predict which solidifying materials are able to push a certain particulate material. In metal systems consisting of lead and lead alloys it has been shown (10) that an advancing planar interface does not reject particles. For interfaces with cellular and dendritic morphologies the segregation of particles in intercellular or interdendritic regions is accounted for on the basis of particle velocity and fluid flow.

Considering ceramic particles in aluminum alloys, SiC particles have been reported to be rejected by an advancing solid/liquid interface during directional Bridgeman solidification (11) and segregated to the cell walls in cast composites (3).

This investigation was undertaken to determine whether alumina and silicon carbide particles are rejected by an advancing solid-liquid interface in aluminum and aluminum alloys by direct observation using a zone refining solidification procedure.

Experimental

Directional solidification of pure Al, Al-Si and Al-Cu alloys containing alumina or silicon carbide particles was carried out by moving a liquid zone along a rod. Cast alloys were prepared from metal/matrix composite material provided by Alcan, pure aluminum, high purity copper (99.999%) and Al-7%Si. Cylindrical samples 7 or 9mm in diameter and 150mm in length were prepared from weighed amounts of constituents which were melted and cast into graphite coated quartz tubes. The samples were directional solidified vertically upwards or downwards by positioning the quartz tube in a small furnace which produced a short liquid zone in the rod of 20mm length. The liquid zone was moved along the rod by moving the rod at a constant rate with respect to the furnace. As the liquid zone moved, particles were continuously released at the melting interface of the zone and interacted with the solidifying interface of the zone. After approximately 50mm of growth, the sample was quenched in water. The sample was then cut, polished and etched to determine the solidification structure and the particle distribution in the sections parallel and perpendicular to the growth direction. The alloy compositions and growth conditions are listed in Table I.

Table 1: Composition and Growth Conditions

Sample Number	Composition Wt %	Volume %	Solidification	
			Rate 10^{-3} cm/s	Direction
1	Al-7%Si	15SiC	2.5	down
2	Al-7%Si	15SiC	2.5	up
3	Al-7%Si	5.7SiC	cast	
4	Al	$2.5Al_2O_3$	2.5	up
5	Al-3%Si	$5.0Al_2O_3$	3.53	down
6	Al-3%Si	$5.0Al_2O_3$	2.26	down
7	Al-5%Si	$3.0Al_2O_3$	2.26	down
8	Al-5%Si	$3.0Al_2O_3$	1.13	down
9	Al-5.2%Si	$2.5Al_2O_3$	1.13	down
10	Al-1.4%Cu	$1.9Al_2O_3$	2.26	down
11	Al-4.7%Cu	$2.8Al_2O_3$	2.26	down
12	Al-4.7%Cu	$2.7Al_2O_3$	2.26	down

An Al-7%Si alloy containing 5.7 volume percent silicon carbide particles prepared in a manner similar to that used in the directional solidification experiments was melted in a graphite crucible. The melt was stirred with a graphite rod, held at 760°C for 20 minutes and solidified in the furnace. This produced a cylindrical sample 20mm in diameter and 100mm in length. The sample was sectioned longitudinally, polished and etched to determine the solidification structure and silicon carbide particle distribution.

The alumina and silicon carbide particles in the sectioned and polished samples were identified using SEM/WDX analysis. The relationship of the particle positions and concentrations of segregated Si and Cu in the casting was also determined using SEM/WDX analysis.

Results

Macroetching of all directional solidified (DS) samples revealed a sharp transition from a columnar to a fine equiaxed structure corresponding to the DS and quenched alloys respectively. Metallographic observations revealed that the density of particles in both the DS and quenched regions are similar in all cases. This is particularly significant in the case of sample 4 consisting of pure aluminum matrix and alumina particles. This sample solidified with a flat interface and the result suggests that the particles are not pushed by the solidifying interface. For the alloyed composites the results are as follows.

Al-7%Si and SiC Particles

Two DS samples were examined, one solidified downward (sample 1) and one upward (sample 2). The cast structure on a vertical plane of sample 1 which includes the interface is shown in Figure 1. The quenched interface plane is clearly delineated. The SiC particles appear dark in the figure, and there is no evidence of particles piled-up ahead of the downward advancing interface. A careful examination of particle distribution in the quenched liquid gave no indication that the particles clumped or were not uniformly distributed in the melt. The particles are observed to be segregated into elongated bands in the DS region of Figure 1. Longitudinal and transverse sections of the samples showed that the particles were segregated into the regions between cells in the longitudinal cell structure. This is shown clearly in a longitudinal section of sample 1, Figure 2(a) and a transverse section Figure 2(b). In Figure 2 the cell centres are marked C and the cell walls W. The SiC particles are observed to be mostly in the cell walls, with a few in the cells. The smallest particles (A,B Figure 2(a)) are in the centre of the cell, SEM-WDX confirmed that particles A and B were silicon carbide particles and not silicon precipitates.

The sample which was solidified upward (sample 2) exhibited the same cast structure and carbide particle distribution as that observed in sample 1 in the interface region.

In one test (sample 3) the entire liquid was held molten for 1.2×10^3s and slowly solidified nondirectionally. Polished and etched sections of the sample at different heights showed that the SiC particles sank in the Al-7%Si melt during the holding period. This is shown in Figures 3(a),(b) and (c). The structure near the top of the sample (c) shows large dendritic regions (D) relatively clear of the dark SiC particles, and clumps of SiC particles between the dendrites. At the mid-section of the casting (b) and near the bottom (a) the dendritic areas free of SiC particle density increases by a factor of about 1.5 from (c) to (b) and from (b) to (a). The dendrite size is largest at the top of the sample, decreasing with distance down the sample, indicating the local solidification time decreases with distance below the top surface. The SiC particles segregated between the dendrites form non-uniform clusters in these regions, larger clusters being associated with larger dendrites. This indicates the clusters formed during solidification and that considerable movement of particles and liquid occurred during solidification, to corroborate this, a liquid zone during a DS experiment was kept molten for 20 minutes and quenched. Polished and etched sections of this sample showed no sign of particle clumping.

Al-Si and Alumina Particles

In the Al-Si alloys with alumina particles which were directionally solidified and quenched the structure was cellular, with the alumina particles concentrated predominately in the cell boundaries in both the DS and quenched regions. An example of a transverse section of

Al-3%Si (sample 6) containing 5% particles in the DS region is shown in Figure 4. The particles marked (A) are mostly segregated to the cell boundaries. The particle size ranges between 4 to 20 microns and they tend to be clumped. The clumps are present in the three regions DS, quenched (molten zone) and unmelted cast rod. The size, shape and density of clumps in the three regions are similar. These indicate that clumps were present in the composite material and the vigorous stirring during sample preparation did not disperse the particles in the clump. The estimated volume fraction of alumina particles in the clumps is about 20% which is higher than the average volume fraction of 15% in the Alcan composite.

Reducing the rate of solidification from 2.26×10^{-3}cm s^{-1} (sample 7) to 1.13×10^{-3}cm s^{-1} (sample 8) or increasing the Si concentration from 3%Si (sample 6) to 5%Si (sample 7) did not significantly change the particle distribution in the alloy or the DS structure, other than the cell size increasing with a decreasing freezing rate.

Al-Cu Alloys and Alumina Particles

Alloys containing 1.4 and 4.7wt% Cu were examined. Both alloys had a cell structure in the DS region with the cell walls not being fully developed in the 1.4% alloy and the cells being irregular in the 4.7% alloy. The particle distribution was generally similar to that in Al-Si alloys. With Al-Cu alloys the number of clumps was larger than with Al-Si alloys, consequently fewer isolated particles were observed. Figure 5 corresponds to a longitudinal section of sample 12. The correlation between the position of the particle clumps and the cell walls was not as clear as in Al-Si, particularly in the Al-1.4%Cu alloy. The distribution of Cu in the structure was determined by SEM/WDX. The results showed that for clumps at the cell walls the copper concentration between the particles was appreciably lower than that at the cell walls without clumps. On the other hand, for clumps within the cells the copper concentration was higher than that at the surrounding matrix material.

Discussion

The results show that alumina particles have a much stronger tendency to clump than silicon carbide particles. Most of the alumina clumps were present in the starting material and dilution or vigorous stirring did not disperse them. The interaction mechanisms of clumps and invidivual particles with the advancing interface are considered below.

Individual Particles

In the case of silicon carbide particles the results showed that the particles segregate to intercellular or interdendritic regions in all cases. For directional solidified samples particle segregation is insensitive to solidification direction, downwards or upwards; solidification velocity; solute concentration and particle size. These results cannot be explained by the mechanism of "particle pushing" which predicts particles are trapped at a critical solidification rate which increases with decreasing particle size. The critical velocities predicted by the theories of Chernov et al. and Bolling and Cisse for a range of particle radius are listed in Table II.

At the growth velocities used in the present experiments of 11.3μm s^{-1} and 35.3μm s^{-1} particles 5.0μm and larger should have been trapped rather than pushed. Note that few small particles were found in the matrix and no particle build up was observed ahead of the advancing interface. This clearly indicates the segregation of particles in the present experiments cannot be attributed to the mechanism of "pushing". The mechanism proposed by Schvezov and Weinberg (10) based on particle velocity and fluid flow can be applied in the present case. In this mechanism particles are segregated by successive collision with the interface (dendrite or cell tips or side walls) until they are trapped in the last solidifying melt or between secondary dendrite arms. The momentum associated in the collision comes from the movement of the particle with respect to the liquid due to the density difference and dragging by the convecting melt.

Table II: Critical Velocities versus Particle Size

Particle Radius	Chernov et al. Theory	Boilling and Cisse Theory
μm	V_c [μms^{-1}] $=50.32R^{-4/3}$	V_c [μms^{-1}] $=143.7R^{-4/3}$
1	50.32	143.7
5	5.88	12.8
10	2.34	4.5
20	0.93	1.6

In the present case the particle Stokes velocity is small. For a silicon carbide particle of 10μm radius the velocity is 0.065μm s^{-1} and for an alumina particle of the same radius is 0.12μm s^{-1}. These velocities are one to two orders of magnitude less than the interface velocity (assumed to be the same as the zone furnace velocity). In this case the momentum involved in each collision must come from the movement of the interface and possibly melt flow. This can explain why there is no effect of solidification direction, upwards or downwards on particle segregation. The momentum associated with the particle relative to the interface is practically the same in both cases. With this mechanism particles colliding perpendicular to a dendrite or cell tip surface are likely to be trapped as observed. This mechanism is also consistent with the results obtained with the cast sample and the alloys containing alumina particles.

Clusters

The interaction of clusters with the interface is more difficult to analyze than the interaction of individual particles due to the cluster structure. The alumina particles in the cluster are not close-packed. In a cross-section approximately 20% of the area corresponds to particles and the remaining area (volume) consists of molten alloy. Clusters can therefore be considered as quasi-liquid particles. The comparable sizes and shapes of clusters in the quenched region and the DS region indicates that the interface does not disperse or deform the cluster. This observation is consistent with a cluster segregation mechanism similiar to the mechanism for particle segregation. By a sequence of collisions with the cell walls or tips the cluster is rejected to the cell wall until a few out particles are trapped by the growing cell. The whole cluster is subsequently trapped without further interaction or deformation. If a cluster is trapped at the cell tip, the interface is severely distorted locally.

Summary

Aluminum based composites containing alumina or silicon carbide particles have been directionally solidified using a zone refining technique. The results can be summarized as follows:

1 Alumina and silicon carbide particles are segregated to intercellular/interdendritic regions.
2 There is no evidence that the particles are pushed during solidification by a planar solid/liquid interface.
3 Particle segregation is insensitive to alloying element, growth velocity, solidification direction and particle size.
4 Alumina particles have a strong tendency to clump in the melt. The clumps which formed in the melt are found in the cell walls.
5 It is proposed that the particles or clums of particles which are segregated at cell walls or interdendritically arrive there by a mechanism which involves fluid flow and collision of the particles or clumps with the advancing curved interface.

Acknowledgements

Financial support from NSERC is gratefully acknowledged. Composite samples provided by D.J. Lloyd from Alcan International Ltd. are appreciated.

References

1. D.K. Rohatgi, R. Asthana and S. Das, International Metals Review 31, 115 (1986).
2. F. Rana and D.M. Stefanescu, Met. Trans. 20A, 1564 (1989).
3. H. Lagace and D.J. Lloyd, Canadian Metallurgical Quarterly, 28, 145 (1989).
4. M.V. Pikunov, Non-Ferrous Metals, Their Treatment and Working, p.55, Metallurgizdat, Moscow, (1957).
5. D.R. Uhlmann, B. Chalmers and K.A. Jackson, J. of Applied Phys., 35, 2986 (1964).
6. S.N. Omenyi and A.W. Neumann, J. of Applied Phys., 47, 3956 (1976).
7. G.F. Bolling and J. Cisse, J. of Crystal Growth, 10, 56 (1971).
8. J. Cisse and G.F. Bolling, J. of Crystal Growth, 10, 67 (1971).
9. A.A. Chernov, D.E. Temkin and A.M. Melnikova, Sov. Phys. Crystallogr., 21, 4 (1976).
10. C.E. Schvezov and F. Weinberg, Met. Trans. 16B, 367 (1985).
11. D.M. Stefanescu, B.K. Dhindaw, S.A. Kacar and A. Moitra, Met. Trans., 19A, 2847 (1988).

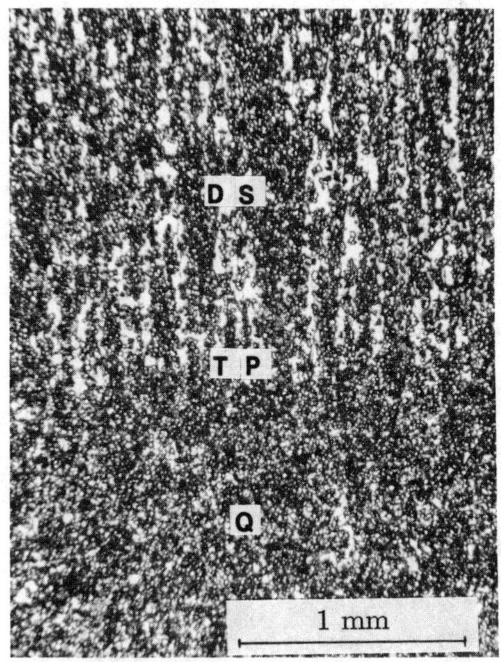

Figure 1: Al-7% Si with SiC solidified downard. Vertical section showing directionally solidified region (DS), quenched region (Q) and transition plane (TP). The dark spots are the SiC particles.

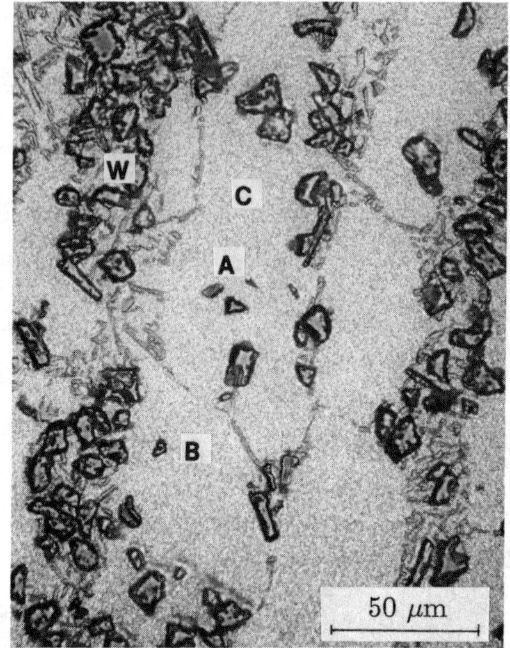

(a)

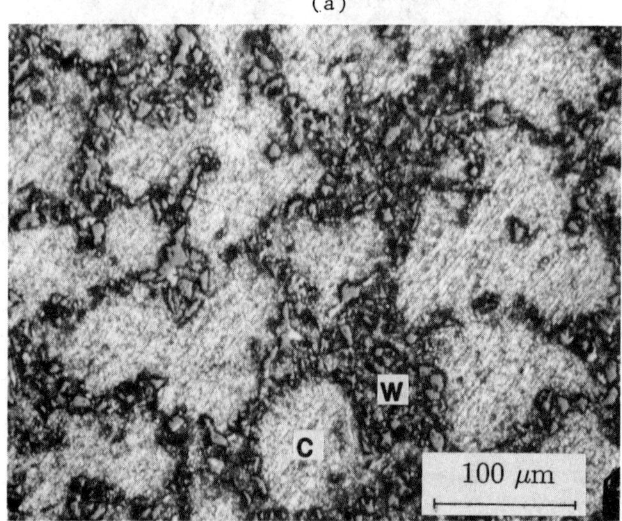

(b)

Figure 2: (a) Longitudinal section of sample 1 with the cell centres indicated by C and cell walls W. The dark SiC particles are mostely concentrated in the cell walls.
(b) Transverse section of sample 1 showing SiC particles concentrated in cell walls.

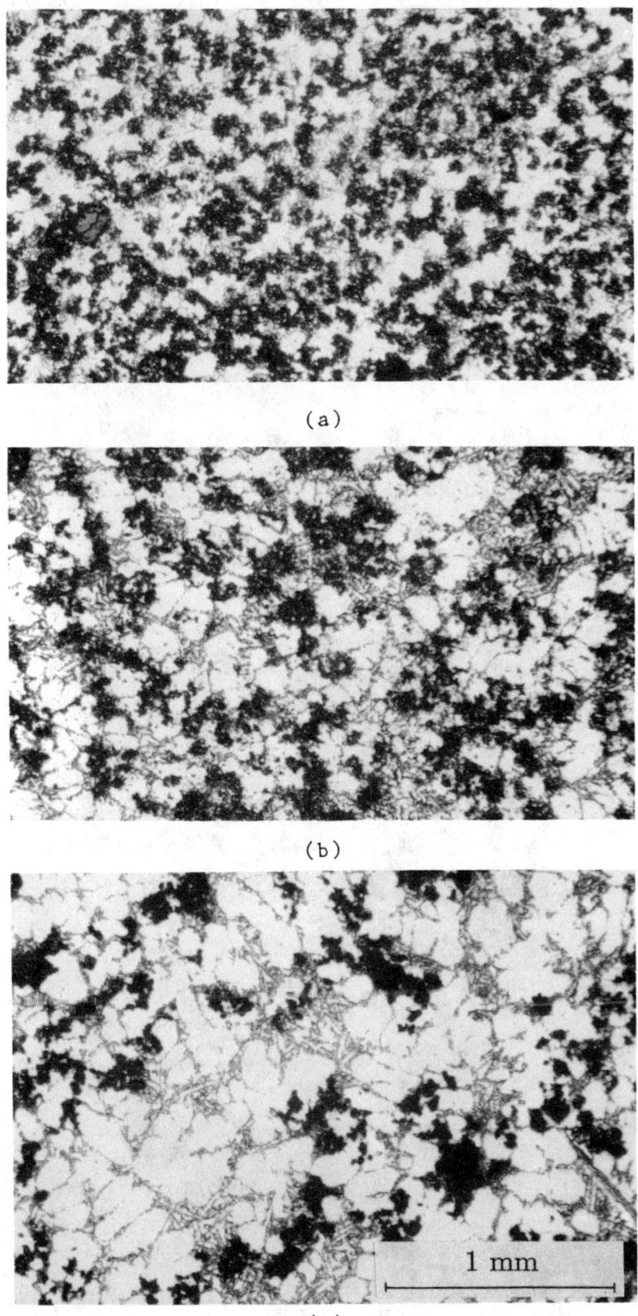

Figure 3: Longitudinal section of bulk sample 3 at three positions: (a) bottom, (b) middle and (c) top of the sample.

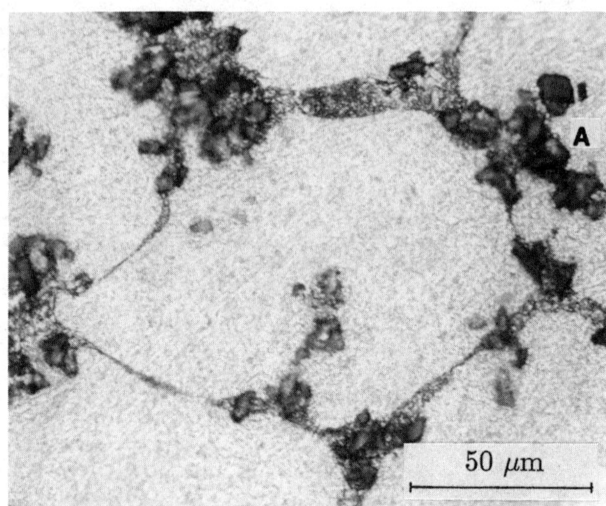

Figure 4: Alumina particles (a) at cell boundaries in sample 6, Al-3% Si with 5% particles.

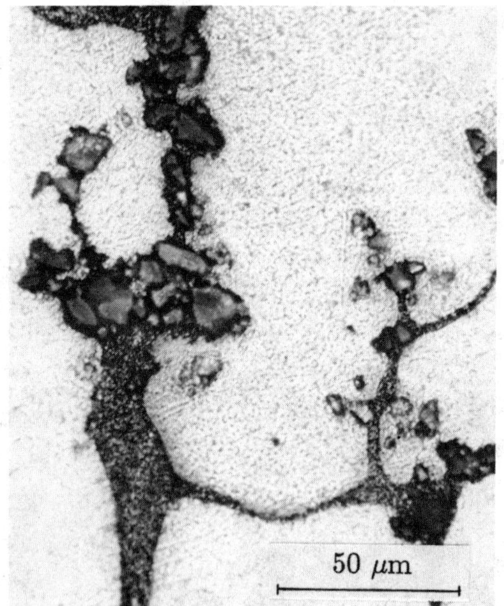

Figure 5: Distribution of alumina particles in longitudinal section. DS region. Sample 12, Al-4.7% Cu with 2.7T particles.

V. Static casting of cast iron

Chairman: H. Biloni
LEMIT-CIC, LaPlata, Argentina

Static casting of cast iron (Keynote Paper)

I. Minkoff
Department of Materials Engineering, Technion, Israel Institute of Technology, Haifa 32000, Israel

1. Introduction

The first part of this paper reviews the experimental and theoretical basis for evolution of structure in the alloys of cast iron which solidify with the graphite phase. The growth mechanisms are described for the three types of eutectic having graphite in flake, compacted and spheroidal form. The appropriate kinetics for each type of graphite eutectic is discussed and compared with the models of micro-structure evolution used for process control by thermal analysis and computer simulation.

Problems of modelling the melt are discussed, and its chemistry, as well as current research on the modelling of shrinkage.

2. A Brief Review of Cast Iron Solidification. Description of Three Graphite Eutectics.

In the cast iron system, growth of the graphite phase proceeds by mechanisms which are specific for a faceted material. Both the growth of graphite as a single phase and its growth as a eutectic phase relate to structural defects, which also determine the growth kinetics. Different graphite forms then grow by different defect mechanisms. The differences in kinetics lead to the possibility of using thermal analysis to differentiate between the types. These differences are also used to characterise aspects of the structure, and to control the process.

The growth mechanisms of the graphite phase in the Fe-C system are determined by the crystallography and defect structure. This distinguishes solidification and growth in cast iron from similar aspects in metallic systems, as for example the dendritic growth of single phases and cooperative growth of eutectics. Instabilities in growth of graphite lead to different forms as single phases and in two phase growth lead to three distinct eutectic types.

These structural types are as follows:-

a) Flake
b) Compacted
c) Spheroidal

The mechanisms are shown in Figure 1. They can be briefly described as follows:-

Flake graphite - The crystal of graphite grows from the steps of rotation boundaries, which are defects in the crystal in the form of rotations of the lattice round a crystallographic axis, in this case (0001). These rotations

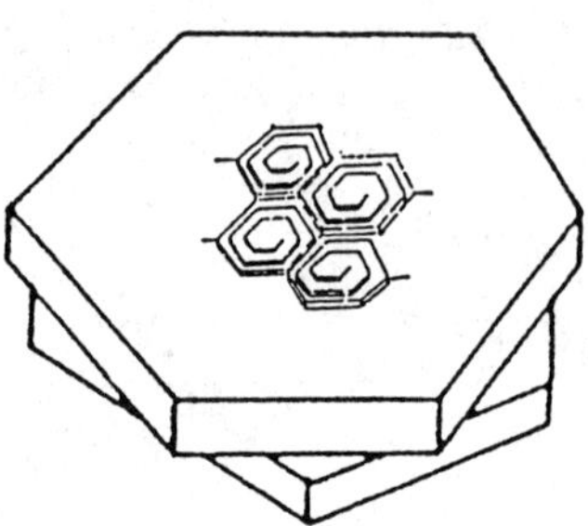

Figure 1a

Mechanism of growth of flake graphite from a rotation boundary which provides steps for the nucleation of $(10\bar{1}0)$ faces.

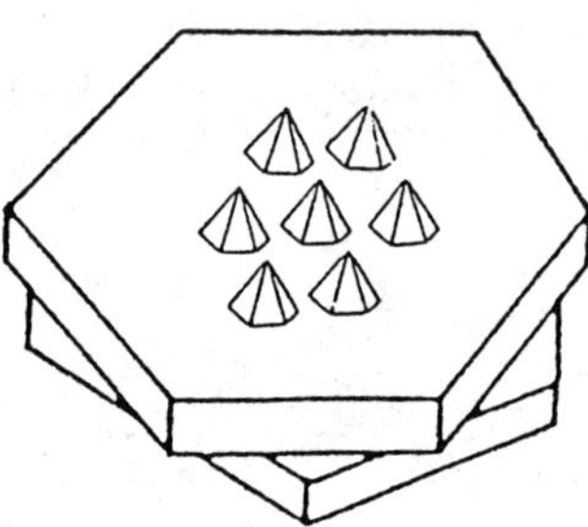

Figure 1b

Mechanism of growth of compacted graphite by development of pyramidal forms on the crystal surface from steps of screw dislocations.

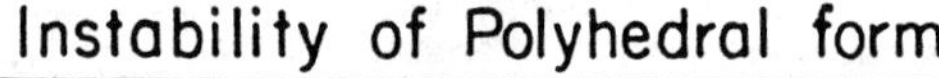

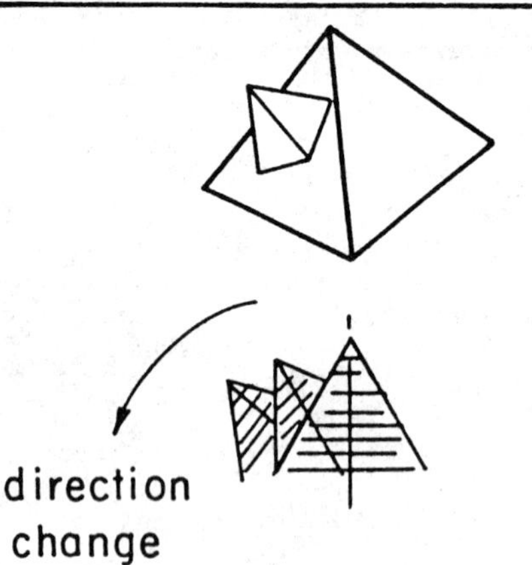

Figure 1c

Mechanism of growth of spheroidal graphite by repeated instability of pyramidal surfaces forming a radial array.

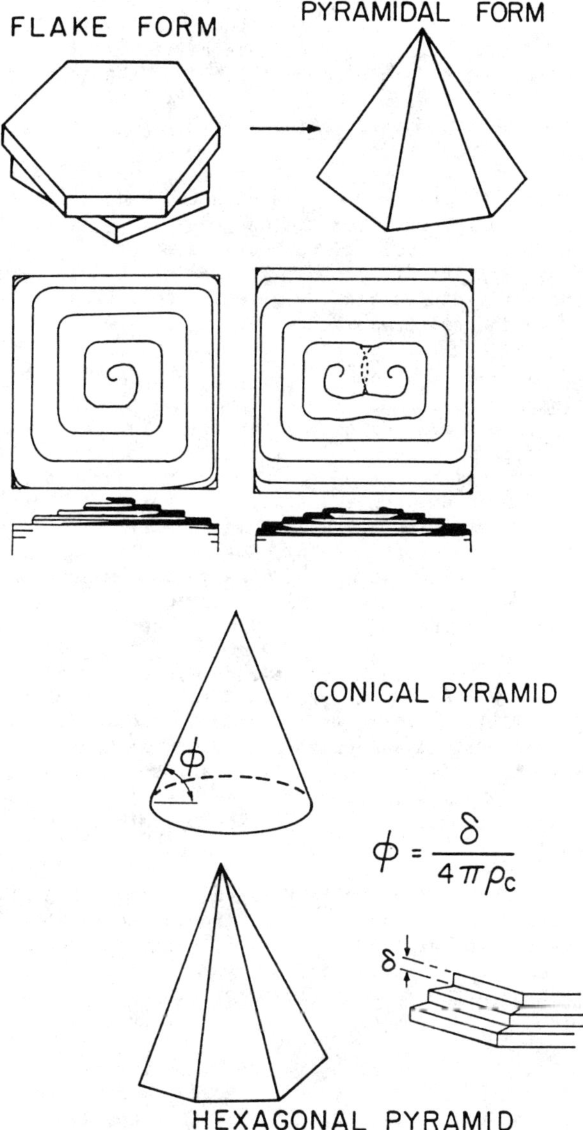

Figure 2. Pyramidal forms can develop from flake forms or pyramids can develop on the surface of flake forms. The basic mechanism is related to the movement of steps centered on screw dislocations. The angle ϕ of the cone or pyramid is given by

$$\phi = \frac{\delta}{4\pi\rho_c}$$

ρ_c is the critical radius for nucleation of a step, and ϕ therefore becomes bigger as ρ_c decreases with undercooling.

are exactly defined by coincidence boundary theory (1). For the case of most rapid growth in graphite they conform to angles having the maximum multiplicity and are equal to 27.8° and 21.8° (2). The steps permit growth on $(10\bar{1}0)$ faces and the related kinetics should be exponential (3) of the form

$V = \text{const}\ e^{-\frac{\Delta Q}{kT}}$ where ΔQ is the activation energy for nucleating a $(10\bar{1}0)$ face on a step.

Compacted Graphite Forms. Thickening of graphite crystals at small values of the undercooling occurs by growth from the steps of screw dislocations which have a Burgers vector in the <0001> direction. The nature of these defects was investigated by Thomas (4), Amelinckx (5). The angle ϕ of the hexagonal pyramid, which grows from the screw dislocation is given by $\phi=\delta/(4\pi\rho_c)$ where δ is the step height and ρ_c is the critical radius for nucleation of a step, Figure 2.

The angle therefore changes as a function of the undercooling, in its influence on ρ_c, and becomes bigger as the undercooling becomes greater. At small undercoolings, the pyramids are scarcely detectable. They become pronounced and observable as the undercooling increases. The growth of graphite is accompanied by recalescence and with the change of temperature the pyramids once more decrease their angle and become congruent with the surface. These pyramids are dislocation related and grow with parabolic kinetics at low undercooling and linear kinetics at higher undercooling (6). The crystal itself in the compact graphite mode propagates lengthwise by the same mechanism as described for the flake. The mode of undercooling in cast iron technology is chemical by the addition of reactive impurities (Mg or Ce). This is described in Section 3.

Spheroids. These grow from pyramidal crystals, by repeated instability of pyramidal surfaces, so that a radial array is formed of pyramids (Figure 1c). This was predicted by Tiller (7) and is now evident from different experimental observations. Hillert (8) also suggested a growth mechanism from screw dislocations which would differ by an angular misorientation due to the absorbed impurities. A spherical surface replaces the pyramidal one as the undercooling decreases with recalescence (9) and pyramidal growth proceeds with smaller pyramidal angles.

Relationship with Melt Chemistry. Growth Form and Undercooling. The forms of graphite in their dependence on the different growth mechanisms of the crystal, are controlled by melt chemistry. In particular, the reactive elements Mg and Ce are used to arrive at different undercoolings of the melt, and it is the undercooling which changes the mode of growth of graphite. By undercooling, the understanding is total undercooling at the interface, Figure 3.

This is made up of a kinetic component ΔT_E and a constitutional undercooling component ΔT_S. The latter is mainly contributed by elements which form a boundary layer important in unstabilising steps. This results in one type of graphite in rod-like form termed coral graphite (10). It is however the elements which make up the kinetic influence, which are largely responsible for the compacted and spheroidal forms.

Their mode of action is different from the elements which influence constitutional supercooling. They act by impeding the movement of steps on surfaces. One such mechanism is by forming a network of atoms which create an obstacle to the passage of steps (Figure 4).

The spacing between impurity atoms must be greater than the diameter of the critical nucleus for formation of a new atomic plane, if a step is to move on the surface. The system undercools to permit this movement. Mg and Ce affect the undercooling differently. This can be related to their distribution on the

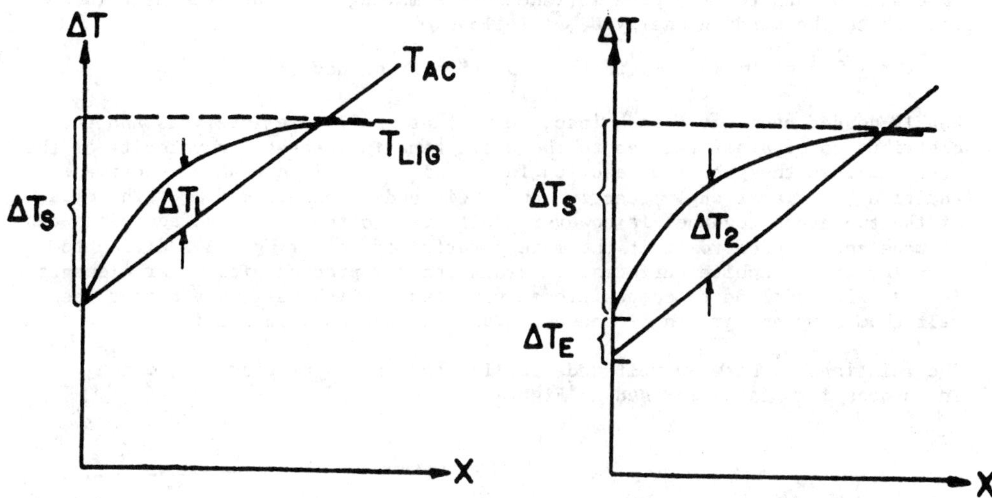

Figure 3. The total undercooling is made up of the constitutional undercooling ΔT_S and the kinetic undercooling ΔT_E . In Figure 3a, the undercooling is ΔT_1 and in Figure 3b, the kinetic component gives the larger undercooling ΔT_2 .

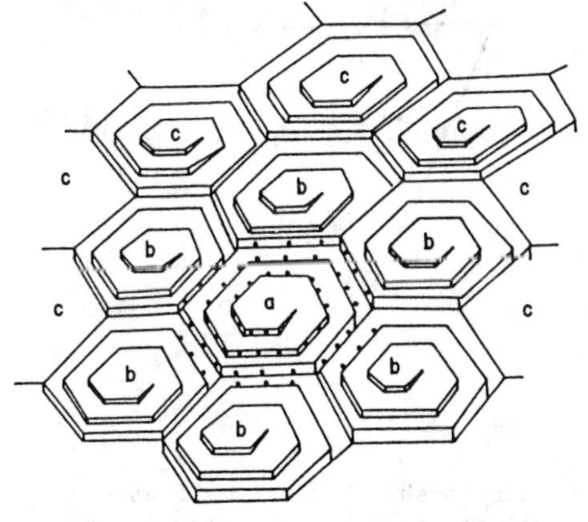

Figure 4.

The elements which influence kinetic undercooling form a network of atoms on the surface, which create an obstacle to the passage of steps.

surface. One possibility for the difference in their action is the difference in bonding energy between different reactive elements and the surface. The time an atom has to stay on a surface before moving back into the melt is related to the bonding energy U as follows:

$t = v^{-1} \exp (U/kt)$ v is the vibration frequency.

Weakly bonded atoms move back into the melt at small time intervals and the distribution of atoms related to the dwell time influences the velocity of the step (12). In the graphite reaction in cast iron, Ce is a weakly reactive addition. Mg is strongly reactive and their undercoolings differ. The effect of the two atoms together is however additional so that combined additions can be made and Ce used to limit the more powerful effect of Mg. In addition to reacting with graphite surfaces, interactions may proceed with other elements in the melt and lead to compositional variations which seriously affect the melt chemistry and growth of phases. This is discussed in Section 6.

The relationship between melt undercooling and growth of graphite, which influences form is summarised in Figure 5.

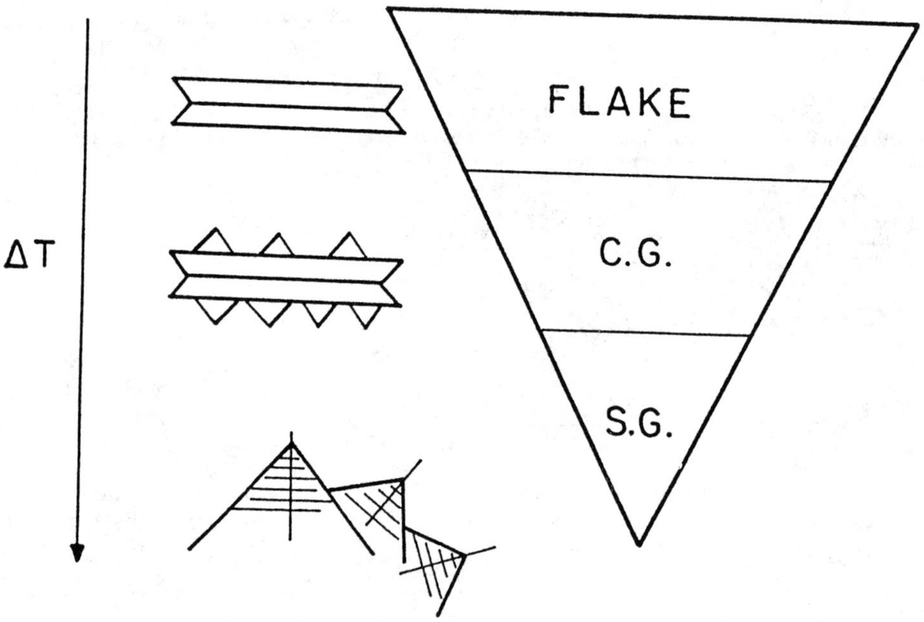

Figure 5. Schematic representation of the relationship between melt undercooling and growth of graphite influencing form. The flake form is representative of growth at the smallest undercooling. Pyramids grow on flake forms and become unstable to form the compacted graphite form. The spheroidal graphite form develops from a pyramid, which becomes unstable on the pyramidal faces.

The stages from flake growth are -

(a) For a critical distribution of impurity, step movement is stopped and the melt undercools. The initial mechanism of growth at small undercooling by steps of rotation boundaries is replaced by pyramidal growth.

(b) The pyramidal crystals undergo instability by growth of pyramids on pyramidal faces. The angular displacements leading to a spherical array are related to low angle boundaries in the structure.

(c) The surface geometry changes with recalescence to attain a more spherical boundary.

Eutectic Growth in the Three Graphite Systems. The three distinctive growth types of graphite lead to three separate eutectic types in graphite cast iron. These three types, which have broad industrial applications, can be characterised by their nucleation and growth kinetics. The structures can be predicted using modelling techniques with the computer, and the casting process controlled.

The nucleation kinetics, as well as the growth kinetics, should relate to the mechanisms observed experimentally and which differentiate the three structures.

These eutectics have been in the main, regarded as extensions of the metallic eutectic systems. Because the flake eutectic is irregular, i.e. the phases have no geometrical order, either lamellar or rod, a model has been used of growth at the extremum (13). In this model, large spacings between lamellae lead to depressions in the curvature of the interface of the widest phase. A construction of minimum and maximum spacing is required to fit the observations into a scheme, which requires diffusion kinetics for growth. In the case of the spheroidal graphite eutectic, it has also been proposed that the graphite grows by diffusion within an austenite envelope. These models are difficult to correlate with the experimental observations. In the computer modelling techniques, factors are introduced into the computations to correlate the calculations with the experimental curves.

In calculating the growth kinetics of the flake graphite eutectic, Hillert (14) compared the experimental values with a Zener type calculation for a two phase structure growing by diffusion. The experimental values were far higher than the computed ones and were reasonably close to those calculated for single phase growth of a disk growing by diffusion using the Ivantsov analysis (15).

It would appear better to model the growth of graphite in the three eutectics by the kinetics of graphite growth as a single phase, developing in each of the three cases by its own mechanism. The austenite phase follows.

In the case of the spheroidal graphite eutectic this gives a different picture of the physical situation than has been generally adopted and might lead to a better modelling approach for avoiding shrinkage. In the case of the lamellar and compact graphite eutectics, the combined volume change for the liquid transformation to graphite and austenite can be taken along a combined boundary advancing into the melt. The volume change at this boundary is approximately zero

$$\Delta V = \Delta V_s + \Delta V_\gamma = 0$$

In the spheroidal graphite case, ΔV is not equal to zero since the graphite grows before the austenite, and when the austenite grows, it does so dendritically, and virtually independently, in the liquid system. The volume increase of the graphite phase is dissipated in the liquid systems before the γ begins to grow. The γ phase grows dendritically in a mushy manner and the feeding problem becomes comparable to that at the centre of a steel casting.

3. Microstructural Modelling

Microstructural modelling is an important advance in applying solidification theory to cast iron. The overall analysis for computing temperature distribution in a solidifying casting is combined with kinetic models for nucleation and growth of phases. A computed curve for temperature with time can then be correlated with the curve for cooling obtained in a sample of the metal used for casting (16),(17),(18).

These curves follow patterns characteristic of the type of alloy being cast, Figure 6, and result from nucleation processes, mainly related to melt composition and inoculation treatments, and to the growth processes which have been discussed (19).

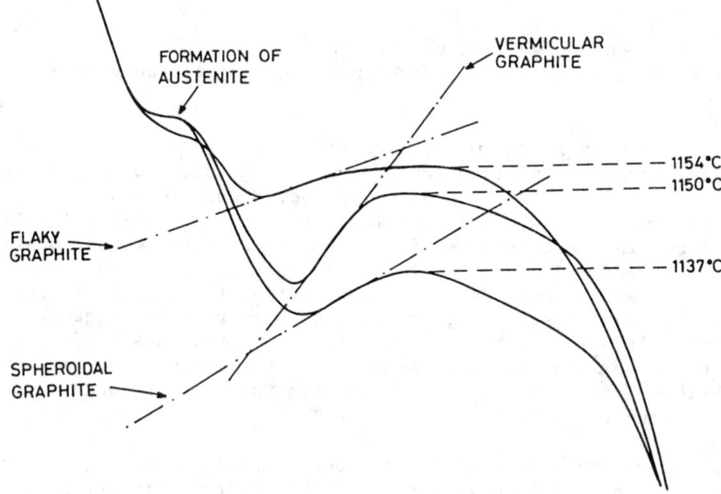

Figure 6. Cooling curves for the three types of structure typical of cast iron (19). The compacted graphite structure is labelled here as vermicular graphite.

The major characteristics of these cooling curves can be roughly analysed on the basis of undercooling, i.e. the curves should demonstrate an increase of undercooling on going through flake, compact and spheroidal graphite forms. The rate of recalescence after undercooling is determined by the nucleation rate and the growth rate. The maximum rate of recalescence is shown by the compact graphite melts and this must be dependent on the higher growth rate of the compact form which is still basically a flake form growing at a higher undercooling.

A disparity between the curves as calculated from theory, and the measured curves, is corrected by introducing factors. It is suggested that further work needs to be done in improving the theoretical approach. As an example the lamellar eutectic is modelled by a diffusion growth equation as follows:-

$$V = \mu(\Delta T)^2$$

μ is a growth constant and is determined experimentally. The equations provide good correlation between the computed and experimental curves for cast iron but there is a discrepancy in the figures for grain size between modelling and measured values.

4. Further Applications of Computing

The cast iron melt needs to be modelled as a system involving a large number of solute elements which undergo complex interactions. These play an important role in solidification and the growth of phases. They interact with themselves, with growing graphite, and with elements in solution important in establishing structure. The concentrations of elements in solution can decrease by interaction or increase by segregation. Important questions like spheroidal graphite iron casting for large sections are dependent on the chemical behaviour of the melt in solidification over extended periods of time. This type of modelling is best tackled by procedures based on Artificial Intelligence and using languages such as Prolog.

In the field of investigation using computers for the analysis of casting problems involving shrinkage in ductile iron, the present approach has been statistical (20, 21) and involves computing a Shrinkage Predictor.

5. References

1. W. Bollmann. Crystal Lattices, Interfaces, Matrices. W. Bollmann, Geneva, 1982.
2. I. Minkoff and S. Myron. Phil. Mag. 379, 19, 1969.
3. I. Minkoff. The Physical Metallurgy of Cast Iron, John Wiley 1986.
4. C. Roscoe and J.M. Thomas. Carbon 389, 4, 1966.
5. S. Amelinckx and P. Delarignette. J. Nucl. Mater. 17, 5(i), 1962.
6. K.W. Burton, N. Cabrera and F.C. Frank. Proc. Roy. Soc. London, A, 299, 243, 1951.
7. W. Oldfield, G.T. Geering and W.A. Tiller. The Solidification of Metals. Iron Steel Inst. Publ. 110 London 1968.
8. M. Hillert and Y. Lindblom. J. Iron Steel Inst. London 388, 176, 1954.
9. I. Minkoff. The Physical Metallurgy of Cast Iron. MRS. Sweden 1985.
10. I. Minkoff and B. Lux. Micron 282, 2, 1971.
11. I. Minkoff. The Physical Metallurgy of Cast Iron. Tokyo, Japan, MRS 1989.
12. I. Minkoff. Phil. Mag. London 1083, 12, 1965.
13. W. Kurz and D.J. Fisher. Int. Metall. Rev. 177, 24, 1979.
14. M. Hillert and V.V. Subba Rao. The Solidification of Metals. Iron Steel Inst. London Publ. 110, 1968.
15. G.P. Ivantsov, Growth of Crystals, Consultants Bureau, New York, 1960.
16. M. Rappaz and D.M. Stefanescu. Metals Handbook.
17. F.J. Bradley, P.F. Bartelt. C.A. Fung, R.W. Heine. The Physical Metallurgy of Cast Iron. Tokyo, Japan, MRS 1989.
18. Liu Jincheng and F. Weinberg. Personal Communication.
19. L. Backerud, K. Nilsson and H. Steen. The Metallurgy of Cast Iron, Georgi Publ. Co., St. Saphorin, 1975.
20. K. Ableidinger and D. Raebus. British Foundryman Aug/Sept p.320, 1986.
21. F.J. Bradley, M. H. Zmerli, and JunCai. The Physical Metallurgy of Cast Iron. Tokyo, Japan, MRS 1989.

The strength of permanent mould grey iron

K.G. Davis and J-G Magny
Metals Technology Laboratories, CANMET, Ottawa, Ontario, Canada

Abstract

The mechanical properties of 25 mm thick plates cast in an iron mould have been compared with those for similar plates cast in sand moulds. The carbon equivalent was maintained at the eutectic value, with silicon content 2.0, 2.5, and 3.0%. It was found that strength was controlled mainly by the iron composition, with variables such as pour temperature and mould type having comparatively minor effects. An expected correlation between graphite flake size and strength was not realised.

Introduction

The production of grey iron in "permanent", i.e. metal, moulds currently accounts for only a small proportion of the total iron castings produced in Western Europe and North America. It is more popular in Eastern Europe, where some 6 to 12% of iron castings are made in metal moulds(1). In sand foundries, maintaining acceptable working conditions is becoming increasingly costly, and the price for dumping used sand and for sand recycling is mounting. When these factors are considered along with the good surface finish and excellent dimensional reproducibility associated with the permanent mould process, it appears likely that a higher proportion of iron castings will soon be made in permanent moulds.

Davis et al(2) have demonstrated that with the liberal use of modern inoculants, chill-free grey iron with carbon content down to slightly over 3.0% and final Si under 2.5% can be produced in rod diameters down to 22 mm, equivalent to a plate with thickness 6 to 12 mm. However, most permanent mould cast iron is made with a carbon equivalent close to eutectic. What are the properties of this material compared with the sand-cast equivalent? An extensive investigation of machinability by the Eaton company (now Grede Permcast) showed a significant improvement for permanent mould iron(3). As to strength, it seems reasonable to expect permanent mould iron to be superior(4). Cast iron strength depends on the nature of the matrix, either pearlitic or ferritic, and on the volume, shape, and distribution of the graphite flakes. Higher solidification rates result in finer graphite flakes, and hence in general to greater strength. One may therefore expect permanent mould iron to be stronger.

Little work has been done comparing strengths in sand cast iron and permanent mould cast iron produced under exactly the same conditions of composition, inoculation practice, and pour temperature. Fig 1 shows some of the published data on fully ferritic castings. The UTS for permanent mould iron is greater than for sand castings at the smaller rod diameters, but at the largest diameter (51 mm, equivalent to a 25 mm thick plate), there was no significant difference. A similar conclusion was reached by Zuithoff et al(6). The aim of the current work is to investigate why the thicker-section permanent mould iron, despite a finer graphite size, failed to show superior strength.

Experimental Procedure

The charge consisted of high purity pig iron and steel, plus 75% FeSi and FeMn. 50 kg heats were melted in air in a medium frequency core-less induction furnace. Three compositions were chosen, designed to have a constant carbon equivalent of 4.25% and final Si contents of 2.0, 2.5. and 3.0%. Mn was kept at 0.2 %, and S and P contents were always low (approx. 0.015% and 0.020% respectively). The pour temperature was either 1250 or 1350 C. Nominal and actual C and Si contents, and pour temperatures, are shown in Table 1. From each heat one plate 8 in by 8 in. by 1 in.

TABLE 1
Final C and Si content of the castings. Obtained by thermal analysis of the base metal combined with emission spectroscopy of quenched samples.

Nominal Composition	Heat No.	Pour Temp, C	Carbon, %	Silicon, %
3.6 C, 2.0 Si	1	1350	3.5	1.9
	2	1250	3.6	1.8
	9	1250	3.7	1.9
3.4 C, 2.5 Si	3	1350	3.5	2.4
	4	1250	3.5	2.4
3.3 C, 3.0 Si	5	1350	3.3	2.9
	6	1250	3.2	2.8
3.4 C,* 2.5 Si	7	1350	3.4	2.6
	8	1250	3.3	2.5
	10	1250	3.4	2.4

* ... Castings ejected from permanent mould after 15 s instead of after 60 s.

(203 X 203 X 25 mm) was poured in an iron mould, and one in a sand mould. The iron mould was a split vertical design, with a single gate at the bottom/side. Before the mould was put into use the cavity surfaces were first sprayed with a refractory wash, then with a graphitic wash. The moulds were also coated with acetylene soot before each pour. Mould temperature at pour was kept to 200 ± 20 C. The sand moulds were made in CO_2 sand, poured horizontally, with a small riser in the centre. For both the sand mould and the permanent mould a single solid inoculant insert weighing approx. 27 g was incorporated in the gating system.

The procedure was as follows. Half the 50 kg heat was tapped into a pour ladle, 310 g (1.24%) of a proprietary inoculant containing 45% Si was stirred in at about 25 C above the pour temperature, and the iron mould was poured. This operation was followed immediately by repeating the procedure with the remaining metal and pouring the sand mould. Previous work has shown that eutectic solidification at the centre of the permanent mould plates is complete after about 50 s(7). Ejection time for the iron mould was normally 1 minute, at which stage it would be fully solidified. For one series of tests the ejection time was reduced to 15 s, at which stage the casting would be about 1/3 solid. In one test the sand mould was instrumented with thermocouples. Pour temperature was 1350 C. The solidification time at the plate half-thickness position varied from 225 s about 12 mm from the edge, to 800 s at the centre (the riser was moved off-centre for this casting).

After casting, a slice was cut across the centre of each plate and metallographic samples were prepared for examination of structure at the centre and at the edge. Three standard ASTM tensile test pieces with reduced sections 13 mm diameter by 70 mm long were machined from each of the remaining half plates, the centres of the test pieces coinciding with the plate half-thickness. To examine the effect of matrix structure, half the test pieces were given a ferritizing anneal of 740 C for 5 h in an argon atmosphere, followed by furnace cooling.

Results and Comments

Microstructure

The microstructures at the plate half thickness did not vary greatly from a position about 25 mm from the outside of the plate to the centre. As the tensile test results will reflect the properties of the central portions of the castings, only microstructures from the plate centres will be described. Fig 2, 3 show typical examples of the graphite structures. The sand cast plates showed coarse type A or type B graphite. There was little evidence of any effect from either pour temperature or silicon content. The permanent mould castings contained a fine type A graphite with areas of very fine type D graphite. Across the section of the permanent mould plates there were often three regions, see Fig 4,5. The outer region had fine ferrite dendrites in a type D graphite, followed by a similar region with coarser, less directional dendrites. At the centre, primary dendrites almost disappeared and the structure consisted of areas of fine type B graphite plus regions of fine type D graphite. Pearlite content increased in the successive layers from the outside to the centre. There was again no marked effect of temperature or of silicon content. The shorter ejection time (15 s against 60 s) produced a coarser type B graphite, Fig 6.

An image analysing microscope was used to measure pearlite contents. Attempts were also made to measure average and maximum flake lengths with the same instrument, but this proved difficult. For most of the castings there was no well-defined cell structure, and flakes often impinged. To test the dependence of mechanical properties on graphite flake size, chord lengths of the longest flakes in 10 fields stretched across the diameters of sectioned and polished tensile test pieces were measured from the screen of a standard metallograph. The results are presented in Tables 2 and 3.

TABLE 2
Max. chord lengths of graphite flakes, taken over 10 fields across a diameter of polished, sectioned tensile test pieces. The UTS for the ferritized test pieces is also given.
Sand Castings:

Heat No.	Flake Lengths μ-m	Av. μ-m	UTS MPa
1	400, 420, 390, 400, 450 350, 280, 380, 270, 370	371	103
2	480, 480, 560, 360, 400 520, 430, 350, 380, 420	438	72
3	510, 410, 530, 540, 450 600, 560, 650, 600, 520	537	98
4	480, 340, 470, 540, 580 400, 510, 750, 600, 450	512	74
5	620, 450, 550, 530, 480 460, 540, 520, 460, 390	500	115
6	480, 420, 500, 460, 460 500, 390, 410, 380, 510	451	97
7	540, 460, 520, 420, 570 460, 480, 470, 380, 450	475	91
8	410, 500, 490, 540, 450 470, 470, 550, 480, 480	484	87

TABLE 3

Max. chord lengths of graphite flakes, taken over 10 fields across a diameter of polished, sectioned tensile test pieces. The UTS for the ferritized test pieces is also given.

Permanent Mould Castings:

Heat No.	Flake Lengths μ-m	Av. μ-m	UTS MPa
1	130, 134, 142, 132, 172 130, 170, 146, 150, 140	145	103
2	80, 150, 100, 124, 116 126, 132, 86, 144, 102	116	82
3	96, 84, 82, 222, 136 130, 122, 200, 104, 144	132	113
4	60, 80, 102, 90, 160 138, 112, 128, 122, 150	114	99
5	176, 150, 142, 162, 110 114, 140, 120, 210, 152	148	117
6	70, 150, 96, 120, 146 114, 130, 130, 112, 130	120	121
7	136, 136, 130, 230, 270 178, 250, 220, 280, 266	210	92
9	88, 130, 98, 130, 131 226, 120, 180, 110, 160	137	89
10	280, 360, 230, 192, 284 296, 324, 302, 230, 260	276	89

Pearlite content:

The ferritized samples had either no pearlite, or else very small regions of highly spheroidized pearlite.

Structures at the centres of as-cast plates are summarized in Table 4. It may be concluded that:

1) Composition has the dominating effect, with less pearlite at the higher silicon contents. A notable exception is the permanent mould plate for heat 9. This casting had a carbon content greater than nominal, and it appears that the small increase in carbon equivalent represented by this was sufficient to suppress the pearlite transformation. Other inconsistencies illustrate the profound effect on pearlite content that small variations in composition can produce.

2) The mould type and pour temperature, and hence cooling rate, had a comparatively minor effect. Only in heats 3, 4 did there appear to be a large effect, with the higher pour temperature stabilising pearlite.

TABLE 4
Image analyzing microscope determination of pearlite content in samples taken from the centres of as-cast plates.

Nominal Composition	Heat No.	Permold/ Sand (P/S)	Pour Temp, C	% Pearlite
3.6 C, 2.0 Si	1	P	1350	88
	1	S	1350	88
	2	P	1250	73
	2	S	1250	75
	9	P	1250	16
3.4 C, 2.5 Si	3	P	1350	90
	3	S	1350	85
	7	S	1350	8
	4	P	1250	8
	4	S	1250	33
	8	S	1250	6
3.3 C, 3.0 Si	5	P	1350	1
	5	S	1350	3
	6	P	1250	2
	6	S	1250	4
3.4 C, 2.5 Si*	7	P	1350	17
	10	P	1250	97

*...Castings ejected from permanent mould after 15 s instead of after 60 s.

Tensile Tests

Tensile test results are shown in Table 5 and in Fig 7, 8. The strong effect of the ferritizing anneal is immediately clear. However, even with the influence of pearlite on strength removed, there is still no clear strength advantage for the permanent mould cast material. The lower silicon irons are stronger in the as-cast material, but not after annealing, showing the entire effect of silicon to be associated with pearlite content.

TABLE 5

Tensile test data for the as-cast and ferritized test pieces.

Sand Castings:

Nominal C, Si%	Heat No.	As Cast		Ferritized	
		UTS, MPa	Av.	UTS, MPa	Av.
3.6 C,	1	190, 188, 189	189	100, 108, 101	103
2.0 Si	2	124, 127, 125	125	75, 71, 70	72
3.4 C,	3	152, 157, 144	151	99, 101, 94	98
2.5 Si	4	93, 96, 94	95	74, 74, 74	74
3.3 C,	5	120, 121, 130	124	118, 128, 118	121
3.0 Si	6	108, 114, 116	113	99, 101, 93	97
3.4 C,	7	110, 119, 122	117	81, 92, 99	91
2.5 Si	8	108, 101, 108	106	89, 89, 83	87

Permanent Mould Castings:

Nominal C,Si%	Heat No.	As Cast		Ferritized	
		UTS.MPa	Av.	UTS, MPa	Av.
3.6 C,	1	196, 206, 195	199	103, 104, 102	103
2.0 Si	2	141, 139	140	79, 88, 81	82
	9	119, 112, 115	115	90, 86, 92	89
3.4 C,	3	150, 159, 155	155	111, 114, 114	113
2.5 Si	4	120, 116, 123	120	99, 100, 99	99
3.3 C,	5	130, 128, 129	129	113, 121	117
3.0 Si	6	138, 132, 137	135	118, 128, 118	121
3.4 C,*	7	106, 114, 117	112	91, 95, 90	92
2.5 Si	10	175, 166	170	91, 90, 88	89

*.. Castings ejected from permanent mould after 15 s instead of after 60 s.

Hardness

Brinell hardness tests were performed on polished surfaces at 500 kg load. For the as-cast material, hardness was measured on the pieces cut from the plates; values are averages of three readings. For the ferritized samples, hardness was measured on polished sections across the 13 mm diam. test pieces; only one measurement could be made per sample. The results are shown in Table 6. In Fig 9 hardness for the ferritized test pieces is plotted against UTS. At the same strength and composition, the permanent mould castings were a little harder.

TABLE 6

Brinell hardness. 500 kg load.

As-Cast:

Heat No.	Sand Mould	Permanent Mould
1	100	130
2	86	126
3	100	119
4	72	93
5	83	93
6	81	100
7	74	84
8	81	---
9	---	93
10	---	123

Ferritized:

Heat No.	Sand Mould	Permanent Mould
1	74	93
2	67	74
3	71	93
4	65	86
5	83	89
6	80	93
7	77	71
8	72	--
9	--	83
10	--	80

"Machinability"

No machinability tests as such were done. However, a qualitative assessment of machinability was made from observations on the threaded portions of the tensile bars. For each bar the surface finish was rated "poor", "fair", or "good" in comparison with the three samples chosen as standards shown in Fig 10. The results are recorded in Table 7. It is interesting to examine this table in the light of the hardness/UTS relationship shown in Fig 9. The permanent mould plates are almost without exception more easily machined than the sand mould plates, and yet they have a higher hardness/UTS ratio. This is opposite to the normal predictions. It seems that the main factor controlling machinability is not the ratio of hardness to strength, but is simply the graphite flake size. Coarser flakes pull out of the surface more easily, and also cause mini-fractures ahead of the tool, to produce a poor machined surface.

TABLE 7
Machinability rating from observations on the threaded portions of the tensile test pieces. "G" = good, "F" = fair', "P" = poor, see Fig 10. There are two assessments for each cast plate, one for the three ferritized test pieces and one for the three as- cast test pieces.

Nominal C%, Si%	Heat No.	Sand Mould		Permanent Mould	
		Ferritized	As-Cast	Ferritized	As-Cast
3.6 C	1	P	P	G	G
2.0 Si	2	P	P/F	G	G/F
	9			G	G
3.4 C	7	F	P	G	G
2.5 Si	8	G	F		
	10			F	F
	3	P	P	G	F/G
	4	P	P	G	F
3.3 C	5	F	F	G	G
3.0 Si	6	F	F	F	G/F

Discussion

Castillo and Baker(8) have expressed fracture toughness in grey cast iron by the simple expression

$$K_{1C} = 0.70 * UTS * sqrt(Pi X d)$$

where d is the eutectic cell size.

Using values for K_{1C} of 17 $MN/m^{3/2}$ for a pearlitic matrix and 11 $MN/m^{3/2}$ for a ferritic matrix, Castillo and Baker found this expression to give reasonably good agreement between measured cell diameter and UTS for sand castings with a carbon equivalent of 4.15% and cell size about 1.5 mm. However, they noted that the predicted strength for finer cell sizes was too high.

One problem with applying this approach to the current results is the lack of well-defined eutectic cells. One might reasonably expect, however, that the values for "max. flake length" as measured here will be related to the critical crack length that opens up during fracture. Assuming a basic cell-like form with flakes radiating from nucleus points, the value for cell diameter to be used in the above expressions should be close to twice the measured max. flake length, multiplied by the geometric factor 1.7 to allow for volumetric effects. For the ferritic irons this leads to

$$UTS = 15.7 / sqrt(3.4 X Pi X FL)$$

where FL is the average max. flake length in metres, and the UTS is in MPa. Examination of Tables 2 and 3 show that the max.

average flake lengths vary from 116 to 537 m. Applying the above criterion, the tensile strength is predicted to vary between 440 and 210 MPa, compared with the actual range of 121 to 72 MPa.

It may be concluded that the actual initiating crack length is greater than that obtained from the maximum measured flake length. In fact, when tensile strength is plotted against the inverse square root of the max. crack length, there appears to be little correlation, Fig 11. On the other hand, Srinivasan(9) has found the tensile strength in permanent mould cast iron plates and cylinders with 3.5% C, 3.0% Si, to be a narrow function of the local solidification time, and hence of the cooling rate. The answer would appear to lie in there being some mechanism whereby the fine flakes join up under fairly low stress, and it is the size of this joined crack, which in turn is a function of the graphite form, that determines the strength.

Acknowledgments

The authors are indebted to K.Kleinschmidt, J.R.Emmett, D.A.Brown, and B.Gracia for metallography, to G.Weatherall for the tensile testing, to B.Casault, who performed the image analysis to determine pearlite contents, and to M.Phaneuf for help in preparing the manuscript for publication.

References

1. "Advantages of permanent mold casting of gray and ductile iron for near-net shapes", A.P.Clark, ASM Near Net Shape Conference, Sept. 1984, Detroit.

2. "The inoculation of permanent mold cast iron", K.G.Davis and J-G Magny, Trans. AFS 1985, vol. 93, pp 939-952.

3. Machinability Evaluation of Eaton Permanent Mold Cast Iron vs. Sand Mold Cast Iron, report on work conducted for the Eaton Corporation Foundry Division (now Grede Permcast), 1982.

4. "Permanent mold casting of gray, ductile and malleable iron, C.A.Jones, J.C.Fisher, and C.E.Bates, Trans AFS 1971, vol. 79, pp 547-559.

5. "Permanent mould iron castings", R.K.Dostal, Design News, Aug. 1976, pp 65-66.

6. "The section sensitivity of cast iron permanent mold castings", AFS Cast Metal Research J., June 1972, pp 83-89.

7. "Modelling the solidification of cast iron in permanent molds", Z.Abdullah, M.Salcudean, K.G.Davis, Computer Modelling of Fabrication Processes and Constitutive Behaviour of Metals, CANMET, 1986, pp 397-415.

8. "The relationship between tensile strength and microstructure in flake graphite cast iron", R.N.Castillo and T.J.Baker, The Physical Metallurgy of Cast Iron, Materials Research Soc. Symposium(Ed: H.Fredriksson and M.Hillert), Vol. 34, 1984, pp 487-495.

9. "Computation and application of local solidification times of grey cast iron cast in metallic moulds", M.N.Srinivasan, Bull. Mater. Sci, 1981, vol. 3, pp 37-55.

10. Cast Iron, Physical and Engineering Properties, H.T.Angus, Butterworths. 2nd edition, 1976.

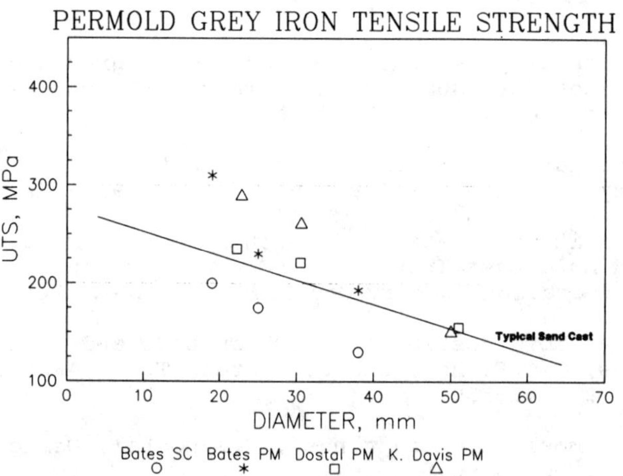

Figure 1. Data from previous work on tensile strength in permanent mould grey iron. All the data shown are for ferritic irons, either as-cast or annealed. Data sources: Bates..ref. 4, Dostal..ref. 5, Davis..ref. 2. The line for sand cast material is taken from ref 10, p 161.

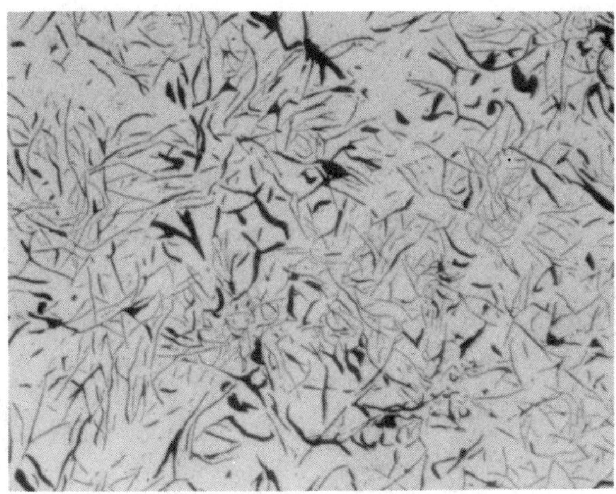

Figure 2. Graphite structure at centre of sand cast plate, heat #4. As-polished. X 50.

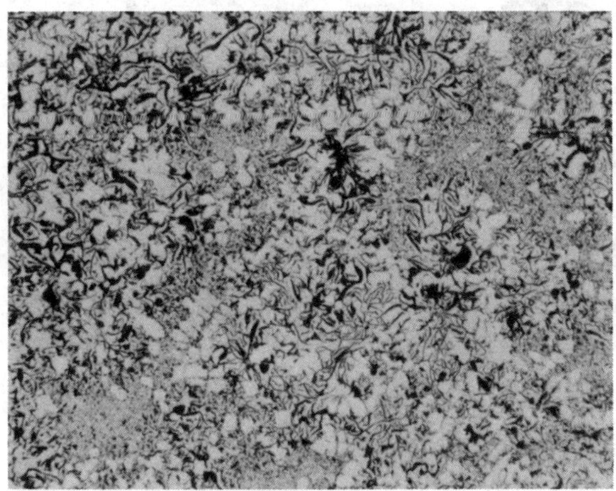

Figure 3. Graphite structure at centre of permanent mould cast plate, heat #4. As-polished. X 50.

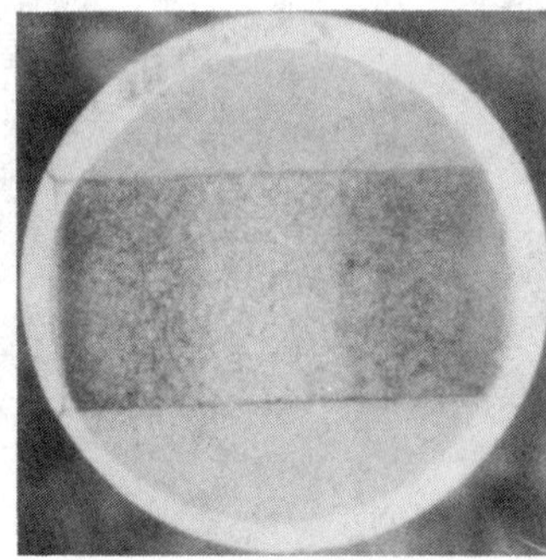

Figure 4. Section through centre of permanent mould cast plate, heat #3. 2% Nital etch. Approx. actual size.

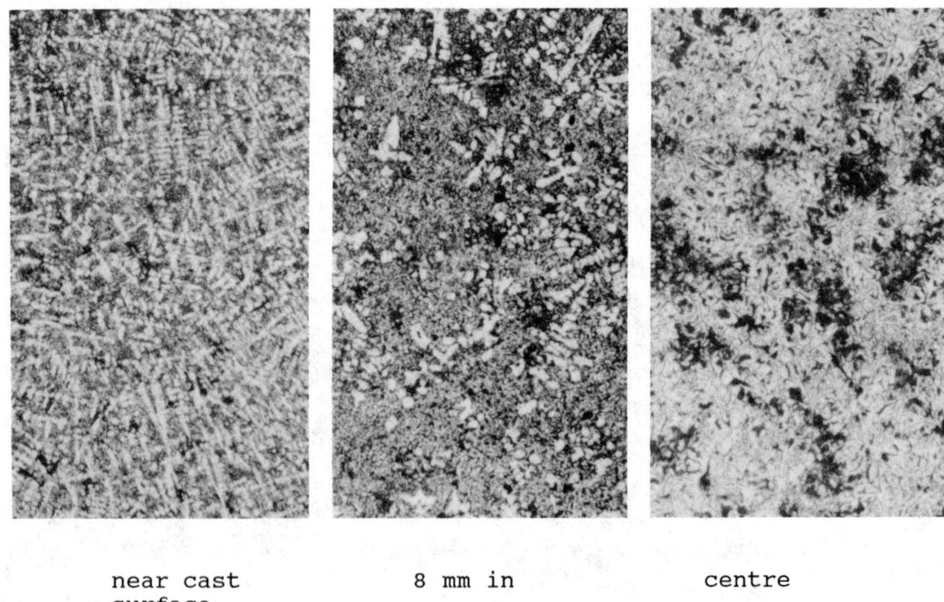

Figure 5. Microstructures in the three zones shown in Fig. 4. 2% Nital etch. X 50.

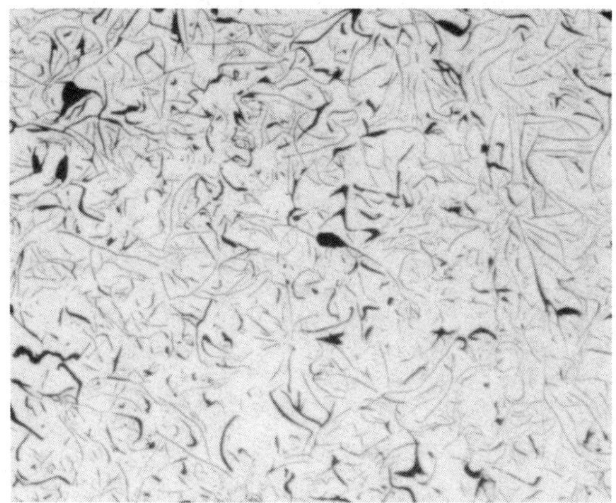

Figure 6. Graphite structure at centre of permanent mould cast plate, heat # 10, ejected after 15 s. As-polished. X 50.

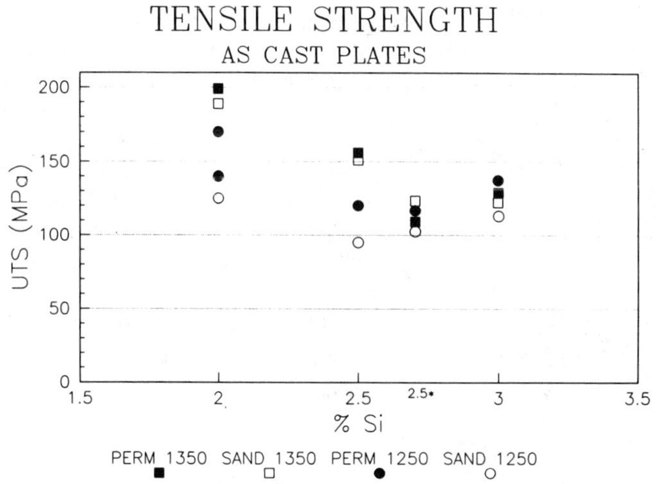

Figure 7. Tensile strength of the unannealed test pieces, as a function of the nominal silicon content. * ... 15 s ejection time from the permanent mould.

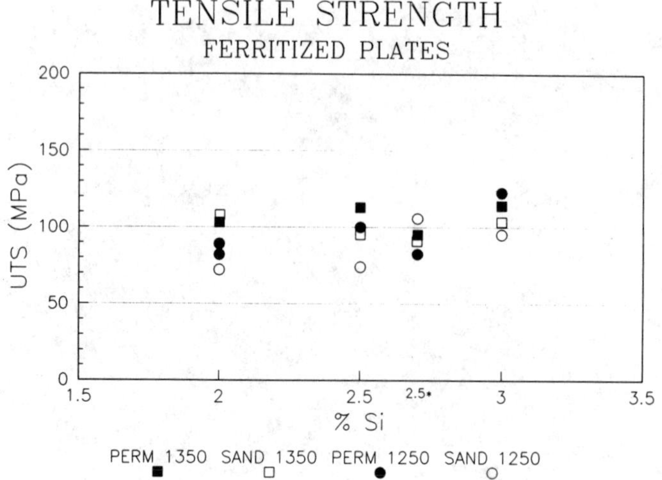

Figure 8. Tensile strength of the ferritized test pieces, as a function of the nominal silicon content. *... 15 s ejection time from the permanent mould.

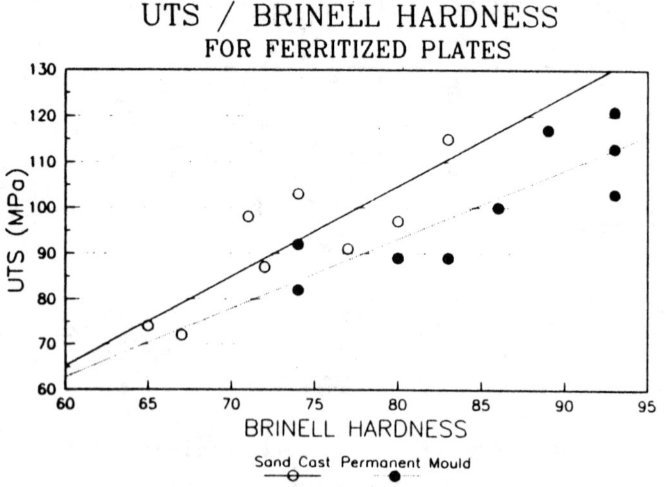

Figure 9. 500 kg Brinell hardness of the ferritized tensile test pieces, as a function of tensile strength.

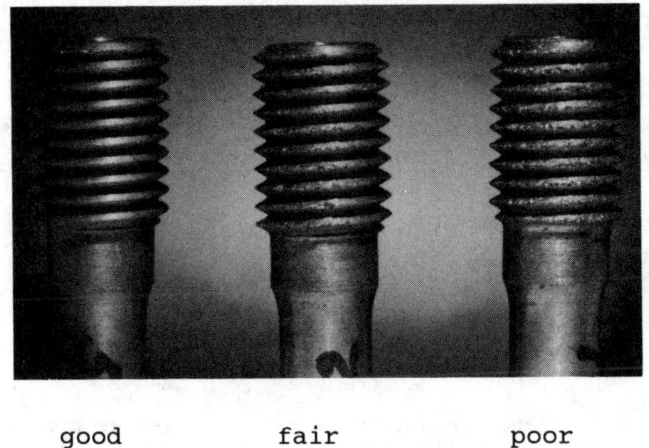

Figure 10. Threaded end sections of tensile test pieces selected as standards for machinability rating.

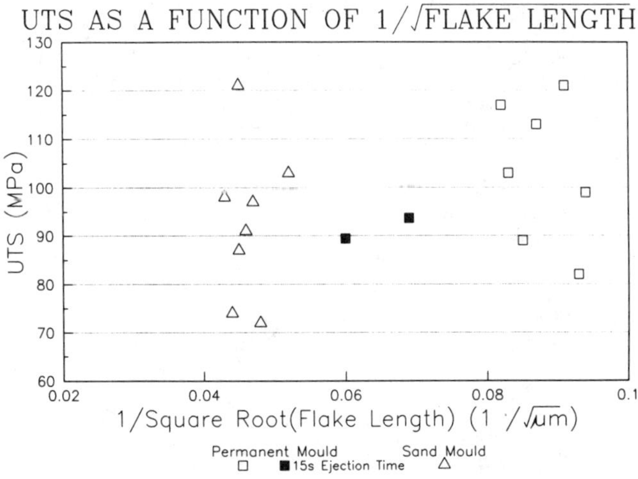

Figure 11. Tensile strength of the ferritized test pieces as a function of the inverse square root of the max. graphite flake chord length.

Metallographic study of the eutectic solidification of gray, vermicular and nodular cast iron

J.A. Sikora, G.L. Rivera
INTEMA-Universidad Nacional de Mar del Plata,
J.B. Justo 4302 (7600) Mar del Plata, Argentina

H. Biloni
LEMIT-CIC, 52 entre 121 y 122 - (1900) La Plata, Argentina

Abstract

A colour map of the segregation pattern can be revealed as a consequence of the development of a film of varying thickness formed by the etchant at the sample surface. When the light incises on this film isoconcentrates in the alloy are revealed in bright colours broght out by double reflection and interference.

The techniques have been used extensively in the past in order to relate the segregation substructures with casting structures. As a consequence several contributions have been done to the understanding of crystal nucleation, instability of the solid-liquid interface, origin and development of chill and equiaxed zones as well as the stability of crystals growing freely in the bulk liquid.

More recently the use of these techniques has been extended to the detection of the microsegregation associated with nodular cast iron solidification.

On this frame, the present paper intents to contribute to the understanding of the solidification mechanisms associated with the eutectic solidification when gray, vermicular and, specially, nodular cast irons are considered.

Introduction

The relationship between segregation substructures and casting structures, as detected by special optical metallographic techniques, has been studied extensively in the past by Biloni and coauthors (1-2). Through out these techniques was improved the understanding of the origin and development of the Solid-Liquid instabilities in controlled unidirectional solidification and, as a result, the microsegregation pattern associated with the solidification substructures in alloys with different crystallography (3-8). In addition , contributions were made to the understanding of crystal nucleation (9),origin and development of the chill and equiaxed zones in ingots (9-12) as well as to the stability of crystals growing freely in the bulk liquid (13). Most of these works were performed using the techniques developed by Lacombe and Mouflard (14).

In essence, the techniques consist in the development of a film of varying thickness formed by the etchant at the sample furnace. When the light incises on this film, isoconcentrates in the alloy are revealed in bright colours brought out by a double reflection and interference. As a result a colour map of the segregation pattern can be revealed and used as an excellent guide for microprobe analysis (15). The coincidence of colours fringes with iso-concentration contours has been documented in the literature, principally for Al-Cu alloys (16-19).

Recently Mortensen et al (15) summarize the research performed at MIT in order to characterize the microstructures in cast metal composites using this type of techniques and Motz (20) stressed the importance of the relationship between a careful colour metallography,detecting qualitatively the micro-segregation map, with a proper quantitative measurement in nodular cast iron.

In the particular case of cast iron many authors reported microsegregation studies, specially in nodular cast irons (21-30) and seems to exist difference of interpretations when eutectic solidification is considered.

On this frame the present paper intents,through out the application of these special metallographic techniques and other complementary, to contribute to the understanding of the origin and development of the eutectic colonies when commercial gray, vermicular and, specially, nodular cast iron structures are examined. The authors consider that a comprehensive microsegregation profile of the samples, eventually combined with quantitative microanalysis may assure a proper correlation between structure and physical, chemical and mechanical properties.

Experimental

Table I gives details of the samples used in the present work. Gray, vermicular and nodular castings with approximately eutectic composition were cast at at LEMIT foundry pilot plant, Moldifer S.R.L. and Metalúrgica Tandil, both medium industries involved in casting technological research with INTEMA.

TABLE I. Characteristics of the Samples

CASTING	C	Si	Mn	Cu	Ni	Cr	Mg	P	S	Gr.Structure
LG	3.5	2.9	0.2					0.12	0.02	flake
LCG	3.5	2.9	0.2					0.12	0.02	compact
ME2	2.99	2.63	0.96	0.88		0.26	0.047	0.022	0.015	nodular
MT	3.22	2.98	0.9	0.81				0.032	0.022	nodular
Mn62	2.68	3.91	6.2							nodular
LV	3.49	1.69	0.89	0.65	0.60			0.035		nodular
LVI	3.43	1.83	0.87	0.64	1.22		0.049	0.038		nodular
ME6	3.40	3.03	0.27					0.02		nodular
LIV	2.97	2.95	0.73	1.27				0.061	0.016	nodular

In order to reveal the primary and secondary structures of the samples the following optical metallographic techniques were used.

1) Conventional Techniques

As first step in the metallographic analysis, secondary structure was always revealed by Nital 2% etching.

2) Colour Metallography

In the Introduction are explained the principles of the colour metallography able to reveal the segregation pattern of elements with $K<1$ or $k>1$. The result is the possibility of a detailed analysis of the primary structures and, as a consequence, the interpretation of the solidification mechanisms governing the liquid-solid transformation (31)

i) Heat Tinting

It consists in a deposition of an oxide film through out thermal annealing in a furnace during 15 to 20 minutes (32).

ii) Chemical Colour Etching

It essentially consists in the deposition of thin films resulting from chemical or electrochemical processes. In the present work were need two chemical etchings named A and B having the following characteristics and compositions, according to Motz (20).

Etchant A: 10g NaOH, 40g OH, 10g picric acid and 50ml destilled water. The etchant is used when boiling at 120°C. It principally detects silicon segregation and the zones with higher silicon contents colour first, so that the colour differences emerge according to the silicon contents.

Etchant B: 50g sodium thiosulphate, 2g potassium metabilsulphate, 100ml destilled water, 20ml aqueous picric acid (0.2 picric acid in 100ml water). It is used at 60°C when boiling. It detects preferentially P segregation and the P rich regions are only very slightly tinted, whereas low-phosphorous zones colour deeply, turning dark.

Background Results and Discussion

1. Gray Cast Iron

In gray cast iron the morphology of the graphite present in the eutectic is laminar or flake like (F.G.). It is well recognized that the austenite (γ)-FG eutectic solidifies with the formation of eutectic colonies, named cells, more or less of spherical shape. Also is generally accepted that each eutectic is the product of a nucleation event. An interconnected network of FG surrounded by austenite with a morphological ramification, depending on undercooling, is the result of a coupled growth, characteristic of a faceted (FG)-no faceted (γ) eutectic growth (21) (23-25) (27-28).

In order to reveal the eutectic cells in gray iron the etchant B proved to be very useful. This etchant shows quite clear the microsegregation. Fig.1 shows clearly the cellular eutectic colonies delineated by the microsegregation, presenting a different colour with respect to the rest of the structure.

2. Vermicular or Compact (CG) Cast Irons

When CG alloys are considered, the current theories and experimental results indicated that the growth of the γ-CG eutectic is quite complicated both for the growing mechanism as well as the origin of the compact morphology (23-25). Nevertheless no doubt seems to exist on the fact that during the whole process of eutectic solidification the tips of the CG are always in contact with the melt, growing coupled with the austenite. The result is the formation of eutectic colonies that can be detected with colour metallography techniques.

Fig.2 corresponds to CG eutectic sample showing the existence of colony borders delineated by microsegregation with a different colour as in the case of γ-FG eutectic colonies. Essentially, the type of coupled growth is the same but it seems that in the case of the γ-CG eutectic the colonies must be more irregulars according to the models postulated. Compare Fig.30 with Fig.16 in reference (23). Fig.2 indicates the more irregular shape of the colonies in CG when compare with gray cast iron (Fig.1)

3. Nodular or Spheroidal Graphite (SG) Cast Irons

Meanwhile the cellular colony structures in gray and vermicular cast iron eutectics are the result of a coupled austenite-graphite growth, it is well recognized that the γ-SG eutectic corresponds to a divorced eutectic (23). As a consequence, in order to explain the formation of eutectic colonies is necessary to understand properly the growing mechanism.

Until recently the most accepted theory was that the growth of the γ-SG eutectic begins with nucleation and subsequent growth of graphite in the liquid,followed by early encapsulation of these graphite spheroids in austenite shells. Once the austenite shell is formed, further growth of graphite can occur only by the diffusion through the austenite. The consequence of this model is that each eutectic colony has as origin the nucleation of one graphite spheroid (2) (22) (24-28). Still many papers of the literature considers correct this model.

However, recent research, using quenching techniques during solidification and/or unidirectional solidification has shown that the solidification mechanism of the γ-SG eutectic is more complicated. Stefanescu (23) describes the sequence of solidification as follows:

i) At the eutectic temperature, austenite dendrites and graphite spheroids nucleate independently in the liquid.

ii)Limited growth of spheroidal graphite occurs in contact with the liquid.

iii) Flotation or convection then determines the collision of spheroids and austenite dendrites.

iv) Further growth of graphite occurs by carbon diffusion through the austenite shell.

When industrial nodular cast irons are considered the growth is multidirectional and the recognition of the solidification growth history during freezing can not be determined using conventional metallography. Nital only reveals clearly the secondary structure and as a result limited details of the primary segregation substructure are obtained. As has been done in Al-Cu alloys (16) (31), the detection of a detailed microsegregation map with special optical metallographic techniques, as colour metallography, may clarify the complicated solidification pattern giving an idea of the mechanisms, determining the origin and development of the eutectic colonies.

On this frame, the following results have been obtained:

i) When γ-SG eutectic structures with perlitic or ferritic matrix are considered, the application of the chemical etchants and thermal tinting reveals:

a) Fig. 3 corresponds to the etching of ME_2 samples with etchant A. The interior of the eutectic cells is tinted with a green colour meanwhile the contours are yellow. The dark areas correspond to the final transient of the phase solidification process. The enrichment of that area in alloy components with $k<1$ has as a consequence the precipitation of carbides and another phases. It is proposed that the SG have been trapped by dendrites during the solidification process as proposed in the literature (21-23). Meanwhile the spheroids only can grow by austenite diffusion the eutectic cells borders are delineated by the segregation associated with the final segregation transient of the dendrites growing in the bulk liquid. The chemical colour etching reveals clearly the size and shape of the cell colonies.

b)Fig.4a) and b) give an example of the potentiality of these metallographic methods in order to detect shape, size and a qualitative degree of microsegregation when different cooling rates control the solidification.

In effect, Fig.4 a) corresponds to a cast bar of Ø = 20mm, meanwhile Fig.4 b) corresponds to a bar of Ø = 80mm.

In both cases the colours are similar to Fig.3

c) Similar results were obtained with thermal tinting, Fig.5. In this case the eutectic cells have a yellow colour meanwhile their limits are brown. It is quite apparent that the SG are inside the colony cells presenting a dendritic substructure.

ii) When Mn 62 alloys with 6.2% Mn were considered etching with Nital 2% was quite useful. In effect this etchant reveals, in this case, the primary structure as a consequence of solid phase transformations. As a result, dendrites with bainitic structures are clearly revealed. It is quite apparent in Fig.6 that most of the SG are situated inside the dendrites. With respect to this rich Mn alloy the following solidification events are proposed:

1) Austenite dendrites and SG are nucleated in the bulk liquid.

2) The austenite dendrites grown trapping SG and rejecting Mn to the liquid.

3) When solidification is over the dendrites transform in bainitic structures meanwhile the interdendritic and intercolony liquid enriched in Mn retain the austenitic structure. At that regions carbides and other phases can be detected as a consequence of positive segregation during the solidification of the multicomponent alloy. When thermal tinting is used, similar results are obtained. In this case, the interdendritic and intercolony areas are detected by a brown colour meanwhile the bainitic structure of the dendrites present a pale yellow colour. Fig. 7. No doubts seems to be in the sense that most of the SG are associated with the dendritic substructure of the cell colonies.

As a result of the present work the authors consider that the mechanism proposed by Rickert and Engler (21), Loper and Heine (22 and, recently, by Stefanescu (24) is essentially correct. The microsegregation map permits to infer the solidification history of the nodular casting. During the process the dendrites trap among its branches the graphite spheroids nucleated independently in the liquid and growing through the diffusion mechanism. The degree of microsegregation detected by the special metallographic techniques permits to detect not only the interdendritic areas but also the impingement of the dendrites with different orientations. The larger microsegregation associated with the final solidification transient permits to detect the eutectic cells or colonies. Then, after freezing, the final structure of nodular cast iron correspond to multinodular eutectic cells with a dendritic substructure. Fig.8 corresponds to alloy MEG; it is considered a good example of the mentioned final structure.

From the present work, some considerations seems to be useful when commercial nodular cast irons structures are analyzed: As a result of homogeneity process and phase transformations occurring during freezing, conventional metallography present, in most of the cases, difficulties to reveal total or partially the eutectic cells. This not only jeopardize the comprehension of the origin and development of the colonies but also may produce mistakes on the degree of microsegregation between two graphite spheroids. In effect, the microsegregation mapping determined by colour metallographic techniques detects quite clearly that two graphite spheroids can be situated inside of a same dendrite or separeted by an interdendritic or an eutectic cell border. In each case the segregation pattern will be quite different and the quantitative values obtained through microanalysis could differ dramatically. These differences may affect the physical, chemical or mechanical properties. As an example, crack propagation in nodular cast iron has been properly determined (33).

Summary and Conclusion

Special metallographic techniques proved to be useful in order to reveal the primary structure in eutectic iron casting. Detecting in detail the micro-segregation substructure are able to dilucidate:

1) The solidification mechanisms in gray , vermicular and nodular eutectic cast irons.

2) The microsegregation associated to the eutectic cells , as a result, the potential influence on physical, chemical and mechanical properties.

Acknowledgements

To the Argentine National Research Council (CONICET), Buenos Aires Province Research Council (CIC), the International Development and Research Centre (IDRC) from Canada and The Organization of American States (OAS) for economical support to LEMIT and INTEMA Research and Development Programs in Casting Technology. To Mr.A.S.Allende for experimental collaboration.

References

1. Biloni,H., Solidification, Physical Metallurgy, Chapter 9, eth.ed. R.W. Cahn and P.Haasen editors. Elsevier Science (1983).
2. Biloni,H., Aluminum Transformation Technology and Applications, p.1, Eds. C.A.Pampillo, H.Biloni and D.E.Embury, ASM, Metals Park, Ohio, (1979).
3. Biloni,H., Canadian J.of Physics, 39, p. 1501, (1961).
4. Biloni,H., Bolling, G.F., Domian,H.A., Trans.Metallurgical So., AIME, 233, p.1926, (1965).
5. Biloni,H., Bolling,G.F. and Cole,G.S., Trans.Metallurgical So., AIME, 233, p.1926, (1965).
6. Biloni,H., Bolling,G.F. and Cole,G.S., Trans.Metallurgical Soc., AIME, 236, p. 930, (1966).
7. Biloni,H., Di Bella,R.D. and Bolling,G.F., Trans.Metallurgical Soc.,AIME, 239, p.2012, (1967).
8. Audero,M.A. and Biloni,H., J.of Crystal Growth, 12, p. 297, (1972).
9. Biloni,H. and Chalmers B., Trans.Metallurgical So., AIME , 233, p. 373, (1965).
10. Biloni,H. and Chalmers B., J.of Materials Science, 3, p. 139, (1968).
11. Biloni,H. and Morando,R., Trans. Metallurgical So., AIME, 242, p. 1121, (1968).
12. Prates,M. and Biloni,H., Metallurgical transaction,AIME, 3, p.1501, (1972).
13. Kiss,F.J. and Biloni,H., The solidification of metals, Iron and Steel Inst., London, Publication 110, p. 74, (1968).
14. Lacombe,P. and Mouflard,H., Metaux-Corrosion-Industries, p. 340, (1953).
15. Mortensen,A., Gungor,M.N., Cornie, J.A. and Flemings, M.C., Journal of Metals, p. 30, (1968).
16. Calvo,C. and Biloni,H., Z.Metallkunde, 62, p. 664, (1971).
17. Cornie,J.A.,Mortensen,A., Gungor,M.N. and Flemings,M.C., Proc. of the First International Conference on Composite Materials, ICCM V, San Diego, Harrimon et al. eds., p. 809, (1985).
18. Buckle,C., Chargainier,C. and Calvet,J., Comptes Rendues de l'Academie de Sciences, 235, p. 1040, (1952).
19. Erdmann-Jestnitzer,F. and Bernhardt,W., Metall.,11,(12),p. 1032, (1957).
20. Motz, J.M.,Practical Metallography, 25, p. 285, (1988).
21. Rickert,A., Engler,S., Material Research So. Symp.Proc., 34, 165, (1985).
22. Loper,C.R. and Heine,R.W.,AFS Trans., 94, p. 547, (1968).
23. Stefanescu,D.M., Cast Iron, Metals Handbook, Ninth ed., 15, p. 169,(1988).
24. Hendrix, J.C., Curreri,P.A., Stefanescu,D.M.,AFS Trans., 99, p.435,(1984).
25. Stefanescu,D.M., Materials Research Symp.Proc., 34, p. 151, (1985).
26. Zhu Zhang., Flower,H.M., Yinyi Niu, Materials Science and Technology,5, p.657, (1989).

27.Fredriksson,H., Svensson,I.L., Materials Research Symp.Proc.,34, p.273, (1985).
28.Stefanesfu,D.M. and Kanetkar,C.S., AFS Trans., 68, p. 139, (1987).
29.Hayrynen,K.L., Moore,D.J. and Rundman,K.B., AFS Trans., 183, p. 619, (1988).
30.Boeri,R. and Weinberg,F., AFS Trans., 106, (1989).
31.Biloni,H., The solidification of metals (Iron and Steel Inst., Publication 110), p. 74, (1968).
32.Beraha,E. and Sphigler,B., Color Metallography, (1977).
33.Villacis,W. and Sikora,J., Cast Metals, 1, p.210, (1989).

Fig. 1.- Sample LG. Ferritic matrix plus FG. Different colour delineates the eutectic cells. Etchant B.

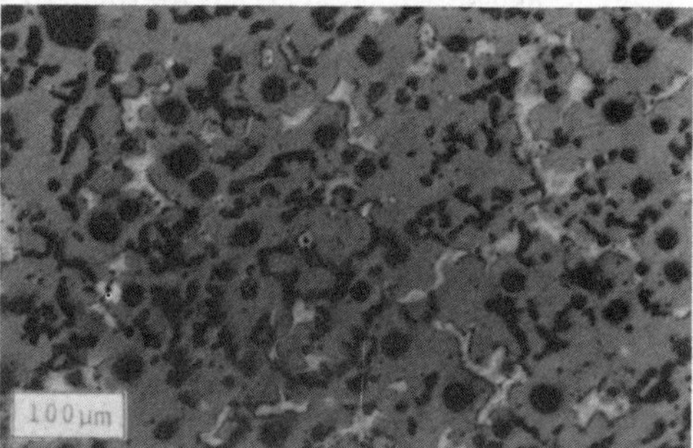

Fig. 2.- Sample LCG. Ferritic matrix plus compact and some nodular ghephite. The eutectic cells are delineated by different colour than the matrix. Etchant B.

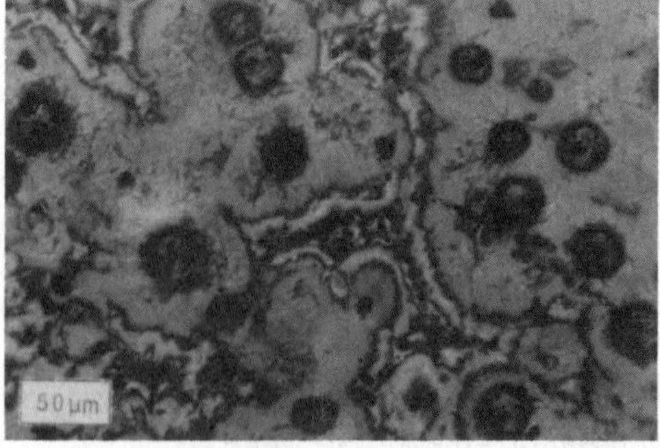

Fig. 3.- Sample ME2. Dendrites with perlitic structure presenting green colour delineated by yellow colour and precipitation of phases. Notice the SG inside of the dendrites. Etchant A.

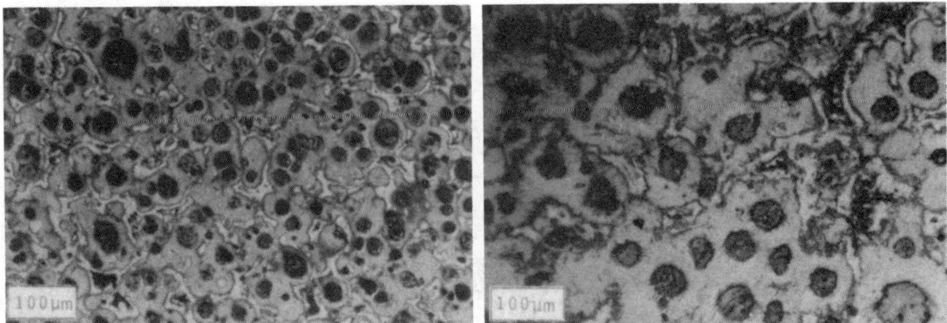

Fig. 4.- a) Sample ME2; Ø = 20 mm. b) Sample ME2; Ø = 80 mm. Notice the difference in size of the eutectic colonies as a function of difference in freezing rate. Etchant A.

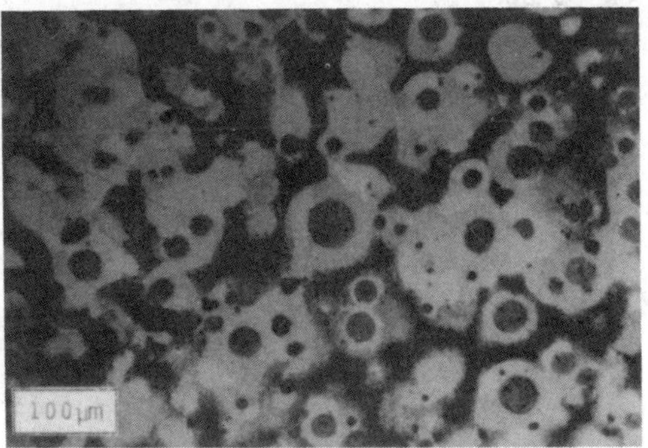

Fig. 5.- Sample MT. Structure as revealed by thermal tinting. Eutectic colonies presenting dendritic substructre and yellow colour delineated by brown colour. Notice the multinodular characteristic of the eutectic colonies.

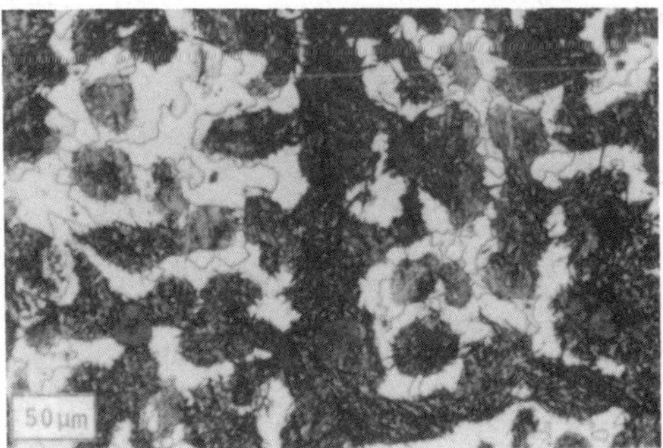

Fig. 6.- Sample Mn 62. Brown dendrites with barnitic structures associated to SG. The interdentrictic areas correspond to retained austenite and carbides both white. Etching: Nital 2%.

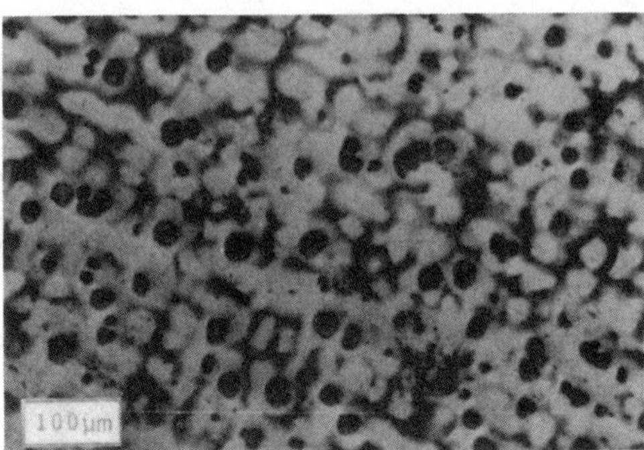

Fig. 7.- Sample Mn 62. Thermal tinting etching revealing interdendritic and intercolony areas.

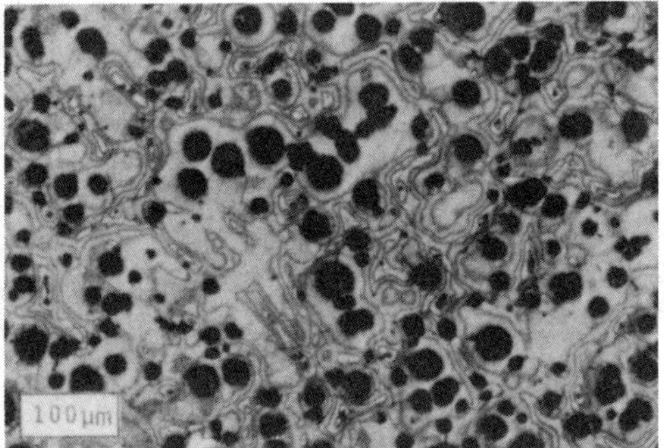

Fig.8.- Simple MEG. Nodular cast iron with ferritic structure. Plurinodular eutectic colonies presenting dendritic substructure. Etchant A.

Graphite morphology control in cast iron

S.V. Subramanian and G.R. Purdy
Department of Metallurgy and Materials Science, McMaster University, Hamilton, Ontario, Canada

Abstract

Graphite morphology in cast iron is analysed in terms of nucleation and growth kinetics of graphite crystals in liquid iron. The high interfacial energy between the graphite/melt interface does not allow the homogeneous nucleation of graphite in the melt. The efficacy of the heterogeneous substrates influences the growth undercooling.

At small driving forces, i.e., low supersaturation or small kinetic undercooling, graphite growth is characterised by faceted growth resulting in flake, compacted and spherulitic graphite morphologies. However, at large driving forces, there is a transition from faceted to non-faceted growth, resulting in a dendritic growth morphology.

Flake morphology is rationalised in terms of impurity dependent crystal growth mechanisms, whereas a spherulitic morphology is attributed to a defect controlled spiral growth mechanism. Compacted graphite morphology is considered a transition between flake and spherulitic morphology.

A thermodynamic approach is used to inter-relate the residual concentrations of impurities of technological interest, i.e., S and O as a function of the residual concentration of the reactive elements, Mg, Ca, and Ce in a typical cast iron melt at 1500C and atmospheric pressure. Such a diagram that quantitatively relates graphite morphology in thick cast iron sections to soluble concentrations of impurities is referred to as a graphite morphology control diagram.

In thin sections with characteristic large undercoolings, the defect controlled spiral growth mechanism dominates over impurity dependent growth mechanisms, leading to deviations from predictions based on graphite morphology control diagram. Estimates of the growth kinetic curves for cementite, austenite and graphite (flake and spherulitic) are presented. The nucleation and growth of cementite can be prevented in thin sections through the provision of an adequate number of heterogeneous nuclei for graphite, and a selective driving force for graphite growth through the addition of graphite stabilising elements. There is an optimum dispersion of graphite nuclei for a given section size to promote the required degree of interconnected growth of graphite in compacted morphology.

Nucleation

The interfacial energy between graphite and iron melt is high, which precludes the possibility of homogeneous nucleation. The interfacial energies of basal and prism planes with iron are reported to be 1460 and 1720 in magnesium treated iron and 1270 and 845 ergs/cm^2 in untreated iron respectively (1). Experimental work involving high purity silicon with liquid iron diffusion couple has confirmed that the increase in carbon supersaturation associated with local silicon concentration transients favour the nucleation of graphite (2). Invariably graphite nucleates on deoxidisation and desulfurisation reaction products. Figs. 1a and 1b show an example of cerium oxysulfide occurring in the core of spherulites. The desulfurisation reaction products from magnesium treatment are found to act as heterogeneous substances for nucleating graphite nodules (3). The potency of the heterogeneous nucleating substrate influences the degree of undercooling for solidification and thus the growth morphology. The local dispersion of graphite nuclei influences the degree of interconnected growth of graphite associated with flake and compacted graphite morphology (4).

The Control of Graphite Growth Morphology in Cast Iron

Cast iron structures can be rationalised in terms of growth competition between austenite, graphite and cementite (5-10). Graphite growth, in turn, can be predominantly on the prism face, as in flake graphite, or on the basal face as in nodular graphite.

The growth velocity of each interface can be analysed in terms of:

(i) its mobility, i.e., the growth response of the interface (its velocity for a given driving force), and

(ii) the actual driving force available.

The force can be related to the growth undercooling (ΔT), which is the difference between the growth arrest temperature in the thermal analysis curve and the equilibrium phase boundary temperature, expressed in degrees Celsius, or the growth supersaturation (ΔC), expressed as a mole fraction or as a fractional quantity. The driving force can be influenced by the melt constitution (the equilibrium effect), and nucleation and solidification growth variables (the kinetic effects).

Interface Mobility

Interface mobility during solidification is determined by the mechanism of crystal growth. Non-faceted interfaces are considered to be atomically 'rough' and are characterised by a normal or continuous growth mechanism (11). Such interfaces exhibit high mobility and grow under small negligible driving forces. (Metallic interfaces are good examples of high mobility interfaces). Growth rates of such interfaces are governed by the rate of transport of solute to the interface. In the absence of fluid flow in the liquid, growth is controlled by diffusional transport in the liquid. The growth of austenite and cementite are generally characterised by 'normal' growth mechanisms.

Once an interface is faceted, as in pure graphite, lateral growth mechanisms operate on the crystallographically smooth interface, which is generally of low mobility. Thus, larger driving forces are involved, depending upon the specific growth mechanism. In the absence of impurities, i.e., in pure crystal growth, the two well known crystal growth mechanisms of faceted interfaces are:

(i) 2-dimensional nucleation controlled growth in which a new crystal layer has to be nucleated before growth spreads and covers the layer, and

(ii) defect controlled growth, wherein crystal defects serve as sources of steps for growth of the new layer.

Amongst the different crystal growth mechanisms, it is clear that two-dimensional nucleation controlled growth requires by far the largest driving force, and neither prism nor basal growth of graphite requires such a large driving force. In the absence of impurities, however, spherulitic graphite grows at a rather large under-cooling; this has been attributed to a spiral growth mechanism (12). In the presence of the impurities that are normally present in untreated technical cast iron, flake, i.e., prism growth is characterised by a high interfacial mobility, and therefore a basic understanding of the role of impurities on the mobility of faceted interfaces is required for the control of the relative rates of graphite growth on basal and prism faces.

The Effect of Impurities on the Interface Mobility of Faceted Interfaces: Theoretical Predictions

The effect of impurities on the growth rates of faceted crystals has been analysed in an elegant manner by Gilmer and co-workers (13,14) by computer modelling the dynamics of crystal growth. The results, though not specific to graphite growth, give a useful insight into the dramatic influence of impurities, even when present at very low concentrations, on interface mobility.

Fig. 2 compares results of simulated growth rates in the presence of impurities, with the growth rates of the pure crystal. The hypothesis here is that an impurity that forms strong bonds with the host species will adhere to the crystal surface and afford favourable sites or steps for the crystallisation of new layers. Such an impurity is designated as an ordering impurity. Even a minute concentration of such an impurity can cause a large relative increase in the nucleation rate. A single impurity atom can make the difference between an unstable cluster, and one that will persist and expand until the new layer has crystallised. The strong bond energy between carbon and oxygen would point in favour of such a mechanism rendering the prism face highly mobile, even at small driving forces.

Under large driving force, a gradual change from very anisotropic to nearly isotropic growth is predicted. Under these conditions, it should be possible, in principle, to produce dendritic morphology. During solidification, such a large driving force is not normally accessible.

Phenomenological Observations on Graphite Grown from Iron Melts

A detailed study of graphite grown under conditions of controlled chemistry in iron melts confirmed that the removal of soluble concentrations of surface active elements like sulphur and oxygen promotes the faceting of the prism faces of the graphite, thereby affecting their mobility. Fig. 3 shows a TEM picture of graphite grown from an Fe-C melt of low sulphur content. The faceting of the prism face is clearly seen. On the further addition of impurity, i.e., sulphur, the prism face in Fig. 4 shows a distinct tendency to form a rounded, 'sinusoidally' perturbed interface. Even when faceting was detected in high sulphur samples, the prism facets were separated by smooth rounded segments at their junctions; these junctions evidently acted as sources of growth steps, as predicted by Chernov (15).

In the absence of impurities, pure graphite crystal growth is characterised by prism faceting and the driving force needed for two-dimensional nucleation to form new layers on the prism face far exceeds the driving force required for spiral growth normal to the basal plane to form spherulites. In the presence of impurities, such as sulphur and oxygen, that occur in untreated technical cast iron, the mobility of the prism face is dramatically increased. The result is flake graphite growth that can compete with austenite growth, establishing a weakly coupled eutectic. These phenomenological observations are thus in accord with the computer predictions, lending support to the hypothesis that impurities like sulphur and oxygen promote prism growth at low undercooling.

As silicon raises the activity of carbon, the driving force for graphite growth can be increased by increasing the silicon concentration. A silicon concentration gradient was obtained in a liquid column of cerium treated cast iron by contacting it with high purity silicon and the iron was frozen in situ. A dendritic morphology resulted in the silicon enriched region in accordance with the predictions of growth under large supersaturation (16).

Physical Chemistry of Impurity Control in the Melt

Graphite crystal growth mechanisms are sensitive to soluble impurity concentrations at the parts per million level, and therefore, control of these impurities at such low concentration levels is critical for graphite morphology control. A thermodynamic approach to this problem is an obvious choice. Since oxygen and sulphur are the impurities of technological interest, soluble concentrations of these impurities in cast iron melt can be related to residual concentrations of reactive elements like Mg, Ce and Ca. The diagram shown in Fig. 5 quantitatively relates the soluble concentrations of sulphur and oxygen with residual concentrations of each of the reactive elements Mg, Ce and Ca, computed for a typical cast iron of composition 3.5%C, 2.0%Si at 1500°C, using the Henrian activity formalism (17).

In the absence of probes based on solid electrolytes to measure the soluble concentration of elements as predicted in the diagram, experimental investigations were undertaken by making programmed additions to iron melts in order to relate graphite morphology to the concentration ranges in the diagram at different cooling rates.

At slow cooling rates, such as those obtained in thick sections, where the kinetic undercooling is not significant, the graphite morphology can be directly related to the diagram.

At rather low sulphur contents, oxygen as an impurity, because of its strong bond energy with carbon, can promote flake growth, even at the very low concentrations (less than 10ppm) present in untreated, carbon saturated iron. However, impurities other than sulphur and oxygen can influence crystal growth mechanisms and thus graphite morphology.

The Application of the Graphite Morphology Control Diagram to Large Section Sizes

The graphite morphology control diagram can be readily applied in making programmed additions of treatment alloys to obtain a compacted graphite morphology in large section sizes, such as in ingot moulds, where the cooling rates are very slow and growth is characterised by small driving forces or kinetic undercoolings (18).

The successful application of the graphite morphology control diagram to consistently produce compacted graphite iron in large sections is dependent on impurity controlled crystal growth mechanisms at small driving forces. At large kinetic undercooling, characteristics of thin sections, impurity growth mechanism are dominated by the defect controlled spiral growth mechanism, thereby increasing the nodularity. Competitive growth of cementite has to be suppressed also. These factors are analysed in the following section.

Compacted Graphite Morphology Control in Thin Sections

In an engineering casting, where a range of section sizes are involved, such as in an automotive engine block, a wide range of cooling rates occur, the faster the cooling rate, the greater the undercooling. Therefore, nucleation and growth of phases in thin sections are attended by larger undercooling, there are two important factors that affect graphite morphology control:

(i) Spiral growth mechanisms are faster than impurity growth mechanisms (14), leading to deviations from predictions based simply on the graphite morphology control diagram,

(ii) Cementite is characterised by a highly mobile interface (6) and therefore, the prevention of nucleation and growth of cementite at large undercooling, is an essential aspect of graphite morphology control.

Solidification Analysis

Estimates of Growth Kinetic Curves

The observed promotion of basal growth in preference to prism growth indicates that the kinetic barrier for growth of the faceted prism face exceeds that required for faceted basal growth. Numerical calculations based on typical growth undercoolings, obtained from thermal analysis observations (29) during spherulitic growth, further indicate that the basal growth mechanism is probably defect-controlled.

Estimate of Growth Constants for Spherulitic Growth from Thermal Analysis Data

W. Oldfield and J.G. Humphreys (19) have attempted to measure the undercooling for each type of graphite morphology, as a function of cooling rate, by varying section diameters. The results give a comparison of growth arrest temperatures for ledeburite eutectic (coupled growth) and spherulitic graphite-austenite eutectic, for each section thickness under comparable chemistry. Their analysis, however, failed to take into account the divorced eutectic reaction involved in the case of spherulitic graphite, resulting in a gross under-estimate of undercoolings for spherulitic growth. Their data-base is re-analysed in Table 1 incorporating the divorced eutectic concept of

Table 1

BAR Dia. in inches	Cooling Rate °C/min.	$\Delta T_{spherulite}$ (Divorced Eutectic) °C	$\Delta T_{cementite}$ (Coupled Growth) °C	$\frac{\Delta T_{S.G.}}{\Delta T_{cem.}}$
0.6	350	265	11	24
0.75	200	150	7.5	20
1.0	110	96	4	24
1.3	80	72	3	24
2.5	18.5	46	2*	23

* estimated

$$\frac{K_{cem}}{K_{S.G.}} = \left(\frac{\Delta T_{S.G.}}{\Delta T_{cem}}\right)^2$$ where K represents the kinetic growth constant

spherulitic growth. Over a range of cooling rates obtained under varying section sizes, the ratio of the undercooling of spherulitic growth to that of cementite growth at carbide transition temperature for each section size is a constant (i.e., $\Delta T_{S.G.}/\Delta T_{cem} \simeq 25$ for carbide transition) in agreement with the theoretical prediction. This implies that the parabolic constant for spiral growth on basal faceted interface is 0.0017 times that of the diffusion controlled growth constant for cementite, i.e., spherulitic growth is 580 times slower than cementite growth at a given undercooling. Since the driving force is partitioned between interface reaction and diffusion, with diffusional component dominating as spherulite grows to a micrometre size, the growth constant for interface controlled regime is expected to be even lower.

Figure 6 shows the estimated growth kinetic curves for interface-controlled spiral growth normal to the basal face (spherulitic growth) in comparison with the diffusion controlled growth curves for austenite, cementite, and flake graphite (20). The metallic phases austenite and cementite are characterised by a normal growth mechanism that involves the lowest kinetic barrier. Flake growth is characterised by high mobility prism growth brought about by an impurity mechanism. The growth of the high mobility prism interface is transport limited and, in the absence of fluid flow, is diffusion controlled. The diffusion controlled growth constants are extracted from the estimates of Hillert and Subba Rao (21) and Carlberg and Fredriksson (22). The growth constant for flake graphite is an order of magnitude slower than austenite, thus accounting for the weakly coupled growth between austenite and flake graphite in grey iron.

Clearly a large driving force is required by the faceted basal interface in comparison with the non-faceted interfaces involved in the cases of cementite, austenite and flake graphite with impurity adsorption.

The Effect of Low Interface Mobility of One Phase on Eutectic Solidification

A consequence of the low mobility of one of the phases is the divorced eutectic reaction. The basal growth of graphite lags far behind that of austenite. This results in the isolated growth of spherulites, essentially decoupled from the austenite.

Relative Undercooling for Coupled and Divorced Eutectic Growth

Coupled growth is easily established when both phases possess high mobility interfaces. The liquid and solid are thus in local equilibrium at each interface and the rate of growth is then determined by the rate of transport in the liquid, which is often diffusion controlled. The driving force involved then, is that required to support the diffusion appropriate to the interface velocity, and the corresponding undercooling is reflected in the isothermal growth arrest temperature analysis curve. Thus the growth arrest in the thermal analysis curve occurs at a temperature close to the eutectic temperature, at an undercooling characteristic of diffusional transport.

In the case of a divorced eutectic, in which one of the phases has a low mobility interface, the growth arrest temperature is well below the eutectic temperature such that the large driving force required by the low mobility interface is provided. The kinetics of the low mobility interface require that a substantial departure from local equilibrium be maintained during growth.

Figures 7a and 7b show schematically, the divorced eutectic case for spherulitic growth. Irrespective of the initial composition, a large supersaturation of carbon is established for the basal growth, by the continued growth of austenite below the eutectic. This enriches the residual liquid with respect to carbon, until the carbon supersaturation is sufficient to sustain the interfacial reaction as well as the diffusional transport of carbon from the bulk liquid to the growing graphite. Thus, austenite grows with negligible kinetic undercooling since a normal crystal growth mechanism is involved; growth of austenite is limited by the rate of transport of carbon away from advancing austenite interface. Basal faceted growth requires a large driving force, and from the typical undercoolings that characterise spherulitic growth, it is postulated that a defect-controlled spiral growth mechanism operates. The large kinetic undercooling required for the basal spiral growth in comparison with the negligible undercooling required by austenite leads to the divorced eutectic reaction. As basal growth progresses, the driving force for diffusion is progressively increased; this is considered in the next section.

Growth Model for Spherulitic Graphite

The present growth model for spherulitic graphite is based on the assumptions of heterogeneous nucleation and free growth of graphite in direct contact with the melt, followed by a later stage of growth in which the spherulite is encapsulated by austenite. The model incorporates the observations of several previous workers (3,10,23-25).

Quantitative analyses were carried out in the present work for typical isothermal growth arrests in order to quantify:

1) the effect of curvature on the total driving force

2) the partitioning of the total driving force between interfacial reaction and diffusion during the free growth of spherulites in the melt, and

3) the contact angle required for heterogeneous nucleation of austenite on growing spherulites for typical size ranges observed, and supersaturations inferred from the model. These quantitative results are incorporated in the description of spherulitic graphite growth model.

Growth of a Spherulite Directly in Contact With the Melt

The free growth of spherulites in direct contact with the melt has two stages, the first stage being predominantly controlled by the interface reaction and the second stage by diffusion.

The first stage starts off by being totally interface controlled. Any sharp curvature of the graphite interface alters the phase diagram in such a way as to reduce the total driving force available for basal spiral growth. At a stage when the spherulite is 0.1 micron radius (10^{-5}cm), the total driving force in terms of carbon supersaturation is reduced by about 50 per cent on account of this curvature effect. Two thirds of this reduced driving force is required for the interfacial reaction and the balance, one third, for the diffusional supply of carbon from the liquid. Because of the slow response (growth rate) of the graphite interface to the driving force in comparison to that of austenite, coupled growth is not possible and a fully divorced eutectic is developed.

The second stage of the free growth of spherulites in direct contact with the melt, is the diffusion controlled isothermal growth, corresponding to the growth arrest temperature in the thermal analysis curve. As the size of the spherulite becomes enlarged, approaching 1 micron radius, the curvature effect on the phase diagram becomes negligible, thus increasing the total driving force, of which about 80 per cent is required for the diffusional supply of carbon to the graphite interface. As the spherulite grows isothermally, carbon from the liquid in the vicinity of the spherulites is depleted, and the transport of carbon to the graphite interface becomes rate controlling. Thus an ever increasing fraction of the total driving force is used to ensure diffusional supply of carbon, as the sphere becomes enlarged. By the time the sphere radius has reached 5 microns, approximately 85 percent of the total driving force is used by diffusion and the remainder by interfacial reaction. The total driving force, ΔC, is thus divided between that required to drive the interface, ΔC_R, against the kinetic barrier, and that required to sustain the diffusional supply of carbon ΔC_D so as to give an unique velocity, V. Experimental work, based on interrupted solidification through quenching, by Welterfall, Fredriksson and M. Hillert (10) indicates that the radii of spherulites, at the end of the free growth stage are typically in the range 5-14 microns.

Growth of spherulites after the formation of an austenite envelope - As the isothermal growth of spherulites progresses, the continued depletion of carbon in the liquid at the graphite interface, increases the driving force for the heterogeneous nucleation of austenite on graphite. Nucleation calculations yield a low contact angle ($\theta \simeq 25°$), suggesting that graphite is an efficient heterogeneous substrate for austenite. However, fresh nucleation of austenite may not be warranted. Phenomenological observations suggest that there are growing austenites in the vicinity of spherulites; the nearest neighbouring austenite invariably outgrows and encompasses the spherulite.

Once the austenite shell forms, the graphite growth is determined by the transport of carbon and the total driving force can be considered as essentially that required for diffusional transport, see Fig. 7b. An implicit assumption here is that the mobility of a graphite-austenite interface is approximately the same as that a graphite-liquid interface. Hence the diffusion model developed by E.Scheil and J.D. Schobel (25) and Welterfall, Fredriksson and M. Hillert (10) becomes an adequate description of spherulitic growth in an austenite envelope.

The advantage of the present model compared with previous models which have assumed diffusion controlled growth throughout, ignoring interfacial kinetics, is that it accounts for the large undercoolings characteristic of spherulitic growth (27) and for the extended period of free growth prior to austenite shell formation. Interface kinetics dominate at the onset of spherulitic growth and it is only in the later stages of growth that the diffusion controlled local equilibrium model is strictly applicable. However, it must be appreciated that the bulk of spherulitic growth is diffusion controlled as proposed in previous models (10,24).

Prevention of Cementite Growth

Cementite is characterised by a highly mobile interface and therefore its growth can compete favourably even at low undercooling. Thus the prevention of nucleation and growth of cementite is an essential aspect of graphite morphology control.

Ternary addition to increase selectively the driving force for graphite growth - Any ternary addition to a binary Fe-C alloy that will stabilise graphite growth in preference to that of metastable carbides, by means of its equilibrium diagram effect, can be used to suppress carbides. Silicon additions are commonly used to selectively increase the driving force for graphite growth. Silicon increases the stable (graphite) eutectic temperature and lowers the metastable (carbide) eutectic temperature as measured by Oldfield (26,27). Carlberg and Fredriksson (22) have calculated eutectic temperatures as a function of Si content for stable and metastable Fe-C-Si systems.

With a 2 wt% Si addition, the separation between the graphite eutectic and the carbide eutectic (Ledeburite) equilibrium temperature is increased from 5°C to 39°C and consequently the limit on the graphite growth velocity is increased from 0.0003 cm/sec to 0.169 cm/sec (characteristic of thin sections). Carlberg and

Fredriksson (22) have pointed out the beneficial effect of aluminum in promoting graphite growth. By the same token, carbide stabilising elements (e.g. chromium) will decrease the driving force for graphite growth and increase that of cementite growth, the notable exception being P (28).

Nucleation Effects on Growth Undercooling

The effect of increased nucleation is to raise the isothermal growth arrest temperature to well above the cementite liquidus. The greater the number of nuclei, i.e., nodule count, the greater the volume fraction of graphite and the greater the release of latent heat. Thus, with an increased number of nuclei, the rate of release of latent heat is increased for any undercooling. A smaller growth undercooling is required for a high nodule count, compared with that required for a low nodule count, to balance a given rate of heat abstraction imposed on the system. The isothermal growth arrest temperature is thus raised, in accordance with the smaller undercooling requirement for the high nodule count case. Reduced undercooling has the effect of lowering the local velocity of graphite growth. However, graphite formation is ensured as the isothermal growth temperature is well above that required for the nucleation and growth or carbides.

A typical example, taken from data in the literature (29) is given in Fig. 8. For practically the same final chemistry, there is a distinct difference in the end of freezing in the two cases which can be related to the nodule density. For a low nodule count of $n^*=33$, the end of freezing temperature is 1107°C, i.e., 8°C below the cementite liquidus line. Consequently, the structure showed carbide formation. Increasing the nodule count n^* to 212, the end of freezing is raised to 1121°C, i.e., 6°C above the cementite liquidus line (1115°C), thus preventing any carbide formation.

The Effect of Large Kinetic Undercooling on the Impurity Mechanism of Prism Growth

Impurity mechanisms dominate at small driving forces, because of the linear dependence of growth rate on driving force, see Fig. 2. However, the parabolic dependence of the spiral growth rate on driving force results in the spiral mechanism overtaking the impurity mechanism, at larger driving forces. Thus, at the larger undercoolings characteristic of thin sections, spiral growth predominates. The implication of the computer predictions is that the degree of nodularity in compacted graphite morphology will be increased at the faster cooling rates, characteristic of thin sections, as the spiral mechanism dominates over the impurity mechanism at large undercooling. In other words, section sensitivity for compacted graphite castings cannot be eliminated at large undercoolings.

Technological Options to Minimize the Section Sensitivity in Compacted Iron Castings

In order to reduce the nodularity problem in thinner sections, it is essential to reduce the growth undercooling; this, in turn, can be achieved by increasing the degree of nucleation, but that there exists an optimum degree of nucleation to promote the required degree of interconnected growth of compacted graphite under a reduced degree of undercooling (30). Far too many nuclei tend to promote nodularity and suppress the interconnected vermicular graphite growth. On the other hand, far too few nuclei can promote carbide formation. Since deoxidation and desulfurisation reaction products are potential sites for the heterogeneous nucleation of graphite, it is important to control the base metal sulfur to consistently low levels for better reproducibility of results. Even though it is not possible to control the nucleation density and uniformity of dispersion of graphite nuclei per se, it is possible to exert control over the nucleation step by carrying out the inoculation for each individual mold as the melt enters the mold cavity. This process step eliminates the holding and handling of the melt after inoculation, and therefore eliminates the variables associated with the coalescence and flotation of embryos and any possible impairment of nucleating substrates on account of reoxidation.

Figure 9 summarises the compacted graphite results from plant trials of an automotive casting in which the section size ranged from 4mm to 15mm. The rate earth addition rate was calculated on the basis of residual Henrian oxygen activity of 10^{-6}, appropriate to compacted graphite growth morphology. Since the same desulfurised ductile base iron was used in all the cases, the rare earth addition rate was kept constant. But the sprue inoculant (75% grade ferro-silicon) addition rate was increased from 0 to 0.3 wt% by an increment of 0.1 wt% each time. The % age compacted graphite was estimated from metallographic studies using the grids. The % age compacted graphite decreased with the section sizes. In the 4mm section, the compacted graphite increased to a maximum value of 50% corresponding to a 0.2 wt% addition of sprue inoculant. Further addition of sprue inoculant to 0.3% decreased the % age compacted graphite to 20%, confirming that there is an optimum nucleation density of graphite to promote the required degree of interconnected growth of graphite associated with compacted graphite morphology.

Acknowledgements

Financial support of this research by Microalloying International Inc., Houston, Texas, U.S.A. and the Natural Sciences and Engineering Research Council of Canada is gratefully acknowledged. Grateful thanks are expressed to Dr. D.A.R. Kay of McMaster University, Dr. D.S. Ghosh, Alberta Research Council, and Dr. Weizhen Zhong, Beijing University, who collaborated in the research. Thanks are due to Algoma Steel Corporation Ltd., Canada, General Motors Foundry, St. Catharines, and Union Carbide Metals Division, now El Kem Metals Company, U.S.A., for experimental evaluation.

References

1. R.H. McSwain and C.E. Bates, The Metallurgy of Cast Iron, Ed. B. Lux, I. Minkoff and F. Mollard, Georgi Pub. Co., Switzerland (1975), pp. 431-438.
2. S.V. Subramanian, D.A.R. Kay and G.R. Purdy, Compacted Graphite Morphology Control, A.F.S. Trans., 90, 589 (1982).
3. M.H. Jacobs, T.J. Law, D.A. Melford and M.J. Stowell, Metals Technology, Nov., pp. 490-500 (1975).
4. S.V. Subramanian, "Compacted Graphite Morphology Control in Shaped Castings", Proc. 4th Int. Symp. on Physical Metallurgy of Cast Iron, Tokyo, Sept. 4-6, 1989 (to be published).
5. I. Minkoff, "Graphite Crystallization", Chapter 1 in Preparation and Properties of Solid State Materials, Vol. 4, Marcel Dekker, Inc., New York, NY (1979), pp. 1-46.
6. M. Hillert, ASM Symposium on Cast Iron, Detroit, Ed. H.D. Merchant, Gordon and Breach, New York, NY (1964), pp. 101-127.
7. R. Brigham, G.R. Purdy and J.S. Kirkaldy, Int. Conf. Crystal Growth, Boston, MA, June 1966, pp. 161-169.
8. H. Morrogh and W.J. Williams, J.I.S.I., March 1947, pp. 321-371.
9. E. Scheil and J.D. Schobel, Giesserei, Techn. Wiss. Bieh, (1963), p. 203.
10. S.E. Wetterfall, H. Frederiksson and M. Hillert, J.I.S.I., May (1972), ASM Source Book on "Ductile Iron", pp. 162-171.

11. K.A. Jackson, "Current Concepts in Crystal Growth from the Melt", Chapter 2 in Progress in Solid State Chemistry, Vol. 4, Ed. Reiss, Pergamon Press, New York, NY (1967).
12. I. Minkoff and W.C. Nixon, J. Appl. Phys., 37 (1966), p. 4848.
13. J.D. Weeks and G.H. Gilmer, Adv. Chem. Phys. 40 (1979), pp. 175-227.
14. G.H. Gilmer, Proceedings of Symposium on "Modelling of Casting and Welding Processes". Ed. H.D. Brody and D. Apelian, AIME Publications, July 1981, pp. 385-401.
15. A.A. Chernov, "Stability of Faceted Shapes", J. of Crystal Growth, Vols. 24-25 (1974), pp. 11-31.
16. S.V. Subramanian, D.A.R. Kay and G.R. Purdy, "Graphite Morphology Control in Cast Irons", Proc. of 3rd Int. Symp. on the Physical Metallurgy of Cast Iron, Sodertalje, Sweden, August 1984; Mat. Res. Soc. Symp. Proc., Vol. 34 (1985), Elsevier Science Publishing Co., Inc., pp. 47-56.
17. C. Wagner, "Thermodynamics of Alloys", Addison-Wesley Publishing Company, Reading, MA (1952).
18. S.V. Subramanian, D.S. Ghosh, G.R. Purdy, D.A.R. Kay and J.M. Gray, "Process for the Production of Vermicular Cast Iron", U.S. Patent No. 4,227,924 (1980).
19. W. Oldfield and J.G. Humphreys, B.C.I.R.A. J1., 10, May 1962, pp. 315-325.
20. S.V. Subramanian, D.S. Ghosh, D.A.R. Kay and G.R. Purdy, Iron and Steelmaker, March (1980), Vol. 7, No. 3, p. 18.
21. M. Hillert and V.V. Subba Rao, "The Solidification of Metals", ISI Publ., No. 110 (1968), pp. 204-211.
22. T. Carlberg and H. Frederiksson, Proceedings of Int. Conf. on "Solidification and Casting of Metals", Sheffield, July (1977).
23. H. Morrogh, J.I.S.I., 1954, Vol. (1976), p. 378-382.
24. E. Scheil and J.D. Schobel, Giesserei, Techn. Wiss. Beih. (1961), 3, p. 203.
25. J.D. Schobel, ASM Seminar on "Recent Research on Cast Iron", Ed. H.D. Merchant, Gorden and Breach, New York, NY (1968), p. 303.
26. W. Oldfield, BCIRAJ, 1962, 10, pp. 17-27.
27. W. Oldfield, ASM Seminar on "Recent Research on Cast Iron", Ed. H.D. Merchant, Gorden and Breach, New York, NY (1968), pp. 374-362.
28. M. Hillert and P. Soderholm, "The Metallurgy of Cast Iron", Georgi, St. Saphorin, Switzerland (1975), pp. 197-207."The Metallurgy of Cast Iron", Georgi, St. Saphorin, Switzerland (1975), p. 644.
29 C.R. Loper, R.W. Heine and M.D. Chaudhari, "The Metallurgy of Cast Iron", Georgi, St. Saphorin, Switzerland (1975), p. 644.
30. S.V. Subramanian, U.S. Patent No. 4,806,157, Feb. 21, (1989).

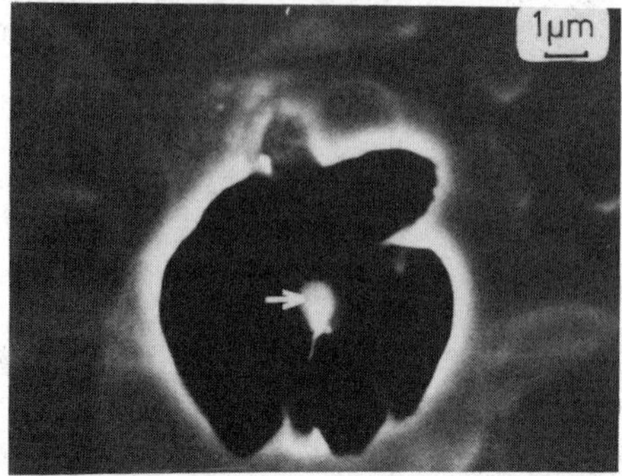

Fig. 1a SEM picture of ion-thinned spherulite showing cerium oxysulfide inclusion in the core.

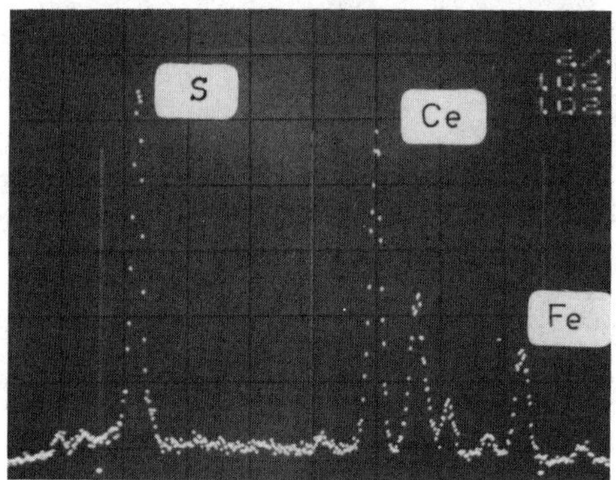

Fig. 1b Energy dispersive analysis of the inclusion

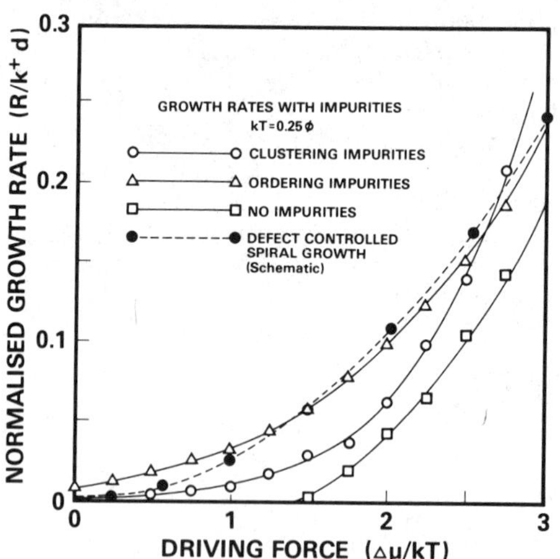

Fig. 2 Growth rates of faceted crystals with and without impurities after Gilmer (14).

Fig. 3
TEM picture of graphite, equilibrated with a pure Fe-C melt, showing facetting of the prism face.

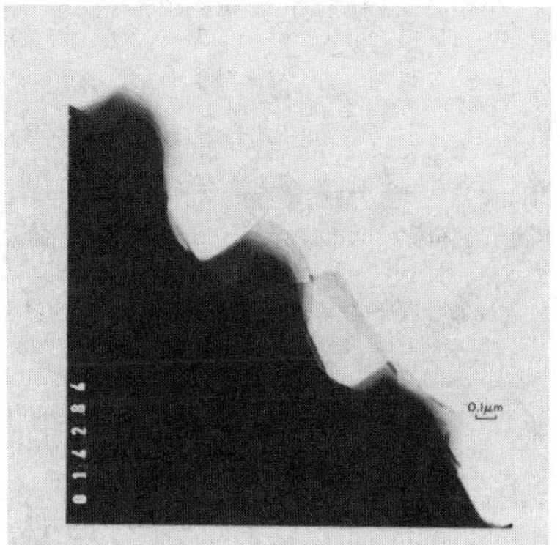

Fig. 4
TEM picture of graphite, equilibrated with a pure Fe-C melt containing sulphur, showing the tendency for interface roughening.

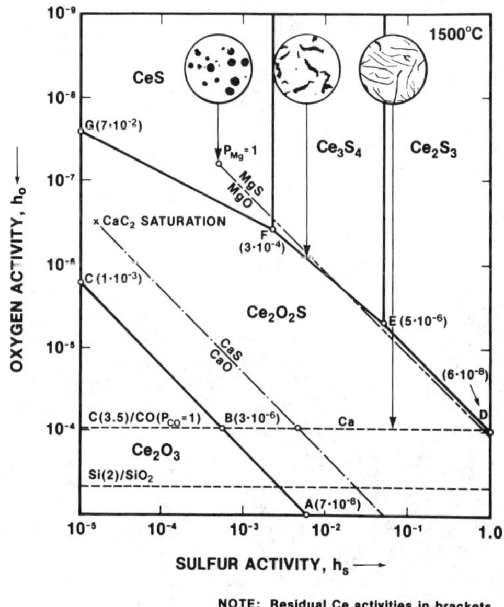

Fig. 5 Graphite morphology control diagram.

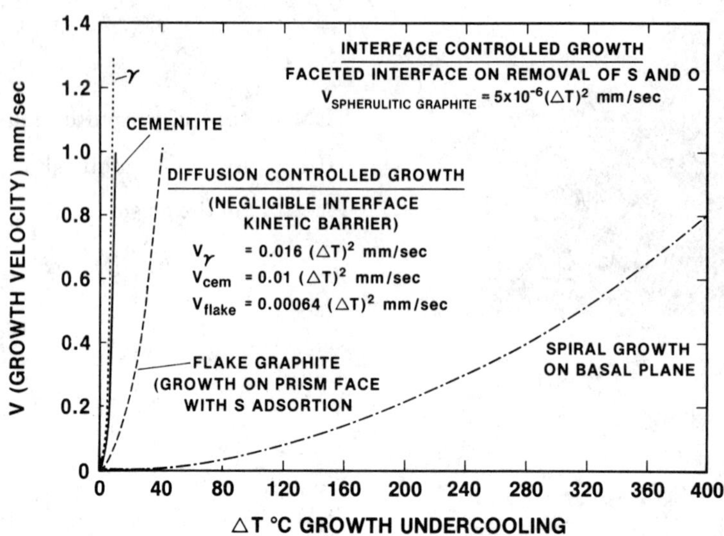

Fig. 6 Estimated growth kinetic curves of the competing phases in cast iron.

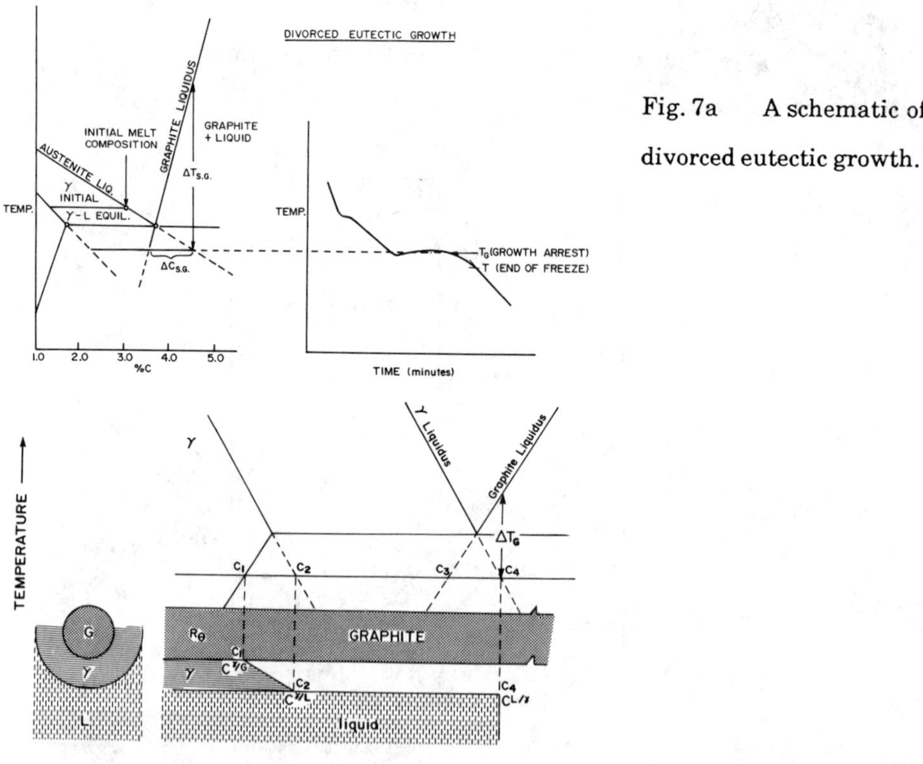

Fig. 7a A schematic of divorced eutectic growth.

Fig. 7b A schematic diagram of spherulitic graphite growing with an austenite shell in the melt.

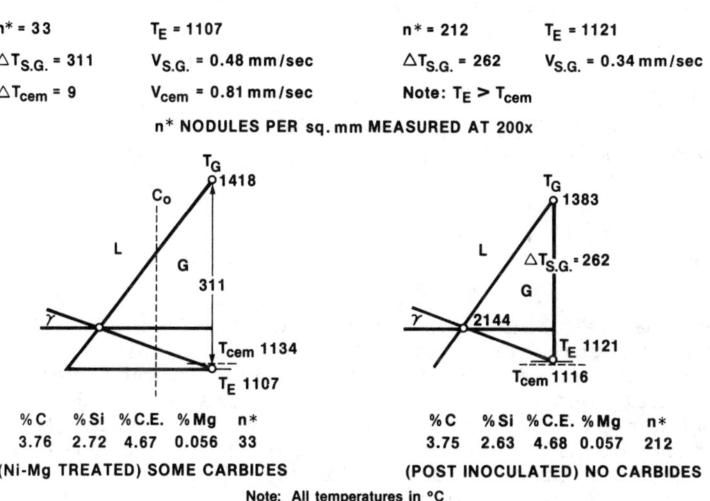

Fig. 8 Nucleation effect on growth undercooling in spheroidal graphite iron.

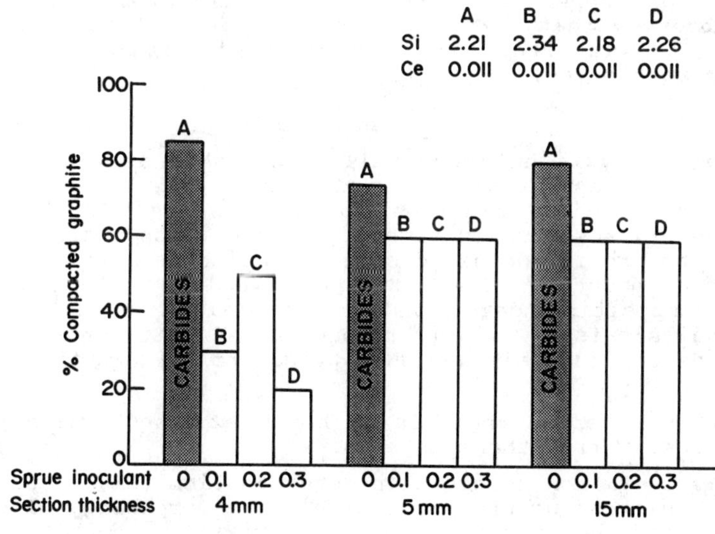

Fig. 9 The effect of inoculation addition rate on the percentage of compacted graphite as a function of casting section size.

The use of thermal analysis to predict the microstructure of cast iron

P. Zhu and R.W. Smith
Department of Metallurgical Engineering, Queen's University, Kingston, Ontario, Canada K7L 3N6

Abstract

During the last decade, with the development of computer technology, computer-aided casting design/solidification modelling techniques has advanced significantly. However, much remains to be done, especially to predict microstructure evolution. In this study, Computer-Aided Differential Thermal Analysis (CADTA) and solidification modelling techniques are incorporated together to predict the composition and microstructural features of cast iron from the data obtained from a test sample of the melt before the pouring the real casting. The composition prediction is accomplished by detecting the austenite liquidus and eutectic temperature by thermal analysis before any treatment. The microstructure prediction is carried-out by the incorporation of CADTA, which determines the amount of latent heat liberated and its release rate, and microstructure evolution modeling. It is shown both theoretically and experimentally that the formation of white iron is associated with less latent heat than is grey iron; this is reflected in the cooling curve by a reduced eutectic transformation temperature and a shorter transformation time. It is also observed that for grey iron the latent heat release rate during the eutectic reaction varies in a distinct way for various graphite morphologies.

1. Introduction

Thermal analysis has been well-established as a technique to study phase transformations and to obtain data for the construction of phase diagrams for many years. In more recent years, however, researchers have begun to develop thermal analysis techniques to provide detailed solidification data for cast irons. Further developments of the technique have led to its use for the chemical composition determination of cast iron (1 - 8) and thermal analysis is now widely recognized as an important tool for on-line chemical analysis and quality control within the cast iron foundry industry.

Currently, thermal analysis is being explored for use in the foundry industry from three aspects:

(1). cast iron composition determination (CEV, C & Si);
(2). microstructure prediction (the morphology of graphite and the structure of the matrix);
(3). based on (1) and (2), mechanical property prediction.

2. Literature Review

2.1. Composition Prediction

Composition prediction mainly involves carbon equivalent, carbon and silicon content prediction. The basic principle involved is to monitor the temperature during the solidification process, and note the presence or otherwise of certain characteristic "arrests" or "kinks" in the cooling curve, since these have been associated with the onset of the transformation.Empirical relationships between composition and these characteristic temperatures can be derived.

For the carbon equivalent value(CEV*), all the previous researchers found that the CEV has a direct relationship to austenite liquidus temperature (T_{AL}) and has the following form.

$$T_{AL} = a - b * CEV \quad [1]$$

However different *a* and *b* values are reported by the various researchers(1,5,8). Therefore, after the temperature measurement, the CEV can be predicted.

Moore (8) found that for the Fe-C (gray iron) system, the eutectic temperature is dependent not only on composition but also on several additional factors, which caused the reaction to occur over a temperature range (recalescence). Consequently, no precise relationship exists between eutectic temperature and metal composition in the gray cast iron system. In the Fe - Fe_3C (white iron) system, on the other hand, the eutectic reaction is dependent only on composition. Therefore the use of tellurium-coated cups to promote the white iron solidification of an otherwise gray iron allows the relationship between the white iron eutectic temperature(T_{EW} °C) and the metal composition to be determined. Moore obtained the following equation:

$$T_{EW} = 1104 + 9.8(\%C - 1.23\%Si - 3.00\%P) \quad [2]$$
$$r = 0.989$$

Solving equations [1] and [2] by neglecting (or presuming a certain) phosphorus content, the carbon and silicon contents can be predicted. It is self-evident that the accuracy of the carbon and silicon predictions by thermal analysis is dependent on the actual phosphorus content in the iron. The results were obtained when the phosphorus level was no more than 0.05%P, the accuracy being quoted as ±0.14%Si (3).

2.2. Microstructure Prediction

Changes in the kind and magnitude of a solidification event during the freezing of a liquid metal charge will be reflected in the cooling curve. This principle suggests that cooling curve might be a measure and predictor of the kind of macrostructure and microstructure generated during solidification. In practice, the so-called cast iron quality control generally refers to composition control (prediction) and microstructure control (prediction). In the literature, there are many reports of studies directed towards this goal(9 - 27), but the results are scattered, especially for ductile iron.

2.2.1. Distinguishing between Gray and White Iron

From a thermodynamic point of view, because the phases formed are different, the solidification process of white iron should result in a cooling curve different from that of gray iron. Typical cooling curves for white and gray iron with same CEV are shown in fig.1.

---

* CEV = %C + (%Si + %P)/3 or CEV = %C + %Si/4 + %P/2

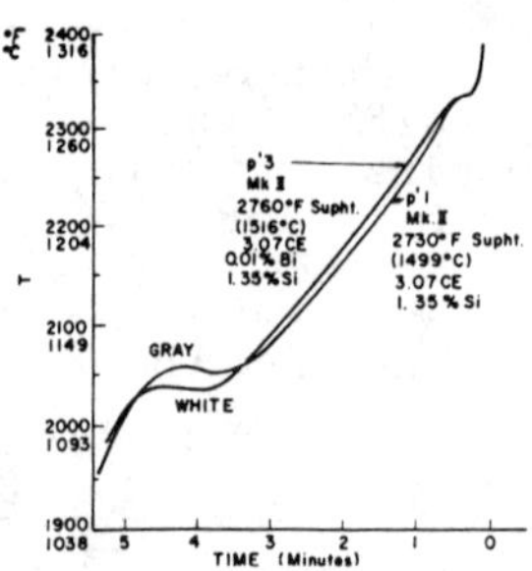

FIG. 1 Cooling curves of grey and white iron (28).

The liquidus arrest temperature (T_{AL}) of white iron is not much different from that of gray iron, but the bulk eutectic arrest temperature is much lower. Generally, a completely carbidic structure is associated with a eutectic temperature (T_{EU}) lower than 1221.1 °C (23). It was observed that the recalescence for a completely white iron is much smaller than that of a grey iron(28).

2.2.2. Distinguishing Among Various Graphite Shapes

Many attempts have been made to correlate the data obtained from a cooling curve with nodularity and nodule count, in order to develop a method of controlling the liquid treatment of ductile iron. Unfortunately, the data reported by different authors are quite scattered and have low reproducibility. From this work, it might be concluded that:

- If the eutectic recalescence > 10 °C, a compacted graphite iron is almost certain to be present (10,24); Otherwise either spheroidal or flake-type iron will present.
- The eutectic temperature of A-type flake iron is generally higher than that of spheroidal iron(10,24). However, for the D-type or other flake irons, there are not enough data available at present to make particular associations.
- There is little difference of latent heat evolution amongst flake, vermicular and spheroidal cast irons(26,29). However, there is a significant difference in latent heat evolution between white iron and any of the gray irons(10). The latent heat of gray iron (~ 56cal/g) is greater than that of white iron (~ 49cal/g).

2.3. Mechanical Properties

It is not difficult to believe that, if the composition and microstructure can be predicted, and the section size is known, then mechanical properties such as strength, hardness, elongation, *etc.*, might also be predicted.

Work by Jeffery, *et al*,(30) Schneidewind and MacElwee,(31) and Watmough, *et al*(32), has resulted in algorithms that can be used with section size information to predict the approximate strength of gray cast iron. One such expression is as follows (31):

$$\sigma_s = 10{,}000 \text{ psi} \times 11.6 - 2.1 \times (\%C + 0.25Si + 0.5P) - \log D_c \times [1 + 0.45 \times (\%Mo)] \times [1 + 0.23 \times (\%Cr)] \times [1 + 0.57 \times (\%V)] \times [1 + 0.065 \times (\%Ni)]$$

$$\times [1 + 0.05 \times (\%Cu)] \times [1 + 0.10 \times (\%Mn)] \quad [3]$$

Where D_c = equivalent diameter of the cast section (33).

This equation takes into account the carbon equivalent of the iron, the presence of elements that harden or strengthen the matrix, and the section thickness. In addition to the equation [3], there are several equations available to predict mechanical properties. The results from ref.(34) suggest:

$$HB = - 1105.2 + 1.14 * T_{AL} - 132 * \beta - 1.15 * B \quad [4]$$

$$\sigma_s \ (N/mm^2) = - 552.6 + 4.95 * 10^4 * T_{AL} + 8.4 * B - 0.16 * B^2 \quad [5]$$

where HB is the hardness; B is the the absolute value of first minimum point on the derivative of the cooling curve; and β is the angle of the derived cooling curve corresponding to its cooling curve after recalescence. The results from ref.(35) are

$$\sigma_s \ (psi) = - 388{,}447 + 357 \ (\text{Liquidus Temperature, } ^\circ C) \quad [6]$$

$$\sigma_s \ (psi) = 24{,}776 + 142(\%\text{Dendrite Interaction}) \quad [7]$$

$$\sigma_s \ (psi) = 248{,}504 + 234(\text{Liquidus Temperature, } ^\circ C) + 93(\%\text{Dendrite Interaction}) \quad [8]$$

where the dendrite interaction is a visual estimate of the percentage area of dendrites intersected in the field of a view. Equations [6] and [7] indicate that both liquidus temperature and dendrite interaction influence the tensile strength. Equation [8] indicates the combined effect of liquidus temperature and dendrite interaction.

2.4. Other Factors Affecting the Cooling Curve

2.4.1. Experimental Conditions

Döpp (36) observed a decrease of pouring temperature would result in a decrease of both liquidus and solidus, due to the higher cooling rate involved. A reduction in sample size is considered to have the same effect as a lower pouring temperature. The cup conditions, i.e. heat transfer characteristics, will also exert an influence on the cooing curve obtained.

2.4.2. Effects of Some Elements

C and Si: C and Si are the main graphitizing elements. They affect the cooling curve by changing the CEV and the potential to form graphite. It is essential to study the effect of C and Si on the cooling curve, and most of the previous researches have done this.

Mg and Ce: The importance of Mg and Ce as nodulizers for ductile iron is well known. For spheroidal graphite cast iron, Titze and Dichtl (37) showed that the solidus temperature is decreased by additions of Mg and is increased with holding time and after treating with silicon-containing inoculants.

S and P: Phosphorus changes T_{AL} because it changes the CEV as mentioned earlier. P changes T_e. Both S and P are detrimental to the spheroidal graphite formation, when the amounts present exceed some critical values.

Bi: For ductile iron, Morrogh's results(38) indicated that Bi in the amount of 0.003% begin to have a harmful effect, and that 0.006% Bi can completely inhibit the nodular structure. Experience suggests that the amount of Bi which can be tolerated depends on the cooling rate of the casting, e.g. larger amounts of Bi can be tolerated in rapidly cooled sections than in slowly cooled sections.

Sb: Antimony is a useful element in producing thick-section a ductile iron castings if either a pearlite matrix or avoidance of float-out of graphite is required, or both. However the amount of Sb must be carefully controlled otherwise Sb will act like Bi to impair spheroidal graphite formation.

Pb: Pb will not affect the shape of the spheroidal graphite in ductile when its amount is kept less than 0.1%(39). Schellerg(40) found that Pb promotes the formation of spheroidal graphite if a Mg-Ce treatment is used. However Pb promotes the formation flake graphite if the melt does not contain Ce. Lin(39) found that, with increasing amount of Pb (max<0.1%), the number of nodules is raised.

Te: Crucibles with a coating or spots of tellurium lead to a significant decrease in the solidus temperature, if pouring temperature is higher than 1300°C. The reproducibility of tellurium test is fairly high. The main purpose of tellurium is to promote the white structure. This leads to a better determination of silicon content, according to Oldfield (41) and Moore (8).

Sn: It is well known that Sn promotes a fine pearlite matrix for both gray and ductile iron. As far as the relation of its presence with thermal analysis data is concerned, however, few data are available.

2.5. The Main Problems of Previous Thermal Analyses and the Objectives of This Study

As mentioned earlier, there are some factors other than composition and the solidification structure of cast iron, which will affect the thermal record of solidification. These factors are pouring temperature, the mass of the sample and the heat transfer characteristics of the system. These interfering factors will give rise to difficulties in applying thermal analysis. Perhaps that is the one of reasons why various researchers under a variety of experimental conditions have obtained different results and so suggest that the led to experimental results have poor reproducibility. Therefore one of the objectives of this research was to develop a technique to eliminate these shortcomings.

In producing ductile iron, some trace elements may play an important role in determining the final microstructure, i.e. the graphite morphology in the cast iron. To understand how and possibly why these element affect the thermal analysis was the second objective of this study. To fully correlate cooling curve data to composition, microstructure and mechanical properties has been the third objective of this study.

3. Theoretical Analysis

3.1. Computer-Aided Differential Thermal Analysis

Classical quantitative thermal analysis of solidification has been done with Differential Thermal Analysis (DTA) on many metallurgical systems to determine either the phase transformation temperature or the amount of latent heat associated with the transformation. This is accomplished by using of a dummy sample which truly represents the test sample but without a phase transformation. By monitoring the temperature difference between the sample and the natural body under the same cooling(heating) conditions, the heat evolution due to any phase transformation can be measured. By using this principle, the limitations in previous techniques are completely eliminated. However, this technique has practical limitations in foundry use because of the requirement of a natural body, small sample size, and complicated measuring instruments (42).

With the development of computer techniques, a new method called Computer-Aided Differential Thermal Analysis (CADTA)* may be used. This was first put forwarded by Ekpoom and Stefanesco (42). Their method consists of using a "simulated natural body" to represent the standard involved in the classical DTA with a computer data-acquisition system, which also records the cooling curve of the sample. In this way, the same types of information obtainable with classic DTA can be gathered, but using a cheaper and more versatile instrument on the foundry floor. However their work had two limitations. First, the simulation of the natural body was carried out by a using simplified analytical method, which in our experience cannot well represent the real cooling of the natural body; and second, their research was limited to measuring the amount latent heat for the eutectic transformation to white iron and was never extended to microstructure prediction for other cast irons. In our study, a numerical method is used to simulate the natural body cooling, and besides that, CADTA is incorporated into the modelling of microstructure evolution.

3.2. Modelling of Microstructure Evolution

Although the modelling of solidification processes has received great attention during the last decade, the number of studies which encompass microstructure evolution is relatively limited (43). The main shortcomings of the rare previous microstructural evolution studies (43) is that many assumptions were made in order to model, and these models lack the flexibility to reflect what is really going on in the sample.

The solidification process of cast iron is controlled, like any other solidification process, by nucleation and growth phenomena. Thus, using the proper growth model and its associated kinetics for the nucleation and growth of solidification, heat transfer considerations are used to simulate the microstructure of the sample casting (43,44). The ultimate goal of modelling microstructural changes during solidification is to predict the room-temperature microstructures and mechanical properties.

3.3. Austenite-Flake Graphite Eutectic Growth Model

It is generally accepted that the eutectic growth of compacted/vermicular graphite iron is identical to that of flake iron whereas the nucleation of eutectic cells is the same as that of spheroidal graphite iron. Therefore, austenite-flake graphite eutectic growth modelling is suitable for both flake and compacted graphite irons.

It may be shown that the linear coupled growth rate of a eutectic cell of F.G (flake graphite iron) and C.G (compacted graphite iron) can be written as (see Appendix I)

$$dR_e/dt = \mu' * (T - T_e)^2 \qquad [9]$$

---

*Note: The method used to generate natural body cooling differs from that reported in this paper.

where dR_e/dt is the growth rate of eutectic cell; μ' is the constant; and $T - T_e = \Delta T$ is the undercooling.

If an eutectic reaction starts from a point in an undercooled melt, the eutectic phase will grow in a spherical form until a critical spacing (λ) between adjacent lamellae of the two phases has developed; this spacing will be maintained by a branching mechanism (10). According to this branching mechanism, the rate of volume formation of solid, dV/dt, can be written as:

$$dV/dt = \mu * R_e * (dR_e/dt) \quad [10]$$

where R_e is the radius of the eutectic cell and μ is the constant.

Equation [10] can be readily converted to the amount of latent heat liberated and its release rate in the grey iron eutectic reaction.

3.4. Austenite-Spheroidal Graphite Eutectic Growth Model

The well-known eutectic growth mechanism for the formation of spheroidal graphite iron is that its graphite nuclei are considered to be encapsulated in a austenite shell right after their formation, so that further growth occurs by diffusion through the shell. According to Fick's first diffusion law and the conservation of energy (see Appendix II for details), it may be shown that the eutectic volume ratio increase rate is given by

$$dF_{etu}/dt = 4\sqrt{2}\,\pi(1-F_{etu}-F_{\gamma})*N*\{D_c^{\gamma}\frac{K+1}{\sqrt[3]{K+1}-1}(C^{\gamma/L}-C^{\gamma/G})\}^{3/2}*\sqrt{t}$$

....[11]

Where F_{eut} is the fraction of solidified eutectic; F_{γ} is the fraction of solidified primary austenite, which can be calculated from the phase diagram and CEV; K is Volume ratio of austenite and graphite in an eutectic cell (the value of K can be known by calculation); D_c^{r} is the diffusion coefficient of carbon in γ (austenite shell); and $C^{\gamma/L}$, $C^{\gamma/G}$ are the carbon concentrations at phase boundaries, which can be derived from the Fe-C phase diagram.

3.5. Computer Program Flow Chart

Fig. 2 is the flow chart which was used to carry-out the composition and microstructure analyses. For accuracy, the temperature arrests in the cooling curve were determined by using the derived cooling curve. The modelling of the natural body cooling has been made using a finite difference method (FDM) at this stage.

4. Experimental Procedure

The temperature information can be obtained from the sensor, which is actually a K-type thermocouple positioned in the standard mold (cup). The EMF which is obtained from the sensor is several mV, and thus, a transmitter (EXP/16), which can magnify the voltage information, is placed between the sensor and the A/D converter (DAS/16). The converter is connected to an IBM computer in the computer slot. The speed of collecting data (controlled by NoteBook Software) is 4Hz, i.e., 4 data per second. The total solidification time is set to 6 minutes, which allows the sample to cool below the eutectoid temperature. If only composition and microstructure (excluding mechanical properties) information is required, it is not necessary to cool the sample below the eutectoid temperature, and thus the time required for this experiment is only about 4 minutes. After recording the temperature, then we use our program to analyze the data obtained, a process taking less than one minute.

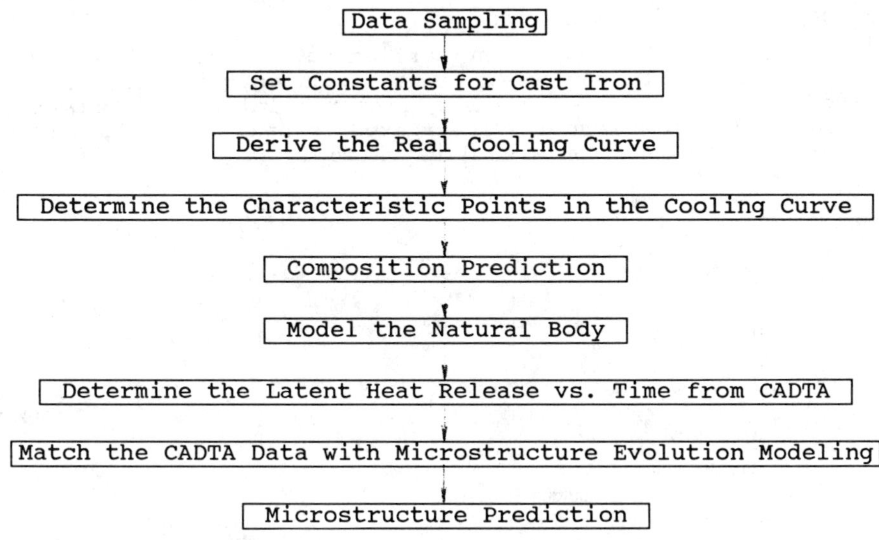

FIG. 2. Computer program flow chart

The metals are melted in a 10 kg induction furnace, then treated or not, based on the different sample requirements. The composition of pig iron used is as follows: C: 3.945%, Si: 1.384%, Mn: 0.010%, S: 0.007% and P: 0.020%.

4. Preliminary Results

4.1. Modeling of The Natural Body

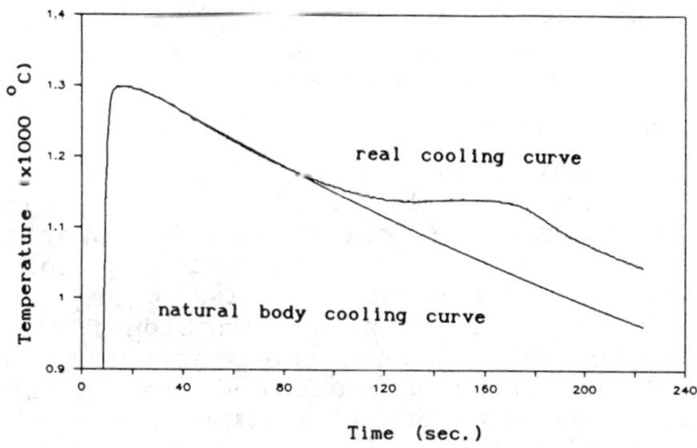

FIG. 3 The Modeling of natural bodies and real cooling curves

Because of the symmetry of the sampling cup, a 2-D finite difference method has been used to model the natural body. As expected, it is found that the heat transfer coefficients for various samples are dependent on the pouring conditions. Therefore, instead of making any assumptions for the heat transfer parameters for the system, as was done in previous studies, the heat transfer coefficient is assumed to be a variable and is determined from the cooling curve of a real sample. The typical curves for both the real sample and the natural body are shown in fig. 3.

4.2. Composition Determination

Some experiments have been carried-out to date. The cooling curves are directly analyzed by a computer program. The data obtained show that there exists a linear relationship between T_{AL} and carbon equivalence. The experimental data and the linear-regressed equation are illustrated in fig. 4.

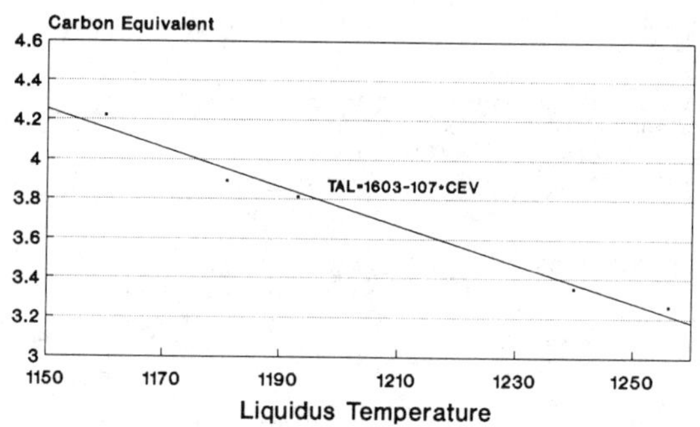

FIG. 4 The relationship between CEV and T_{AL}

It was also found that, under the experimental conditions used, the eutectic transformation temperatures for either "plain cup" or "tellurium-coated cup" are relatively constant (less than 5 °C recalescence was observed in either case). However, no useful relationship between Si content and the eutectic temperature in the Fe-C system was observed. The relationship between the white eutectic temperature and the Si content is shown in figure 5.

If a plain sampling cup is used to do the composition prediction for hypereutectic cast iron, an eutectic reaction temperature slightly above the equilibrium temperature (1154 °C), namely approximately 1160 °C, was observed.

4.3. Microstructure Determination

As noted earlier, the presence of white iron in eutectic microstructure may be indicated by the thermal analysis data, since its eutectic transformation temperature is lower than 1125 °C*, and the eutectic reaction time is generally shorter than that of the comparable gray iron. However CADTA may be used to show that the latent heat of the eutectic reaction for white iron is less than that for grey iron. Hence, the presence of white iron is confirmed by the eutectic temperature and the latent heat of the eutectic reaction. The latent heat liberated for grey and white iron in their eutectic reaction are about $2.68*10^5$ J/kg and $2.05*10^5$ J/kg, respectively.

---

*Note: From some studies, it appears that the eutectic reaction temperature of the vermicular /compacted graphite cast iron occasionally also goes below 1125°C, but the recalescence for CG iron is generally over 10°C.

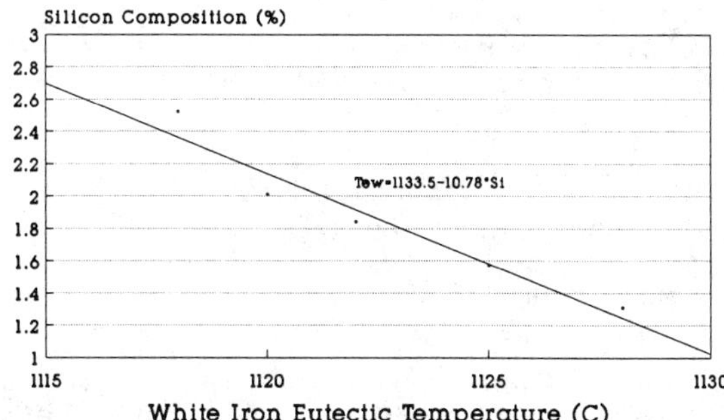

FIG. 5 White eutectic temperature and silicon composition

For flake graphite (FG) iron, the curves of the latent heat release rate vs time for the eutectic reaction as revealed by CADTA and by the microstructure evolution method are shown in fig 6.

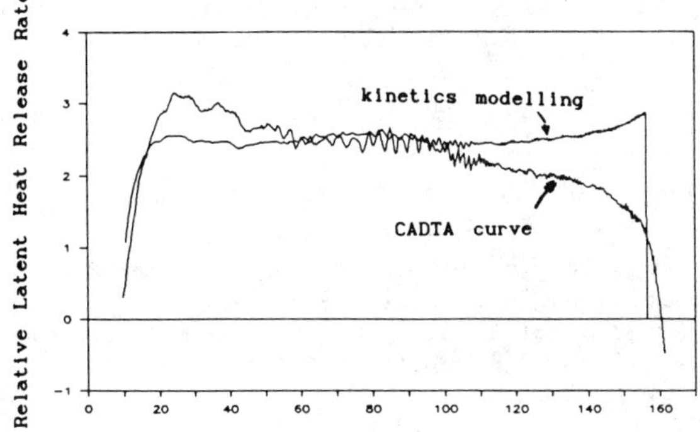

FIG. 6 The comparison between CADTA and microstructure evolution modeling for FG iron.

For spheroidal graphite(SG) iron, the curves of the latent heat release rate vs time for the eutectic reaction by CADTA and by the microstructure evolution method are shown in fig 7.

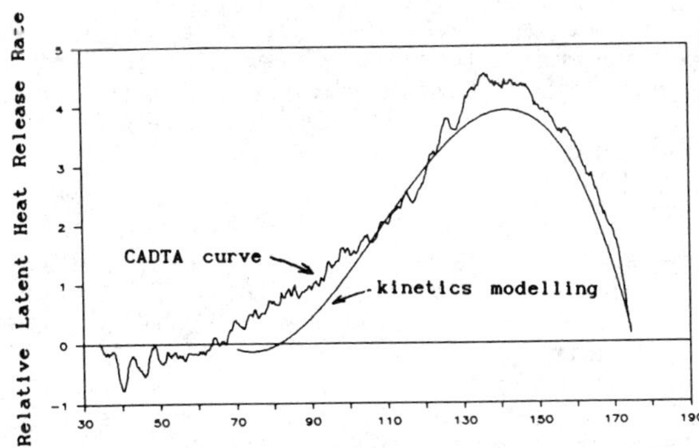

FIG. 7 The comparison between CADTA and microstructure evolution modeling for SG iron

It is seen from fig 6 and 7 that the eutectic growth kinetics models used for grey and spheroidal graphite iron provide a reasonably good fit. However in the present study, the formation of primary austenite has not been considered. It is scheduled to be considered later.

5. Preliminary Conclusions and Current Work

This paper must be considered as the first step towards the accomplishment of the aim to model the microstructure evolution of cast iron from thermal analysis of test sample of an iron melt. However, some preliminary conclusions can be reached:

1. For the composition prediction, a "plain-cup" can be used to predict the CEV up to 4.3 (hypoeutectic) without any problem. For hypereutectic (CEV > 4.3) cast iron, the "tellurium-coated cup" should be used. In either case, if the carbon and silicon content are required to be known, the tellurium-coated sampling cup is recommended.

2. White iron can be distinguished from grey iron quite easily by using two parameters, namely, the temperature of the eutectic and the latent heat liberated during the eutectic reaction.

3. Computer-aided differential thermal analysis, coupled with microstructure evolution modeling is an effective method to predict microstructure evolution during solidification processing.

As mentioned above, the paper reports a study in progress. Current activity is directed towards:

1. correlating graphite morphology changes from spheroidal to flake with the latent heat release rates and hence with the influence of trace elements on nodulization; and thereupon derive a mechanism suitable for the eutectic growth of vermicular /compacted graphite iron;

2. quantifying the nodularity of spheroidal cast iron by using a shape factor, and so, when several microstructures are present together, being able to quantify the amount of each in the microstructure; and

3. extending the work to other alloy eutectic systems, such as Al alloys, to check whether it may be used extensively.

Acknowledgement

This work forms part of the research program of R.W. Smith and is supported by the National Science and Engineering Research Council of Canada, the International Development Research Center, Ottawa and Queen's University through the School of Graduate Studies and Research.

References

1. J.G. Humpreys, *BCIRA Report* No.605(1961), 609.
2. A. Moore, *Foundry M & T.*, (July, 1974), 80.
3. M. Booth, *British Foundryman*, (March, 1983), 35.
4. R.W. Heine, *Trans. AFS* 81(1973), 462.
5. R.W. Heine *Trans. AFS* 85(1977), 537.
6. A. Moore, *AFS Cast Metal Research J.*, Mar. 1972, 15.
7. A. Alagarsamy, et al, *Trans. AFS* 92(1984), 871.
8. A. Moore, *Foundry Tr. J.* 131, 1971, 885.
9. C.R. Loper Jr., R.W. Heine & M.D. Chaudhari, *The Metallurgy of*

Cast Iron, St. Saphorin, Switzerland, Georgi Publishing Company (1975), P640.
10. L. Backerud, K. Nilsson & N. Steen, *The Metallurgy of Cast Iron,* Switzerland, Georgi Publishing Company (1975), P625.
11. A.N. Snigir, *Soviet Castings Technology*, No.10, 1987, 1.
12. A.N. Snigir, *Soviet Castings Technology*, No.3, 1986, 61.
13. L.E. Menawati, R.W. Heine & C.R. Loper, Jr, *Trans. AFS* 78 (1970), 363.
14. D. Hui, Y. Jingxiang & K.G. Davis, *Trans. AFS* 93(1985), 917.
15. C.R. Loper, Jr, & R.W. Heine, et al, *Trans. AFS* 75(1967), 541.
16. A.N. Snigir, et al, *Thermochimica Acta,* 93(1985), 657.
17. B.R. Hrohn, Modern Casting, March, 1985, 21.
18. G.R. Strong, G.R., *Trans. AFS* 91(1983), 151.
19. R.W. Heine, C.R. Loper, Jr, & M.D. Chaudhari, *Trans. AFS,* 79 (1971), 399.
20. R. Monroe & C.E. Bates, *Trans. AFS* 90(1982), 307.
21. E.F. Ryntz & J.F. Janowak, et. al., *Trans. AFS* 79(1971), 141.
22. M.D. Chaudhari & R.W. Heine, *AFS Cast Metals Research J.*, June 1975, 52.
23. M.D. Chaudhari & R.W. Heine, *AFS Cast Metals Research J.*, Sept., 1975, 77.
24. D.M. Stefanescu, C.P. Loper Jr, et al *Trans. AFS* 90(1982), 333
25. I-G Chen & D.M. Stefanescu, *Trans.AFS* 92(1984), 947.
26. U. Ekpoom & R.W. Heine, *Trans. AFS* 89(1981), 27.
27. R. Döpp, *53rd International Foundry Congress,* Prague, Czechoslovakia, Sept.10, 1986.
28. U. Ekpoom & R.W. Heine, *Trans.AFS* 85(1977), 537.
29. H. T. Angus, *Cast Iron: Physical and Engineering Properties*, London, Butterworths(1976), p122.
30. W.C. Jeffery, E.E. Languer, W.G. Mitchness & G.D. Azizi, *Trans. AFS* 62(1954), 568.
31. R. Schneidewind & R.G. McElwee, *Trans. AFS* 68(1960), 312.
32. I. Watmough, W.F. Shaw & F.C. Buck, *Trans. AFS* 79(1971), 225.
33. B.R. Patterson & C.E. Bates, *Trans. AFS* 89(1981), 369.
34. S. Jura, H. Borek & J. Sakwa, *46th International Foundry Congress* Madrid, Spain, 1979
35. D. Glover, C.E. Bates & R. Monroe, *Trans. AFS* 90(1982), 745.
36. R. Döpp, *The Metallurgy of Cast Iron,* St. Saphorin, Switzerland, Georgi Publishing Company 1975, p235.
37. E. Titze & H.J. Dichtl, *Giesserei* 61, 1974, Nr. 20, 619.
38. H. Morrogh, *Trans. AFS* 60 (1952), 439.
39. C.G. Lin, et. al, *J. Chinese Foundryman's Association*, June, 1986, 1.
40. R.D. Schelleng, *Trans. AFS* 74 (1966), 700.
41. W. Oldfield, *BCIRA J.* 10, 1962, 17.
42. U. Ekpoom, Ph.D Thesis, University of Wisconsin-Madison, 1980
43. M. Rappaz, *International Mat. Rev.*, 34, 1989, 3, 93.
44. C.S. Kanetkar, Ph.D Thesis, U. of Alabama, 1988.

Appendix A

Eutectic Growth Rate for Gray and Compacted Iron (coupled growth)

The eutectic growth mechanism of grey iron is schematically shown in fig.A1, where λ and dx are graphite spacing and distance grown per unit time respectively.

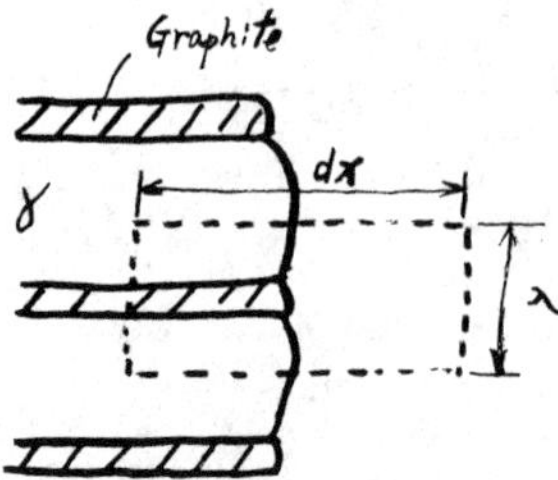

Fig. A1. The eutectic growth mechanism of grey iron

Assuming lamellae growth and considering unit thickness of the eutectic cell, then the eutectic volume (V_{eut}) transformed during unit time can be expressed as:

$$V_{eut} = \lambda * dx * 1 \quad [A-1]$$

Hence the mass transformed in unit time correspondingly is

$$M_{etu} = \rho * \lambda * dx \quad [A-2]$$

where ρ is the density of the eutectic cell. Assuming the phase transformation energy is proportional to the energy spent on creating new γ/graphite interface, we have:

$$\Delta G * \rho * \lambda * dx / M_w = k * 2 * dx * \sigma \quad [A-3]$$

where ΔG is the free energy change (per mole) associated with the phase transformation; σ is the surface energy per unit area; M_w is the weight per mole; and k is a constant. Thus from equations [A-1] and [A-3], we have

$$\lambda = \frac{2\ \sigma\ M_w\ k}{\Delta G\ \rho} \quad [A-4]$$

On the other hand, the free energy change during a phase transformation is

$$\Delta G = (\Delta H / T_e) * \Delta T \quad [A-5]$$

where ΔG is the free energy change (driving force) in a phase transformation; ΔH is the enthalpy change during a phase transformation; T_e is the equilibrium phase transformation temperature; and ΔT is the undercooling. The spacing λ is proportional to the $(dR_e/dt)^{-1/2}$, i.e.

$$\lambda = k' * (dR_e/dt)^{-1/2} \quad [A-6]$$

Substituting [A-4] in [A-5] and combining with [A-6], we obtain that:

$$dR_e/dt = \mu' * \Delta T^2 \quad [A-7]$$

where μ' is a constant

Appendix B

Eutectic Growth Model of Spheroidal Graphite Iron

The diffusion rate of carbon through the austenite shell is, according to Fick's first law:

$$J_D = (4\pi R^2)\, D_c^{\gamma}\, (dC/dR) \qquad [B-1]$$

where D_c^{r} is the diffusion coefficient of carbon in γ (austenite shell); dC/dR is the carbon concentration gradient in γ shell; and R is the radius of arbitrary selected γ shell. By integrating equation(B-1), we can get

$$J_D = 4\pi D_c^{\gamma} (C^{\gamma/L} - C^{\gamma/G}) * (R_e * R_g)/(R_e - R_g) \qquad [B-2]$$

where $C^{\gamma/L}$, $C^{\gamma/G}$ are the carbon concentrations at the phase boundaries; and R_e, R_g are the radii of eutectic cells and graphite nuclei respectively. The graphite nuclei growth rate can be written as:

$$dR_g/dt = J_D/(4\pi * R_g^2) \qquad [B-3]$$

And the corresponding eutectic cell growth rate is

$$dR_e/dt = (K+1)^{1/3} * (dR_g/dt)$$

$$= D_c^{\gamma} \frac{K+1}{\sqrt[3]{K+1} - 1} (C^{\gamma/L} - C^{\gamma/G})/R_e \qquad [B-4]$$

Where K $[=(R_e^3 - R_g^3)/R_g^3]$ is the volume ratio of austenite and graphite in an eutectic cell; the value of K can be obtained by equilibrium phase diagram. By integrating equ.[B-4], finally, we obtain

$$R_e^2 = 2 \frac{K+1}{\sqrt[3]{K+1} - 1} (C^{\gamma/L} - C^{\gamma/G}) * t \qquad [B-5]$$

The eutectic cell volume increase rate (dV_e/dt) thus becomes

$$dV_e/dt = N * 4\pi R_e^2 (dR_e/dt)$$

$$= N * 4\pi D_c^{\gamma} K' (C^{\gamma/L} - C^{\gamma/G}) R_e$$

$$= N * \sqrt{2}\, \pi (D_c^{\gamma} * K')^{3/2} * (C^{\gamma/L} - C^{\gamma/G})^{3/2} * \sqrt{t}$$

....[B-6]

where $K' = \dfrac{K+1}{\sqrt[3]{K+1} - 1}$

N is the the number of eutectic cells

The latent heat release rate (dQ/dt) during the eutectic reaction can be determined as follows:

$$dQ/dt = L * \rho * dV_e/dt$$

$$= \alpha' * N * (C^{\gamma/L} - C^{\gamma/G})^{3/2} \sqrt{t} \qquad [B-7]$$

Where $\alpha' = L * \rho * 4\sqrt{2}\, \pi (D_c^{\gamma} * K')^{3/2}$; L is latent heat per kg. By balancing the energy involved in the system,

Heat Evolved - Heat Given Away = Energy Change in System

$$\alpha' * N (C^{\gamma/L} - C^{\gamma/G})^{3/2} \sqrt{t} * dt - h * A(T - T_\infty) dt = c\rho V * dT$$

....[B-8]

where h is the heat transfer coefficient; A is the area; T_∞ is the temperature of surroundings; V is the sample volume; and c is the specific heat of cast iron. According to the phase diagram of Fe-C, we assume:

$$(C^{\gamma/L} - C^{\gamma/G}) = K_1 (T - T_e) \quad [B-9]$$

where K_1 is a constant; T_e is the equilibrium eutectic temperature (~ 1150 °C); T is the actual eutectic temperature, which varies with pre-eutectic cooling conditions and the eutectic reaction time. Combining equations [B-8] and [B-9], we have:

$$dT/dt = \alpha * N (T - T_e)^{3/2} * \sqrt{t} - h * A(T - T_\infty) \quad [B-10]$$

where $\alpha = \alpha' * K_1$

In order to obtain the equation [B-10], we have made the following assumptions, namely, the number of eutectic cells growing at any moment during the eutectic reaction are same; the eutectic cells start growing at the same time and grow at the same rate. To take note of impingement, we therefore use a modified expression for the Johnson-Mehl equation which takes into account eutectic grains impingement. The fraction of eutectic solid can be calculated from:

$$F_{eut} = 1 - F\gamma - \exp(-4/3 \, \pi \, N \, R_e^3) \quad [B-11]$$

$$dF_{eut}/dt = 4\pi (1 - F_{eut} - F\gamma) R_e^2 * N \, dR_e/dt \quad [B-12]$$

Where F_{eut} is the fraction of solidified eutectic; and $F\gamma$ is the fraction of solidified primary austenite, which can be calculated by using the phase diagram and the CEV. Substituting equation [B-12] with equations [B-4] and [B-5], equation [B-12] can be written as:

$$dF_{etu}/dt = 4\sqrt{2} \, \pi (1 - F_{etu} - F\gamma) * N * \{ D_c^{\gamma} \frac{K + 1}{\sqrt[3]{K+1} - 1} (C^{\gamma/L} - C^{\gamma/G}) \}^{3/2} * \sqrt{t}$$

....[B-13]

From equ.[B-13], we can easily relate dF_{eut}/dt to dT/dt, and thus we may show that the larger the number(N) of eutectic cells, the higher the recalescence will be due to a larger dF_{eut}/dt.

A physical and mathematical model for prediction of microstructural features in cast iron parts

D.K. Banerjee, G. Upadhya and D.M. Stefanescu
Solidification Laboratory, The University of Alabama, Tuscaloosa, Alabama, U.S.A.

Abstract

Significant progress has been achieved recently in modeling microstructural evolution during solidification of castings by coupling macroscopic heat transfer with microscopic solidification kinetics models. This paper describes a 2-D heat transfer-solidification kinetics model which was developed for predicting the *thermal history* and *microstructural data* for shaped castings. Basic models for dendritic and eutectic solidification, as well as their coupling are discussed.

In particular, a physical model for the evaluation of chill formation, i.e. of the gray/white structural transition in iron castings was developed. This model is based on the concept of the existence of a critical cooling rate at which a gray to white transition occurs. The critical cooling rate for the gray/white structural transition was experimentally determined on a step-casting. This rate was then used in the heat transfer-solidification kinetics model to determine the time-temperature history of the test casting, and to predict the region of chill formation in the casting. The model was also used to predict the interlamellar spacing, graphite flake size and eutectic grain size in hypoeutectic gray iron as a function of cooling rate. Predicted results were then compared with experimentally obtained data. Validation of calculated results was performed against experimental data in terms of temperature history, evolution of fraction of solid, chill pattern and other microstructural characteristics. Other information such as solute element profile, undercooling, equivalent grain radius as a function of time are also discussed.

Introduction

Recent developments in the field of computer simulation of solidification of castings support the idea that a combination of both macroscopic and microscopic models are necessary in order to accurately describe the process. Required macroscopic models include models that describe heat transfer from metal to mold, fluid flow of liquid metal during mold filling, and stress field in the casting. At the microscopic level, the models should include solidification kinetics and fluid flow in the mushy zone.

This paper describes only work involving macroscopic heat transfer (HT) and microscopic solidification kinetics (SK) models. As discussed in previous papers(1,2), HT modeling for a given casting-mold combination requires solving the energy conservation equation for heat conduction:

$$\nabla . \left[K(T) \ \nabla \ T(\vec{x},t) \right] + \dot{Q} = \rho \ C_p(T) \ \frac{\partial T(\vec{x},t)}{\partial t} \qquad (1)$$

where K: thermal conductivity; T: temperature at a given location in casting; ρ: density; C_p: specific heat; t: time. The source term $\dot{Q}$ describes the rate of latent heat evolution during the liquid-solid transformation and may be written as:

$$\dot{Q} = L\frac{\partial f_s(\vec{x},t)}{\partial t} \qquad (2)$$

where L: constant volumetric latent heat of fusion during the phase transformation; f_S: volume fraction of solid. The heat evolution problem is solved via the *latent heat method*(2,3). This method consists in calculating $f_s(x,t)$ by a solidification kinetics approach, e.g. from nucleation and growth laws. If f_s in eq. (2) is known, solving of eq. (1) becomes trivial. The coupling of the macroscopic heat flow to solidification kinetics is quite apparent through eq.(2).

The rate of formation of fraction of solid is obtained from the following equation:

$$\frac{\partial f_s}{\partial t} = 4\pi R^2 N \frac{\partial R}{\partial t}(1-f_s) \qquad (3)$$

where N: number of grains per unit volume (grain density); R: average equiaxed grain radius. The effective area of the interface between the solid grains and the liquid was weighted by the factor $(1-f_s)$, which is the well known Johnson-Mehl(4) approximation. This equation has the advantage that it accounts for grain impingement. N(x,t) and R(x,t) must be calculated from nucleation and growth laws.

Theoretical Work

In order to describe the solidification of hypo-eutectic and/or of eutectic cast iron one needs appropriate models for the solidification of primary austenite, of the stable austenite-graphite eutectic and of the metastable austenite/iron carbide eutectic.

1. Model for equiaxed dendritic growth

Several models for growth of equiaxed dendrites have been proposed before(5,6). An improved model, which uses both perturbation analysis and the Ivantsov solution is introduced in this paper.

The basic assumptions are as follows (see notations in FIG. 1):

- the interface composition at the tip of the dendrite (C^* at R_d) is the same with that at the circumference of the sphere that incorporates the total mass of the dendrite (C^* at R_s);
- when the tip of the dendrite reaches the available grain size (i.e. $R_d = R_g$) the sphere R_s advances with the velocity of the tip of the dendrite, V;
- constitutional supercooling in the interdendritic space is small, which means that the dendrite is a paraboloid of revolution (no secondary arms).

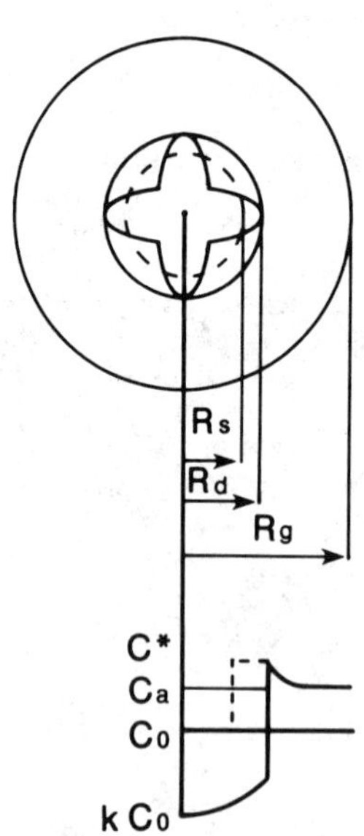

FIG. 1 Composition profile within the volume of a dendritic grain.

Calculation of fraction of solid f_s

The fraction of solid can be calculated as:

$$f_s = \left(\frac{R_s}{R_g}\right)^3 \qquad (4)$$

If it is assumed that the volume of the dendrite is compressed into a sphere, and that the tridimensional (3D) dendrite is made of three paraboloids of revolution, from geometrical considerations it can be calculated that the fraction of solid is:

$$f_s = 3\frac{rR_d^2}{R_g^3}\left[\frac{3}{2} - \left(\frac{8r}{R_d}\right)^{\frac{1}{2}}\right] \qquad (5)$$

where r is the tip radius of the dendrite. In this equation R_g is known from the nucleation law, while R_d can be calculated as a function of the dendrite tip velocity, V, with:

$$R_d = \int_0^t V(t)\ dt \qquad (6)$$

Obviously to calculate f_s, one needs to know the dendrite tip velocity, V, and tip radius, r.

Calculation of the Tip Velocity of Dendrite, V

Following the approach suggested by Lipton, Glicksman and Kurz(7), the Ivantsov function of the Peclet number can be calculated as:

$$I(P_c) = \frac{\Delta T_c}{(k-1)(mC_o - \Delta T_c)} \qquad (7)$$

where k: partition coefficient; ΔT_c: solutal undercooling; m: slope of the liquidus line; C_o: initial composition in the liquid; P_c: solutal Peclet number; $I(P_c) = \Omega$; Ω: supersaturation at the tip; $P_c = Vr/2D$. Using perturbation analysis:

$$r^2 = \frac{4\pi^2\Gamma}{mG_c - G} = \frac{4\pi^2\Gamma}{\frac{P_tL}{c}\frac{1}{r} - \frac{P_c\ m(1-k)C*}{r}}$$

where Γ: Gibbs Thompson coefficient; G_c: constitutional gradient; G: thermal gradient; C^*: interface composition; P_t: thermal Peclet number. Since $P_t = Vr/2\alpha$:

$$Vr^2 = \frac{8\pi^2\Gamma D\alpha c}{DL - m(1-k)C^*\alpha c} \qquad (8)$$

where D: liquid diffusivity; α: thermal diffusivity. If ΔT_c and C^* are known, eqs. (7) and (8) can be solved simultaneously numerically for V and r.

For the first approximation of $I(P_c)$ i.e. $I(P_c) = P_c$, an analytical expression for the tip velocity can be derived.

Calculation of Liquid Composition at Interface, C^*

Writing a flux balance at R_g as suggested in ref.(5), it can be shown that the interface temperature can be calculated as:

$$T^* = \int dt^* = \frac{1}{\rho c}\int_0^t\left(\frac{3}{R_g}\ dQ_{ext} + L\ df_s\right)dt \qquad (9)$$

and:

$$C^* = \int dC^* \qquad (10)$$

where Q_{ext}: external heat flow. If a numerical solution for the thermal field is available, Q_{ext} is known from the previous time step as:

$$\Delta Q_{ext} = R_g/3\rho c \ \Delta T$$

Calculation of Solutal Undercooling, ΔTc

Assuming complete mixing in the interdendritic liquid, the solutal undercooling can be written as:

$$\Delta T_c = m\,(C_a - C^*) \tag{11}$$

where C_a is the average composition in the liquid. To find C_a one must solve the diffusion equation:

$$\frac{\partial C}{\partial t} = D\left[\frac{\partial^2 C}{\partial r^2} + \frac{1}{r}\frac{\partial C}{\partial r}\right] \tag{12}$$

This equation is solved numerically with the following conditions:

Initial condition:	$C(r,o) = C_o$
BC1	$C(R_s,t) = C^*$
BC2	$(\partial C/\partial r)_{r\,=\,Rg} = 0$

The solution of eq. (12) is the solute profile in the liquid C(r,t) for $r = R_s$ to $r = R_g$.

The total solute in liquid is:

$$C_{tot} = \int_{Rs}^{Rg} 4\pi r^2 C(r,t)\,dr \tag{13}$$

The average composition in the liquid is:

$$C_a = \frac{C_{tot}}{\frac{4}{3}\pi\,(R_g{}^3 - R_s^3)} = \frac{3\int_{Rs}^{R_g} r^2 C(r,t)\,dr}{R_g{}^3 - R_s^3} \tag{14}$$

With C^* known from eq. (10) and C_a from eq. (14), ΔT_c in eq. (11) can be calculated.

2. <u>Model for equiaxed eutectic growth</u>

The treatment of the nucleation problem has been discussed in detail in ref. (3). The eutectic grain size or the grain density can be predicted with that model, and compares well with experimental data.

Based on the classic Jackson-Hunt model for regular eutectic solidification, the growth velocity of the planar interface can be calculated as:

$$V = \mu\,(\Delta T^*)^2 \tag{15}$$

where ΔT^* is the interface undercooling. In most of the published work on HT-SK modeling, this interface undercooling is substituted by the bulk undercooling, ΔT_b, when calculating the velocity, i.e.:

$$\Delta T^* = \Delta T_b \tag{16}$$

This approximation implies first, that $\Delta T^* = \Delta T_t$, and second that $\Delta T_t = \Delta T_b$, where ΔT_t is the thermal undercooling at the interface.

A more complete approach should also consider solutal undercooling, ΔT_c, and curvature undercooling, ΔT_r. Then eq. (16) becomes:

$$\Delta T^* = \Delta T_t + \Delta T_c + \Delta T_r \quad (17)$$

The calculation of the three terms on the right hand side of the above equation is discussed in the following paragraphs.

Thermal Undercooling, ΔT_t

As mentioned before, it is common practice to assume that $\Delta T_t = \Delta T_b$. As shown in an earlier paper (1), this is not necessarily true.

The calculation of ΔT_t can be done from a heat balance over the final volume of a grain at some intermediate time when both liquid and solid coexist. This volume is defined as 1/N. The radius of the solid sphere is calculated as:

$$R_S = [3f_s/(4\pi N)]^{1/3}$$

$$R_L = [R_S^3 + 3(1-f_s)/(4\pi N)]^{1/3}$$

The heat transfer problem is solved as suggested in ref.(8), by using the energy equation in cylindrical coordinates:

$$\frac{1}{r}\frac{\partial}{\partial r}(rq_r) = \dot{Q} \quad (18)$$

The boundary conditions for the solid are:

$$r = 0 \qquad q^S \neq 0$$

$$r = R_S \qquad T^S = T_t$$

where q^S is the radial heat flux in the solid, T^S is the temperature in the solid, $T_t = T_E - \Delta T_t$, and T_E is the equilibrium transformation temperature.

The boundary conditions for the liquid are:

$$r = R_S \qquad T^L = T_t$$

$$r = R_S \qquad q^S = q^L$$

where q^L is the radial heat flux in the liquid, and T^L is the temperature in the liquid.

By solving eq.(18) with the appropriate boundary conditions it is calculated that:

$$T^S = T^* + \frac{\dot{H}}{4k_s} R_s^2 \left[1 - \left(\frac{r}{R_s}\right)^2\right] \quad (19)$$

where $\dot{H}$ is the rate of change of enthalpy for the solid and k_s is the thermal conductivity of solid. Also:

$$T^L = T^* + \frac{\dot{M}}{4k_L}(R_s^2 - r^2) + \frac{(\dot{H} - \dot{M})}{2k_L} R_s^2 (\ln R_s - \ln r) \quad (20)$$

where M is the rate of change of enthalpy for the liquid, and k_L is the thermal conductivity of liquid. These two temperatures are related to the bulk temperature, T_b, through:

$$T_b = \frac{\int_0^{R_s} T^s \, dr + \int_{R_s}^{R_L} T^L \, dr}{R_L} \tag{21}$$

From this last equation one can obtain the T^* as a function of T_b.

Solutal Undercooling, ΔT_c

To calculate ΔT_c one needs to know the solutal profile in the liquid, C(r,t), between R_S and R_L. This can be calculated under different sets of assumptions.

For example it can be assumed that mixing in the liquid occurs only by diffusion (i.e. no convection). In this case the diffusion equation (12) must be solved numerically with the boundary conditions:

$$r = R_S \qquad C(r,t) = C_o/k$$
$$r = R_L \qquad \partial C(r,t)/\partial r = 0$$

A simpler approach will be to assume complete mixing in liquid, in which case the diffusion equation can be solved analytically (Scheil equation). With C(r,t) known, the profile of the actual transformation temperature can be calculated as:

$$T_c = T_E + m' C(R_L,t) \tag{22}$$

where m': constant related to the influence of a particular element on the equilibrium temperature. Then the solutal undercooling is:

$$\Delta T_c = T_E - T_c \tag{23}$$

Curvature Undercooling, ΔT_r

Since the basic assumption of the SK model is that the grains grow as spheres, an undercooling due to the radius of curvature will occur:

$$\Delta T_r = 2\,\Gamma / R_S \tag{24}$$

Eutectic Interlamellar Spacing

The austenite-lamellar graphite eutectic is an irregular eutectic. Graphite grows as a faceted phase. According to Magnin and Kurz (12), for this type of eutectics, two different spacings must be considered:

- a minimum spacing resulting from growth at extremum, λ_{ex}, as proposed by Jackson and Hunt
- a maximum (branching) spacing resulting from the instability criterion, λ_{br}

The minimum spacing can be calculated as:

$$\lambda^2_{ex} \cdot V = \mu_1 \tag{25}$$

where μ_1 is a material constant which can be calculated (see ref. 12).

The maximum spacing can be calculated as:

$$\lambda_{br} = \left(\frac{2\Gamma_{xF}\cos(\Theta_{xF})}{f_{xF}\left(\frac{|m_{xF}|C_0}{D\tan(\Theta_{xF})}\left(\Pi_{xF} - \frac{P}{f_{xF}} \right) V + \frac{f_{xF}}{24} G \right)} \right)^{\frac{1}{2}} \mu_2 \tag{26}$$

where xF: index for faceted phase; Θ: the wetting angle; Π and P: functions of volume fraction; and μ_2: material constant, which for cast iron can be considered to be 1.9. Then the average spacing is:

$$<\lambda> = (\lambda_{ex} + \lambda_{br})/2 \tag{27}$$

It must be noted that this approach is valid for steady state solidification.

3. Coupling of dendritic and eutectic growth models

Scheil Equation for Solidification with Undercooling

For equilibrium solidification the amount of eutectic resulting at the end of solidification can be calculated from the classic Scheil equation as:

$$f_E = (C_{SM}/k\, C_o)^{1/k-1} \tag{28}$$

The symbols are defined in FIG. 2. If solidification occurs with undercooling both primary dendritic and eutectic solidification will be delayed with ΔT_P and ΔT_E respectively. From FIG. 2 it can be calculated that the fraction of eutectic for solidification with undercooling is:

$$f_E' = \left(\frac{m_s C_{SM} - \Delta T_E + \Delta T_P}{k C_o m_s} \right)^{\frac{1}{k-1}} \tag{29}$$

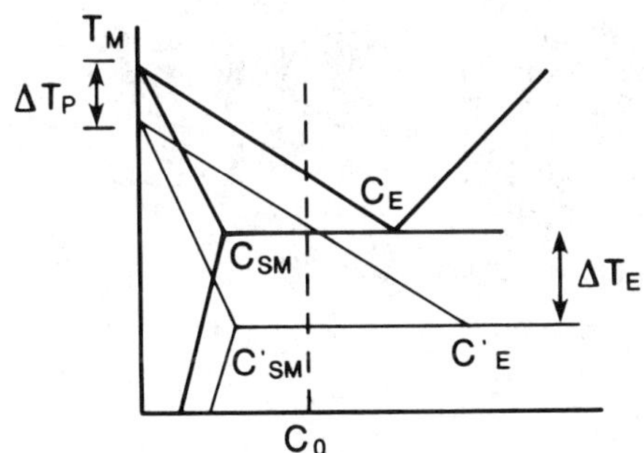

FIG. 2 Shift of temperature-composition lines as a function of undercooling.

Analysis of eq.(26) shows that:

if $\Delta T_E = \Delta T_P$ then $f_E' = f_E$
if $\Delta T_E > \Delta T_P$ then $f_E' < f_E$
if $\Delta T_E < \Delta T_P$ then $f_E' > f_E$

If the assumption of local equilibrium is relaxed, the fraction of primary dendrites, f_s, will continue to grow even when the interface temperature becomes smaller than the eutectic temperature. Therefore, one should expect that:

for $\Delta T_E > \Delta T_P$ $\quad f_E'{}_{\text{non equil}} < f_E'{}_{\text{equil}}$.

Note that ΔT_E and ΔT_P are functions of dT/dt through the nucleation law.

4. Model for the Structural Gray (Stable)/White (Metastable) Transition in Cast Iron

When an iron casting cools and solidifies, the eutectic solidification can take place either according to the stable iron-graphite eutectic transformation, leading to gray structure, or according to the metastable iron-iron carbide (cementite) transformation, resulting in white structure. The eutectic temperature for the stable transformation (T_{st}) of a binary Fe-C alloy is 7°C higher than that of the metastable reaction (T_{met}). This difference can be considerably increased or decreased by the addition of alloying elements. The type of structure to form during solidification depends on whether solidification occurs above or under T_{met}.

The basic variables affecting the gray/white transition in cast iron are cooling rate of casting, nucleation potential (inoculation), and chemical composition of melt. When the casting solidifies at high cooling rates excessive undercooling under T_{st} may occur, and the liquid metal may cool below the metastable transformation temperature before solidifying. This can result in a white or partially white structure. Inoculation increases the nucleation potential of the iron and can significantly reduce undercooling, and thus the chilling tendency. Alloying elements can significantly affect the potential for chill and intercellular carbide formation through kinetic and thermodynamic influences. In general, graphite promoters, such as Si and Ni simultaneously raise the stable transformation temperature and reduce the metastable transformation temperature. Carbide promoters such as Cr and V have the opposite effect. The relation between T_{met} and various alloying elements can be linearly expressed as (ref.9):

$$T_{met} = 1148°C - 19\ \%Si + 3\ \%Mn - 2.3\ \%Cu \qquad (30)$$

Thus for an alloy of given composition, T_{met} can be calculated. Experimental or calculated cooling curve data for various positions in the casting can then be compared with the calculated T_{met}. It can be assumed that at locations where the temperature of eutectic start, TES, is below T_{met}, the casting solidifies with a chilled structure provided there is no significant recalescence.

The physical model proposed for the evaluation of the gray/white transition in cast iron (ref.10) is based on the concept of a critical cooling rate $(dT/dt)_{cr}$, as shown in FIG.3 which is a plot of TES vs. cooling rate. The intersections between the T_{met} lines and the TES - dT/dt lines determine the critical cooling rates at which the gray/white structural transition occurs as a function of composition (silicon content) and inoculation. If the casting or a particular section of the casting cools faster than the critical cooling rate, $(dT/dt)_{cr}$, white structure will result. It must be noted that in fact T_{met} is also a function of cooling rate (ref.11), although to a much lesser extent.

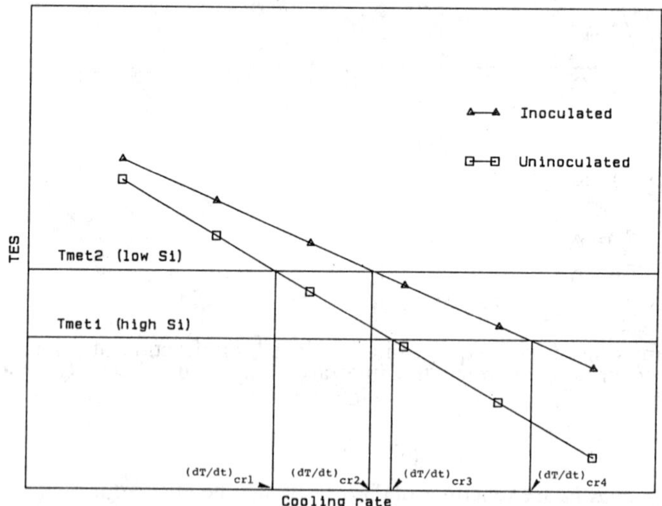

FIG. 3 Schematic representation of the influence of cooling rate on the temperature of start of eutectic solidification and of the critical cooling rate for the occurrence of the gray/white transition.

Experimental Work

In order to obtain a range of structures from white through mottled to gray in the same casting, a step casting was selected. The geometry of the casting is shown in FIG. 4. The thickness of the steps were 25, 18, 12.5, 6, and 3 mm. The mold was made with pep-set resin bonded sand, and was instrumented with K-type thermocouples to record the temperature history at the center of the steps. A cylindrical casting, 5 cm. in diameter and 28 cm. long, molded in sand, and instrumented with a thermocouple, was also used for validation purposes.

A gray iron heat was produced. Details are given in ref.(10). The cooling curve data for the casting at locations were collected with an IBM PC computer interfaced with a Dash-8 board. From eq.(30), T_{met} for the particular composition used in this work can be calculated as 1114°C. A complete representation of the microstructural pattern in the experimental casting is given in FIG. 4. It will be compared later with calculated results.

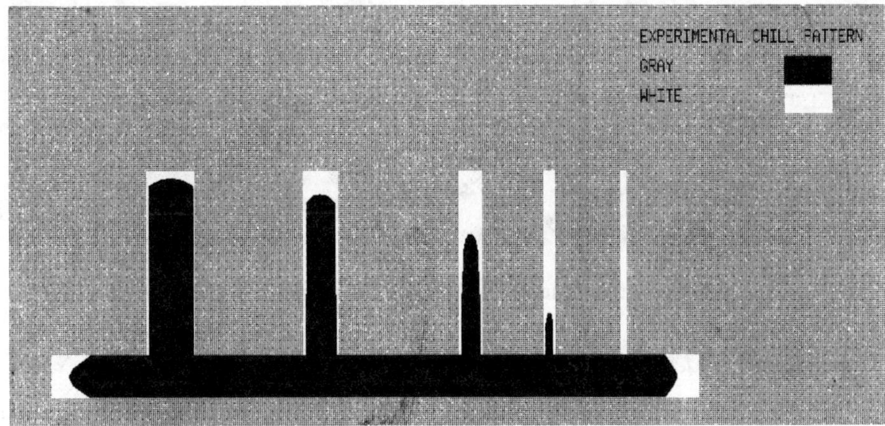

FIG. 4 Map of experimental gray and white structural regions on the casting.

As explained in FIG. 3, the critical cooling rate for the gray/white transition was obtained by taking the cooling rate for which the TES equals T_{met}. This gives a critical cooling rate of 13.3°C/sec. for the composition poured. For validation purposes this critical rate has been used in the HT-SK model as will be discussed later.

In order to collect data for the interlamellar spacing in the graphite-austenite eutectic, a sample was taken from the area in the vicinity of the thermocouple from the 25 mm. thick step. The sample was deep etched to remove the austenite and examined under SEM. A typical eutectic grain is shown in FIG. 5. The measurement of the interlamellar spacing included the following steps: 1) find the center of the grain; 2) use circular grids of increasing diameters, centered in the middle of the grain; 3) count the number of intercepts for each particular grid; 4) measure the volume fraction of graphite inside the grid; 5) use standard stereological equations to obtain mean interlamellar spacing, λ_{av}. This experimental spacing was plotted against the radial distance from the center of the eutectic grain in FIG. 10.

Validation and Discussion

The HT-SK model developed for the prediction of chill formation during the solidification of cast iron was run on an Apollo Domain Series 10000 computer with color graphic display facilities. The HT code is named ALCAST 2D. It is an FEM code described elsewhere (3).

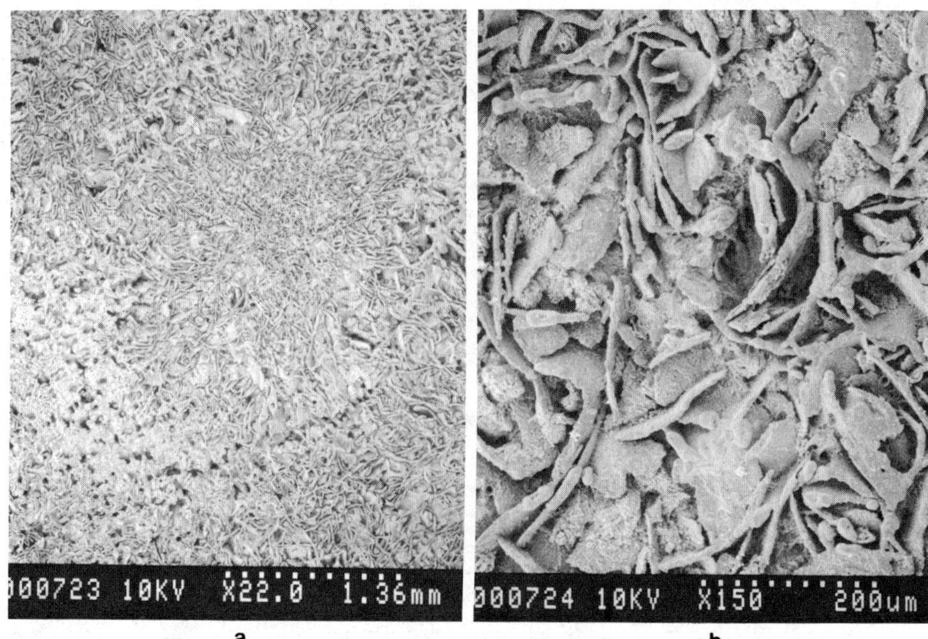

FIG. 5 Microstructures of deep etched eutectic grains in cast iron at different magnifications.

Interface vs. Bulk Undercooling

The interface undercooling was calculated using eqs. (19) through (21) and then compared to the bulk undercooling obtained from the HT-SK code. The difference $T^* - T_b$ was then plotted against time as a function of cooling rate in FIG. 6. The cooling rates corresponded to three different steps in the casting. It can be seen that the difference can be as high as 1.1 °C for the 9 °C/s cooling rate, and decrease to 0.6 °C for the 2.8 °C/s cooling rate. This difference did not seem to have a significant influence on the cooling curve, as shown in FIG. 7.

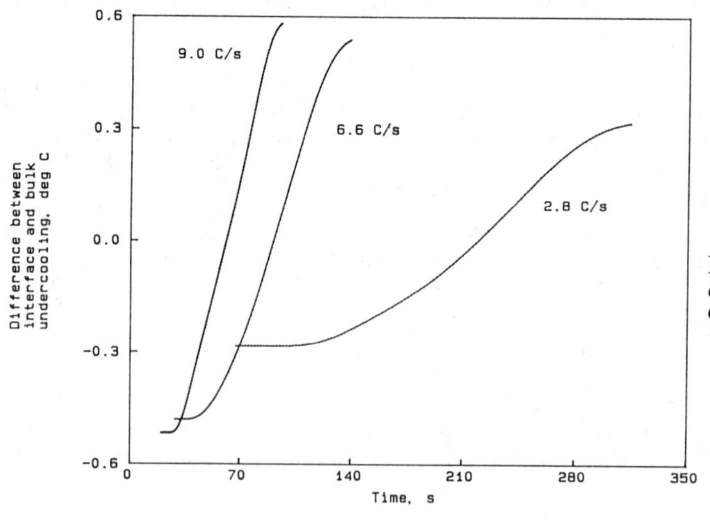

FIG. 6 Influence of cooling rate on the difference $(T^* - T_b)$.

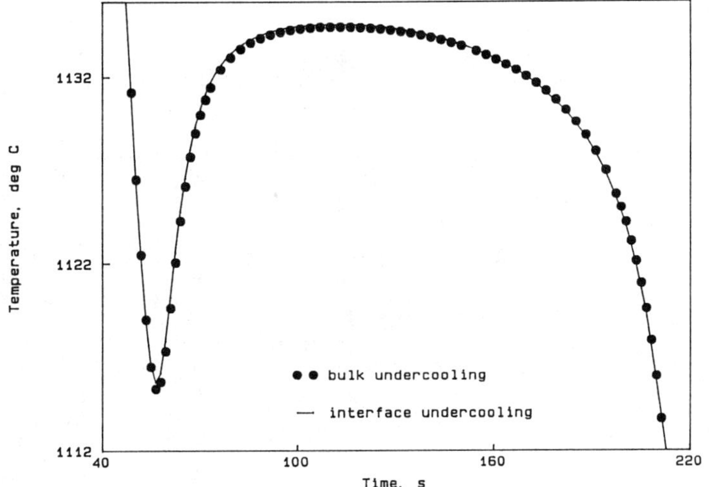

FIG. 7 Comparison of calculated cooling curves using interface or bulk undercooling for a 3 cm. diameter bar molded in sand.

Coupling of Primary and Eutectic Solidification

Experimental and calculated cooling curves for a hypoeutectic gray iron poured into the experimental cylindrical casting (5 cm. dia., 28 cm. long) are shown in FIG. 8. Both a primary and an eutectic arrest were predicted, and match rather well with the experimental curve. Some noticeable discrepancy is obvious at the end of solidification. Probably this discrepancy is mostly due to the correction factor used to account for grain impingements at the end of solidification (see eq. 3). Indeed when the correction factor was changed rather arbitrarily from (1-fs) to $(1\text{-fs})^3$ the agreement improved considerably (see FIG. 8).

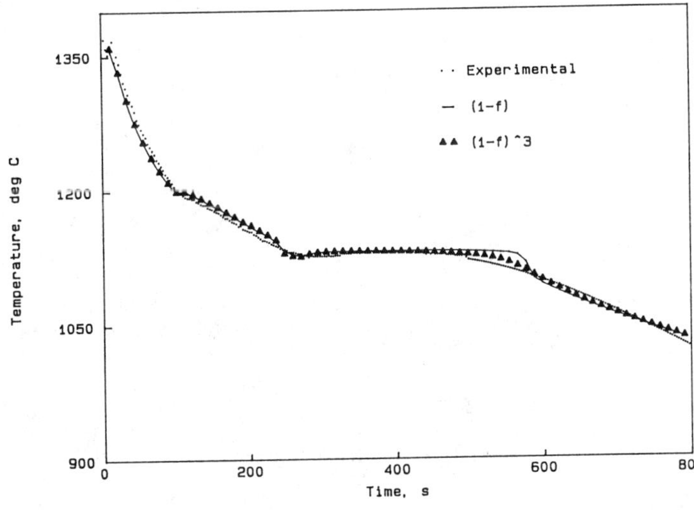

FIG. 8 Experimental and modeled cooling curves, using different corrections for grain impingement.

Lamellar Spacing in Gray Iron

The steady-state average interlamellar spacing for the austenite-graphite eutectic was calculated for the case of the 25 mm. step in the experimental casting. The casting poured was hypoeutectic and had a carbon equivalent of 3.9. The growth velocity was calculated based on the interface undercooling as given in eq. (17). The average spacing was obtained by using eq. (27), using constants from ref. (12). The result of this calculation is shown in FIG. 9 together with the experimental measurements. Calculations show that the maximum spacing occurs immediately after the beginning of solidification. This maximum results from the influence of the radius of curvature on the undercooling. Nevertheless this maximum obviously contradicts experimental data. An additional constraint must be considered. This is the fact that $<\lambda>$ cannot be larger than the instantaneous circumference of the grain, $2\pi R_S$. This constraint is also plotted in FIG. 9. After the first maximum of $<\lambda>$, the spacing decreases, following the trend in ΔT^*, until a second maximum occurs. This maximum results from the contribution of constitutional undercooling calculated under the assumption of complete mixing in liquid. The discrepancy between the steady state model and the experimental data is noticeable. The steady state model may give better agreement if only mixing by diffusion is considered (eq. 12).

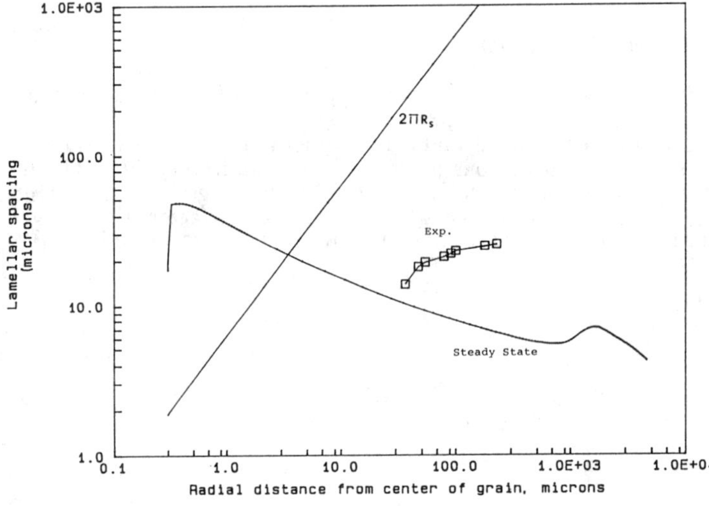

FIG. 9 Comparison between experimental and calculated interlamellar spacing in gray iron.

The Gray/White Transition in Cast Iron

The gray/white transitions in the step casting, as calculated through the model, is shown in FIG. 10. It compares rather well with the experimentally obtained structural map shown in FIG. 4.

Summary

An improved model for the growth of equiaxed dendrites, which uses both perturbation analysis and the Ivantsov solution, is introduced in this paper. Also a more complete model for equiaxed eutectic growth was proposed. This model uses interface undercooling rather than bulk undercooling. An energy balance approach has been used for the calculation of interface temperature from bulk temperature. The difference between the interface temperature and the bulk temperature was found to vary in the range 0.6 to 1.1 °C for the particular casting used in these experiments.

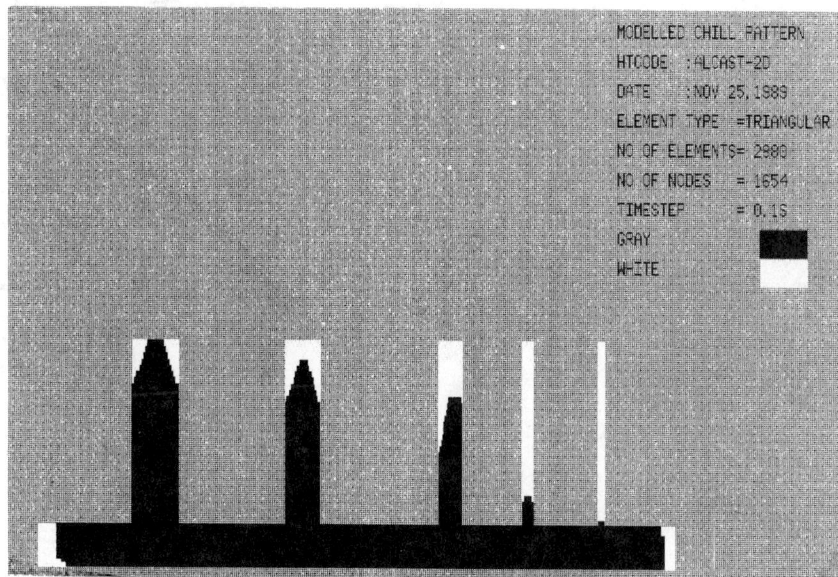

FIG. 10 Map of calculated gray and white structural regions on the casting.

The contributions of solutal undercooling and curvature undercooling have also been incorporated in the eutectic model. Calculations with the coupled primary and eutectic models have been successfully validated for a hypoeutectic cast iron.

Prediction of interlamellar spacing has also been attempted by incorporating the existing steady state model for faceted/non-faceted eutectics. Experimental work involving SEM and deep etching was done, in order to measure the interlamellar spacing in an eutectic grain along the grain radius. While at the present time the agreement with experiment is not very good, there are reasons to believe that a more complete treatment of the solutal undercooling could improve this agreement.

A physical model for prediction of chill formation (i.e., gray/white structural transition) in gray iron castings, previously proposed, has also been discussed. The model is essentially based on the determination of a critical cooling rate above which the solidified structure is white. It was shown that this model can describe rather accurately the experimental map of structural transition over the whole section of an experimental casting.

Acknowledgements

This work was supported by NSF EPSCoR Alabama, and NSF CADCAST, Contract No. DMC - 8612019. The authors are grateful to Dr. Dongkai Shangguan for many helpful discussions.

References

1. D.M. Stefanescu and C. S. Kanetkar, State of the Art of Computer Simulation of Casting and Solidification Processes, H. Fredriksson, editor, p.255, Les Editions de Physique, Paris, (1986).
2. C.S. Kanetkar,I.G. Chen, D.M. Stefanescu, and N. El-Kaddah, J. of ISI of Japan, 28, p.860, (1988).
3. D.M.Stefanescu, G.Upadhya, and D.Bandyopadhyay, Met.Trans., 21A, 997 (1990).

4. W.A. Johnson and R.F. Mehl, Trans AIME, 135, 416 (1939).
5. M. Rappaz and P. Thevoz, Acta Met., 35, 7, 1487 (1987).
6. C.S. Kanetkar and D.M. Stefanescu, Modeling of Casting and Welding Processes IV, A.F. Giamei and G.J. Abbaschian, editors, p. 697, TMS, Warrendale, PA, (1988).
7. J.Lipton, M.E.Glicksman, and W.Kurz, Mat.Sci.and Eng. 65, 1, 57 (1984).
8. R.B.Bird, W.E.Stewart, and E.N.Lightfoot, Transport Phenomena, p. 274, John Wiley & Sons, New York (1960).
9. D.M.Stefanescu, Metals Handbook vol.15: Casting, p. 61, ASM International (1988).
10. G.Upadhya, D.K.Banerjee, D.M.Stefanescu, and J.L.Hill, AFS Trans., paper no. 156 (1990) in print.
11. K.Nakamura, H.Sumimoto: Japan Foundrymen's Society, report no.47 (1987).
12. P.Magnin and W.Kurz, Acta Metall.35, 5, 1119 (1987).

On thermal analysis for shrinkage prediction in commercial ductile iron castings

F.J. Bradley
Department of Materials Science and Engineering, University of Wisconsin-Madison, Madison, Wisconsin, U.S.A.

C.A. Fung
Department of Industrial Engineering and Center for Quality and Productivity Improvement, University of Wisconsin-Madison, Madison, Wisconsin, U.S.A

Introduction

This paper summarizes results to date from two complementary ongoing research projects concerning shrinkage prediction in ductile iron castings: one an industry-sponsored project on statistical modeling for process control, and the other an NSF-funded project on finite element computer-aided design of castings. Each project represents a distinct application of thermal analysis. In the statistical modeling project, thermal analysis variables are utilized directly as shrinkage predictors for process control; whereas in the numerical simulation project, thermal analysis is utilized to obtain a heat generation function which can be used to model the solidification behavior of ductile iron. For the latter case, previous studies have suggested the use of solidification parameters such as thermal gradient, cooling rate, solidus velocity, and fraction solidified for shrinkage prediction [1-6]. Integration of these two approaches to the shrinkage problem provides a direct link between computer-aided design based on numerical simulation, and process control in the foundry based on thermal analysis.

Riser design and feeding considerations for ductile iron are aggravated by the complexity of solidification mode and the extreme sensitivity of solidification behavior to processing practice, which combine to make the prediction of volume changes occurring during solidification very difficult. In particular, the specific volume of the solidified eutectic may be greater than that of the liquid from which it forms because of the precipitation of graphite, which has a significantly lower density than iron. This complicates riser design as expansion during solidification can lead to the development of significant contact pressure at the mold/metal interface which tends to distort the shape of the mold cavity. Any resulting mold enlargement leads to an increased demand for feed metal which must be accounted for in the initial casting design. The extent of mold enlargement is determined by the contact pressure at the mold/metal interface and the deformation response of the green sand.

Another important consideration with regard to shrinkage is the 'feeding capability' of a given iron. The ease with which an iron can feed a thicker section through a thinner section is dependent on the mode and sequence of solidification, both in terms of how solidification behavior affects the physical nature of the feed paths between the riser and the hot spot and how it influences 'self-feeding', a term which refers to the capability of molten cast iron to compensate for solidification contraction of the austenite during the formation of the eutectic via expansion associated with graphite precipitation. Thus, with sufficient self-feeding, sound castings may be produced even in the case that feeding of a hot spot from a riser has been terminated due to closure of the feed channels. Self-feeding is often exploited to obtain more efficient risering of gray iron castings, but this same beneficial effect is not typically found in commercial ductile iron castings — with the result that scrap rates due to shrinkage defects are often much higher in ductile iron castings than in gray iron castings.

Fundamental considerations in numerical simulation-based shrinkage prediction research are (i) the development of a kinetic model of ductile iron solidification which realistically accounts for the effects of metallurgical processing on the mode and sequence of solidification, and (ii) application of the model to estimating the volume changes occurring during freezing. Thermal analysis and microstructural modeling [7,8] based on fundamental nucleation and growth considerations are two approaches that have been taken in simulating the solidification behavior of ductile iron. The microstructural models of ductile iron solidification developed to date have required input of unrealistic values of nodule count in order to obtain cooling curves consistent with experimental results [9-11]. The difficulties in modeling the solidification of ductile iron are associated with limited data on nucleation of the phases, the possibility of the phases growing independently during a portion of the solidification cycle, and segregation effects [9,12].

In addition to the complexities associated with the formulation and development of microscopic solidification models, practical considerations currently limit the application of this approach to computer-aided design. A direct coupling of microstructural and macroscopic models not only greatly increases computational requirements and simulation time, but also significantly increases the degree of nonlinearity of solidification heat transfer analysis and may lead to convergence problems. While this paper is primarily concerned with the application of thermal analysis to shrinkage prediction in ductile iron castings, research utilizing both thermal analysis-based and micro-macroscopic modeling approaches to the problem is ongoing at UW-Madison.

Thermal Analysis

A more applied approach to studying solidification characteristics of cast iron and other alloys than micro-macroscopic modeling is thermal analysis [13-26], a well-established technique which essentially involves recording time-temperature data of a sample as it solidifies under standard conditions. Figure 1 shows a typical ductile iron cooling curve, its first and second derivative curves, and the nomenclature used to identify characteristic cooling curve parameters. The premise of thermal analysis for process control is that the nature of cooling curve data obtained under standard conditions is inherently reflective of the effects of metallurgical processing on solidification behavior.

Most thermal analysis research to date has been concerned with evaluating the effects of molten metal processing on solidification behavior, and correlating cooling curve parameters with alloy chemistry, as-cast microstructure, and other processing variables. For example, in early work Heine, Loper, and co-workers [13] developed prediction charts for nodularity, nodule count, and carbide content of select cast irons based on a number of cooling curve parameters, including eutectic recalescence, eutectic undercooling, average recalescence rate, and cooling time between a specified temperature range. Ryntz et al [14] developed a 'visual comparator overlay' method for the prediction of ductile iron microstructures. They noted that the technique was particularly effective in detecting the effect of magnesium treatment and postinoculation effects on graphite nodularity and carbide content; and concluded that it is preferable to base microstructure prediction on relative shape of cooling curves rather than on absolute values of temperature and time, which were found to be dependent on sampling technique and sample cup type. Backerud et al [15] also found absolute values of cooling curve parameters to be less reliable as microstructure predictors than relative or derivative parameters, which were found to correlate well with graphite morphology in spheroidal and vermicular cast irons.

Much of the recent work in thermal analysis has been concerned with the

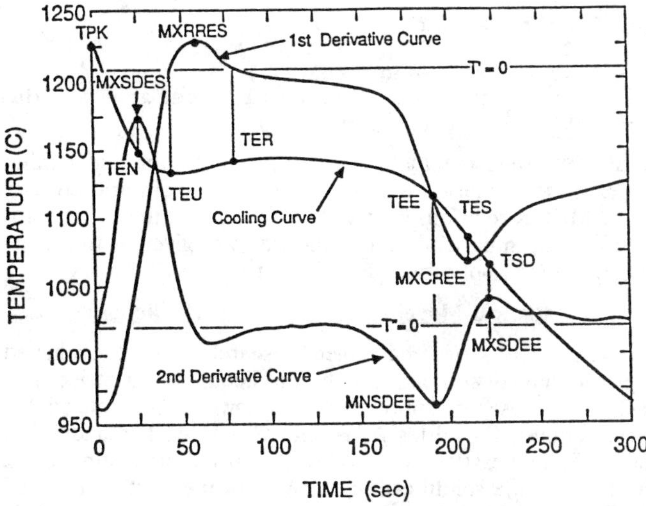

FIG. 1. Typical Eutectic Ductile Iron Cooling Curve

development of computer-based systems for process control. In particular, Chen and Stefanescu [18] developed a technique termed computer-aided differential thermal analysis (CA-DTA) for the prediction of microstructure and graphite morphology of cast iron. While good correlations between nodularity and select first and second derivative parameters were observed for certain conditions, it was noted that variations in carbon equivalent, pouring temperature, and nodule count tended to increase the spread in the results. Louvo and co-workers [21] at the Technical Research Center of Finland developed a computer-based derivative thermal analysis (DERTA) method to assess the quality of ductile iron melts. They conducted directional solidification studies and found nodularity to correlate well with the minimum in the cooling rate curve at the end of eutectic solidification, and nodule count to correlate well with the minimum second derivative at the end of eutectic solidification. Multiple regression analysis was used to develop predictive formulas for both nodularity and nodule count.

Recently, there has been much interest in thermal analysis for shrinkage prediction in ductile iron castings. In particular, Ableidinger and Raebus [20] suggest that the feeding behavior and shrinkage tendency of ductile iron melts can be characterized using thermal analysis criteria based on the following temperature, derivative, and integral cooling curve parameters : $T_{eut\ max}$, the maximum eutectic temperature [TER]; $T_{sol\ corr}$, the corrected solidus temperature [TES]; negative $peak_{corr}$, the negative peak in the first derivative curve at the solidus [MXCREE]; the maximum of the second derivative before $solidus_{corr}$ [MNSDEE]; and $Q_{tot\ corr}$, the total heat of solidification in Joules per gram [QSOLID]. The variable names of cooling curve parameters identified in Figure 1 most closely corresponding to the Ableidinger and Rabeus parameters are given in square brackets. Ableidinger and Raebus suggest that ductile iron feeding characteristics are optimized when the following criteria are met:

$$T_{eut\ max} > 1152\ ^{\circ}C \quad (1)$$

$$T_{sol\ corr} > 1109\ ^{\circ}C \quad (2)$$

$$\text{neg peak at solidus} < -4.0\ ^{\circ}C/s \quad (3)$$

$$\text{max. 2nd deriv. at solidus} < -0.40\ ^{\circ}C/s^2 \quad (4)$$

$$Q_{tot} > 250\ J/g \quad (5)$$

They also presented a shrinkage predictor MEGK, which is a combination of all of the five thermal analysis parameters noted above, computed using the following equation:

$$\begin{aligned} MEGK = &\ 0.0575 \times (T_{sol\ corr} - 1096) \\ &+ 0.033 \times (T_{eut\ max} - 1145) \\ &+ 0.11 \times (-neg\ peak_{corr}) \\ &- 2.0 \times (|max.\ 2nd\ deriv.\ before\ solidus_{corr}|) \\ &+ 0.02875 \times (Q_{tot\ corr} - 232) \end{aligned} \tag{6}$$

Ableidinger and Raebus reported that good feeding behavior and minimum shrinkage tendency correspond to the criterion MEGK > 2.80. They did not, however, publish the data from which the MEGK relationship was derived, nor did they offer statistical substantiation for the coefficients; in particular, no indication was given as to the possible need to recalibrate the equation for applications in different foundries.

Statistical Modeling for Shrinkage Prediction

In an ongoing industry-sponsored project, research is being conducted to develop a thermal analysis-based statistical model for the prediction of shrinkage defects in ductile iron castings [24-26]. The research approach is as follows. Plant-scale trials are conducted at three foundries, denoted Foundries A, B, and C. At each foundry, sets of specially designed end-risered T-plate castings are produced. During the casting of each set, cooling curve data to characterize the solidification behavior of the melt is collected concurrently from the same ladle from which the castings are poured. Regression analysis is used to obtain predictive relationships for shrinkage in terms of thermal analysis parameters, alloy chemistry, and other processing variables.

The design of the test castings (see Figure 2) is similar to that used in a previous investigation by Mikkola and Heine [27] of the effect of molding variables and casting design on the formation of shrinkage defects in white cast iron. The T-plate design incorporates a hot spot junction which varies in ease of feeding, depending on its dimensions. The relative ease of feeding of a given casting in the set is determined by the thickness of the horizontal feed section connecting the riser and the end of the T-plate. Except for the thickness of the horizontal feed section: 19mm (0.75in), 25mm (1.00in), 32mm (1.25in), and 38mm (1.50in), respectively; and the ingate area (which is 85% of the cross-sectional area of the horizontal section of the T-plate), all other dimensions of each of the test castings shown in Figure 2 are the same. Results to date indicate that the design of the T-plate test castings provides a reasonably sensitive indicator of the effects of both geometry and metal quality on shrinkage tendency.

In the initial series of plant trials conducted to characterize the current operating practice of each foundry, shrinkage was first assessed using a relatively fast and inexpensive procedure in which the T-plate test castings are sectioned to locate shrinkage which, if present, is measured as the average of the cross-sectional areas of the shrinkage voids on the two halves of a sectioned casting. Since the 0.75in test casting had the highest shrinkage defect rate, the shrinkage response of this test casting (denoted A075) was used to represent shrinkage in the preliminary statistical analysis of initial plant trial data.

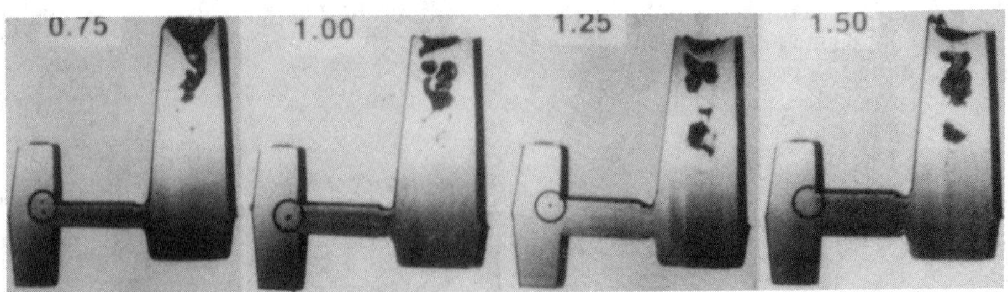

FIG. 2. Set of T-Plate Shrinkage Test Castings

The results from these initial plant trials were examined in a preliminary study to assess the general validity of the Ableidinger and Rabeus MEGK 'shrinkage tendency' predictor. The data were examined both as individual sets per foundry and as a combined set. In addition, since Foundries A and B had relatively similar processes that were significantly different from the metallurgical process at Foundry C, the data from Foundries A and B were also grouped as a separate set. This data set displayed the most significant correlations overall between shrinkage (A075) and the five thermal analysis parameters which appear in the MEGK relationship shown above (TER, TES, MXCREE, MNSDEE, and QSOLID). This is apparent in Figure 3, a scatter plot of A075 versus TER for the

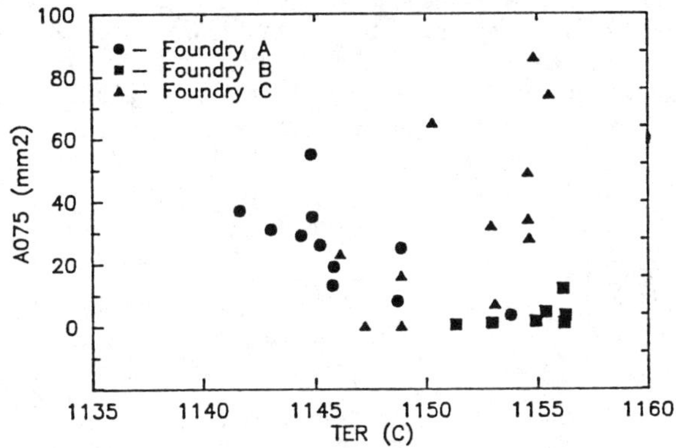

FIG. 3. Shrinkage Area (A075) Versus Recalescence Temperature (TER)

combined data set. Consequently, the data set comprising Foundries A and B was chosen as the basis for development of a first order prediction model using regression analysis. Multiple regression yielded the following relationship for ASP, the "area-of-shrinkage predictor":

$$\begin{aligned} ASP = 6790 &- 4.7 \times TER \quad (^\circ C) \\ &- 1.4 \times TES \quad (^\circ C) \\ &+ 23 \times MXCREE \quad (^\circ C/s) \\ &- 500 \times MNSDEE \quad (^\circ C/s^2) \\ &+ 0.29 \times QSOLID \quad (J/g) \end{aligned} \tag{7}$$

A correlation coefficient of 0.864 indicated that ASP is an effective predictor of shrinkage. However, proper interpretation of this predictive model depends on first considering two issues of statistical analysis to be discussed below.

First, when a regression model involves parameters that differ significantly in magnitude and/or variability, it is difficult to understand the actual effect of each variable on the response by direct comparison of the regression coefficients in the predictive equation. In the case of the ASP predictor, the coefficients for the five predictor variables differ by three orders of magnitude. To account for this, a linear transformation of the predictor variables, referred to as "centering and scaling", was applied. Each predictor variable X_i is centered about its own mean $\bar{X}$ and scaled by its own standard deviation σ as shown in equation (8),

$$\frac{(X_i - \bar{X})}{\sigma} \tag{8}$$

The "centered and scaled" predictor variables (indicated by the prefix CS) all have mean 0 and standard deviation 1. This transformation does not affect relationships between the predictor variables and the response. Multiple regression of the transformed data results in equation (9).

$$
\begin{aligned}
\mathrm{ASP} = 39 &- 44 \quad \mathrm{CSTER} \\
&- 26 \quad \mathrm{CSTES} \\
&+ 24 \quad \mathrm{CSMXCREE} \\
&- 54 \quad \mathrm{CSMNSDEE} \\
&+ 17 \quad \mathrm{CSQSOLID}
\end{aligned}
\tag{9}
$$

The coefficients in equation (9) each represent the magnitude of change in ASP due to an unit change in the given variable within in its observed range. If the observed ranges in the predictor variables are characteristic of plant practice, then these coefficients directly measure the sensitivity of ASP to each of these factors.

A second important statistical issue is the following: in observing the correlations among the variables in the ASP regression, it was apparent that several of the variables were significantly correlated with one another, and hence "collinear". Figure 4 illustrates the strong collinearity between the derivative parameters MXCREF and MNSDEE. A

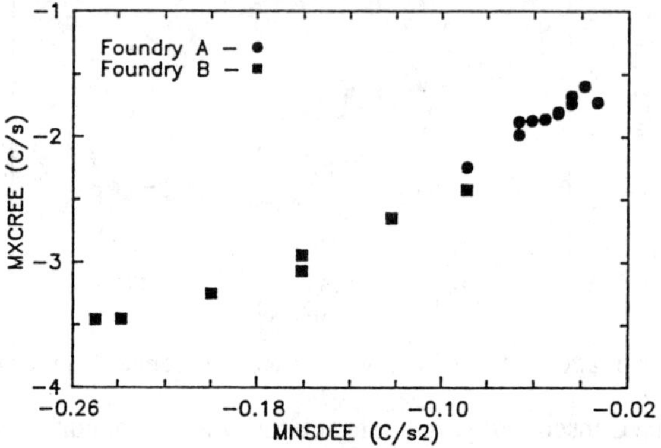

FIG. 4. Cooling Rate at Solidus (MXCREE) Versus Minimum 2nd Derivative (MNSDEE)

moderate collinearity was also observed between the temperature parameters TER and TES. Highly collinear variables do not improve the predictive ability of a regression equation and often lead to incorrect estimates of the effects of the individual variables. Taking this into consideration the ASP relationship was fitted again using only a subset of the predictors: TER, TES, MXCREE, and QSOLID for equation (10):

$$
\begin{aligned}
\mathrm{ASP4} = 24 &- 25 \quad \mathrm{CSTER} \\
&- 2.6 \quad \mathrm{CSTES} \\
&+ 2.0 \quad \mathrm{CSMXCREE} \\
&+ 2.0 \quad \mathrm{CSQSOLID}
\end{aligned}
\tag{10}
$$

and only TER, MXCREE, and QSOLID for equation (11):

$$
\begin{aligned}
\mathrm{ASP3} = 22 &- 22 \quad \mathrm{CSTER} \\
&+ 1.3 \quad \mathrm{CSMXCREE} \\
&+ 1.1 \quad \mathrm{CSQSOLID}
\end{aligned}
\tag{11}
$$

From plots of these revised ASP relationships it was apparent that deletion of the collinear parameters MNSDEE and TES did not seriously degrade the ability of the ASP relationship to predict shrinkage.

In the course of carrying out the cross-sectional area measurements, it was observed that the geometry of the shrinkage cavities tended to vary from casting to casting. Consequently, the volume of shrinkage (denoted V075, etc.) was subsequently accurately determined for each casting using a titration method. Figure 5 is a plot of volume of shrinkage (V075) versus area of shrinkage (A075) for Foundries A, B, and C, with the dotted line corresponding to the theoretical volume-to-area relationship for a sphere. In

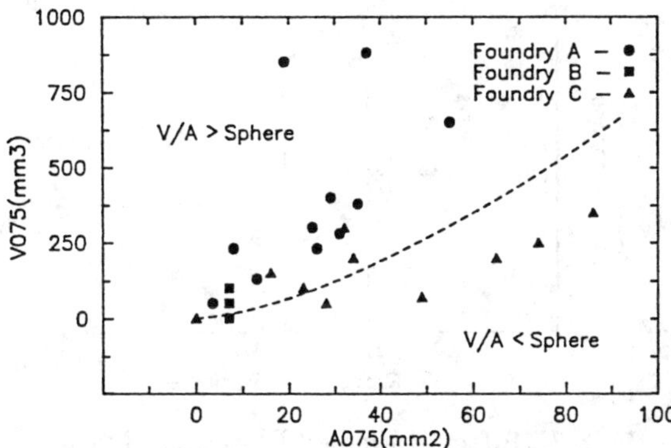

FIG. 5. Shrinkage Volume (V075) Versus Shrinkage Area (A075)

contrast to Figure 3, a scatter plot of A075 versus TER, the scatter plot of V075 versus TER in Figure 6 shows the Foundry C data to be less anomalous, although closer inspection reveals that the trend in the Foundry C data appears to be opposed to that in the data for Foundries A and B. From Figure 5, it is apparent that it is the aspect ratio of the shrinkage defects observed in the test castings that causes Foundry C data to appear different when cross-sectional area of shrinkage is analyzed.

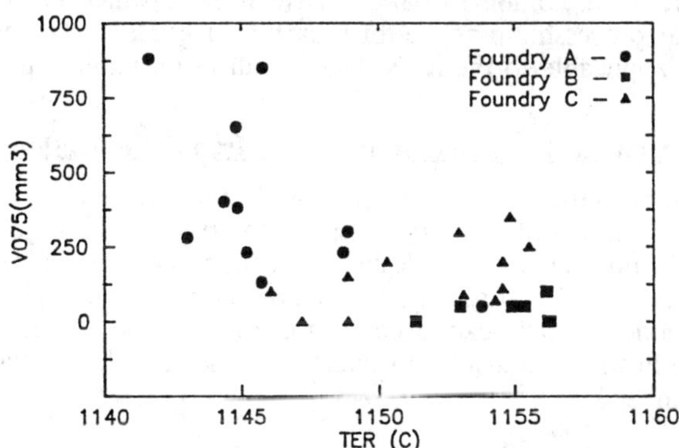

FIG. 6. Shrinkage Volume (V075) Versus Recalescence Temperature (TER)

The validity of any empirical model depends directly on the quality of the data that is used to develop it. It is emphasized that statistical analyses conducted to date have been based on data from normal plant production, rather than from designed experiments; these preliminary data were gathered simply to characterize the current operating practices at each of the participating foundries. Production data may be subject to limitations such as those illustrated in Figure 7, where the ranges of the process variables are limited because of production needs. As a result, analysis of such data may not accurately reveal the existing relationships among the variables. Furthermore, historical data are vulnerable to inconsistencies due to unrecorded process changes or process drifts — both of which would invalidate any model developed from such data. Finally, the use of data from undesigned experiments may result in confounding effects, wherein a variable may appear to depend on one cause when, in fact, it is dependent on a different factor or operational policy.

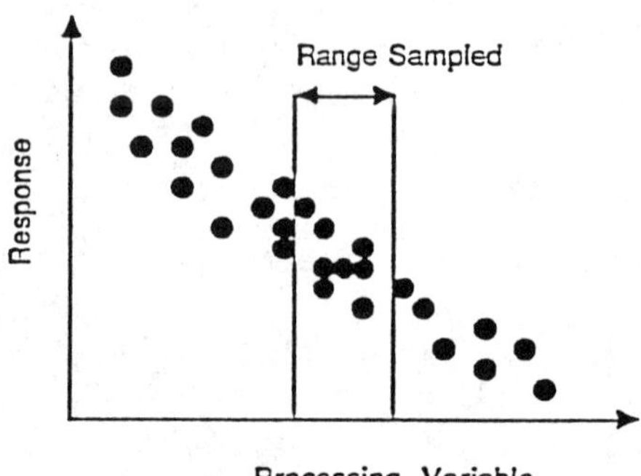

FIG. 7. Effect of Limited Processing Range

Since the data in this study were obtained during normal plant production rather than from designed experiments, the hazards mentioned above (as well as the procedure for quantitatively determining shrinkage) must be kept in mind when interpreting the data. For example, the plot of A075 versus TER (Figure 3) shows three foundries each occupying a distinct zone. This clustering of the data was initially interpreted to reflect the different processing practices at each foundry. The plot of V075 versus TER (Figure 6) suggests that the clustering of data is related to the aspect ratio of the shrinkage defect, which indirectly may reflect the effect of distinct metallurgical processing practice at Foundry C in general, or for the particular grades of ductile iron considered in the initial trial. If a model were to be developed from the combined data, each cluster of which represents a dissimilar process or dissimilar processing practice for a particular grade, the resulting model would not be legitimately applicable to the individual foundries or within a given foundry for all grades produced.

Numerical Simulation of Ductile Iron Solidification

The ultimate objective of the NSF-sponsored finite element solidification modeling project is the development of numerical simulation-based shrinkage defect prediction criteria for optimization of casting design of commercial ductile iron castings. A fundamental requirement for shrinkage prediction based on numerical analysis is the development of models which can account for the complex effects of metallurgical processing on solidification behavior. In numerically solving the partial differential heat conduction equation,

$$\nabla \cdot (k\nabla T) + g = \rho c \frac{\partial T}{\partial t} \tag{12}$$

the release of latent heat is generally handled by expressing the heat generation source term g in terms of the rate of change of fraction solidified as follows

$$g = \rho L \frac{\partial f_s}{\partial t} \tag{13}$$

where T is temperature, t is time, k is thermal conductivity, ρ is density, c is specific heat per unit mass, g is the rate of heat generation per unit volume, L is the latent heat of fusion per unit mass, and f_s is fraction solidified. While much progress has been made in developing efficient algorithms for solving the heat transfer with phase change problem [28-37], conventional methods of handling latent heat in macroscopic models utilizing simple temperature-dependent fraction solidified relationships and current versions of micro-macroscopic models based on fundamental nucleation and growth considerations are as yet incapable of realistically simulating the effects of metallurgical processing on the solidification behavior of commercial eutectic ductile iron.

Ekpoom and Heine [17], Chen and Stefanescu [18], Ableidinger and Raebus [20], and Sivula, Rantala, and Louvo [21] have all applied a form of lumped system analysis of raw cooling curve data using a Newtonian or constant heat transfer coefficient heat extraction model to calculate latent heats associated with cast iron solidification. In this research project, lumped system analysis of cooling curve data is applied to obtain the heat generation function g. The Newtonian cooling case is considered to illustrate the procedure for determining g. A lumped system heat balance of the thermal analysis specimen shown in Figure 8 yields:

$$-hA(T-T_\infty) + gV = \rho cV \frac{dT}{dt} \quad (14)$$

where h is the constant heat transfer coefficient for the system, T_∞ is ambient temperature, A is the effective surface area over which heat transfer occurs, and V is the volume of the thermal analysis cup. Rearranging equation (14) gives g for the Newtonian cooling case:

$$g(t) = \rho c \left[\frac{dT}{dt} + \theta(T-T_\infty) \right] \quad (15)$$

where the Newtonian cooling curve parameter θ is defined by

$$\theta = \frac{hA}{\rho cV} . \quad (16)$$

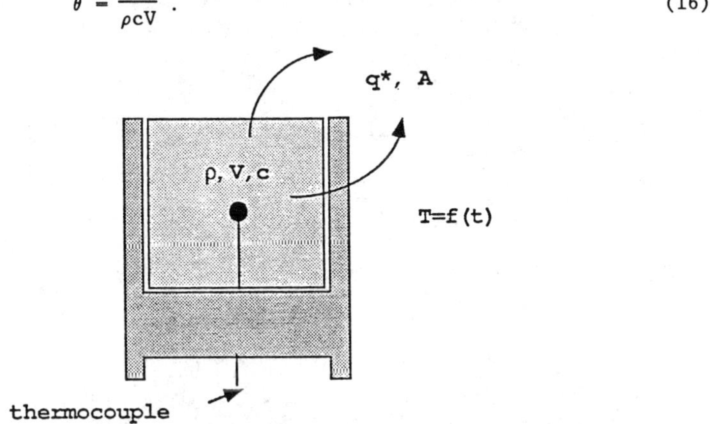

FIG. 8. Schematic of Lumped System Model of Thermal Analysis Cup

The procedure for determining the heat generation function is described in detail in references [38] and [39]. A key aspect of the thermal analysis-based approach is that the cooling curve and its derivatives are used to delineate the range of solidification. In this example the nominal start of eutectic solidification (TEN) is taken as the temperature corresponding to the maximum peak in the second derivative curve at the beginning of eutectic solidification (MXSDES). The nominal end of eutectic solidification (TES) is taken as the temperature corresponding to the minimum in the first derivative curve at the end of the eutectic plateau (MXCREE).

Figure 9 shows the cooling curve and corresponding heat generation function of a commercial eutectic ductile iron. The latent heat of fusion L per unit mass is obtained by integrating the heat generation function g(t) over the duration of solidification τ_s, that is

$$\rho L = \int_0^{\tau_s} g(t)\, dt \tag{17}$$

or, on substituting the expression for the heat generation function,

$$L = c \int \left[\frac{dT}{dt} + \theta (T - T_\infty) \right] dt \tag{18}$$

We note that only the value of the specific heat c is needed to obtain an estimate of the heat of solidification per unit mass. The lumped system analysis of cooling curve data provides both an estimate of the latent heat of fusion as well as the rate of liberation of latent heat. The latent heat can then be utilized in conventional solidification heat transfer models as L and in statistical models for shrinkage prediction as QSOLID. An alternative approach with regard to numerical simulation is to incorporate the heat generation function g directly into the finite element model to account for the time-dependent evolution of latent heat. We refer to this approach as the G-method for handling latent heat.

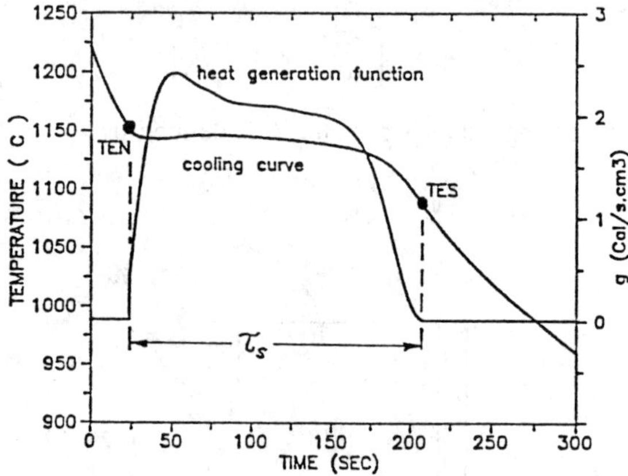

FIG. 9. Cooling Curve and Heat Generation Function for a Commercial Eutectic Ductile Iron Thermal Analysis Specimen

Finite element equations are commonly expressed in the form

$$[C]\ \{\dot{T}\} + [K]\ \{T\} = \{R\} \tag{19}$$

where [C] is the capacitance matrix, [K] is the conductance matrix, and {R} is the thermal load vector. In the G-method, the nonlinearity due to the release of latent heat during solidification is incorporated in the source term $\{R_g\}$ in the thermal load vector on the right hand side of equation (19),

$$\{R_g\} = \int_\Omega g\{N\}\, d\Omega \tag{20}$$

where {N} is the vector of shape functions, and Ω is the volume of the element. The G-method algorithm for handling latent heat is as follows. Each node has associated with it a local solidification time clock which is activated when the node temperature falls to the solidification start temperature T_s, which corresponds to TEN in Figure 1. Once activated, the solidification time clock for each node runs for a duration τ_s, the total solidification time as determined from the cooling curve data; i.e. the time interval delineated by TEN and TES. Once the temperature of a given node falls within the freezing range, the nodal contribution to the heat generation vector for a given time interval is computed using equation (20) and the appropriate range of the heat generation function g(t) corresponding to the current local solidification time of the node.

The finite element model has been used to simulate cooling curves at various locations of thermal analysis cups and T-plate test castings [38,39]. Figure 10 [38] shows a measured cooling curve and two simulated cooling curves corresponding to the thermocouple location (node 2 in Figure 11); one simulated curve computed using the G-method and the other obtained with a temperature recovery algorithm (Q-method) which also handles the release of latent heat using the thermal load vector rather than the capacitance matrix [29]. The G-method simulated curve is quite similar to the measured curve in that it exhibits significant undercooling with respect to the eutectic start temperature and slight recalescence. In the temperature recovery algorithm, solidification is constrained to occur at the specified freezing temperature (ie. the eutectic start of solidification temperature TEN), resulting in the horizontal plateau in the Q-model cooling curve shown in Figure 10. In contrast, the heat generation function is set a priori in the G-method, which permits the nodal temperatures to 'float' in a manner determined by the local rate of heat extraction in the vicinity of a particular node, thereby enabling the simulation of undercooling and recalescence.

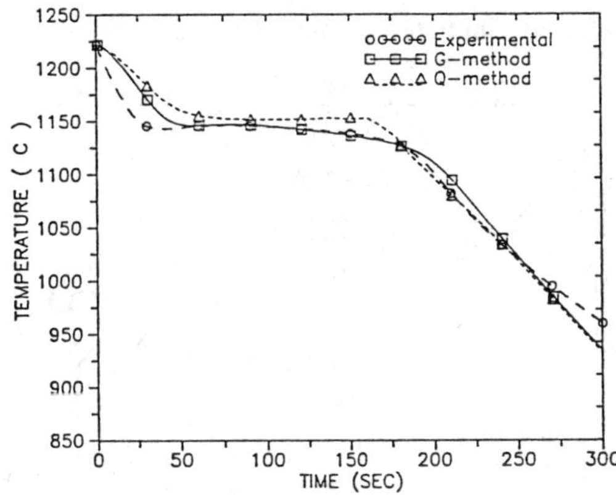

FIG. 10. Experimental, G-Method, and Q-Method Cooling Curves at Node 2

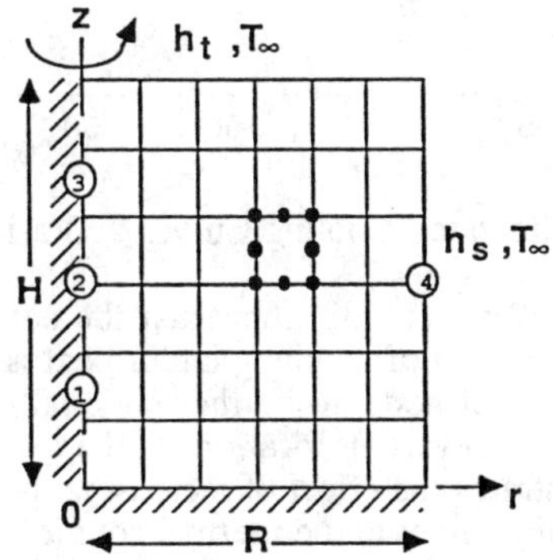

FIG. 11. Finite Element Mesh and Nodes at Which Cooling Curves are Obtained

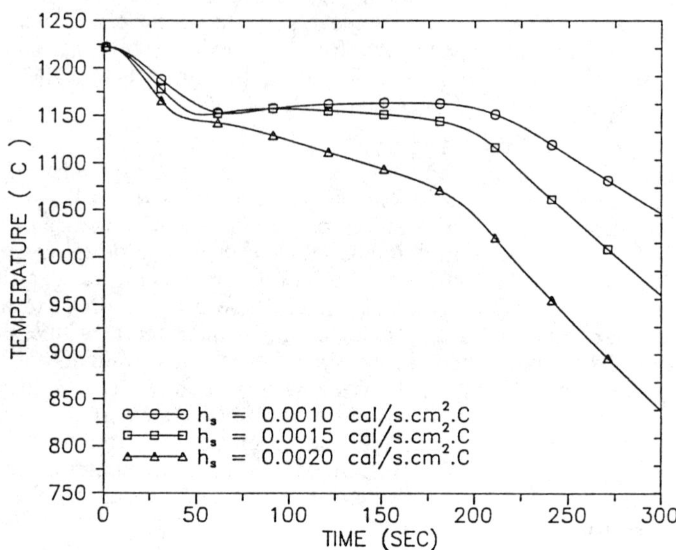

FIG. 12. G-Method Simulated Cooling Curves at Node 2 for Several h_s

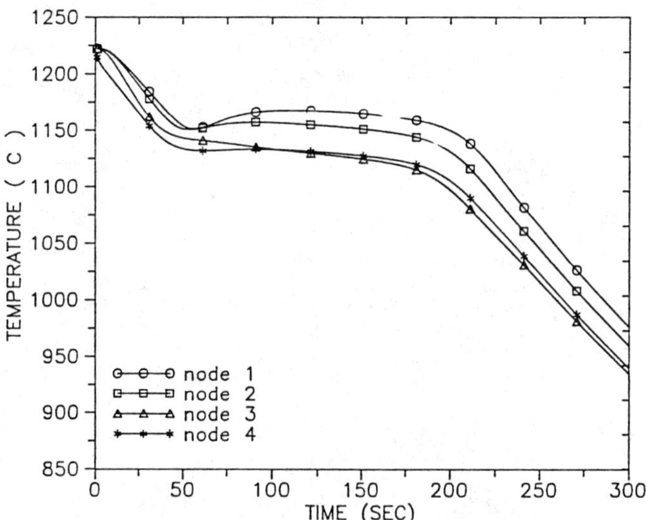

FIG. 13. G-Method Simulated Cooling Curves at Various Locations of Cup

The results in Figures 12 and 13 [38] show that the nature of the simulated cooling curves obtained using the G-method of handling latent heat is strongly dependent on heat transfer coefficient and location of node in the thermal analysis cup. Figure 11 shows G-method simulated cooling curves at various locations of the thermal analysis cup (h_s=0.0015 cal/s/cm^2/C). Figure 12 illustrates the effect of increasing the heat transfer coefficient on the G-method simulated cooling curves at the thermal couple location (node 2). In general, the overall effect of increasing rate of heat extraction at a node is greater undercooling with less recalescence.

Summary

The validity of the research approach to the development of a statistical model for the control of molten metal processing of ductile iron has been demonstrated, particularly in regard to the applicability of thermal analysis to shrinkage prediction. Operating practice at each of the participating foundries has been characterized and processing and thermal analysis variables that affect shrinkage have been identified. Statistically designed plant experiments are being conducted to obtain data for the development of a more reliable model for shrinkage prediction.

A thermal analysis-based heat generation function algorithm (G-method) has been developed for the finite element simulation of the solidification of commercial ductile iron. An important advantage of the G-method is its ability to simulate undercooling and recalescence, characteristic features of ductile iron cooling curves that relate to the effect of metallurgical processing on alloy solidification behavior. The G-method is able to simulate undercooling and recalescence by decoupling rate of heat extraction and rate of heat generation. While in real metallurgical systems heat extraction and heat generation are coupled via temperature and cooling rate dependence of nucleation and growth phenomena, the G-method of modeling latent heat liberation provides a practical means of incorporating the effects of metallurgical processing on solidification behavior. Furthermore, the numerical technique is computationally efficient and, therefore, well suited for the computer-aided design of castings. In the current phase of this project, we are focusing on model refinements and model verification using instrumented castings. Once a reliable computer model has been developed, output from solidification simulations of the T-plate shrinkage test castings such as cooling rate, thermal gradient, solidification time, rate of solidification, and fraction solidified will be correlated with observed shrinkage in an effort to develop numerical shrinkage prediction criteria.

Acknowledgements

The authors gratefully acknowledge the support of National Science Foundation Grant Nos. DDM-8910418 and DDM-08808138; Grede, Neenah, and John Deere foundries; and MINCO; and wish to thank the technical representatives of these companies and Emeritus Professor R.W. Heine for the stimulating discussions that have taken place in the course of the project. In addition, we also gratefully acknowledge the contribution of research engineer Paul Bartelt.

References

1. V. de L. Davies, "Feeding Range Determination by Numerically Computed Heat Distribution", AFS Cast Metals Research Journal, June, 1975, 33-44
2. E. Niyama, T. Uchida, M. Morikawa, and S. Saito, "A Method of Shrinkage Prediction and Its Application to Steel Castings", 49th International Foundry Congress, Chicago, April, 1982, 1-12
3. S. Minakawa, I.V. Samarasekera, and F. Weinberg," Centerline Porosity in Plate Castings", Metallurgical Transactions B, 16B, 1985, 823-829
4. J.C. Moosbrugger and J.T. Berry, "Calculation of Feeding Range Data for Hypoeutectic A-357 Alloy Using FEM Solidification Model Results", AFS Transactions, Vol. 94, 1986
5. K. Kubo and R.D. Pehlke, "Mathematical Modeling of Porosity Formation in Solidification", Metallurgical Transactions, 16B, (1985) 359-366
6. D.R. Poirier, K. Yeum, and A.L. Maples, "A Thermodynamic Prediction for Microporosity Formation in Aluminum-Rich Al-Cu Alloys", Metallurgical Transactions, 18A, (1987) 1979-1987
7. M. Rappaz and D.M. Stefanescu, "Modeling of Microstructural Evolution", Metals Handbook, 9th Edition, volume 15, Casting, (1988) 883-892
8. M. Rappaz,"Modeling of Microstructure Formation in Solidification Processes",

International Materials Reviews, vol 34, No. 3, (1989) 93-123
9. K. Su, I. Ohnaka, I. Yamauchi, and T. Fukusako, "Computer Simulation of Solidification of Nodular Cast Iron", Mat. Res. Soc. Symp. Proc., vol. 34, 1985, 181-189.
10. D.M. Stefanescu and C.S. Kanetkar, "Computer Modelling of the Solidification of Eutectic Alloys", Computer Simulation of Microstructural Evolution, Ed. by D. J. Scrolovitz, The Metallurgical Society-AIME, 1986, 171-188.
11. D.M. Stefanescu and C.S. Kanetkar, "Modeling of Microstructural Evolution of Eutectic Cast Iron and of the Gray White Transition", AFS Transactions, vol. 95, 1987.
12. D.M. Stefanescu and D.K. Bandyopadhyay, "On the Solidification Kinetics of Spheroidal Graphite Cast Iron", 4th Intl. Conf. on the Physical Metallurgy of Cast Iron, Tokyo, Japan, Sept (1989) in press
13. C.R. Loper, Jr., R.W. Heine, and M.D. Chaudhari, "Thermal Analysis for Structure Control," 2nd International Symposium on the Metallurgy of Cast Irons. Eds. Lux, Minkoff and Mollard, Geneva, Switzerland, (1987) 640-657
14. E.F. Ryntz, Jr., J.F. Janowak, A.W. Hochstein, and C.A. Wargel, "Prediction of Nodular Iron Microstructure Using Thermal Analysis", AFS Transactions, vol 79, (1971) 141-144
15. L. Backerud, K. Nillson, and H. Steen, "Study of Nucleation and Growth of Graphite in Magnesium-Treated Cast Iron by Means of Thermal Analysis," 2nd International Symposium on the Metallurgy of Cast Irons. Eds. Lux, Minkoff and Mollard, Geneva, Switzerland, (1974) 625-635
16. R. Monroe and C.E. Bates, "Thermal Analysis of Ductile Iron Samples for Graphite Shape Prediction", AFS Transactions, vol 90, (1982) 307-311
17. U. Ekpoom and R.W. Heine, "Thermal Analysis (DHA) of Cast Iron," AFS Transactions, vol 89, (1981) 27-38
18. I.G. Chen and D.M. Stefanescu, "Computer-Aided Differential Thermal Analysis of Spheroidal and Compacted Graphite Cast Irons," AFS Transactions, vol 92 (1984) 947-964
19. D.M. Stefanescu, "Solidification of Flake, Compacted/Vermicular and Spheroidal Graphite Cast Irons as Revealed by Thermal Analysis and Directional Solidification Experiments", Mat. Res. Soc. Symp. Proc., vol 34, (1985), 151-162.
20. K. Ableidinger and Dr. Raebus, "Thermoanalytical Computer Control and Correction of SG Iron Melts with Constant Feeding Conditions," British Foundryman, August/September (1986) 321-324
21. A.J. Sivula, T. Rantala, and A. Louvo, "Possibility of Controlling the quality of SC-Irons by Computerized Thermal Analysis," FOCOMP '86 Conference, Krakow, Poland, June (1986).
22. D.M. Stefanescu, D. Bandyopadhyay, and G. Upadhya, "Coupling of Solidification Kinetics Models With Heat Transfer Models for Casting", Casting of Near Net Shape Products, Eds. Y. Sahai, J.E. Battles, R.S. Carbonara, and C.E. Mobley, TMS. (1988) 153-165
23. K.T. Upadhya, D.M. Stefanescu, K. Lieu, and D.P. Yeager, "Computer-Aided Cooling Curve Analysis, Principles and Applications in Metal Casting", AFS Transactions, vol. 97, (1989)
24. F.J. Bradley, P.F. Bartelt, C.A. Fung, and R.W. Heine,"On the Application of Thermal Analysis to the Control of Molten Metal Processing of Ductile Iron", 4th Intl. Conf. on the Physical Metallurgy of Cast Iron, Tokyo, Japan, Sept (1989) in press
25. C.A. Fung, P.F. Bartelt, F.J. Bradley, and R.W. Heine,"On the Development of a Thermal Analysis-based Empirical Model for Shrinkage Defect Prediction in Ductile Iron Castings", 4th Intl. Conf. on the Physical Metallurgy of Cast Iron, Tokyo, Japan,

Sept (1989) in press
26. C.A. Fung, P.F. Bartelt, F.J. Bradley, and R.W. Heine, "Statistical Modeling for the Prediction of Shrinkage in Ductile Iron Castings", AFS Transactions, Detroit, (1990)
27. P.H. Mikkola and R.W. Heine, "Mold Aggregate and Casting Design Interactions Causing Defects in White Iron Castings" AFS Transactions, vol. 78, 265-268. (1970).
28. G. Comini, S. Del Guidice, R.W. Lewis and O.C. Zienkiewicz, "Finite Element Solution of Nonlinear Heat Conduction Problems With Special Reference to Phase Change," International Journal for Numerical Methods in Engineering, vol. 8, (1974), pp 613-624.
29. W.D. Rolph and K.J. Bathe, "An Efficient Algorithm for Analysis of Nonlinear Heat Transfer With Phase Change," International Journal for Numerical Methods in Engineering, Vol. 18, (1982) 119-134.
30. J. Roose and O. Storrer, "Modelization of Phase Changes by Fictitious Heat Flow," International Journal for Numerical Methods in Engineering, vol. 20, (1984) pp 217-225.
31. B.G. Thomas, I.V. Samarasekera, and J.K. Brimacombe, "Comparison of Numerical Modeling Techniques for Complex, Two-Dimensional, Transient Heat-Conduction Problems," Metallurgical Transactions B, Vol. 15B, 307-318, June (1984).
32. M. Samonds, K.Morgan, and R.W. Lewis, "Finite Element Modeling of Solidification in Sand Castings Employing an Implicit-Explicit Algorithm," Appl. Math. Modeling, Vol. 9, (1985), 170-174.
33. J.A. Dantzig and S.C. Lu, "Modeling of Heat Flow in Sand Castings: Part I. The Boundary Curvature Method," Metallurgical Transactions B, 16B, 195-202, June (1985).
34. M.J. Beffel, J.O. Wilkes, and R.D. Pehlke, "Finite Element Simulation of Casting Processes," AFS Transactions, Vol. 94, 757-764 (1986)
35. V.R. Voller, "A Numerical Method for Analysis of Solidification in Heat and Mass Transfer Systems," Numerical Methods in Thermal Problems, Vol. V, Part I, (1987) 693-704.
36. K. Morgan, R.W. Lewis, and O.C. Zienkiewicz, "An Improved Algorithm for Heat Conduction Problems With Phase Change," Int. J. Numerical Methods in Engineering 12 (1987) 1191-1195.
37. R.W. Lewis and P.M. Roberts, "Finite Element Simulation of Solidification Problems," Applied Scientific Research 44 (1987) 61-92.
38. F.J. Bradley, Cai Jun, and M.H. Zmerli, "Numerical Simulation of the Solidification of Eutectic Ductile Iron Casting Alloys",6th Intl. Conf. on Numerical Methods in Thermal Problems, Lewis and Morgan (Eds), Swansea, Pineridge Press, (1989) 320-330
39. F.J. Bradley, M.H. Zmerli, and Cai Jun,"Finite Element Modeling of Ductile Iron Solidification in Thermal Analysis Cups and T-Plate Shrinkage Test Castings",4th Intl. Conf. on the Physical Metallurgy of Cast Iron, Tokyo Japan, September, (1989) in press

VI. Semiconductor and optoelectronic crystal growth

Chairman: A.G. Elliot
Optoelectronics Division,
Hewlett-Packard Company,
San Jose, California, U.S.A.

Toward complete models for bulk crystal growth of semi-conductor materials (Keynote Paper)

R.A. Brown
Department of Chemical Engineering and Materials Processing Center, Massachusetts Institute of Technology, Cambridge, Massachusetts 02139, U.S.A.

In the last decade tremendous effort has gone into the analysis and optimization of processes for crystal growth of bulk semiconductor materials with the aim of improving the crystallographic and composition homogeneity of these materials. This effort has been partially successful. Mathematical modelling and state-of-the-art numerical analysis of complex heat transfer in furnaces, melt and crystal, coupled with studies of the fluid mechanics and dopant transfer in the melt has given new understanding of the mechanism for dopant segregation in a host of crystal growth systems. In this presentation, studies of heat transfer and dopant segregation in the vertical Bridgman growth of gallium-doped germanium and the Czochralski growth of silicon are used to exemplify the depth of analysis that is presently available.

The missing link in the modelling of crystal growth is the connection between macroscopic heat transfer and crystal quality, measured in terms of the densities of dislocations and other defects. Progress is reviewed toward the development of self-consistent models for defect formation in crystal growth. This picture includes the formation of microdefects by clustering of point defects and interstitials, the evolution of microdefects into mobile dislocation loops, and the dynamics of dislocations leading to plastic slip in the crystal.

Research supported by IBM and by NASA.

Observations on mechanisms of poly and twin formation in the LEC growth of GaAs and GaP

A.G. Elliot, K.-H. Huang and D.A. Vanderwater
Hewlett Packard Co., Optoelecronics Division, San Jose, California 95131, U.S.A.

Abstract

Large crystals of GaAs and GaP grown by the liquid encapsulated Czochralski (LEC) technique have strong tendencies to form polycrystalline or twinned regions. In this report we will describe the results of our systematic study of the development of polycrystalline and twinned regions of 2" and 3" diameter GaAs and GaP crystals grown by the LEC technique. Close examination of boules of both GaAs and GaP that have experienced polycrystalline growth reveals that regions of polycrystal result when dislocations form and migrate into low angle grain boundries. With subsequent growth, polygonization into macroscopic cells occurs, finally leading to interface breakdown and the formation of gross polycrystalline regions. The various circumstances under which this occurs suggest that the mechanism is driven by thermal gradients and by segregation phenomena that result during growth from either doped or non-stoichiometric melts. The same techniques used to examine twinned ingots reveal that the twinning phenomena occur most often in reqions that experience excessive melt-back and regrowth and is most influenced by the cone angle and/or the presence of dopants.

Introduction

The liquid encapsulated Czochralski (LEC) growth of III-V single crystals, particularly with a <100> growth axis, has long been plagued by a strong tendency to form polycrystals. Various contributing factors have been cited in the literature: including water content of the B2O3 encapsulent [1], melt stoichiometry [2], excessive thermal stress due to changes of the crystal diameter [3], abrupt shouldering of the crystal [4], and "junk" floating on the melt surface. This last factor, "junk" on the melt surface, can be a pervasive problem if improper handling of grower furniture, charge, crucible, etc., occurs. This, however, is a housekeeping item and is not the subject of this report. Poly formation arising from a change in growth conditions such as encountered in the scale up of a process to larger crystal sizes or greater volumes can have deleterious effects. The underlying physical phenomena driving poly formation has not been understood, therefore efforts at minimizing it have been based on speculation. In this paper we report the results of our work systematically examining the development of polycrystalline regions during the LEC growth of <100> and <111> oriented single crystals of GaAs and GaP. From this work, we have deduced that polycrystalline regions arise during LEC growth of single crystals due to thermoelastic stress. This stress generates dislocations which pile up forming low angle grain boundaries that subsequently become high angle grain boundaries forming polycrystalline regions.

Experimental

The crystal growth experiments were carried out in graphite resistance heated furnaces shown schematically in Figure 1. The only significant difference between the GaAs and GaP growers is the ambient required to prevent decomposition of the crystal and melt. The former are operated at 3 atm of nitrogen while the latter require 80 atm of argon. In either case, the heat shielding can be arranged to give various temperature gradient configurations. Typically, axial gradients are in the 25 to 40 °C per cm range at the solid/liquid interface. This gradient has been lowered to 6 °C per cm at the solid/liquid interface in one GaAs grower configuration by suitable adjustments in the heat shielding.

Most crystals were grown fully encapsulated in B2O3 in order to prevent their decomposition. This was especially necessary when low temperature gradients were used. A variety of pulling rates, seed and crucible rotation rates, and crucible diameters were used. The orientation of the growth axis was either <100> or <111>. The crystals were cooled typically at 1-2 °C/min after growth.

The grown crystals were cut for analysis into (100) or (111) oriented wafers sliced perpendicular to the growth axis and/or into (110) oriented slabs sliced parallel to the growth axis, as shown in Figure 2. These samples were lapped with 5 um Al2O3 grit and polished with sodium hypochlorite solution to remove all surface damage. They were then etched with a specific defect revealing etch. GaAs samples either were etched in molten KOH, or photoetched using the procedures outlined by van de Ven, et al. [6]; GaP material was etched either with a 5:5:2 H2SO4:H3PO4:H2O2 solution for (111) material [6], or with a 3:1:1 HF:H2O2:Acetic acid solution heated to 60 °C for (100) or (110) oriented material which is a modification of procedures described by Hayes, et al. [7], and Gibb and Augustus [8]. After etching, the samples were examined using differential contrast microscopy.

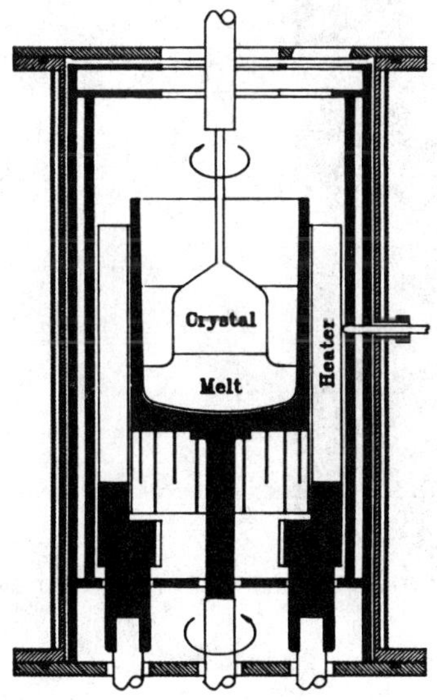

Figure 1.
Section view of a typical LEC crystal grower

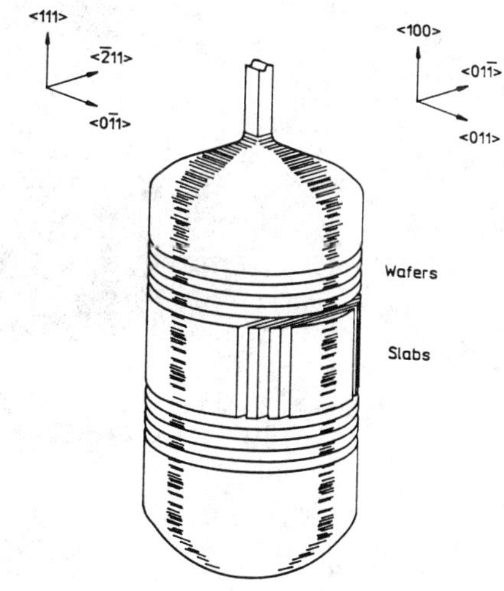

Figure 2.
Schematic diagram showing how samples were cut from a grown ingot

Results

Experience has taught that 100% single crystals of GaAs and GaP can be obtained more easily (more frequently) when the growth axis is <111> rather than <100>. With axial thermal gradients reduced from ~100 °C/cm to 30 or 50 °C/cm, this remains true, but with still lower gradients, there seems to be little difference. However, upon reducing the ratio of the grown crystal diameter to crucible diameter, poly formation is enhanced, whether <111> or <100> growth axes are used.

The type and concentration of added dopant also significantly affects the propensity for poly to form. Both GaAs and GaP doped with Si grow predominantly as 100% single crystals. However, when Te or S dopant, or no dopant is used, there is a stronger tendency for polycrystalline regions to form (given the same growth conditions).

Polycrystalline regions appear to initiate in a very small region at the surface of the crystal and increase in size down the length of the ingot. With GaAs, the poly regions extend inward as well, quickly taking over the entire bulk of the ingot. With GaP, however, the poly often does not extend inward residing instead only at the surface. For both GaAs and GaP the macroscopic regions of poly are always preceded by the formation of several straight traces (lines) roughly parallel to the growth direction as shown in Figure 3. These traces always start ([A] in the figure) at a significant distance above the point where well-defined poly is visible ([B] in the figure). These regions always form on or very closely adjacent to the facet, never between the facets.

Figure 3.
A Photomacrograph of the surface of a grown crystal showing at A the beginings of traces due to the intersection of lineage with the crystal surface and at B the beginning of observable poly.

The development of polycrystalline regions in GaAs is shown graphically in Figure 4 which depicts the dislocation distribution down the length of a crystal. Before poly appears on the surface, dislocations have moved to form low-angle grain boundaries, observed initially as more or less straight lines parallel to <110> directions ([A] in the figure). The intersection of these linear arrays of dislocations with the surface of the crystal correlates with the straight line traces shown in Figure 3. At points remote from the surface dislocations pile up to form randomly oriented sub-grain boundaries ([B] in Figure 4) prior to the formation of extensive networks of large-angle grain boundaries (polycrystalline regions - [C] in Figure 4).

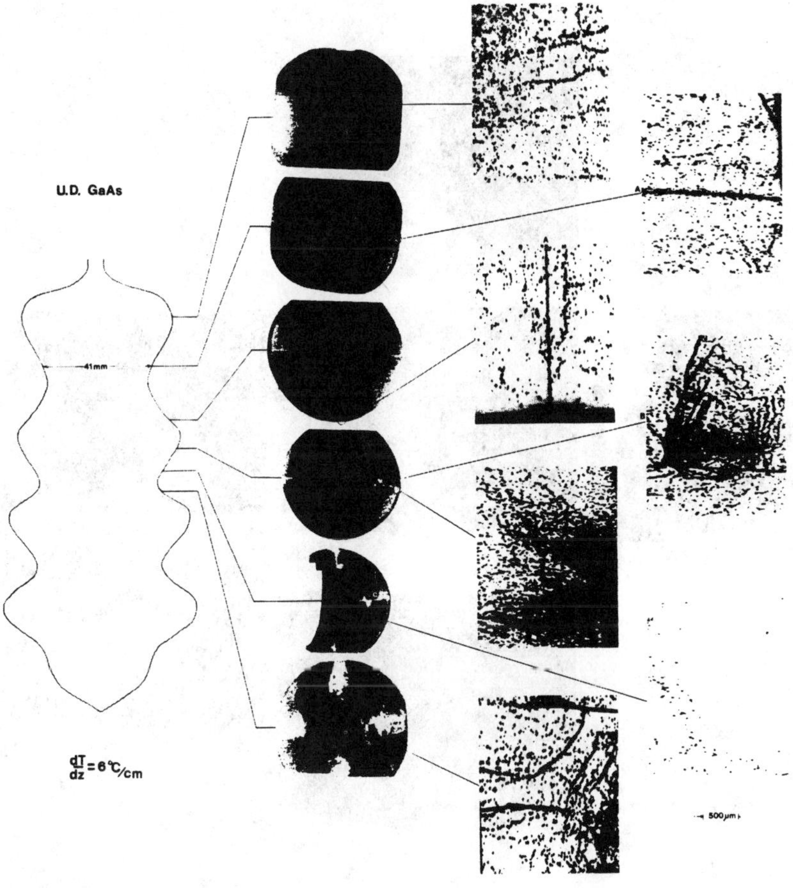

Figure 4.
A series of photomacrographs and photomicrographs from wafers sliced along the length of a grown, undoped, GaAs ingot showing the creation of lineage at A, the formation of areas containing low angle grain boundaries at B, and the subsequent formation of polycrystalline regions at C.

Very similar poly formation behavior occurs in GaP as shown in Figure 5. Figure 5 (a), (b), and (c) are photomacrographs of three, successive, etched wafers cut 1.7 mm apart starting just above where the poly is observed on the crystal surface. Figure 5 (d) shows the entire wafer corresponding to (b). Figure 5 (e) and (f) are enlargements of the region circled in (b), showing the onset of poly through the formation of a large-angle grain boundary. In this series of photomicrographs note the high density of dislocations in the region where the poly occurs, and the formation of the straight boundaries that correlate with the surface traces shown in Figure 3.

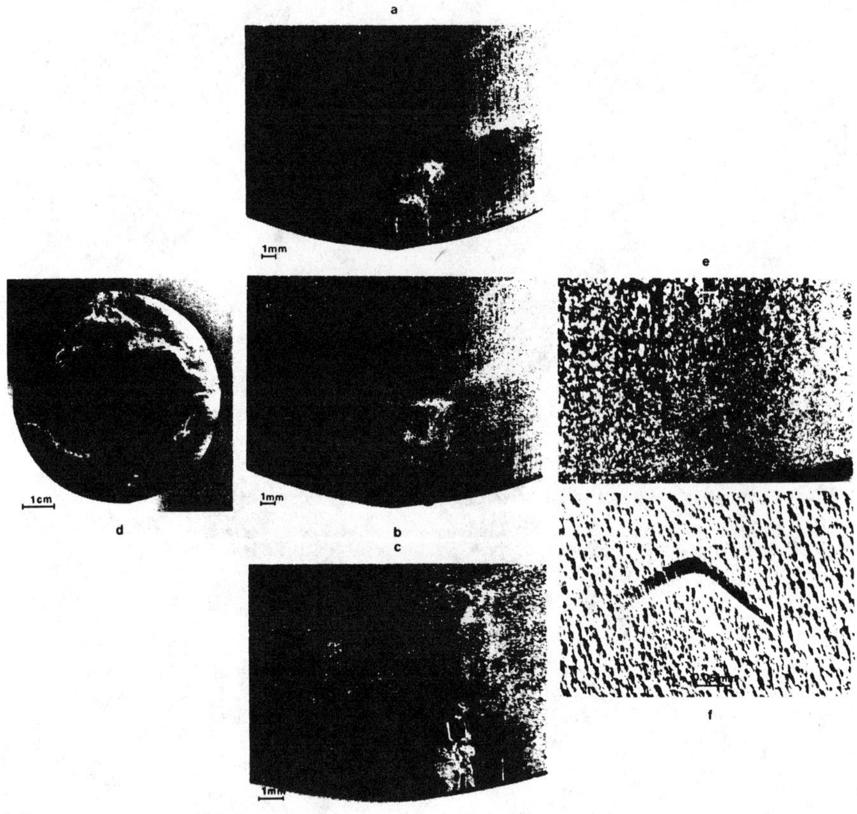

Figure 5.
(a), (b), (c) A series of photomacrographs of successive wafers cut, starting just above the point of observable poly formation, from a GaP ingot; (d) A photomacrograph of the entire wafer shown in (b); (e) A photomicrograph of the circled region in (b); (f) An enlargement of the grainboundary region shown in (e).

The pile-up of dislocations and the formation of low-angle grain boundaries can (and usually does) occur much before the onset of polycrystalline growth, as suggested by Figure 4. This is indicated further by the photomicrographs shown in Figure 6 taken of slabs cut from a Te-doped GaAs crystal of irregular diameter. The point of observable poly formation occurs below the first reduced diameter segment, essentially isolating the region examined in (a). However, extensive dislocation pile up and low-angle grain boundary formation already have occurred in this portion of the crystal. This continues in the regions examined in Figure 6 (b), (c), and (d) close to the point of observable poly formation.

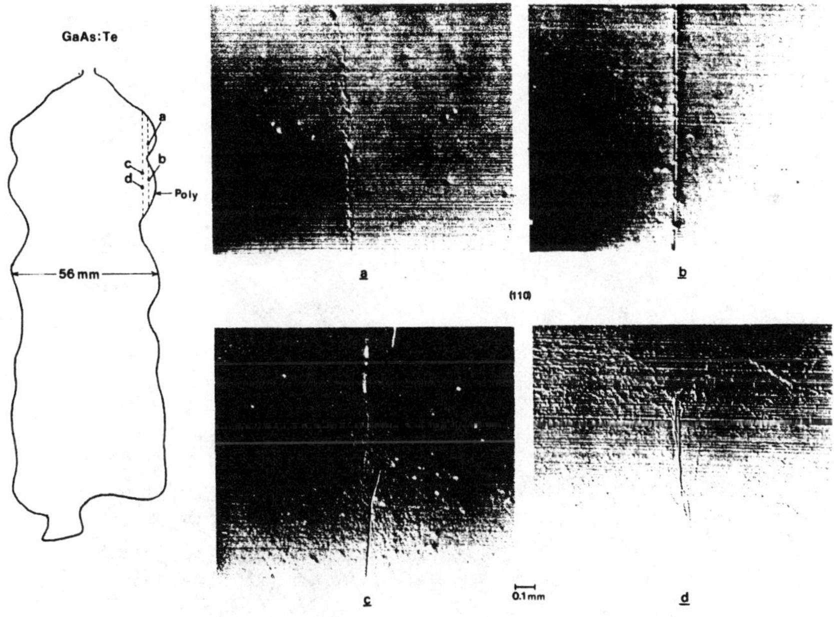

Figure 6.
A selection of photomicrographs taken from slabs cut from the labeled regions of a GaAs:Te crystal.

Twin formation in III-V crystals is related to their strong tendency to facet during growth. This faceting is affected by the dopant added to achieve the desired electrical properties. Figure 7 shows this effect dramatically. Figure 7 (a) is from the top cone region of a GaAs:Te crystal. The faceting is not well-delineated by the etch in this instance, possibly due to a significant difference in dopant incorporation in this region. Evidence for extensive melt-back and regrowth can be seen by the intersection of the growth striae at [A]. A twin line exists as marked by [B]. The surface of the crystal can be seen in the upper right hand corner.

Figure 7 (b) is from the top cone region of a GaAs:Si crystal. Note the vastly different striation pattern and more extensive faceting. Melt-back and regrowth occurs as before. The surface of the crystal can be seen in the upper right hand corner. The frequency of twinning is significantly less for Si-doped crystals compared to Te-doped crystals. However, in both cases it is closely tied to physical misalignments among the crucible, the heater, and the growing crystal.

The region shown in Figure 7 (c) is several millimeters in from the surface of the cone section of a GaAs:Si crystal grown under very low thermal gradient conditions. Very little melt-back and regrowth occurs for this configuration.

Figure7.
Photomicrographs of {111} faceting in the cone region of (a) GaAs:Te, showing evidence for melt-back and re-growth at A and twin formation at B, (b) GaAs:Si, and (c) GaAs:Si grown under low thermal gradients.

Discussion

The salient feature of the studies presented above is that polycrystal formation is preceded by extensive dislocation formation, dislocation pile up, formation of low-angle grain boundaries and their subsequent evolvement into high-angle grain boundaries (polycrystalline regions). In general, dislocation formation during crystal growth of III-V materials occurs as a result of thermoelastic stress within the growing crystal, a model originally proposed by Billig [9] and developed by Jordan, et al. [10,11]. Detailed mathematical modeling of the heat flows and stresses generated in the Czochralski configuration using finite element techniques [12,13,14], have been able to successfully reproduce the dislocation distributions that are observed in III-V crystals. These analyses show that the point of maximum resolved shear stress occurs about 5 to 20 mm above the solid/liquid interface at the periphery of the crystal on the [110] nodes (facet formation locations). If that stress exceeds the critical resolved shear stress then dislocations will form causing slip.

In essence, dislocations arise at the surface of the crystal to relieve the hoop stress generated by temperature gradients within the growing crystal. These dislocations glide down their slip planes ({111} in III-V compounds) until the back stress generated by a linear array of dislocations balances the applied stress, or they become pinned by cross slip and/or climb, interact with another dislocation or interact with an impurity or precipitate. Under continued applied stress these pinned dislocations themselves can become sources for additional dislocation generation.

For <100> oriented growth there are four non-equivalent {111} planes on which the hoop stress may be resolved and along which dislocations may glide. These planes can intersect at a point somewhere within the bulk of the crystal as shown schematically in Figure 8. At this point extensive dislocaton pile-up can occur. Given that growth occurs at the melting point (1235 °C for GaAs and 1455 °C for GaP) and that the time available for motion to occur is great

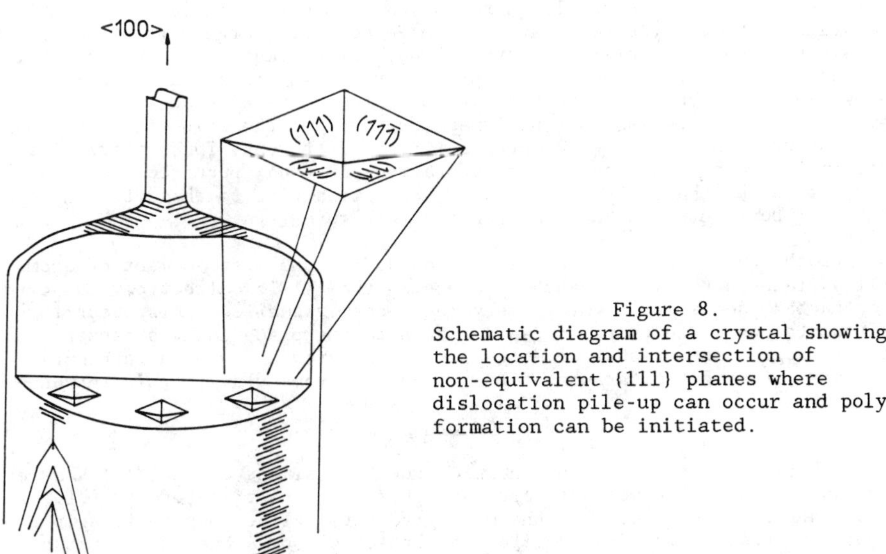

Figure 8.
Schematic diagram of a crystal showing the location and intersection of non-equivalent {111} planes where dislocation pile-up can occur and poly formation can be initiated.

(the crystals grow at a rate of about 6 mm/hr), extensive glide, climb and multiplication can result. Continued application of the stress then gives rise to grain boundary formation (a work hardening-related effect). Since the resolved shear stress is maximum at the [110] nodes, this is the region where the poly most likely will form.

For <111> oriented growth there are 6 non-equivalent {111} planes upon which to resolve the shear stress. Thus, for the same thermal gradient situation (same growth environment), poly formation by the above mechanism will be less for <111> oriented growth than for <100> oriented growth.

Within the thermoelastic model, it is the radial temperature gradients that give rise to the hoop stress. In the Czochralski configuration the radial gradients are inextricably linked to the axial gradients. However, there is a significant effect of crucible diameter relative to crystal diameter on the radial gradients, since close to the crucible wall the radial gradients are non-linear and rise steeply. This is particularly true for quartz crucibles which have very low thermal conductivity. These radial gradients near the crucible wall can be used to control the crystal diameter since the increasing temperature will limit radial growth. However, the associated increase in the resolved shear stress can give rise to increased poly formation as has been observed.

The effect of added dopant on the tendency to form poly can be understood in terms of the dopant's effect on dislocation formation. Si and In both have been shown to be effective agents for dislocation reduction in GaAs [15,16,17]. Si appears to behave similarly in GaP [18]. While the mechanism for this reduction in dislocation density is not yet well understood, the fact that these dopants are effective at reducing dislocation formation explains the dopant effect on poly formation, since reduced dislocation formation will reduce the tendency to form poly.

While there is an effect of dopant on the tendency for twinn formation, the mechanism involved here is not understood. Faceted growth is planar growth occurring on the {111} planes in contact with the liquid. Growth takes place atomic-layer by atomic-layer through a ledge-kink mechanism. For GaAs it is well known that growth on (111)A faces (Ga terminated) is faster than growth on (111)B faces (As terminated). Simple geometric arguments show that under this condition any arbitrary crystalline, III-V material surface will be reduced to intersecting (111)A planes, resulting in a "puckered" structure. Growth of the puckered surface will be slower than that of the faceted surface. Also, because of the irregular surface, the composition of the melt in contact with that growing surface will be different. Thus, there will be a different level of dopant incorporation in the puckered surface region (crystal bulk) than in facet region. This gives rise to the different etching behavior between facet and bulk regions in Figure 7 (a).

A growth twinn in GaAs or GaP arises due to the misplacement of atoms on a (111) plane in the "abcabcabc" sequence in the <111> direction to give an "abcbacba" sequence. Intuitively, one can vizualize this occurring under circumstances where there is rapid growth taking place such as happens in the facet region as a result of melt-back and re-growth as shown in Figure 7 (a). However, it is not clear what role the dopant atoms play in this instance.

Conclusions

Polycrystal formation during the LEC growth of single crystal III-V materials is the result of thermoelastic stress. Dislocations that form to relieve the hoop stress generated by thermal gradients glide down their slip planes, interact, become pinned and pile up creating low-angle grain boundaries.

Continued action of the thermal stress causes further dislocation generation and rearrangement to form high-angle grain boundaries, thus creating polycrystalline regions. These regions form on or just slightly off the facets ([110] nodes) of the growing crystal, since this corresponds to the point of maximum resolved shear stress.

The <111> oriented growth axis has more non-equivalent {111} planes over which to distribute the resolved shear stress compared to the <100> oriented growth axis. Thus, there will be a lower tendency for poly formation in <111> oriented growth.

Dopants that reduce the formation of dislocations, e.g. Si, also reduce the tendency for poly formation. However, the role of dopants in affecting the tendency for twinning to occur is unclear.

The thermal gradients that exist in the growth system dictate dislocation formation and, hence, poly formation. Steep gradients such as those that exist near the crucible wall enhance poly formation. Reducing gradients so as to minimize dislocation formation will result in the elimination of poly.

Acknowledgments

The authors would like to recognize the contributions of Darwin Boblet, Tony Anatra and Mark MacDonald to the growth of the crystals, and that of John Unsworth in their characterization. The encouragement and support of George Craford is gratefully acknowledged.

References

1) T.R. Aucoin, R.L. Ross, M.J. Wade and R.O. Savage, Solid State Technology, 22 (1979) 59
2) A. Steinemann and U. Zimmerli, Solid State Electron., 6 (1963) 597
3) H. Kotake, K. Hirahara and M. Watanabe, J. Cryst. Growth, 50 (1980) 743
4) H.M. Hobgood, T.T. Braggins, D.L. Barrett, G.W. Eldridge and R.N. Thomas, 5th Am. Conf. Cryst. Growth, San Diego, CA. (1981)
5) J. van de Ven, J. L. Weyher, J.E.A.M. van den Meerakker and J.J. Kelly, J. Electrochem. Soc., 133 (1986) 799
6) J. van Dongen, R.R. Tijburg, and J. Bakker, Electrochem. Soc. Spring Meeting (1978) Abstr. No. 253
7) T.J. Hayes, A. Rasul and S.M. Davidson, J. Electron. Mat., 5 (1976) 351
8) R.M. Gibb and P.D. Augustus, J. Electron. Mat., 5 (1976) 585
9) E. Billig, Proc. Royal Soc. (London), A235 (1956) 37
10) A.S. Jordan, A.R. Von Neida and R. Caruso, Bell Syst. Tech. J., 59 (1980) 593
11) A.S. Jordan, R. Caruso, A.R. Von Neida, and J.W. Nielsen, J. Appl. Phys., 52 (1981) 3331
12) M. Duseaux, J. Cryst. Growth, 61 (1983) 576
13) C.E. Schvezov, I.V. Samarasekera and F. Weinberg, J. Cryst. Growth, 84 (1987) 212
14) F. Dupret, P. Nicodeme and Y. Ryckmans, J. Cryst. Growth, 97 (1989) 162
15) A. G. Elliot, C.-L. Wei and D.A. Vanderwater, J. Cryst. Growth, 85 (1987) 59
16) Y. Seki, H. Watanabe and T. Matsui, J. Appl. Phys., 49 (1978) 822
17) G. Jacob, J. Cryst. Growth, 59 (1982) 669
18) K.-H. Huang, unpublished results

Vertical Bridgman and gradient freeze growth of III-V compound semi-conductors

E.D. Bourret
Center for Advanced Materials (2-200), Materials and Chemical Science Division, Lawrence Berkeley Laboratory, University of California, Berkeley, California 94720, U.S.A.

Abstract:

Major improvements in the structural and electrical perfection of single crystals of III-V compound semiconductors have been achieved by using new vertical Bridgman-type and vertical gradient freeze techniques. A general review of experimental set-ups used for growth of large diameter crystals of GaP, InP and GaAs is presented. Crystals properties and characteristic features are discussed to illustrate advantages and disadvantages of the vertical Bridgman-type growth techniques.

1. Introduction:

The Czochralski technique (1) has been the preferred method for growth of many semiconductor single crystals since 1950 when it was first used for growth of single crystals germanium. The first successful growth of GaAs single crystals was carried out using a modified Czochralski technique, the Gremmelmeier technique (2). In this technique, a sealed system with magnetic coupling for the rotation and pulling mechanisms is used to maintain an arsenic vapor pressure in the chamber which prevents decomposition of the GaAs melt. Mullin et al.(3) in 1965 developed the liquid encapsulation technique, an adaptation of the work of Metz et al. (4) for growth of III-V compounds. The Czochralski (LEC) technique quickly became the industry standard for production of semi-insulating GaAs particularly after it was demonstrated that PBN crucibles could be used to produce undoped semi-insulating crystals.

In the mean time, the horizontal gradient freeze and horizontal Bridgman techniques were widely used to grow doped conducting crystals as well as chromium-doped semi-insulating crystals. It was assumed that the horizontal configuration was necessary in order to limit confinement of the crystal and to allow free expansion of the crystal upon solidification. Based on this assumption, the vertical gradient freeze and vertical Bridgman techniques were not fully investigated for growth of III-V compounds, despite encouraging early results obtained for GaP as reported in a review of crystal growth techniques for II-VI and III-V compounds by Fischer (5). Fischer describes vertical Bridgman / gradient freeze growth of GaP in a pressure-relieved ampoule using pyrolytic boron nitride crucibles (figure 1). Blum et al. (6) successfully grew 1.5 cm diameter GaP single crystals in pyrolytic boron nitride crucibles by the liquid encapsulated vertical gradient freeze technique (LE-VGF). These authors concluded that "crystals grown by the LE-VGF technique are appreciably less strained, lower in dislocation density, excellent in diameter control and easier to grow than the conventional LEC GaP". Woodbury (7) tried to scale-up the LE-VGF technique for seeded growth of up to 35 mm diameter GaP crystals. Difficulties in reproducibly growing single crystals were encountered due to spurious nucleation at the crystal/seed-melt junction. Several recommendations were presented to solve these problems and

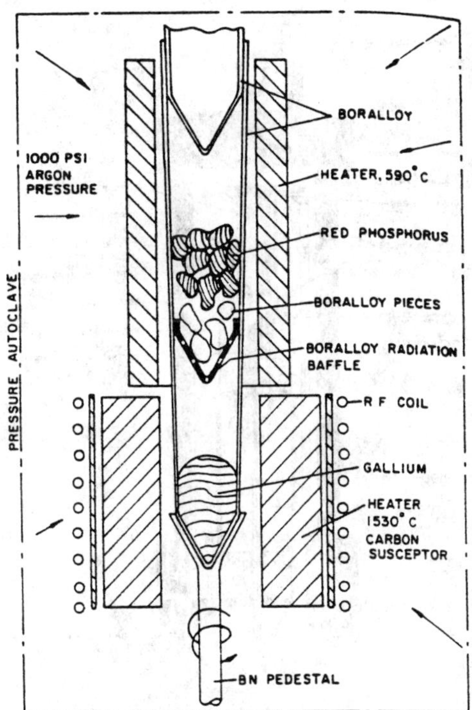

Figure 1. GaP synthesis and crystal growth demontable pyrolytic BN ampoule (reproduc- of fig. 7 from ref. 5).

the author stated that this technique might be the easiest way to grow large, strain-free ingots of materials with high vapor pressure directly from their melt as long are they are chemically compatible with molten boron oxide.

Chang et al.(8) have used GaAs to illustrate the flexibility of the vertical gradient freeze technique in terms of control of the growth parameters. They used high freezing rates to solidify unseeded ingots 10 mm diameter and 17 cm long in silica crucibles and concluded: "we see no reason why large diameter useful GaAs cannot be produced by this technique, at least at low doping levels". It was only in 1986 that the first study of growth of high quality III-V crystals of large diameter was published (9) which started a renewed interest in the technique.

In this paper, a review of the experimental set-ups used for growth of large diameter crystals of GaP, InP and GaAs is presented. Crystal properties and characteristic features are discussed to illustrate advantages and disadvantages of the vertical Bridgman-type techniques.

2. GaP:

GaP was the first III-V compound to be grown by the vertical gradient freeze technique (5). in a Crystals up to 5 cm diameter have now been tion obtained. The largest crystals (9) were grown in a water-cooled, stainless steel lined pressure vessel very similar to the chamber first described by Fischer (5). The chamber assembly is shown in figure 2. Two resistance heaters were used in the vessel. The upper heater (graphite picket-fence) heats a bottom seeded PBN crucible which contains the GaP charge and is used to control the directional solidification process. The lower heater is used to control the temperature of a phosphorus reservoir which provides regulation of the vapor pressure of phosphorus above the melt and prevent its decomposition. The crucible and phosphorus source are placed in a PBN container with a hot-pressed boron nitride cap. This container can be moved vertically up and down and can be rotated during growth. The paper does not specify if these features where used during growth. The phosphorus loss through the cap appears to be small enough that the required partial pressure of phosphorus can be maintained during growth. The axial temperature gradients during growth were about 40ºC/cm over the solid and about 8ºC/cm over the melt. The pressure of the chamber was maintained at 65 atm during growth. The crystals grown were 1300g, <111> seeded and either sulfur or tellurium n-type doped. The dopant concentrations are not stated in the publication. The main feature of these crystals was low dislocation densities ranging from 800 to 2000/cm^2. Green light emitting diodes (LED) made by liquid phase epitaxy on wafers from these low dislocation density VGF substrates were 23% more efficient than LEDs made on LEC substrates.

3. InP:

The same furnace as described for GaP was used to grow large diameter InP crystals (9,10). The axial temperature gradients during growth were mentioned to be lower than for GaP.

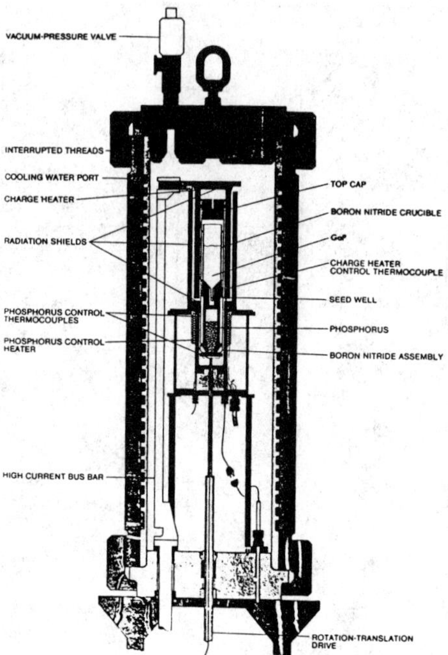

Figure 2. Cross sectional view of vertical gradient freeze crystal growth equipment as configured for GaP (reproduction of fig.1 from ref. 9).

Figure 3. (a) Temperature profile along the axis of the load: experimental (dashed lines) and numerical (full line) results (reproduction of fig.11 from ref.13),

(b) Temperature field throughout the furnace: the increment is 100K (335 K) for the global view while the increment is 10 K in the load (1 K for dashed line), (reproduction of fig.12 from ref. 13).

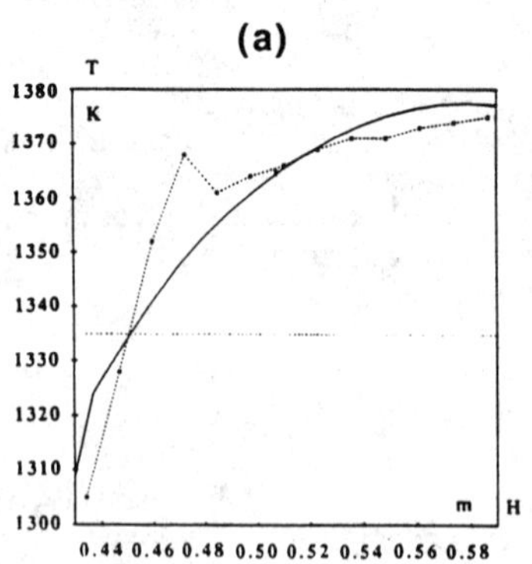

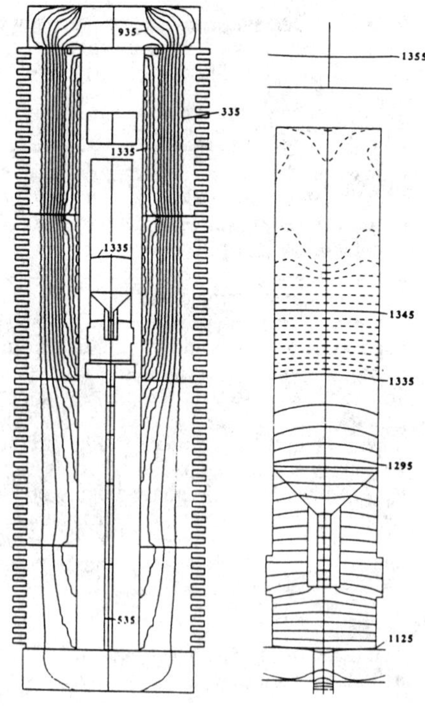

The pressure of the chamber was maintained at 27.5 atm during growth. The grown crystals were 750g, <111> seeded and sulfur doped in the range 2 x 10^{17} to 5 x 10^{18} /cm^3. The average dislocation density is about 350/cm^2 and is not affected by the dopant concentration in the range used. The low level of uniformly distributed dislocations and the absence of slip lines at the periphery of the wafers indicate solidification and cooling under very low thermal stresses. The electrical characteristics reported were similar to those of LEC crystals.

Monberg et al. (11) designed a dynamic gradient freeze furnace for growth of 50 mm diameter InP. The furnace has 23 heating zones grouped into four sections. The pressure vessel which contains the heating elements can be operated under up to 1500 psi pressure. The crystals grown in PBN crucibles with boron oxide encapsulation were 550g, <111> seeded undoped. The dislocation density lies in the range 500-1000 /cm^2. The crystals were n-type with a carrier concentration in the 3 x 10^{15} /cm^3 range and with mobilities comparable to those of crystals grown by the LEC technique. Monberg et al. (12) also obtained zinc doped p-type InP, 50 mm diameter, with dislocation density in the range 300-1200 /cm^2 for a carrier concentration between 1 and 5 x 10^{17} /cm^3. The previous studies, however, show that InP grown confined in pyrolytic boron nitride crucibles has a strong tendency to twin.

A global simulation of Monberg's furnace was developed by Crochet et al. (13). The model includes coupling between the various heat transfers by conduction and radiation, latent heat of solidification and natural convection in the melt and the furnace is realistically represented by a finite element mesh. Comparison of the calculated axial temperature profile and measured profiles in a dummy load of boron nitride are in relatively good agreement (figure 3a). The calculated isotherms in the furnace are reproduced in figure 3b. The crystal-melt interface is slightly convex at the altitude shown. Unfortunately, the experimental interface shapes are not shown for comparison with the calculated shapes.

4. GaAs:

As mentioned in the introduction, the first use of a vertical gradient freeze technique for solidification of GaAs was reported by Chang et al. (7). This study was not aimed at producing high quality large diameter single crystals but indicated that the technique could be used for that purpose. It is again in the paper of Gault et al. (9) that results on the first large diameter (50 mm) GaAs single crystals grown by a seeded VGF technique were reported. The growth conditions (axial temperature gradients, pressure and growth rates) were not reported. The crystals grown were 1200g, <100> seeded and silicon doped in the range 3 x 10^{17} to 3 x 10^{18}/cm^3. The average dislocation density was reported to be about 450/cm^2 for a dopant concentration of about 1.5 x 10^{18}/cm^3. The dislocation density increases significantly when the cooling rate is 100ºC/h. Clemans et al. (14) subsequently published results on 75 mm diameter undoped semi-insulating GaAs using the Gault technique. The average dislocation density is about 2500/cm^2 and exhibits the four-fold symmetry pattern induced by thermal stress during growth. Crystals with undetectable (<10^{14}/cm^3) carbon concentration have a very high resistivity (>10^8 Ω cm) and crystals with a carbon concentration of 2 x 10^{15}/cm^3 at the seed end and 2 x 10^{14}/cm^3 at the tail end have a lower resistivity (ranging from 2 x10^7 Ω cm at the seed end to 7 x10^7 Ω cm at the tail end). The variations in EL2 concentrations are 25%, about half of what they are in LEC crystals. The crystals exhibit mobilities of about 7000 cm^2/Vs.

Abernathy et al. (15) described a VGF system in which the PBN crucible is located in a sealed quartz tube. The crystals grown were 400g to 1000g, 50 mm diameter, <100> seeded, undoped and indium-doped at 2 x 10^{19}/cm^3. The average dislocation density varies from 2000/cm^2 at the seed end to 6000/cm^2 at the tail end for undoped crystals. The dislocation patterns show strong lineage features indicative a growth under significant thermal stresses. Semi-insulating crystals with resistivities >$10^7\Omega$ cm were obtained by addition of gallium oxide to the charge. These crystals have a dominating deep level defect at 0.5 eV.

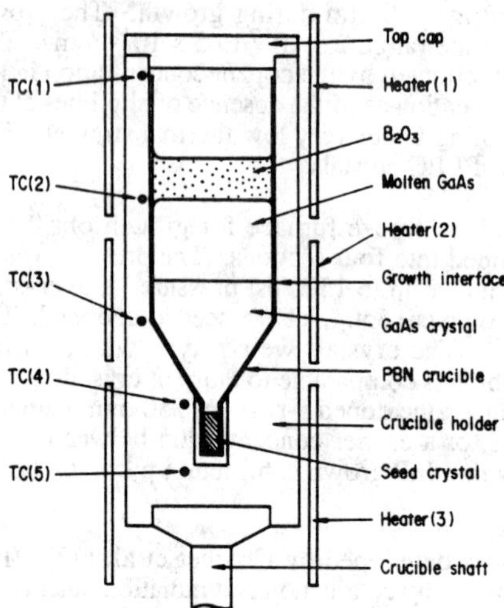

Figure 4. A schematic representation of the VB apparatus used to grow a single crystal (reproduction of fig. 1 from ref. 13).

Hoshikawa et al.(16) have used a liquid encapsulated vertical Bridgman technique (LE-VB) for growth of 75 mm diameter crystals (figure 4). Three graphite heaters are used to control the temperature profile. The crystals grown were 1700g, <100> seeded. About 200 g of boron oxide was placed above the GaAs to encapsulate the melt and an argon pressure of $8kg/cm^2$ was maintained in the furnace. The crucible lowering rate was 3mm/h and they estimated the actual growth rate to be about twice that value. A rotation rate of 2 rpm was used during growth. Undoped semi-insulating crystals have dislocation densities ranging from 5000 to $40000/cm^2$. The electrical characteristics were reported similar to those of undoped semi-insulating LEC crystals. The authors stated three critical issues for reproducible growth: 1) to insure proper seeding by using a seed well of slightly larger diameter than the seed, 2) to have a continuous layer of boron oxide between the LE-crucible and the GaAs, 3) to prevent cracking by keeping a relatively uniform temperature of GaAs distribution in the crystal during cooling.

We have grown GaAs single crystals, 37, 50 and 62.5 mm in diameter, by the dynamic gradient freeze technique (17). Both quartz and pyrolytic boron nitride crucibles have been used successfully. Figure 5 is a schematic of the experimental arrangement. The sealed ampoule configuration is very similar to the one used by Abernathy et al.(15). The ampoule support structure and the temperature control, however, are significantly different. The small diameter part of the quartz ampoule that forms the arsenic reservoir is surrounded by a hot-pressed Boron Nitride (BN) cylinder which acts as a heat-pipe. A thermal gradient of 10ºC/cm, as measured by the control thermocouples on the outside of the quartz ampoule, is maintained over the solid side of the crystal-melt interface. Very shallow axial thermal gradients are maintained over the melt: 1.2ºC/cm within 5 cm of the interface and isothermal zone above that region. Growth rates of 3,4, and 5 mm per hour have been used. After growth the crystals are cooled at 1ºC/min down to 900ºC then at 2ºC/min down to room temperature. Using very shallow temperature gradients over the melt and slow growth rates, the crystal-melt interface can be maintained flat over about 75% of the diameter. Through control of stoichiometry and with a thermal environment designed to reduce thermal stresses, about 50% of the crystals have a dislocation density of less than $1{,}000/cm^2$.

We have used total liquid encapsulation with B_2O_3 to grow 50 mm diameter GaAs single crystals in PBN crucibles in a vertical gradient freeze configuration (18). The B_2O_3 layer efficiently prevents wetting of the crucible by the GaAs charge. The effect of the water content of B_2O_3 on the structural and electrical characteristics of the crystals was investigated. Water vapor can be trapped between the crystal and the crucible affecting the surface morphology of the crystals but not its crystalline perfection.

Numerous theoretical heat transfer analyses of the vertical Bridgman-Stockbarger design have appeared in the literature in the last fifteen years. Most of them were directed toward the understanding of a simple one or two zone design for growth of germanium. Brown recently reviewed the main concepts and results of these analyses (19). Kim et al. analyzed the effect of a magnetic field applied during VB growth of doped germanium (20) and modeled growth of the more complex system HgCdTe also grown in a vertical Bridgman configuration (21). These fundamental studies are mentioned here because the results help the understanding of the process and are useful for growth of GaAs as well. Kim (22) used the same concepts to model growth of GaAs in a closed ampoule. The low thermal conductivity of GaAs is responsible for a concave crystal-melt interface. Such analyses as well as the global analysis mentioned for InP (13) should

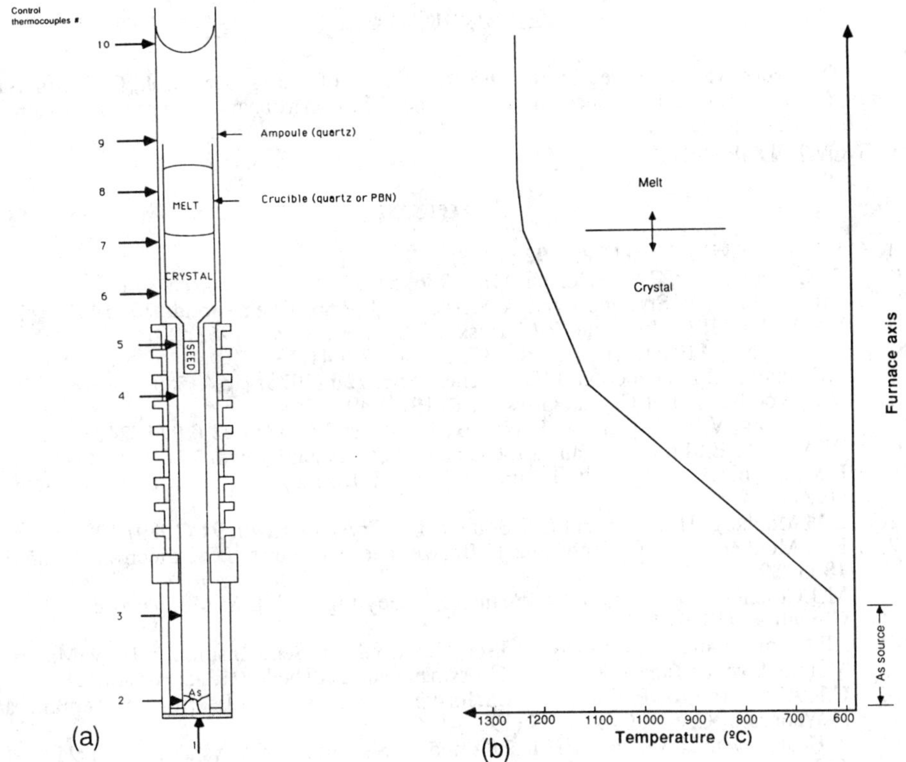

Figure 5: (a) Schematic of the experimental set-up showing the growth ampoule and the position of the control thermocouples
(b) temperature profile as measured on the outside of the growth ampoule during growth.

be done for each experimental set-up since they have reached a level of accuracy which allows optimization of the growth process.

5. Summary:

It has now clearly been demonstrated that vertical Bridgman-type techniques are a viable technique for growth of III-V compounds. These techniques involve a relatively simple technology compared to the Czochralski technique. More importantly crystals can be grown with excellent diameter control which reduces waste from subsequent grinding operations. There appears to be no theoretical limits to the size, length and diameter of crystals which can be grown. However in very large systems the control of heat flows will be more difficult due to the very low thermal conductivities of the III-V compounds. Since the main feature of the technique is growth in a crucible, it is mandatory that there be no reaction between the crucible and the charge. Chemical reactions induce contamination of the melt which can be detrimental to the electrical properties of the crystal. Sticking between the crucible and the charge creates random nucleation sites on the crucible wall and generation of low angle grain boundaries and twins. Pyrolytic boron nitride appears well suited for growth of GaP. Total liquid encapsulation with boron oxide is necessary for reproducible growth of GaAs. Total liquid encapsulation also provides an efficient barrier to impurity contamination, in particular silicon impurities arising from partial decomposition of quartzware that might be present in the furnace. Reproducible growth of fully single crystals of InP has not yet be demonstrated even though very promising results have been obtained.

6. Acknowledgments:

This work was supported by the Director, Office of Energy Research, Office of Basic Energy Sciences, Materials Science Division, of the U.S. Department of Energy under Contract No. DE-AC03-76SF00098.

ACKNOWLEDGEMENT

References:

(1) J.Czochralski, Z.Phys.Chem. **92** (1918) 219.
(2) R.Gremmelmeier, Z.Naturforsch.,11a (1956) 511.
(3) J.B. Mullin, B.W.Straughan and W.S.Brickell, J. Phys. Chem. Solids **26** (1965) 782.
(4) E.P.A.Metz, R.C.Miller and R.Mazelsky, J.Appl. Phys., **33** (1962) 2016.
(5) A.G.Fischer, J.Electrochem.Soc., **117**, No2 (1970) 41C.
(6) S.E.Blum and R.J.Chicotka,J.Electrochem.Soc. **120** (1973) No.4, 588.
(7) H.H.Woodbury, J.of Crystal Growth **35** (1976) 49.
(8) C.E.Chang, V.F.S.Yip and W.R.Wilcox, J.of Crystal Growth **22** (1974) 247.
(9) W.A.Gault, E.M.Monberg and J.E.Clemans, J.of Crystal Growth **74** (1986) 491.
(10) E.M.Monberg, W.A.Gault, F.Simchock and F.Dominguez, J.of Crystal Growth **83** (1987) 174.
(11) E.M.Monberg, H.Brown and C.E.Bonner, J.of Crystal Growth **94** (1989) 109.
(12) E.M.Monberg, P.M.Bridenbaugh, H Brown, and R.L.Barns, J. of Electronic Materials **18** (1989) 549.
(13) M.J.Crochet, F.Dupret, Y.Rickmans, F.T.Geyling and E.M.Monberg, J.of Crystal Growth **97** (1989) 173.
(14) J.E.Clemans and J.H.Conway,in: Proc. 5th Conf. on Semi-Insulating III-V Materials, Malmö, Sweden June 1988, Eds. J.Grossmann and L.Lebedo (Hilger,Bristol,1988).
(15) C.R.Abernathy, A.P.Kinsella, A.S.Jordan, R.Caruso, S.J.Pearton, H.Temkin and H.Wade, J.of Crystal Growth **85** (1987) 106.
(16) K.Hoshikawa, H.Nakanishi H.Kohda and M.Sasaura, J.of Crystal Growth **94** (1989) 643.
(17) E.D. Bourret, M.L. Galiano, R.D. Mih, J.B. Guitron and E.E. Haller, J.Crystal Growth (1990) in press.
(18) E.D.Bourret and E.Merk in: Proc. 8th American Conf. on Crystal Growth, Vail, Colorado July 1990, to be published in J. Crystal Growth.
(19) R.A.Brown, AIChE J. **34** (1988) 881.
(20) D.H.Kim, P.M.Adornato and R.A.Brown, J.of Crystal Growth **89** (1988) 339.
(21) D.H.Kim and R.A.Brown, J.of Crystal Growth **96** (1989) 609.
(22) D.H.Kim, PhD Thesis, Department of Chemical Engineering, Massachusetts Institute of Technology (1990).

Modelling the Vertical Bridgman growth of cadmium telluride

C. Parfeniuk, F. Weinberg, I.V. Samarasekera, C. Schvezov
Department of Metals and Materials Engineering, The University of British Columbia, Vancouver, British Columbia, Canada, V6T 1W5

L. Li
Johnson Matthey Electronics, Trail, British Columbia, Canada V1R 4S5

Abstract

The temperature and stress fields for the Vertical Bridgman growth of CdTe were calculated using the finite element method. The critical resolved shear stress (CRSS) of CdTe was measured at elevated temperatures. The measured CRSS and the calculated shear stress were compared to determine the onset of dislocation formation in the crystal.

Introduction

Cadmium telluride (CdTe) is a II-VI compound used in γ-ray and X-ray spectrometers and as a substrate for $Hg_{1-x}Cd_xTe$ infrared detectors [1,2]. These applications require single crystals of low defect density since grain boundaries, twins and dislocations are detrimental to the device. Dislocations are a direct result of thermal stress. The thermal stresses that occur during crystal growth will cause dislocation formation if the resolved shear stress is greater than the critical resolved shear stress (CRSS).

During the crystal growth of CdTe there are large axial thermal gradients near the solid/liquid interface which result in high thermal stresses in the crystal. There are also radial temperature gradients which lead to a concave interface as shown experimentally [3] and mathematically [4,5,6]. These thermal gradients may contribute to dislocation generation.

The CRSS of CdTe has been measured at temperatures between 100K and 723K [7,8] well below the melting point of CdTe (1364K). Values of the CRSS above 723K are required, since dislocation generation is related to the amount of local thermal stresses exceeding the CRSS. High thermal stresses occur well above 723K, particularly near the solid/liquid interface.

The present investigation was undertaken to determine the thermal and stress fields in a growing CdTe crystal using a finite element mathematical model. In addition measurements were made of the CRSS of CdTe at temperatures well above 723K. Combining the model results with the CRSS values enabled an estimate of the dislocation density and distribution in the crystal to be made. From this the growth conditions can be modified to reduce the dislocation density.

Mathematical Model

The heat transfer and thermal stress models were developed for CdTe, grown in a sealed quartz container, lowered vertically through a thermal gradient at a slow rate. A sketch of the growth system, Figure 1, shows the coordinate system used, the vertical temperature distribution provided by the furnace, and the air gap between the solid CdTe and the quartz due to volume shrinkage of the CdTe during cooling. The inside of the quartz tube is coated with graphite, to prevent the CdTe from reacting and sticking to the quartz. In order to solve the heat and thermal stress equations a number of assumptions were made. It was assumed that the heat and stress fields were symmetric about the z axis, the CdTe acted as a simple linear elastic material, and the solid CdTe was a single crystal with isotropic thermal-physical properties.

The heat and stress models were solved using quadratic isoparametric elements [9]. The model used 1184 nodes and 340 elements, each element being of constant size at 2.5 mm by 3.0 mm. The thermal-physical properties of CdTe used in the models are listed in Table 1 being obtained from the literature [1,10,11]. The thermal transient was calculated using the Dupont three point time stepping technique [12]. The latent heat release at the interface as solidification progressed was incorporated in the thermal model using the enthalpy method [13]. Since CdTe contracts more than quartz on cooling, a gap is formed between the solid CdTe and the quartz across which heat is transferred by radiation. Before solidification, heat transfer between the liquid CdTe and quartz is by conduction. Conduction across the air gap was approximated by using an inflated heat transfer coefficient [14].

Two boundary conditions were used for the thermal model: radiative heat transfer and forced convection. Forced convection was used in the case of a gas which moved through the furnace during crystal growth. The temperature of the gas was assumed to be 10 degrees below the furnace temperature. Radiation heat transfer from the furnace is absorbed by two surfaces. The outside of the quartz will absorb a fraction of the radiation that is equal to the absorptivity (α). The portion of the rest of the radiation ($1 - \alpha$) will be absorbed at the CdTe surface. The radiation that reaches the surface also depends on the orientation of the absorbing surface. For a vertical surface that is parallel to the furnace wall, all of the radiation reaches the surface. For the top and bottom surfaces the radiation view factors were calculated [15] and used to determine the radiative heat transfer to these surfaces.

The thermal stress model, calculates the Von Mises stress at each integration point within an element during growth. This stress is then resolved into twelve components defined by the operating slip planes {111} and slip directions <110> in CdTe [16]. Youngs modulus and Poissons ratio within the {111} plane were used for all stress calculations [17,18]. The component which had the maximum resolved shear stress (MRSS) was selected and compared to the CRSS of CdTe at the temperature being considered. If the MRSS was greater than the CRSS it was assumed that dislocations would be generated locally, the number of dislocations roughly increasing with increasing excess stress, to a first approximation. The stress model assumes that the CdTe does not stick to the quartz during and after solidification

Experimental Procedure

The critical resolved shear stress was measured in CdTe single crystals deformed in tension in the temperature range 723K to 1023K. The testing apparatus is shown in Figure 2. A single crystal plate 1.28 mm thick was formed into the test specimen (A), by first cutting a sample of rectangular section (22.2 mm by 7.94 mm), mounting the section between glass plates using

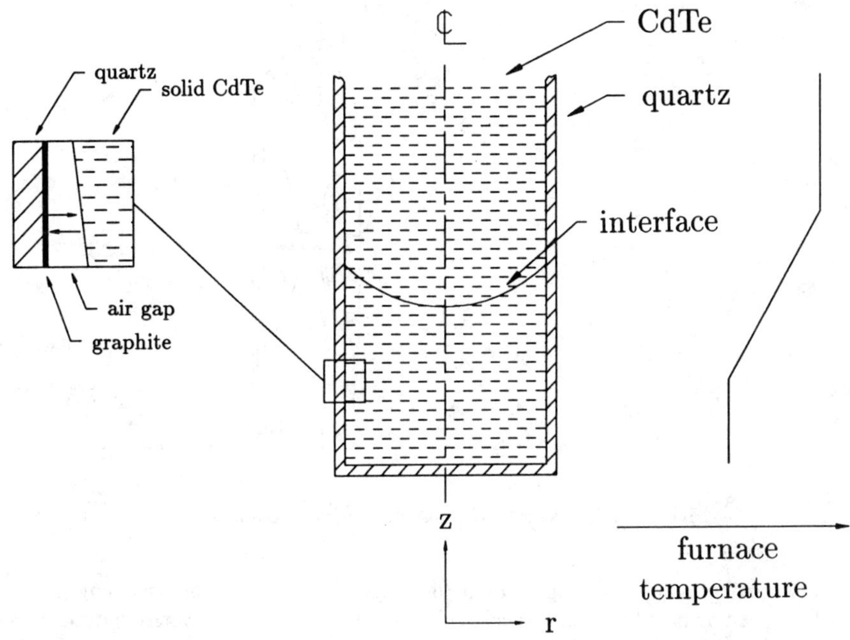

FIG. 1. Vertical Bridgman growth.

TABLE 1: CdTe Properties.

Property	Units	Temperature	Polynomial
Conductivity	W / K m	T < 500K	$10.701 - 1.3 \times 10^{-2}$ T
		500K < T < 1200K	$5.337 - 2.263 \times 10^{-3}$ T
		1200K < T < 1364K	$-38.31 + 6.78 \times 10^{-2}$ T $- 2.811 \times 10^{-5}$ T2
		T > 1364K	$9.987 - 5.676 \times 10^{-3}$ T
Heat of Fusion	J / g	T =1364	209.2
Specific Heat	J / g K	298 < T < 1600K	$0.1778 + 1.663 \times 10^{-4} T - 6.607 \times 10^{-8} T^2 + 1.703 \times 10^{-11} T^3$
Quartz Absorptivity	-	-	0.3
HTC of Gap*	J / m^2 s K	-	8000.0
Forced Convection HTC*	J / m^2 s K	-	50.0
Poissons Ratio	-	-	0.459
Young Modulas	MPa	-	3.98×10^5
Expansion Coefficient	-	T < 525K	$1.031 \times 10^{-6} + 2.054 \times 10^{-8} T -19.779 \times 10^{-12} T^2$
		T > 525K	6.3629×10^{-6}

* HTC = heat transfer coefficient

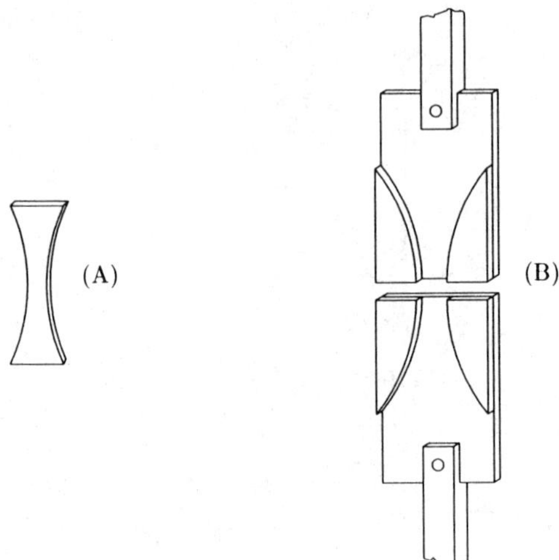

FIG. 2: Tensile sample(A) and tensile test apparatus(B).

lakeside wax, and grinding the long faces of the edge to a circular contour using a 51 mm diameter fast speed diamond grinding wheel. After grinding the samples were removed from the jig and cleaned in ethyl alcohol. The orientation of the samples was determined by the X-Ray back-reflection Laue technique.

The test samples were inserted into the tensile holder (B) shown in Figure 2. The holder and sample were designed to allow the components to expand as the assembly was heated to the test temperature without straining the sample. The assembly was heated in a small resistance tube furnace in a helium atmosphere. Temperatures were measured with chromel/alumel thermocouples attached to the tensile grips on either side of the sample. The samples were deformed in an Instron tester with a cross head speed of 0.85×10^{-3} cm/s. As the sample was loaded, the crystal self-aligned in the grips, then deformed elastically and plastically.

Results

Mathematical Model

Calculations were made for the growth conditions listed in Table 2. A series of furnace gradients were considered, between 0.05 and 4.0 degree/mm and ampoule velocities between 1.39×10^{-4} mm/sec and 5.56×10^{-4} mm/sec as listed. The calculated shape and distribution of the isotherms for crystal growth at 2.78×10^{-4} mm/sec with a furnace gradient of 1 degree/mm after 2.88×10^{5} seconds is shown in Figure 3(a). The isotherms are concave upwards in the bulk of the sample including the solid/liquid interface, indicating that radial heat loss occurs. Near the top of the sample, in the melt, the isotherms become convex upwards and are widely spaced. The vertical central gradient is relatively steep in the solid below the interface, decreasing with distance in the solid away from the interface, and shallow in the melt. The discontinuity in the isotherm at the solid CdTe/quartz interface is due to the presence of an air gap between the solid and the quartz.

The shape and position of the solid/liquid interface with time from the start of solidification is shown in Figure 3(b). The interface is observed to be concave upwards throughout growth tending to become slightly flatter towards the end of solidification. The distance along the axis between the interface positions shown for equal time increments, increases towards the end of solidification. This indicates that the axial velocity increases in this region.

TABLE 2: CdTe Model Operating Parameters.

Parameters	Units	Values
Ampoule Diameter	mm	50.0
Charge Height	mm	90.0
Quartz Wall Thickness	mm	2.75
Hot Zone Temperature	Kelvin	1380.0
Furnace Gradient	degrees / mm	0.05, 0.1, 0.5, 1.0, 2.0, 4.0
Cold Zone Temperature	Kelvin	1335.0
Ampoule Velocity	mm / sec	1.39×10^{-4}, 2.78×10^{-4}, 5.56×10^{-4}

The calculated MRSS contour curves corresponding to the isotherms shown in Figure 3(a) are shown in Figure 3(c). The largest values of the MRSS (18 MPa) are observed to occur adjacent to the quartz near the solid/liquid interface. The stress is observed to drop rapidly both with distance radially from the outside to the centre of the crystal, and with distance longitudinally into the crystal from the interface.

The effect of changing the furnace gradient and the growth velocity on the curvature of the solid/liquid interface and the interface velocity was examined. The curvature of the concave solid/liquid interface is arbitrarily defined as the vertical distance between the interface position at the centre of the crystal and the outside surface. The growth velocity is calculated at the center of the crystal.

The effect of temperature gradient for three ampoule velocities on the interface curvature, at the vertical midpoint of the crystal, is shown in Figure 4. Large curvatures are observed at gradients below 0.5 degrees/mm. Above 0.5 degree/mm the curvature is essentially constant at near 8 mm. The effect of ampoule velocity on the curvature is small except at very shallow gradients where the curvature increases with increased ampoule velocity.

The change in interface curvature with distance from the start of solidification, at different furnace gradients is shown in Figure 5. For a small furnace gradient of 0.05 degrees/mm, the curvature is large (18mm) at the start of solidification decreasing slowly and progressively to near 12 mm at the end of solidification. Increasing the gradient to 0.1 degree/mm results in the same initial curvature of 18 mm at the start, but decreases more rapidly to near 8 mm at the end of solidification. At higher gradients of 0.5 to 4 degrees/mm the curvature is effectively constant at 8 mm for most of the crystal length, dropping to 2 mm near the end of solidification.

The dependance of interface velocity at the center of the crystal with distance from the start of solidification is shown in Figure 6, for an imposed ampoule velocity of 2.78×10^{-4} mm/sec. The calculated results show that the center velocity is relatively constant between 2.78×10^{-4} and 4.17×10^{-4} mm/sec for most of the crystal. Towards the end of crystal growth the interface velocity increases rapidly, 3 to 5 times the ampoule velocity. At the lower furnace gradients of 0.05 and 0.1 degrees/mm the calculated interface velocity is higher than the ampoule velocity, and the increase in velocity towards the end of freezing is appreciable. At the higher furnace gradients, 1 to 4 degrees/mm, the calculated velocities are the same as the ampoule velocity, and the increase in velocity near the end of freezing is small.

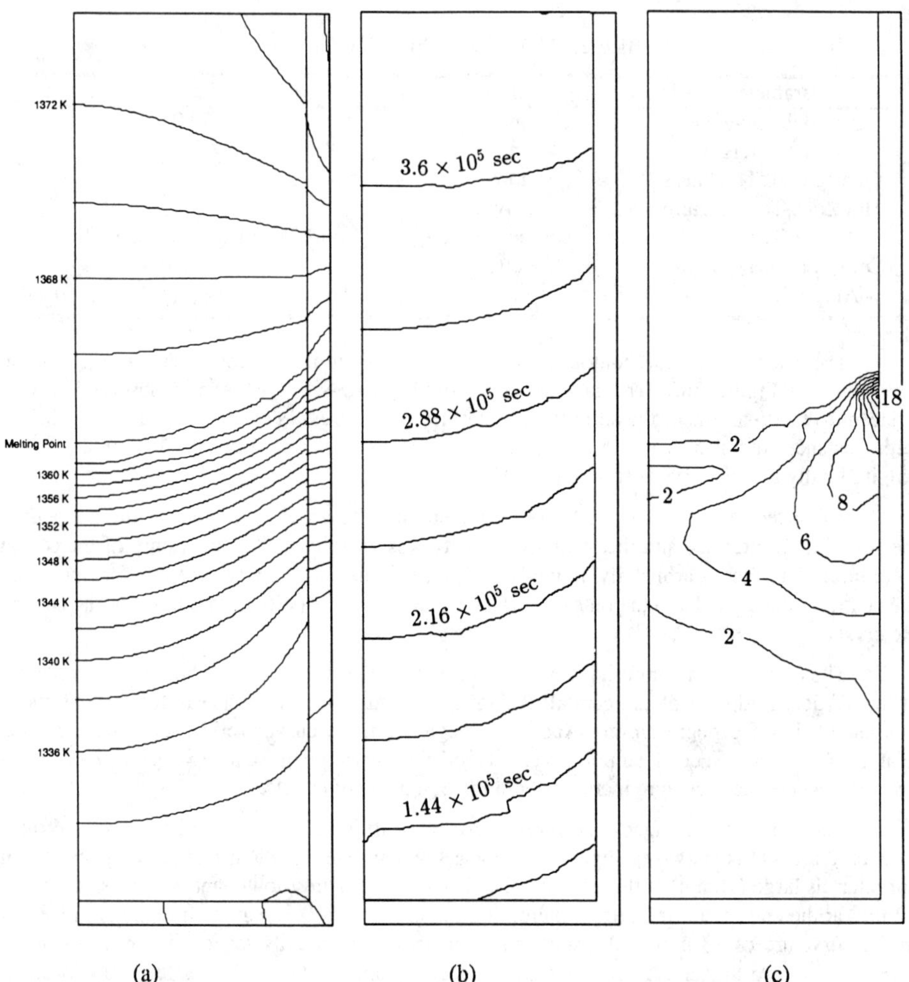

FIG. 3: Crystal grown at 2.78×10^{-4} mm/sec with a furnace gradient of 1 degree/mm. (a) Isotherm contours after 2.88×10^5 sec of growth. (b) Solid/liquid interface shape and position at the lines indicated at intervals of 3.6×10^4 sec of growth. (c) MRSS contours after 2.88×10^5 sec growth, in MPa.

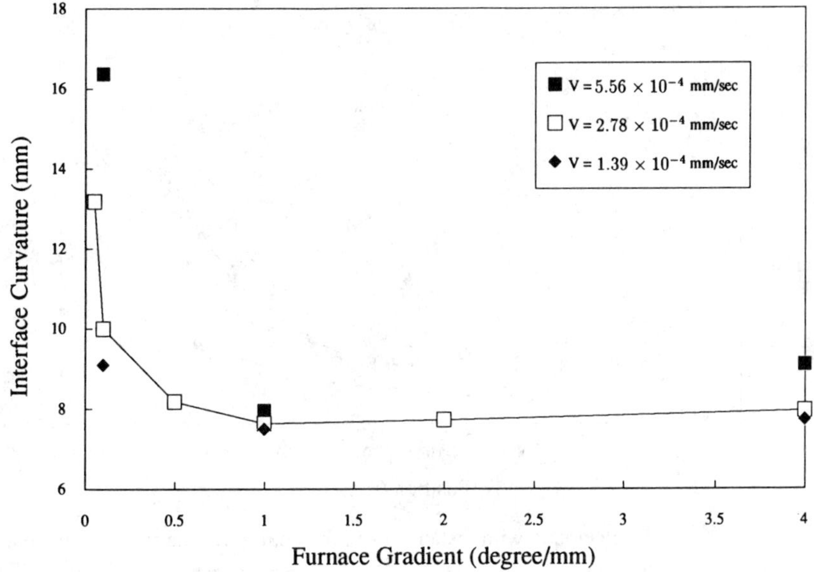

FIG. 4: Dependence of interface curvature on furnace gradient and ampoule velocity (V).

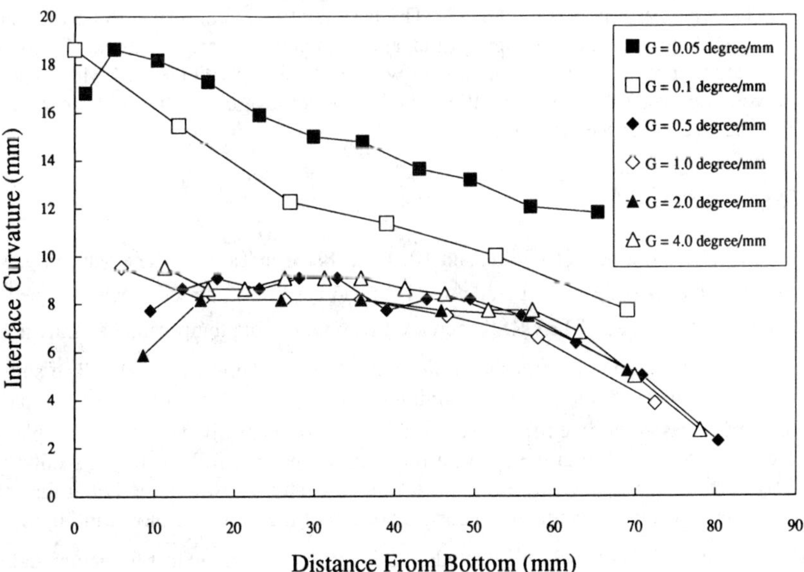

FIG. 5: Variation on interface curvature with distance from the start of solidification for different temperature gradients (G).

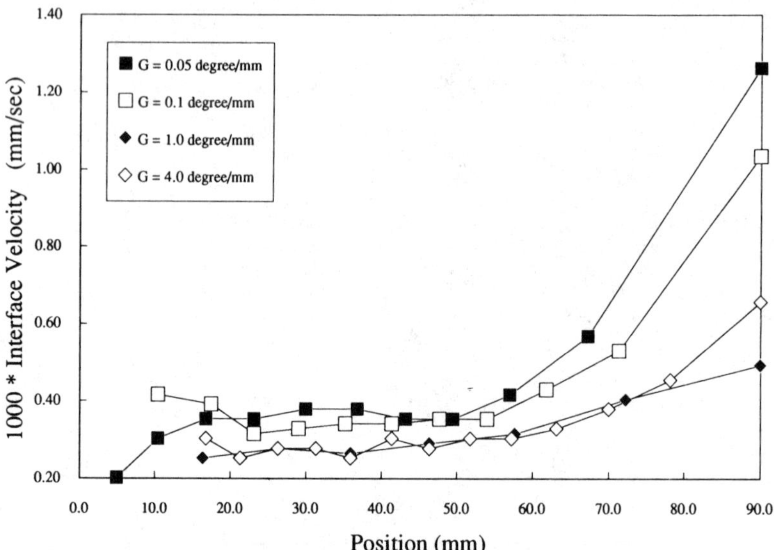

FIG. 6: Variation of interface velocity with distance from the bottom for the furnace gradients (G) shown.

The effect of changing the furnace gradient and growth velocity on the calculated MRSS was determined, using the largest value of MRSS, located at the solid/liquid interface adjacent to the quartz. The results are shown in Figure 7. The largest MRSS is observed to be very small at small furnace gradients, increasing rapidly and approximately linearly to about 18 MPa at a gradient of 1 degree/mm, then increasing slowly with increasing gradient to 24 MPa. The effect of ampoule velocity on the MRSS is small, the higher velocities giving slightly higher values of the MRSS at the larger furnace gradients.

Critical Resolved Shear Stress

Four tests were carried out between 748K and 1024K as listed in Table 3. The test samples were oriented such that the large face was parallel to (111) and the $[\bar{1}\,0\,1]$ direction was inclined to the stress axis at the angles listed. The CRSS decreased with increasing temperature as shown.

A typical stress-strain curve is shown in Figure 8, from the test at 932K. The stress was determined from the measured load and the minimum cross-sectional area of the test sample (3.58 mm^2). The strain was determined from the cross head movement divided by the sample length between the grips (5 mm). The curve shows a linear initial portion followed by a section with a progressively decreasing slope. The CRSS was taken as the stress at which the initial linear curve intersects the line which best fits the initial curved portion of the curve, as shown in Figure 8.

The measured CRSS values are plotted on a logarithmic plot against the reciprocal of the test temperature, giving the results shown Figure 9. The best fit line for the four data points is shown, and corresponds to the following equation:

$$\sigma_{crss} = 0.0136 \exp\left(\frac{4500.0}{T}\right) \qquad \text{(in MPa)}$$

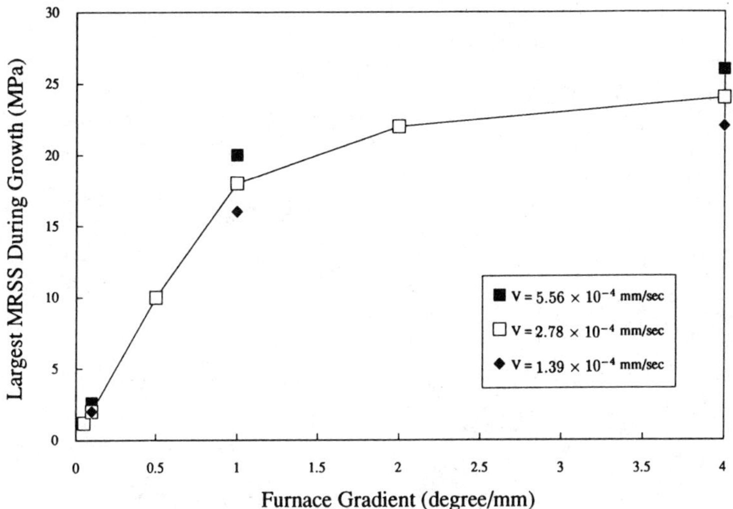

FIG. 7: The largest calculated MRSS as a function of furnace gradient for the growth velocities (V) shown.

The extrapolated value of the CRSS from reported data [7] is shown in Figure 9 by the dotted line, and lies appreciably lower than the measured values at the lower temperatures. The value of CRSS extrapolated to the melting point is 0.4 MPa which is much lower than the CRSS of GaAs at its melting point (approximately 0.9 MPa).

A comparison of the MRSS and excess stress above the critical shear stress is made for a CdTe crystal grown with a thermal gradient of 5 degrees/mm and a growth rate of 2.78×10^{-4} mm/sec at 3.6×10^{5} seconds after solidification started. The results are shown in Figure 10, in which the MRSS contours are shown in (a) and the excess stress contours in (b). In (b), the excess stress is significant in the region of the interface close to the outer wall, as with the MRSS. Since the CRSS is small, subtracting its value from the MRSS produces a relatively small shift in the stress contours. This is the region where dislocations will be generated. In the lower part of the crystal there is no excess stress, therefore dislocations would not be generated in this region during growth. This would indicate that there should be a higher dislocation density in the crystal near the side, compared to the centre, providing dislocation annihilation does not occur. The high dislocation density associated with CdTe indicates there is little recovery during and after crystal growth.

TABLE 3: Critical Resolved Shear Stress of CdTe

Temperature (K)	Orientation		CRSS (MPa)
	Plane*	Angle to $[\bar{1}01]$**	
738	(111)	42	4.78
817	(111)	3	5.22
932	(111)	49	1.72
1024	(111)	0	0.98

*Plane parallel to the surface of the sample

**Angle between the stress axis and the $[\bar{1}01]$ direction

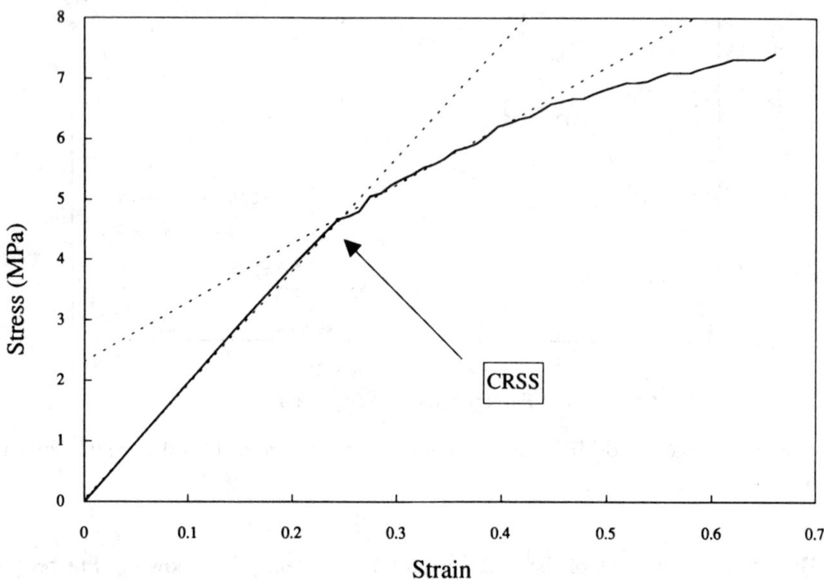

FIG. 8: Stress-Strain curve for a tensile test of CdTe at 932K.

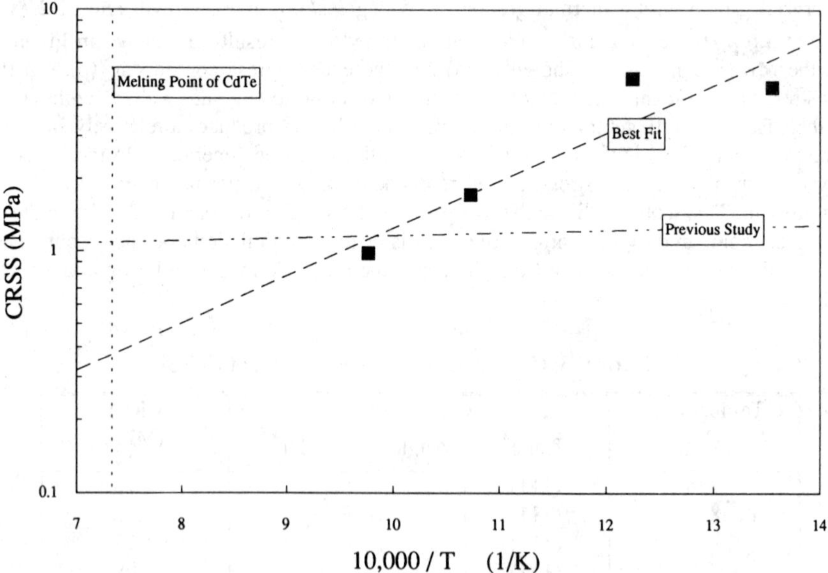

FIG. 9: Dependence of critical resolved shear stress on temperature for CdTe.

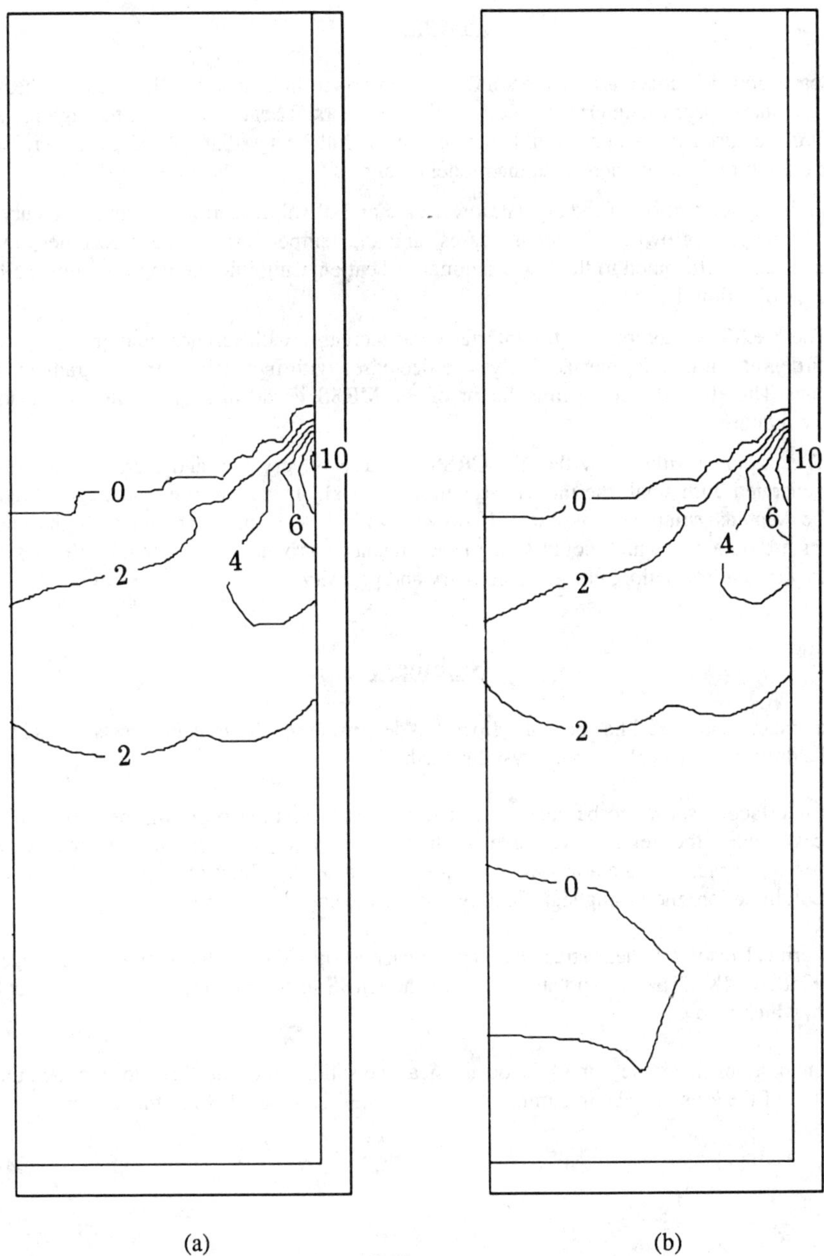

FIG. 10: (a) MRSS contours in MPa (b) Excess stress above the CRSS (MPa). G = 0.5 degree/mm, V = 2.78×10^{-4} mm/sec, 3.6×10^{5} sec from start of crystal growth.

Discussion

The concave upward curvature of the solid/liquid interface indicates that the radiative heat loss through the quartz near the interface is comparable to the axial heat conduction through the solid. Decreasing the furnace gradient would increase the radial heat radiation and thus increase the interface curvature, in agreement with the model results.

In the upper portion of the crystal, toward the end of solidification, the interface curvature tends to flatten, the growth velocity increases, and the temperature gradient becomes smaller. The effects can be attributed to the lower amount of heat entering into the upper part of the boule containing CdTe liquid.

The MRSS is maximum at the interface and increases with furnace gradient for gradients up to 1.5 degrees/mm. The interface curvature decreases with increasing furnace gradient in the same range. Therefore the controlling factor of the MRSS is the furnace gradient, and not the interface curvature.

The present results show that the CRSS in CdTe is very low and therefore dislocations will be generated with small thermal stresses in the crystal. It is therefore evident that gradients should be kept as small as possible. However, with low furnace gradients, high interface curvatures are obtained which could lead to the enhanced production of crystal defects such as twins, stray crystal formation, non-stoichiometry and porosity.

Summary

1) A mathematical model has been developed to determine the thermal and stress fields in CdTe during Vertical Bridgman crystal growth.

2) The interface is shown to be concave to the melt and remains so during the entire growth process under the regime we have studied. The interface curvature increases with decreasing furnace gradient at low gradients. The growth velocity is constant for most of the solidification, increasing significantly towardsthe end of freezing.

3) The critical resolved shear stress has been measured in CdTe in the temperature range of 738K to 1024K. The extrapolated value of the CRSS at the melting point of CdTe is 0.4 MPa, which is low.

4) The maximum shear stress in CdTe occurs near the solid/liquid interface close to the outer surface of the crystal. The maximum stress is greater than the CRSS in this region.

Acknowledgements

The authors wish to thank the British Columbia Science Council for financial support and Johnson Matthey Electronics for providing samples used in this work.

Referances

[1] K. Zanio, *Semiconductors and Semimetal, Vol. 13, Cadmium Telluride*, Academic Press (1978).

[2] O. Oda, K. Hirata, K. Matsumoto and I. Tsuboya, *J. Crystal Growth*, 71, 273 (1985).

[3] A. Tanaka, Y. Masa, S. Seto and T. Kawasaki, *Mat Res Soc Symp Proc*, 90, 111 (1987).

[4] C.E. Huang, D. Elwell and R.S. Feigelson, *Journal of Crystal Growth*, 64, 441 (1983).

[5] T. Jasinski and A.F. Witt, *Journal of Crystal Growth*, 71, 295 (1985).

[6] Y.M. Dakhoul, R. Farmer, S.L. Lehoczky and F.R. Szofran, *Journal of Crystal Growth*, 86, 49 (1988).

[7] E.Y. Gutmanas, N. Travitzky, U. Plitt and P.Haasen, *Scripta Metallurgica*, 13, 293 (1979).

[8] E.L. Hall and J.B. Vander Sande, *Journal of The American Ceramic Society*, 61, 418 (1978).

[9] O.C. Zienkiewicz, *The Finite Element Method*, McGraw-Hill Book Company, NewYork, 1977.

[10] S. Sen et al, *Journal of Crystal Growth*, 86, 111 (1988).

[11] D. Wagman et al, *Selected Values of Chemical Thermodynamic Properties*, National Bureau of Standards Series 270

[12] B.G. Thomas, I.V. Samarasekera and J.K. Brimacombe, *Met. Trans. B*, 15b, 307, (1984).

[13] E.C. Lemmon, *Num. Methods in Heat Transfer*, 201 (1981).

[14] Y. Nishida, W. Droste, and S. Engler, *Met. Trans. B*, 17B, 833 (1986).

[15] E.M. Sparrow, L.U. Albers and E.R.G. Eckert, *Journal of Heat Transfer*, 84, (1962), 73-81

[16] C. Schezov, I.V. Samarasekera and F. Weinberg, *Journal of Crystal Growth*, 84, 219 (1987).

[17] W.A. Brantley, *J. Appl. Phys.*, 44, 534 (1973).

[18] H. McSkimin, D. Thomas, *J. Appl. Phys.*, 33, 56 (1961).

The control and properties of misfit dislocations in OMCVD growth epitaxial layers

G.P. Watson, D.G. Ast
Materials Science and Engineering Department, Cornell University

T.J. Anderson, Y. Hayakawa
Chemical Engineering Department, University of Florida, Florida, U.S.A.

ABSTRACT

It has previously been shown that the density of misfit dislocations formed at the interface of lattice-mismatched epitaxial layers can be controlled by patterning the substrate prior to molecular beam epitaxial growth. This work extends the substrate patterning technique to material grown by organo-metallic chemical vapor deposition. GaAs substrates were first patterned by photolithography and trenches were reactively ion etched to isolate square regions on the wafer surface. Substrates were etched to 70 *nm* to 1200 *nm* deep to observe the effects of trench depth on dislocation isolation. Layers of nominally $In_{0.05}Ga_{0.95}As$, 350 *nm* thick, were then grown on the substrates. Misfit dislocations at the substrate/epi-layer interface were observed by scanning cathodoluminescence. Trench depth results indicate that misfit dislocations can be stopped by trenches that are shallower than the epi layer thickness itself and the layer is continuous across trenches. The dislocation density varies linearly with isolated square size all the way up to 800 μm - indicating that multiplication mechanisms are not active in this material. The defect densities in the 110 and $1\bar{1}0$ directions were found to be identical, which is surprisingly unlike the misfit density asymmetries found in molecular beam epitaxial InGaAs.

INTRODUCTION

The density of misfit dislocations that form at the interface of substrates and lattice mismatched epitaxial layers can be significantly reduced by simply isolating regions of the substrate surface before crystal growth. This effect has been previously been demonstrated with lattice mismatched InGaAs grown by molecular beam epitaxy (MBE) on patterned and etched GaAs substrates (1). Misfit dislocations are energetically favored to form when the epitaxial layer exceeds a critical thickness that can be calculated by theories such as Matthews et al. (2). These dislocations form along the substrate/epitaxial layer interface, relieving some strain in the epitaxial layer that is otherwise forced to conform to the in-plane lattice parameter of the substrate. Misfit dislocations are important to study because they can adversely affect the electronic properties of devices that make use of the interface or properties of the strained material.

Misfit dislocations (hereafter simply called misfits) form by glide and "stretching" of short dislocation segments. These segments extend from the interface to the epitaxial layer surface and form from threading dislocations in the substrate or from other surface defects such as particle contamination on the substrate surface before epitaxial growth. In planar substrates, a segment can propagate a misfit that is on the order of millimeters long. If however the segment is stopped by a deep trench or wall, it can affect only a relatively small region of a wafer. Figure 1 is a schematic showing a cross section of a misfit gliding along a strained epitaxial layer interface grown on a patterned wafer. The gliding segment cannot affect any other isolated region.

One deleterious aspect of patterning and etching before epitaxial growth is that the resulting wafer surface is no longer planar. This could cause problems in subsequent device fabrication steps. In this paper we describe work that determines how deeply a trench must be etched to isolate surface regions. In this work we also extend the isolation technique to strained layers grown by organo-metallic chemical vapor deposition (OMCVD) and study the differences between it and patterned MBE material.

EXPERIMENTAL

Semi-insulating and Si doped GaAs substrates (from Crystal Specialties, Sumitomo and Spectrum Technologies), with etch pit densities (EPD) of 5000 cm^{-2} were solvent cleaned and etched in a solution of NH_3OH, H_2O_2, and H_2O before being coated with SiO_x. The oxide is used as both a mask and as a protective layer covering the GaAs surface until just before epitaxial growth. The SiO_x mask was patterned using conventional photolithography and reactive ion etching processes. The exposed GaAs was then etched by two processes. Most of the material was etched by the chemical assisted ion beam etching (CAIBE) technique using Cl_2 in an Ar ion beam. This technique produces etch features with vertical side walls which for large depths has been found to create discontinuous MBE epitaxial layers (ensuring isolation). Some wafers were etched in a reactive ion process (RIE) using CCl_2F_2 and He. Again the sidewalls were essentially vertical.

Figure 2 shows the pattern etched into the oxide; the narrow channels correspond to removed oxide that outline squares of four different sizes. The channels are aligned along the 110 direction of the GaAs wafer. These channels are disconnected at the square corners so that the epitaxial material forms one continuous layer to permit the making of planar transmission electron microscope specimens.

After patterning, the substrates were etched in HF to remove the masks, and in some cases they were further etched in a very dilute NH_3OH, H_2O_2, H_2O solution to remove about 10 *nm* of material. After rinsing in de-ionized water and blown dry, the wafers were placed in a chamber evacuated to 3×10^{-6} Pa and baked for 1 hour at 950 °C in H_2. OMCVD growth was performed at 650 °C. A 350 *nm* GaAs buffer layer was first deposited, followed by a 350 *nm* InGaAs layer which varied between 3.0 and 5.0% In. Both layers were S doped (n type) at 6×10^{18} cm^{-3} to enhance the cathodoluminescence signal.

The InGaAs thicknesses and compositions were determined by Rutherford back scattering. Compositions were checked by x-ray diffraction rocking curves of the 004 and 224 Bragg conditions. Since the misfit density is small in patterned material, in-plane strain distorts the 004 interplanar spacing of the InGaAs. Therefore an asymmetric Bragg reflection such as the 224 is also needed to accurately determine the In content of the layer.

The misfit dislocations were imaged with a GW Electronics scanning cathodoluminescence detector (CL) attached to a JEOL JSM35 scanning electron microscope (SEM). As a 15 kV electron beam scans a wafer surface electron-hole pairs are created that either recombine radiatively (and picked up with a Si detector) or non-radiatively, such as near the core of a dislocation. Therefore, Misfit dislocations (or groups of them) at the GaAs/InGaAs interface appear as dark lines in the image.

To understand how the epitaxial layers follow the substrate topology, samples were coated with a 800 *nm* of SiO_x, cleaved and stained. It was found that the highly doped, low In, layers were not revealed (stained) by wet chemical etching. To observe the InGaAs profile, the large etch selectivity of InGaAs relative to GaAs in Cl based RIE was exploited, the epitaxial layers (7).

Table 1 indicates that there is a discrepancy at 150 and 170 *nm*. The smaller trench isolated misfits while the slightly deeper one did not. We believe that this is due to the fact that the 170 *nm* layer has a slightly higher In content, which increases the mismatch and therefore the driving force applied to a dislocation segment.

The surface of a wafer that is selectively etched to a depth just beyond the minimum isolation depth and subsequently used as a substrate for OMCVD growth of stained layers is still in effect quite planar as far as subsequent photlithographic processes are concerned. No special treatment is necessary to create devices on this material. At the same time, the density of misfit dislocations at the interface is reduced significantly compared to unpatterned material.

TABLE 1
Depth Necessary for Isolation

Wafer ID	In Fraction	Etch Process	Depth (*nm*)	Isolation? (Yes/No)
064	0.035	CAIBE	70	No
081	0.050	CAIBE	80	No
082	0.050	CAIBE	110	No
065	0.035	CAIBE	150	Yes
080	0.050	CAIBE	170	No
037	0.030	*	310	Yes
083	0.050	CAIBE	340	Yes
066	0.035	CAIBE	500	Yes
030	0.030	RIE	600	Yes
028	0.030	RIE	1200	Yes

* the trenches in wafer 037 were wet etched

Figures 5 and 6 are SEM images of trench profiles after growth. Photos 5 *a* and *b* are of the 110 *nm* deep trench which did not stop misfits from gliding. They appear very similar to the structure of the 340 *nm* trench running in the $1\bar{1}0$ direction, shown in Figure 6 *a*. The shape of the epitaxial layer profile does not determine whether isolation occurs or not; The depth of the trench is the only apparent difference (at least in the $1\bar{1}0$ direction).

The profile of the bottom of the InGaAs layer is much different than the vertical sidewalls obtained in the CAIBE or RIE etch processes. The initial GaAs buffer layer has an important effect on the shape of the trench geometry.

Figures 5 and 6 also show that the InGaAs layer, revealed by etching, is one continuous layer for both trench depths. InGaAs layers on adjacent squares are not strictly disconnected. Misfit isolation is believed to have occurred in the 340 *nm* layer by the abrupt changes in depth. This barrier is much different than that obtained from MBE growth of InGaAs on 2 μm deep trenches. Figure 7 is a SEM image of such material. Note that the InGaAs layer is discontinuous at the trench edge.

The fact that the OMCVD strained material is not broken up by the trenches leads the authors to speculate that the trench isolation "critical" depth, (for the material in Table 1 this value is about 160 nm) is a function of buffer layer thickness, Strained layer thickness, details of which will be described elsewhere (3). The edge profiles were observed in the same SEM used for the CL work.

RESULTS AND DISCUSSION

Figures 3 *a* and *b* are CL images of the patterned material after OMCVD growth of $In_{0.035}Ga_{0.965}As$. The number of misfits per *cm* is clearly much lower in the material isolated into 100 μm squares (*a*) than 800 μm squares (*b*). By confining the gliding dislocations to small regions with abrupt trench steps, the misfit density can be reduced by at least an order of magnitude compared to unpatterned material.

Figure 4 is a plot of the average misfit density varies with square size for a 350 *nm* $In_{0.035}Ga_{0.965}As$ epitaxial layer isolated with 1.2 μm deep trenchs. The substrate in this case was a semi-insulating GaAs wafer with an average EPD of 5000 cm^{-2}. The mean density was taken from the individual square misfit densities of many squares (from 60 100 μm squares to 20 800 μm squares) so that the statistical significance of the relationship can be inferred. The error bar corresponds to one standard deviation on each side of the mean.

The number of misfit nucleation points in this material is estimated to be about 50,000 cm^{-2} - 10 times larger than the EPD. Threading dislocations (assumed to be the same as the EPD) alone cannot account for the observed number of defects. Other defects such as particles on the substrate surface must contribute the majority of nucleation sources. Presumably, improved processing techniques can reduce the density of misfits even further.

The line in Figure 4 is a least squares fit of the data. Even for large squares, the change in misfit density with square size is linear. It has been proposed that at larger densities dislocation multiplication caused by dislocation interactions can occur, leading to an increasing slope at large square sizes (1). No evidence of misfit dislocation multiplication is found in this material.

It is well known that the misfit dislocations that form in the 110 and $1\bar{1}0$ directions possess different core types, due to the non-centrosymmetric nature of III-V compounds. An α type misfit dislocation relieves compressive epitaxial layer stress in the $1\bar{1}0$ direction of the interface plane while a β misfit relieves stress in the 110 direction. Some researchers have found that the density of misfits in MBE grown, Si doped (n type) InGaAs varies with direction, the α misfits being more prevalent (1, 4). They attribute this to the different kinetic properties of dislocation glide. It has been well established, for instance, that the α dislocation glide velocity under stress is unaffected by the Fermi level, while the β dislocation velocity can vary by orders of magnitude in p type and n type GaAs (5). In fact, others have found identical misfit densities in the two directions in undoped MBE InGaAs (6). However, the OMCVD densities in Figure 4 are also identical in the 110 and $1\bar{1}0$ directions to within the expected statistical variation. Since this material is also n type (but S and not Si doped), the carrier concentration is not the cause of the asymmetry; it may be that the other properties unique to the Si dopant lead to the difference. Another possibility is that the misfit nucleation source varies from one set of experiments to another, depending on the substrate EPD and the surface particle density, for instance.

Several wafers, etched to various depths, were prepared to find precisely how deep a trench must be to prevent misfits from gliding from one square to an adjacent one. Table 1 below lists the results. Note that the transition from isolation to non-isolation occurs at a value much less than the total epitaxial layer thicknesses (350 nm buffer and 350 nm InGaAs) and at about half of the strained layer thickness. It is apparently unnecessary to form a discontinuous epitaxial layer to ensure isolation (this point will be brought up again below). Others have come to a similar conclusion in patterned Si substrates with mismatched SiGe and strained layer mismatch. The buffer layer controls the trench profile, a thicker layer probably enhances the sidewall profile change. The strained layer thickness and composition control the stress applied misfit gliding segment. A higher stress should increase the probability that misfits can cross a trench.

CONCLUSIONS

A significant reduction in the density of misfit dislocations is observed in strained, OMCVD grown InGaAs epitaxial layers that are in excess of the critical thickness when grown on patterned substrates. Misfit dislocation densities are as low as 350 cm^{-1} in material that has been grown on 100 μm squares.

Unlike n doped MBE GaAs/InGaAs epitaxial layers, n type OMCVD material possesses equal densities of misfit dislocations in the 110 and $1\bar{1}0$ directions. Since undoped MBE material also has equal densities, the growth technique alone is not the reason for this difference. Since n type OMCVD has equal densities and n type MBE material has different densities, the Fermi level alone cannot explain the discrepancy either. This effect is currently being studied further.

The density of misfits in OMCVD InGaAs increases linearly with the size of isolated squares up to dimensions of 800 μm. Because of the linearity, dislocation multiplication is an unimportant effect.

Although SEM profiles show that the 350 nm InGaAs strained layer is continuously connected across trenches, misfits can be stopped by these trenches. The minimum trench thickness is smaller than the epitaxial layers themselves (about 160 nm deep), so that isolation can be achieved without hindering subsequent device processing.

ACKNOWLEDGEMENTS

This research has been sponsored by the Department of Energy under contract DOE-FG02086ER45278 and by the Defense Advanced Research Project Agency under Navy grant no. MDA 972-88-J-1006. The microscope facility is operated by the Materials Science Center at Cornell under a grant from the National Science Foundation. We also would like to acknowledge the use of the National Nanofabrication Facility at Cornell, which is supported in part by the National Science Foundation. One of us (GPW) is supported by a fellowship from the Department of Defense.

REFERENCES

1 E. A. Fitzgerald, G. P. Watson, R. E. Proano, D. G. Ast, P. D. Kirchner, G. D. Pettit, and J. M. Woodall, J. Appl. Phys. **65**, 2220 (1989).
2 J. W. Matthews, S. Mader, and T. B. Light, J. Appl. Phys. **41**, 3800 (1970).
3 G. P. Watson, D. G. Ast, T. Anderson, and Y. Hayakawa, unpublished.
4 K. L. Kavanagh, M. A. Capano, L. W. Hobbs, J. C. Barbour, P. M. J. Maree, W. Schaff, J. W. Mayer, D. Pettit, J. M. Woodall, J. A. Stroscio, and R. M. Feenstra, J. Appl. Phys. **64**, 4843 (1988).
5 B. C. De Cooman and C. B. Carter, Philos. Mag. A, **60**, 245 (1989).
6 C. Herbeaux, J. Di Persio, and A. Lefebvre, Philos. Mag. Lett., **59**, 243 (1989).
7 E. A. Fitzgerald, Private Communication.

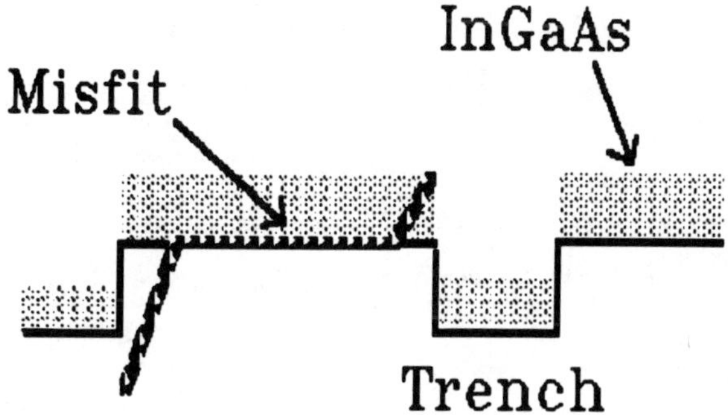

Figure 1. A schematic drawing of a misfit dislocation forming on a patterned substrate.

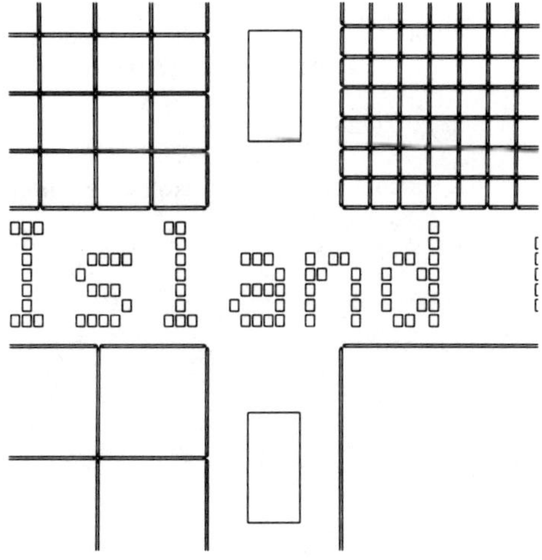

Figure 2. A portion of the photomask pattern used to study misfit densities.

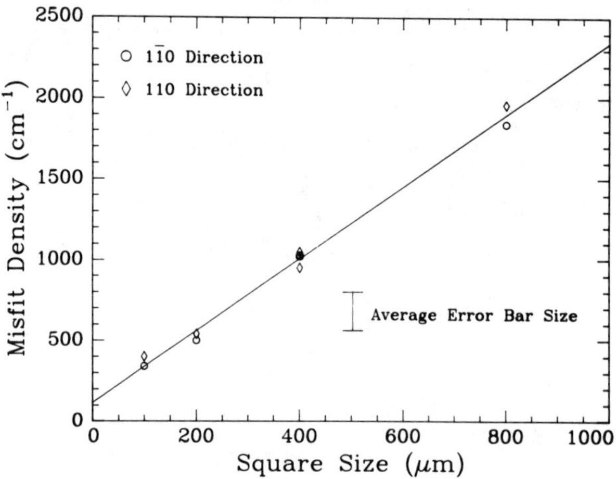

Figure 4. A plot of misfit dislocation density vs. isolating square size for the material shown in Figure 3.

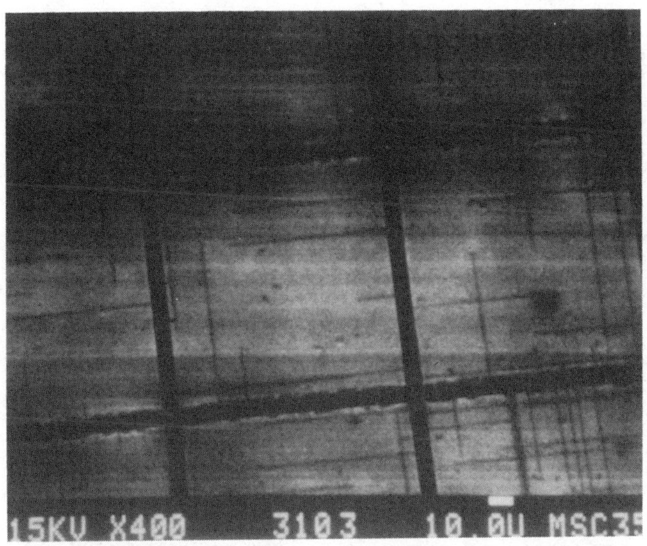

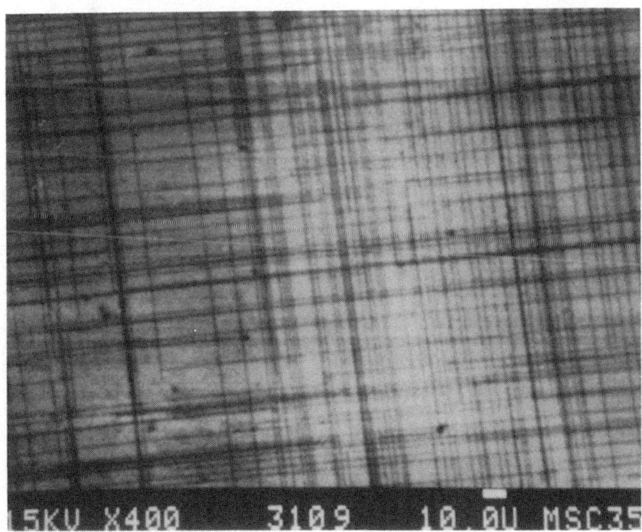

Figure 3. A CL image of misfits formed in 100 μm squares (a), and 800 μm squares (b). The density is significantly lower in (a).

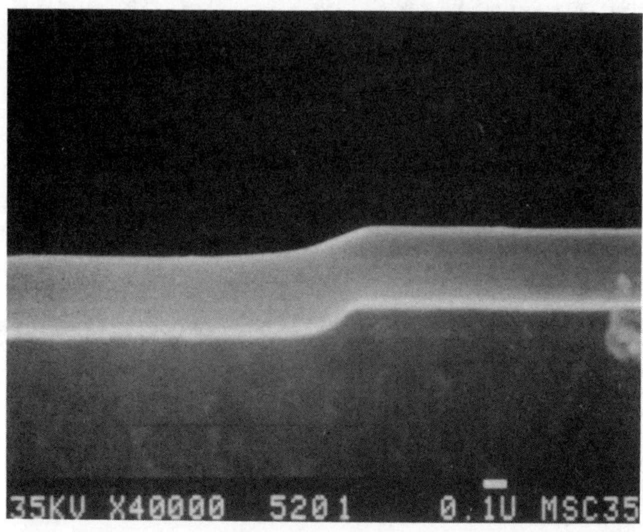

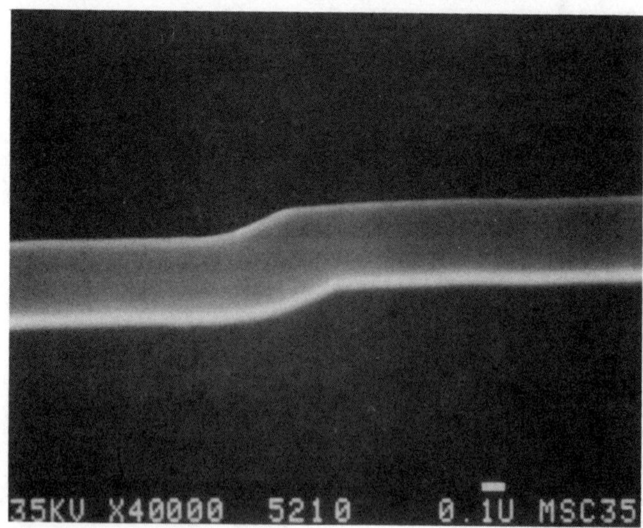

Figure 5. Cross sectional SEM profiles of the InGaAs epitaxial layer (the bright band of material) in the 110 direction (a) and the $1\bar{1}0$ direction (b). The trench is 110 *nm* deep and did not prevent misfits from gliding across.

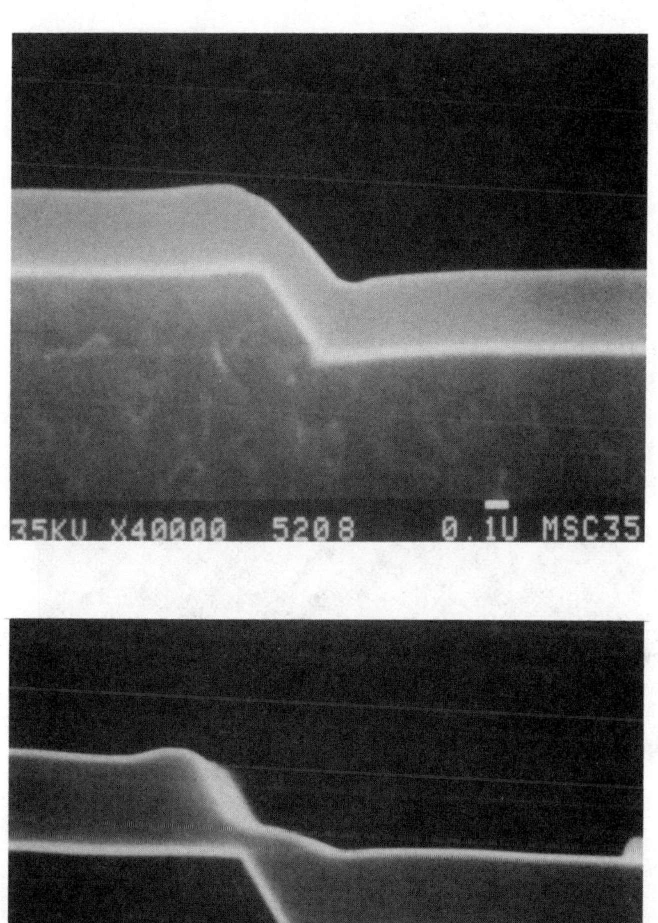

Figure 6. Cross sectional SEM profiles of InGaAs in 350 nm deep trenches. Misfits were blocked in both the 110 (a) and $1\bar{1}0$ (b) directions.

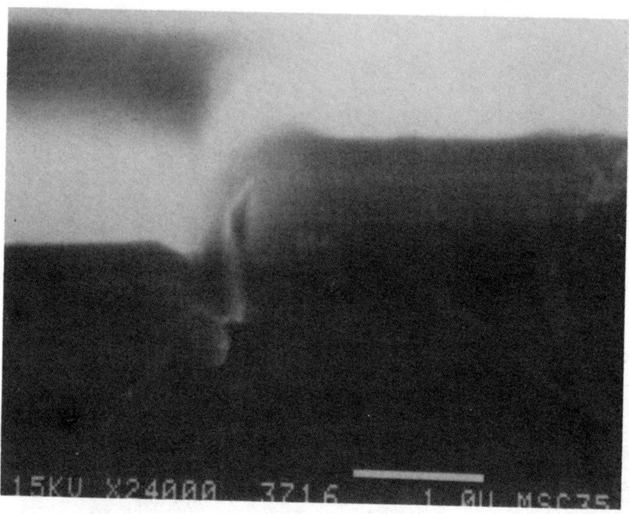

Figure 7. Cross sectional SEM profile of MBE grown InGaAs on a 1.5 μm deep etched wafer. The individual layers are from top to bottom: SiO_x coating, 350 nm $In_{0.05}Ga_{0.95}As$, 350 nm GaAs buffer, and the substrate. The InGaAs layer is discontinuous at the trench edge.

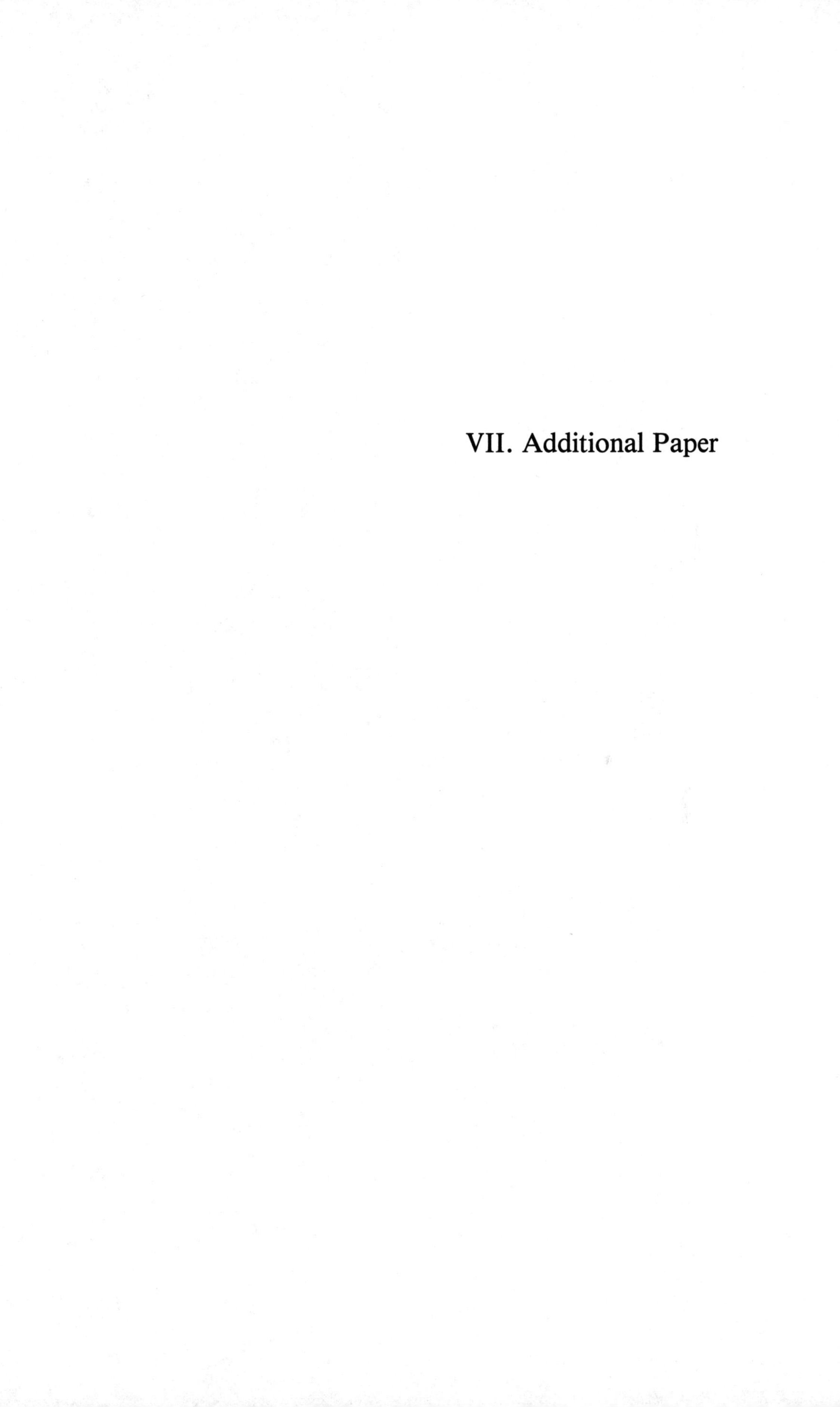

VII. Additional Paper

Micro- and macro-segregation during alloy solidification

A. Hellawell
Department of Metallurgical and Materials Engineering,
Michigan Technological University, Houghton, Michigan, U.S.A.

Introduction

To attempt a comprehensive review of this subject area, with all the ramifications and consequences associated with grain size and distribution is beyond the scope of this presentation. Therefore, to be more selective, there are some general comments about the application of microsegregation analyses, inasmuch as they can be used to model solidification behavior, followed by a discussion of those aspects of macrosegregation which are gravity induced.

Evidently, in solidification processing, as in most fields of activity, there is a current preoccupation with modeling, and if this can be done in a physically realistic and predictive manner for castings, the effort has revolutionary potential for quality control, automation, and innovative design (1,2). This idea is not altogether new and was pioneered for eutectic cast irons by Oldfield (3) some thirty years ago. At present, it is not obvious that predictive modeling of the overall process will be possible but some aspects of the problem are further advanced than others.

Thus there have been model analyses of heterogeneous nucleation processes during initial recalescence (4) and predictions of the columnar to equiaxed transition for simple binary alloys (e.g. 5-7), and most recently, attempts to combine heat flow analyses with nucleation and growth for alloys containing primary and eutectic constituents (e.g. 8-9). As regards analyses of microsegregation, these have probably been advanced to a level beyond our capabilities in handling other aspects of the process as a whole.

I. Microsegregation and Thermal Analysis

The basic equations for fractional crystallization are essentially textbook material; (e.g. 10,11) it is their application to real foundry conditions which requires attention.

With the assumption that the scale of solid + liquid phase distribution during solidification is too coarse to allow significant diffusional mixing in the solid state, but is sufficiently fine to accommodate complete mixing in the liquid, a simple mass balance gives the well known relation attributed to Scheil (12):

$$C_L = C_o f_L^{(k_o-1)}$$

for the composition of a residual liquid fraction, f_L, during the solidification of an alloy, C_o, in a diagram such as Fig. 1. In systems where the solid solubility is negligible, the solid/liquid distribution coefficient, $k_o \rightarrow 0$, and the expression becomes:

$$C_L = C_o/f_L$$

which is the same as the lever rule. In practice, of course, k_o is not exactly a constant, so that with extended solid solubility, using a constant k_o, the Scheil equation becomes increasingly inaccurate at high solid fractions, $f_s = 1 - f_L$, inevitably predicting C_s or $C_L \to \infty$ as $f_L \to 0$. However, if variations of k_o with composition are included, the equation conveniently expresses expected solid/liquid composition traces with respect to some local reference volume and coordinates within a casting, predicting the occurrence or fraction of eutectic or peritectic phases which literal interpretation of the equilibrium diagram cannot. Clearly, also, the relationship can be further developed by introducing density variations with composition and between phases, allowing for expression of the relative volumes to be expected in a microstructure. This also includes the volume change between the liquid and solid phases and the attendant capillarity flow which this causes, of which inverse segregation is an extreme example.

The difference between the Scheil and lever rule predictions then appear as in Fig. 2 as C_s or C_L vs. $f_s = (1 - f_L)$, plotted in this case for a solid solution, terminating at a eutectic reaction. It is important to remember that for the lever rule, the solid composition, C_s, is that of the total, whereas for the Scheil plot, C_s is the local equilibrium solid in contact with liquid of uniform composition at any f_L or equivalent temperature. It is also common practice to plot the mean solid concentration for the Scheil prediction, $\bar{C}_s$, (see Fig. 1) on a phase diagram, in such a way as to express a non-equilibrium solidus which can be used as the basis for a modified lever rule balance where $\bar{C}_s$ only approaches C_o in the limit.

We then recognize, that the basic assumptions of no solid mixing, or of complete liquid mixing, implicit in the Scheil equation may need some relaxation and qualification, leading to segregation profiles intermediate between the two extremes of Fig. 2. Thus, either there may be only partial liquid mixing, or, more probably, there may be some back diffusion in the solid. In the context of microsegregation it is generally unnecessary to consider the case for limited liquid mixing, because the dimensions of local solidification volumes are typically well within $\sqrt{D_L t}$ (D_L = liquid diffusion coefficient and t is local solidification time - i.e. that time taken to cool from the liquidus to the final freezing point). Back diffusion in the solid, however, becomes a more serious problem, especially for alloys containing interstitial solutes - e.g. Fe-C. There have been semi-empirical (13,14) and more rigorous solutions to such back diffusion (15,16) which fit experimental microscopical and EMPA or other data with varying degrees of conformity (17,18,19).

It is pertinent to consider how close a fit to experimental results might reasonably be expected for these modified analyses.

Firstly, let us consider that EMPA data are based on multiple line trace analyses across microsections. However thorough or selective such data may be, they are still only line traces through local volumes which have solidified in three dimensions with continuously (or continually) changing geometrical phase distributions. The problem may be placed within any preferred reference coordinates, cubic or otherwise, but initially, the solidification consists of the thickening of dendrites, essentially the outward growth of cylindrical shapes, and then as these make contact there will be rapid adjustments of the morphology as the growth changes to radially inwards towards the centre of residual liquid droplets within a solid matrix (20,21). The Scheil equation, however modified, can only describe the

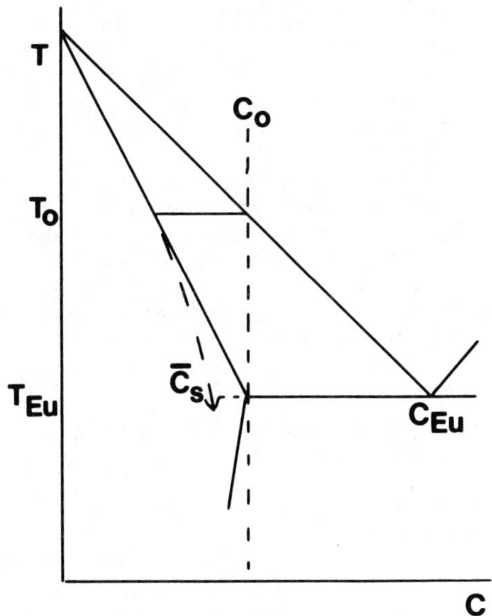

Fig. 1. Partial eutectic phase diagram with compositions and temperatures as in text and Figs. 2 and 3.

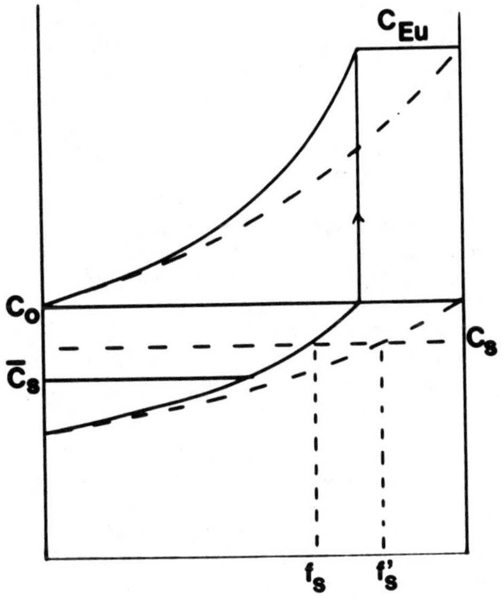

Fig. 2. Composition traces vs. fraction solid, f_s, for solidification of alloy, C_o, in Fig. 1, full lines according to Scheil, broken lines according to lever rule. $\bar{C}_s$ is the mean solid fraction, f_s and local composition, C_s; f's is the corresponding lever rule fraction.

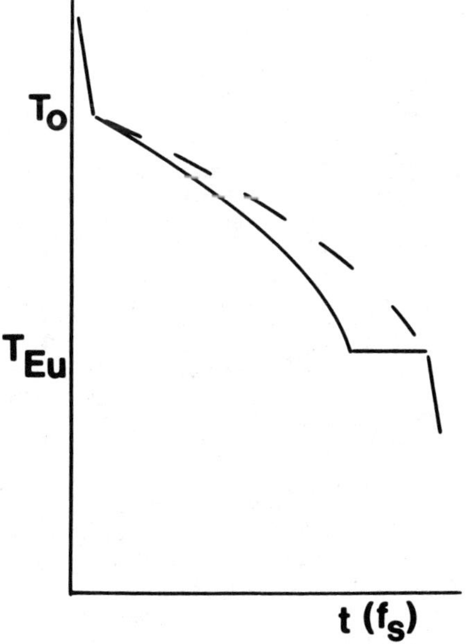

Fig. 3. Schematic cooling curves for a uniform rate of heat abstraction corresponding to alloy C_o, Fig. 1, full line according to Scheil, broken line according to lever rule.

relative masses or volumes of solid and liquid and cannot be expected, on a linear trace, to conform exactly with a changing geometrical distribution, which at longer times is also subject to coarsening and ripening adjustments. Unless this geometrical complexity can be adequately accounted for, precise agreement between analysis and observation cannot reasonably be expected, so that it is somewhat doubtful if this type of experimental data can be used to assess the relative merits of finer modifications to the Scheil equation.

However, in foundry practice, while subsequent and often detailed microanalysis may follow in the laboratory, a more direct, on line control is possible and easily accomplished through thermal analysis, either of isolated melt samples, or more elaborately, on model castings using multiple thermocouples. Given a realistic heat flow analysis, this approach requires that the lever rule or Scheil equations be transformed to a temperature scale vs. phase proportion, using the slope of the liquidus line or trace, if this is established or can be estimated - thus, corresponding to the composition vs. f_s plots of Fig. 2 we have the temperature vs. f_s plots of Fig. 3, i.e. the equilibrium and Scheil equivalent cooling curves for a given rate of heat abstraction. Clearly, also, there can be a continuum of intermediate forms for varying degrees of solid mixing. This is the only obvious direct method available to test solidification models against observation, but even given the ability to calculate local heat flow conditions, there remain other serious problems or unknowns arising from the columnar: equiaxed transition and the nucleation density. These are important, not only inasmuch as they concern the final grain structure but because they determine local growth rates, hence interface shape and the size of representative, local solidification volumes, and hence, the applicability of preferred segregation analyses. Arguments become circular and it is necessary to make arbitrary assumptions about the nucleation kinetics in order to "fit" analyses to observed cooling curves. The predictive capability is then sadly lacking unless nucleation sites can be identified and their surface potentials be quantified. At the present time this capability is generally lacking, except in a few cases where specific grain refining additions appear to operate, and even in these cases representation of the potency of surfaces is open to argument, e.g. for aluminum base alloys with Ti, B, etc. (22). Instead of attempting this imponderable problem, some progress has been made at ‘curve fitting’ to cooling curves of binary alloys, assuming the presence of unspecified nuclei of variable particle density and nucleation potencies, the latter expressed in terms of maximum undercoolings prior to recalescence (8,9). Examples of such comparisons appear in Fig. 4 and relate to (a), hypoeutectic Al-Si alloy and (b), to a cast iron. Both alloys contain primary dendrites and eutectic. The agreement between calculation and observation is similar in either case, but I suspect that by Weinberg standards they would not be hailed as entirely successful, but one might be optimistic and reflect that a start must be made somewhere.

In the absence of specific, identifiable nucleating substrates, the most probable and spontaneously efficient nucleation sites are from particles of the primary solid phase, either generated at low superheats on mold surfaces - so-called ‘big bang’ nucleation (23) - or arising by fragmentation of primary dendrites from the columnar zone at larger superheats (24); in the latter case, a mechanism must be considered for the transport of such fragments into the bulk liquid, where they may develop into equiaxed grains if the thermal conditions allow. Truly predictive modeling would require that such sources and the transport of crystal fragments be not only identified, but also quantified, and this may be the major obstacle to further useful progress. However, in the absence of induced or forced stirring of a melt, natural convection must provide the source of possible transport into and through the melt and it becomes immediately necessary to examine such convective patterns and to recognize that they also determine solute redistribution on a macroscopic scale.

Aspects of Natural Convection and Macrosegregation

In any solidifying (or melting) material which exhibits a freezing range, liquid density gradients arise from both thermal and solutal gradients and interaction between these can lead to so-called thermo-solutal convective patterns (e.g. 25-30). Thus with respect to some reference axis or axes, there are thermal and composition gradients, dT/dz and dc/dz, which have associated density gradients, $d\rho/dz$, the magnitudes and senses of which depend upon the corresponding coefficients, $\alpha = d\rho/dT$ and $\beta = d\rho/dc$. α is always positive but β may be of either sign. Only for certain compositions in multi-component systems will $\beta \rightarrow 0$, in almost every binary system β is finite and is often such as to exceed α by an order of magnitude.

Since β may be of either sign, it is relevant to consider the geometrical configuration, i.e., whether growth is vertically upwards, antiparallel to gravity, $_{\uparrow}{}^{\downarrow}$, downwards, parallel, $\downarrow\downarrow$, or horizontal, i.e. at right angles to the gravitational vector, $\leftarrow\downarrow\rightarrow$. Schematically, Fig. 5, with a developed columnar mushy zone, these alternatives then allow the thermo-solutal contributions to combine or to oppose each other, depending on the sense of the g vector with respect to that of the temperature gradient, and upon the sign (and magnitude) of the solutal coefficient, β.

In a metallurgical context, identification of the buoyancy forces arising from solutal gradients came from directional growth studies (31) and in analogue castings (32,33) some twenty years ago, although it was recognized about the same time in oceanographic contexts (25,28,29) where it causes so-called thermo-haline convection effects. It is a generic effect applying to a wider range of natural phenomena, including geological situations (34).

With constrained growth into positive temperature gradients, composition variations occur between the bulk and interdendritic liquids and at the dendritic growth front itself: the corresponding density gradients then interact with the thermal density variations as shown, Fig. 5(a)-(f). Thus, with horizontal growth (a), β negative, there is a buoyancy force tending to cause flow upwards and with β positive, (b), a reverse tendency, such that in each case, excess solute accumulates at the top or bottom of the ingot, and, with depression of the freezing point, $k_o < 1$, this distorts the growth front accordingly, as indicated. With vertical growth upwards, (c) and (d), the solutal contribution is opposed to the stabilizing thermal field, with β-negative, or reinforces it and maintains a quiescent liquid above and below the growth front, (d). The situation in case (c) is of particular interest because the incidence of convection is not spontaneous, even if the destabilizing solutal density gradient exceeds that from the temperature gradient. this situation has received much experimental and theoretical attention. Lastly, in (e) and (f), the bulk liquid is continuously mixed by the inverted thermal field to which the solutal contribution may be either opposed, (e), or will reinforce such convection, (f), as more dense solute seeks to flow downwards into the bulk liquid.

Convective flow is continuous and circular, i.e., that which goes up must come down since fluids are effectively incomprehensible, although they expand with temperature - a so-called Businesque fluid. Within and without the dendritic array of a mushy zone, the continuity of convective flow is naturally accomplished by the formation of localized channel patterns. In the bulk liquid these localized flows exist as streamlined, solute rich plumes, ~ 1 mm wide and $\geq 10^2$ mm long, emanating from the growth front. Evidence of their existence remains behind in the fully solidified material as channel segregates - well known casting defects, 'A' segregates in steel billets (e.g. 10,35,36) or 'freckles' in directionally solidified E.S.R. ingots or

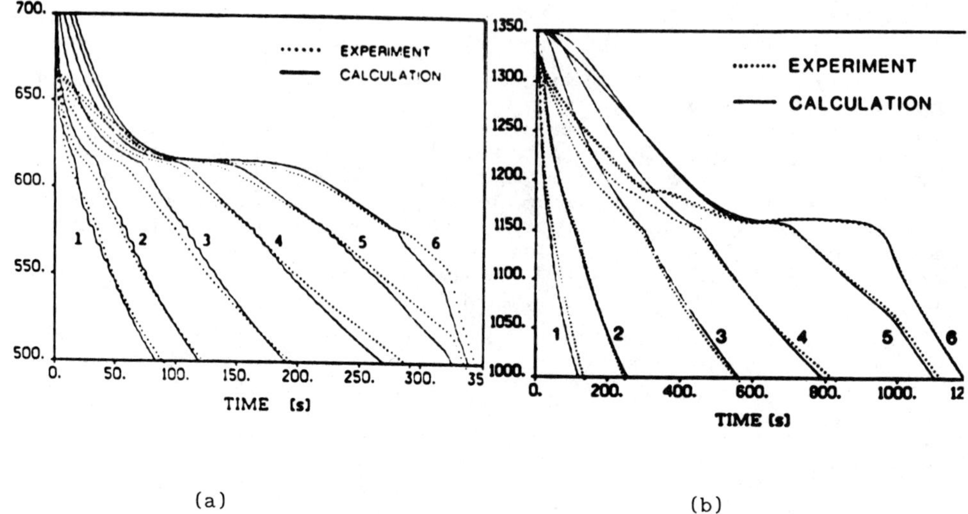

Fig. 4. Calculated and experimental cooling curves for six position in samples of (a) Al-7 wt.% Si and (b) hypoeutectic cast iron, after refs. 8 and 9.

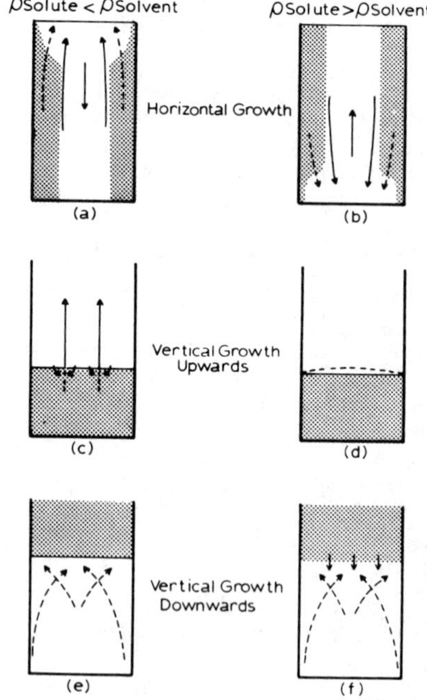

Fig. 5. Schematic convective patterns during alloy solidification with shaded 'mushy zones'. (a) growth normal to gravitational vector, g, ←↓→, $\Delta\rho-$; (b) the same with $\Delta\rho+$; (c) growth upwards, antiparallel to g, ↑↓, $\Delta\rho-$; (d) the same with $\Delta\rho+$; (e) growth downwards, parallel to g, ↓↓, $\Delta\rho-$ and (f) the same with $\Delta\rho+$.

superalloy turbine blades (37). Such channels can be found in the configurations (a), (b), (c), and (f) of Fig. 6 and they certainly offer one mechanism by which dendrite crystal fragments are transported from the mushy zone into the bulk liquid. However, in billet castings, (a) and (b), and with downwards growth, (f), such patterns are combined with other bulk fluid flows and are more difficult to study. That is why most attention has focused upon the particular configuration of Fig. 5(c), where interdendritic liquid exhibits density inversion but is surmounted by quiescent bulk liquid which is stabilized against convection by the positive temperature gradient. It is the manner in which this less dense, interdendritic liquid escapes into the more dense bulk liquid which is particularly interesting and instructive, causing a special type of thermo-solutal convection.

Figure 6 summarizes this situation. An alloy of composition C_o solidifies vertically upwards into a positive temperature gradient, dT/dz. At the dendritic front there is a slightly enriched boundary layer at C_L, depositing solid C_D, while between the dendrites the liquid is in local equilibrium with the solid according to the temperature and composition along the liquidus line. If $\beta \times dc/dz$ exceeds $\alpha \times dT/dz$, along the liquidus, the density gradient below the growth front is positive, i.e. the system exhibits density inversion and is inherently unstable. However, although this positive gradient often exceeds the negative, thermally controlled density gradient above, perturbations leading to some form of convection do not occur spontaneously because of a finite viscosity and the thermal inertia of the system. We note that $D_{thermal} >> D_{solute}$, so that while a perturbation may be able to achieve near thermal equilibrium in a short time, solute diffusion is so slow that there is insufficient time for composition adjustment. It has then been of recent interest to identify where and when these convective channels develop and, subsequently, to understand, what determines their flow rates and spacings. The following relates to the vertical case of Fig. 5(c).

(i) Position of Origin

The choice here is from a site within the partially solid mushy zone or elsewhere, such as at the growth front or above that front in the bulk liquid. Although analyses have been formulated for the former origin (38-40), experimental evidence (41-42) suggests that the convection starts at or immediately ahead of the dendritic front. The observations which indicate this site preference are (a), direct studies of the transparent aqueous system NH_4Cl-H_2O, which reveal that channels within the mushy zone originate from local disturbance of liquid close to the front and then spread back (i.e. downwards) into the partially solid region; (b) artificial creation of channels, by boring holes into the mush region, does not, per se, lead to self perpetuating channel plumes, but, the artificial creation of upward plumes in the bulk liquid, ahead of the growth front does result in the formation of coincident channels and self sustained plume flows; while, (c), mold movements which cause the bulk liquid to sweep across the front will inhibit the development of local convective flows or stop any which have developed. These experimental procedures are summarized in Fig. 7(a)-(c).

In retrospect, it is obvious that channel mouths and established plumes are part of a continuous flow pattern and the one cannot exist without the other, but the most probable level at which these will originate is where the sign of the density gradient reverses. Without perturbation at this level, fluid movements at greater depths cannot be released; it is at this level that the less dense liquid needs to break through into the quiescent, supernatant bulk. In a sense, this level is an interface, although a diffuse one.

There is presently no formal analysis of this break away of the boundary layer fluid. The positions on the front have not been seen to be associated

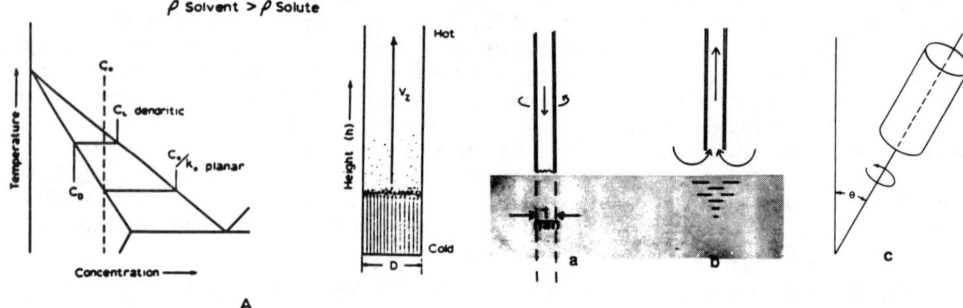

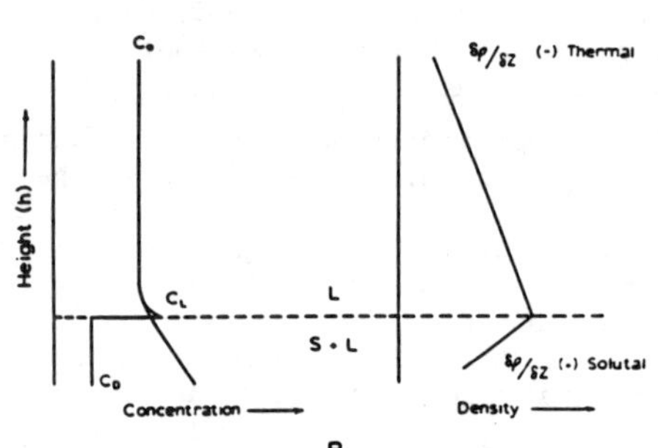

Fig. 6. (A) Conditions as for Fig. 5(c), initial concentration C_o, local liquid concentration at dendrite tips, C_L, (B) expected liquid concentrations above the growth front, $C_L \rightarrow C_o$, and below the front between primary dendrites of solid composition C_D.

Fig. 7. Schematic representation of experiments relating to origin of channels, (a) artificial drill of a channel, (b) artificial creation of a plume above the growth front and (c) mode of precessional movement which eliminates channels and plumes.

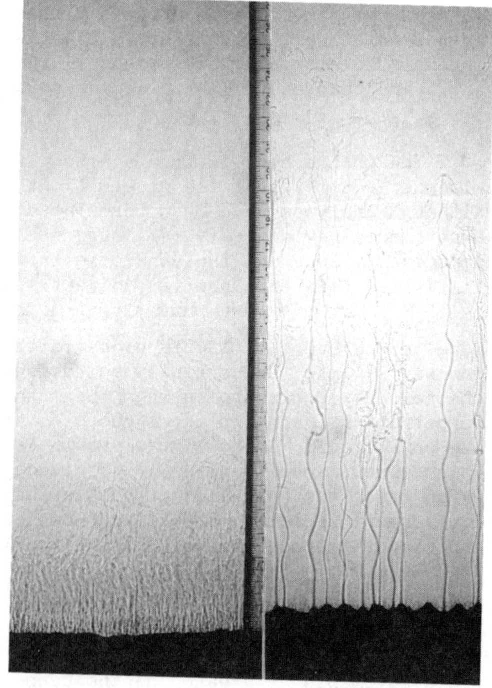

Fig. 8. 'Finger' and 'plume' convection patterns above an aqueous NH_4Cl growth front (courtesy of C. F. Chen).

with defects in the dendritic array, such as grain boundaries. A simple minded analysis of the quantities in a critical Rayleigh number (43) show that if this number were unity, the relevant dimension in the equation, in both metallic and aqueous systems, could be the primary interdendritic spacing. Indeed, since the liquid perturbation is occurring at such a periodic front, this coincidence might well be expected.

For a given alloy, in a given system, the primary dendritic spacing is a function of the imposed temperature gradient, G, and the growth rate, V (approximately proportional to $G^{-1/2}$ $V^{-1/4}$ (e.g. 11)) and while the former also influences the buoyancy gradients, it might be expected, and is qualitatively observed (37,44), that steeper temperature gradients and higher growth rates, which refine the primary spacing, also reduce or inhibit the incidence of segregation channels.

The incidence of channels is also composition dependent, inasmuch as this influences the liquid fraction and permeability of the mushy region; there appears to be a lower composition/permeability limit below which channels do not occur, for a given buoyancy pressure and Prandtl number, and an upper level where single channels and plumes are not observed. Above these upper levels, individual plumes and channels are replaced by a regime of smaller fluctuating 'fingers' on a similar scale to that of the primary dendritic spacing, Fig. 8. The transitional coupling between initial 'finger' perturbations and well defined, wider, quasi-steady state plumes is not well understood. At present, we are not able to predict, except in a very qualitative manner, when long range channel flows will develop.

(ii) Channel Plumes and Their Dimensions

As previously noted, the propagation of quasi-steady-state channel plumes is separate and subsequent to the initial perturbation. There are (at least) six relevant and interrelated, first order quantities involved in this problem; these are listed in Table I, although not necessarily in any selective order.

To these, it may also be relevant to consider second order quantities, including (a) temperature gradient, dT/dz, (b) growth rate, (c) height of mushy zone (d) height of meniscus above the latter and (e) width of container.

Many of these quantities have been measured (Table II) (43,44), including systems from materials having significantly different physical properties. These properties are conveniently defined by the dimensionless Prandtl and Lewis numbers, σ and τ. The systems are metallic Pb-Sb, Pb-Sn and ternary Pb-Sb-Sn, aqueous NH_4Cl and organic succinonitrile (SCN)-ethanol. Some of the quantities cannot be measured in all cases, notably the flow rates in opaque metals, and indeed, it is one of the objects of the research to estimate this quantity by extrapolation from other transparent systems. There are also some uncertainties or margins of error (e.g. $\pm$ 10%) in some of the measurements. Thus, precise channel or plume "widths" need to be defined and measured flow rates in the transparent materials are not necessarily maximum axial rates. Moreover, channel/plume compositions are averages, based on EMPA (Pb-base) or direct sampling and refractive index measurements for the aqueous and organic materials.

The outstanding question which then arises is, "how does the system select these quantities"?

One approach is to examine the plume flow in order to understand how the width, velocity and buoyancy forces are tied together.

The plume flow may be first considered as flow along a tube without constraints at either end. If a 'no slip' condition applied at the tube

TABLE I
Relevant Data

(i) Channel width (to be defined) - radius R_o
(ii) Flow rate (mean or maximum) - w
(iii) ΔC (plume vs. bulk) vs. time and position
(iv) ΔT (plume vs. bulk) vs. time and position
(v) Mean spacing, L
(vi) Permeability of mushy zone (primary and secondary spacings, λ, and fraction liquid, f_L)

TABLE II
Present Data Available

	Metallic	Aqueous	Organic
σ	$\sim 2\ 10^{-2}$	~ 7	~ 30
τ	$\sim 3\ 10^{3}$	$\sim 10^{2}$	$\lesssim 10^{2}$
Width	√	√	√
Velocity	*	√	√
Composition	√	√	√
Temperature	*	√	√
Spacing	√	√	√
Permeability	√	√	√

*No measurements

walls, flow along it would adopt the usual parabolic velocity distribution as in Poiseuille. However, this is not exactly the case because the 'walls' do not have a specific interface but rather represent a region of rapid shear, liquid without the tube being dragged upwards by the internal buoyancy, therefore having an exponential, radial velocity distribution around the plume. At some radius, the internal and external velocity distributions coincide at a common dw/dR Fig. 9; this is one way of identifying the effective plume width, i.e. where $d^2w/dR^2 = 0$. In a transparent material this is the position where the refractive index changes most rapidly and corresponds to a 'width' which is detected optically. The velocity distribution then fits a modified parabolic distribution of the basic form:

$$w = \frac{P R_o^2}{2h\eta}$$

where R_o is the effective plume radius, h is the plume height, P is pressure, of the form Δρ g h, and η is the dynamic viscosity coefficient. It will be seen that the density difference, Δρ is then given by β ρ ΔC, involving the solutal coefficient, β, and the (mean) composition difference, ΔC, between plume and surrounding bulk liquids. This buoyancy pressure may be reduced if the plume liquid temperature remains below that of the bulk (which it does by 0 - 4K) by a term α ρ ΔT.

Combining these contributions for the aqueous system on a plot of w vs. R, Fig. 10(a) and (b), then shows that the parabolic function passes through the small scatter of experimental results which lie around $w \approx 7$ mm s^{-1} at $R \approx 0.6$ mm. Similar agreement is obtained for the organic data, and extrapolating the model for measured ΔC and R values would then indicate extremely rapid metallic plume flow rates exceeding 100 mm s^{-1}!

But, we still lack a criterion for this selection process. Two possibilities might be considered: one, that the plume flow be as rapid as is

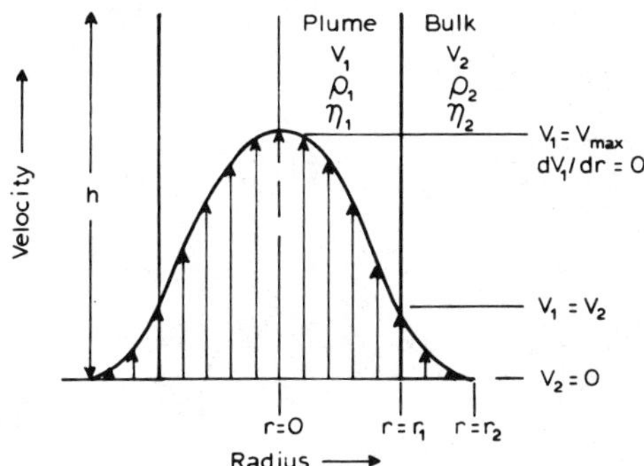

Fig. 9. Schematic velocity profile across a convective plume - refer to text.

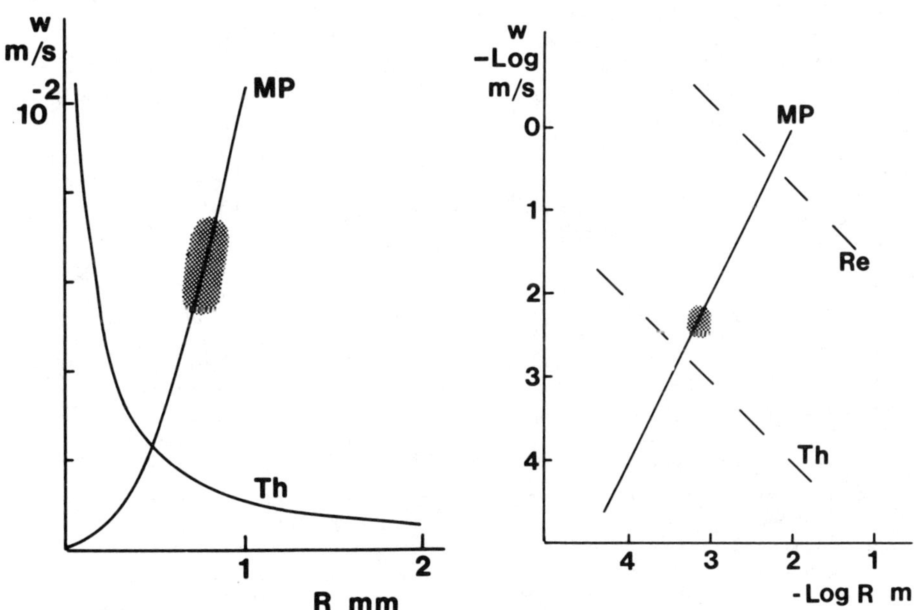

Fig. 10. (a) plot of plume flow velocity, W, vs. radius, R, shaded area includes experimental observations, MP is predicted by a modified Poiseuille equation, Th is a limit for thermal equilibrium, $W = \kappa/R$, (b) corresponding logarithmic plot including turbulence limit for Reynolds #, Re = 1000, $W = Re/R$. Data for NH_4Cl-H_2O.

consistent with streamlined flow - i.e. lies within the onset of turbulence, or, two, that flow occurs at such a rate, along plumes of such a dimension, as to allow the plume liquid to continuously adjust to the temperature of the surrounding bulk liquid. This is the condition which applies to a steady state perturbation in the first place.

For the first of these alternatives, the flow rate, tube radius and kinematic viscosity, ν, should fall within the limits of a critical Reynold's number:

$$Re = w\,R/\nu$$

where ν is the kinematic viscosity, η/ρ. The transition from laminar to turbulent flow is known to occur at a critical value of $Re \tilde{>} 10^3$, so that we may write:

$$w = Re\,\nu/R$$

and trace this upon the same plot of w vs. R in Fig. 10(a). Inserting data for the aqueous solution yields a turbulence limit some three orders of magnitude greater than experimental data, so that it can only be compared on a logarithmic scale, Fig. 10(b). This cannot be a relevant consideration and the observed rates and dimensions are entirely within the streamline regime. The same is true for the organic example and even in the metallic case the turbulent regime would still lie beyond our extrapolated value of velocity, by a factor of ~ x 5.

The second criterion, for thermal equilibrium, would require that the plume radius were close to κ/w, or $w \approx \kappa/R$. This limit also appears on Figs. 11(a) and (b), and as may be seen, it falls below the observed range by about an order of magnitude. This means that the plume flow is not only much too rapid for there to be any composition change, but that it is also too fast to allow thermal equilibrium to be achieved: fine thermocouples, moved through plumes, confirm this conclusion.

Insertion of data for the metallic and organic materials leads to the same conclusion, although in the metal we have no confirmatory temperature measurements - it is virtually impossible to find a plume in an opaque liquid, except by accident and inference. Neither criterion seems to offer a controlling limit.

It is pertinent to reexamine the previous assumption that the plume flow is unconstrained at either end, in particular, the limits on supply of entrained liquid around the channel which exits at the base of a plume, Fig. 11. This is a geometrical problem which is imposed by the fluid viscosity and the permeability of the mushy zone. These control the volume of fluid which can be drawn through (be entrained into) the mushy region surrounding a channel exit, at and below the growth front, i.e., the base of the plume is effectively embedded in a filter.

For a plume to be able to operate it is necessary to consider the number, N, of interdendritic pipes, of radius, r, which feed into the channel circumference, and to what depth below the front this supply is significant. It is also necessary to consider how long the pipes are and how rapid is the entrained flow rate along them. Estimates and measurements can be and have been made (e.g. 42,44). Clearly, there is a lower limit to the porosity/permeability of the dendrite mesh for a given buoyancy pressure, below which the necessary entrainment cannot be sustained. This, in a simplified form, can be expressed in terms of the fraction liquid in the mushy zone at some level, f_L, the channel radius, R, and its ratio to the primary interdendritic spacing, λ, this last dimension being the superimposed grid dimension to which all dimensions can be referred, i.e.

$$f_L \geq R^m/A\ \lambda^n$$

where A is geometrical term, $m \approx 3/2$ and $n \approx 1$. It will be understood that there is some latitude within the term A depending upon the choice of a realistic model for the entrainment pattern. The limiting fraction of liquid also applies at some depth below the growth front, related to the term, A. Since f_L is composition dependent, there is, understandably, a lower critical value for channel/plume formation in any system.

(iii) Other Configurations

We conclude, from the preceding, that the factors controlling perturbation of the boundary layer region are separate from those which allow the larger scale plume/channel flow to be sustained. There can be fine scale fingers without fully developed plumes, Fig. 8. The situation might be compared with conditions allowing plane front breakdown vs. those governing subsequent growth forms.

All this discussion was in the context of a configuration where growth occurs vertically upwards, c.f. Fig. 5 (c), with density inversion below the growth front. Some comments on the other alternatives are in order, particularly with side chill geometry and horizontal growth, as in billet castings, Fig. 5(a) or (b). There is also the problem of what might be expected with continuously changing inclinations, as during continuous casting operations.

With a vertical, or near vertical growth front the situation is modified by convection patterns in the bulk liquid and parallel to the front. These arise, in part, because the solute boundary layer at the front is more or less dense than the bulk and so causes a downward or upward stream which precedes channel formation and is probably more important as a means of transporting solute and crystal fragments within the melt. In the bulk liquid, at some distance from the solute boundary layer, the bulk liquid may also exhibit thermally induced convection patterns with downward flow at the sides, in opposition to the solutal convection for a less dense solute. Channels do develop however, as 'A' segregates and are not only restricted to the upward or downward facing parts of the front which reflect macroscopic solute accumulations. It is not altogether clear whether formation of these channels is assisted or damped by the bulk liquid movements. On the one hand, there is no boundary layer perturbation needed as there was in the previous case, $\uparrow\downarrow$, because the fluid ahead of the front is already being swept along in the same direction as any potential plume flow. On the other hand, we know that in that other configuration a circular or processional motion will inhibit or eliminate plume and channel formation.

Visual examination of transparent analogues is somewhat ambiguous in this respect, but one interesting feature does appear and seems to arise from the upward (or downward) flow over the dendritic front, causing rivulets or streams to develop in the plane of the front, Fig. 12. It is possible that channels develop from these grooves and/or that they become incorporated in the mushy zone and are partially or completely overgrown. This action will enhance the fragmentation of dendrites and be an efficient source of potential equiaxed grains. The effect is enhanced if the growth front is slightly inclined and overhangs the bulk liquid, so it may well assume more importance in continuous casting of steels as the strand bends away from the vertical.

The convective patterns are very sensitive to quite small inclinations away from the vertical - e.g. by as little as 5° - and can then change from symmetrical to asymmetrical distributions. Associated with such inclinations, in continuous steel castings around the peritectic range, notably but not exclusively with curved molds, the columnar:equiaxed grain distribution becomes asymmetric, such that the columnar length on the overhanging, inner

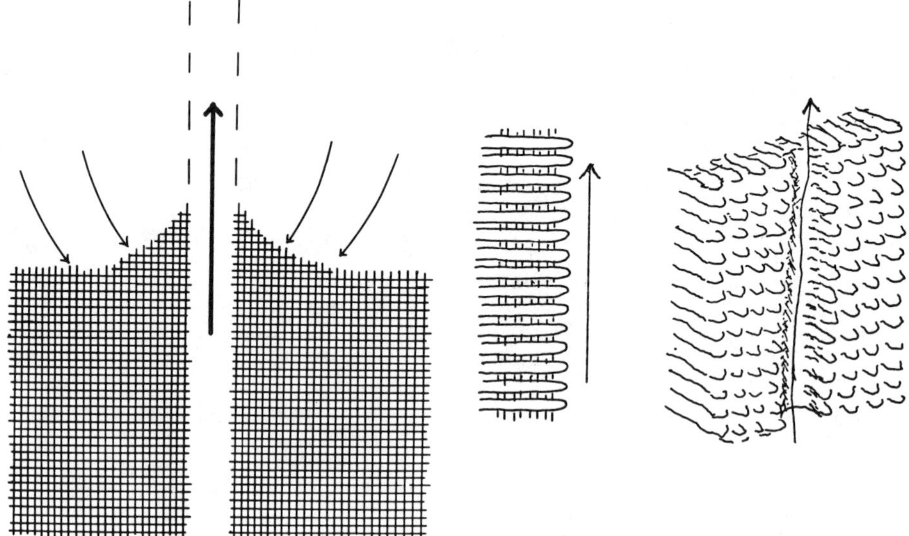

Fig. 11. Schematic representation of local growth front around channel mouth.

Fig. 12. Type of rivulet flow which can develop at a vertical dendritic growth front, $\Delta\rho-$, as in Fig. 5(a).

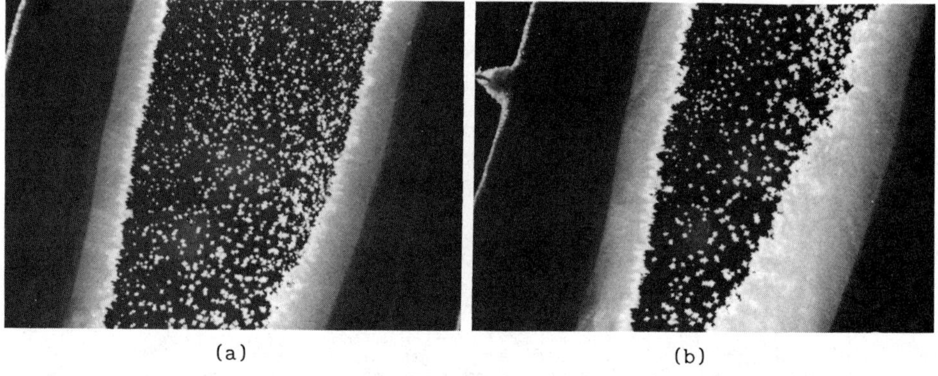

(a) (b)

Fig. 13 (a) and (b). Showing development of an asymmetric grain structure in an NH_4Cl-H_2O analogue casting with an inclined mold, time interval 4 minutes (courtesy of Jinsung Jang - to be published).

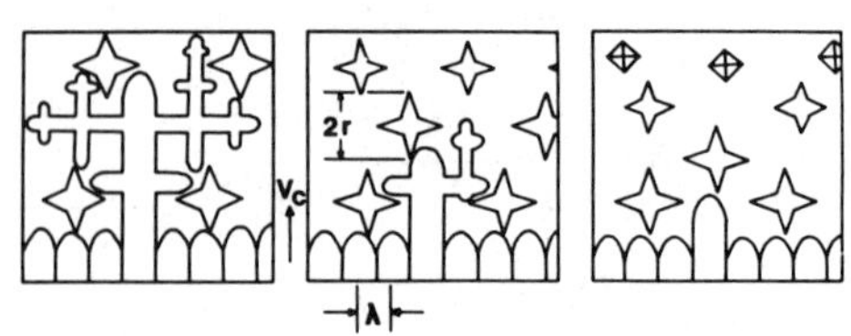

Fig. 14. Schematic representation of the blocking of a columnar front by falling equiaxed crystals.

side of the strand is longer than that on the lower, outer side (e.g. 46). This development has been reasonably attributed (47) to the formation of crystal fragments which grow and fall through the bulk liquid, away from the overhanging front eventually accumulating on the lower front and blocking further columnar growth. This can be visually demonstrated with an analogue casting of NH_4Cl-H_2O, as in Fig. 13(a)-(b). This raises one further problem for which we need a realistic model, namely, what criterion should be accepted for a columnar blocking mechanism? The problem requires consideration of the density of falling crystallites, and therefore their origin at higher levels in the casting, their growth and sedimentation rates and the rate of arrival at the dendritic columnar front, necessary to effectively block that front. There have been a number of models presented for this situation (48,49), essentially involving the arrival of sedimenting, equiaxed crystals upon the front, at such a rate that the columnar dendrites cannot grow between these obstacles in less time than another layer arrives. Fig. 14 depicts the geometrical configuration for a horizontal front, growing upwards. For an inclined, or indeed for a vertical front, further complications arise from a need to invoke a sticking coefficient or 'piling up' arrangement from some lower level, such as from the base of a billet casting or some critical level in a continuously cast strand.

Summary and Conclusions

Necessarily, this has been a rather selective review of various aspects of segregation during alloy solidification. The object was to identify some of the parts of the overall picture which are still in need of interpretation if the casting process is to be modeled in a realistic and predictive manner. Emphasis is laid upon 'predictive', as opposed to 'curve fitting', as of cooling curve data, although it is recognized that the latter practice may represent a necessary preliminary exercise. The following points were discussed in more or less detail:

(i) Analyses for fractional crystallization and solute redistribution during solidification have been developed in some detail for different extents of solid diffusion. These analyses can obviously be applied to describe microsegregation but the complexities of shape changes in the solid-liquid range make precise experimental confirmation difficult.

(ii) Given the necessary information about the relevant phase equilibria (solid-liquid distribution coefficients and liquidus data), coupled with realistic heat flow analysis, the solute mass balance can be transformed to a temperature scale in order to calculate a cooling curve.

(iii) This approach must be set in a context of a casting, incorporating a description of the columnar to equiaxed transition and of the nucleation process leading to the development of equiaxed grains. Information of this type is necessary to identify the scale of local solidification volumes within which the solute redistribution must be predicted.

(iv) The most probable and efficient sources of nuclei are fragments of the solid material, either swept off chill surfaces ('big bang') or transported from the columnar region by convection or other induced stirring. In the latter case it becomes necessary to consider how liquid density variations develop in temperature and composition gradients.

(v) Thermo-solutal convective patterns in the form of channel/plume flows were described, particularly as they develop at a dendritic growth front growing vertically upwards. The origin of these was discussed and the factors controlling their flow rates and dimensions were considered; it was concluded that the permeability of the mushy zone is the major controlling factor. Reference was made to observations of metallic, aqueous, and organic materials.

(vi) In other geometrical configurations with horizontal growth, or with continuously changing inclination to the vertical, bulk liquid convective movements probably override channel patterns as a means of transporting crystal nuclei.

(vii) To complete the picture of the overall process it is necessary to return to the need for a model to describe how falling equiaxed crystals block a dendritic columnar growth front in different configurations.

It will be necessary to develop physically realistic models for all these aspects if truly predictive analysis of ingot solidification is ever to be accomplished.

Acknowledgments

The description of channel convection was largely based on recent work by J. R. Sarazin at MTU, being part of a program supported by NASA, grant #NAG-3-560 and by the NSF Division of Materials Research, grant DMR-8815049.

References

1. Modelling of Casting and Welding Process I, (H. D. Brody, D. Apelian, eds.), The Metallurgical Soc. AIME, 1981.
2. Modelling of Casting and Welding Processes II (J. A. Dantzig, J. T. Berry, eds.) The Metallurgical Soc. AIME, 1984.
3. W. Oldfield, Trans. ASM, 59, 945 (1966).
4. I. Maxwell and A. Hellawell, Acta. Met., 23, 229 (1975).
5. W. C. Winegard and B. Chalmers, Trans. ASM., 46, 1214 (1953).
6. S. C. Flood and J. D. Hunt, Applied Science Res., 44, 27 (1987).
7. R. B. Mahapatra and F. Weinberg, Met. Trans. B, 18B, 425 (1987).
8,9. M. Rappaz and Ph. Thévoz, in Solidification Process, 1987, The Institute of Metals, Book #421, p. 164 and p. 168.
10. M. C. Flemings, Solidification Processing, McGraw-Hill, 1974.
11. W. Kurz and D. J. Fisher, Fundamentals of Solidification, Trans. Tech. Publications, Aedermannsdorf-Switzerland, 1986.
12. E. Scheil, Zeit. f. Met., 34, 70 (1942).
13. H. D. Brody and M. C. Flemings, Trans. Met. Soc. AIME, 236, 615 (1966).
14. T. W. Clyne and W. Kurz, Metall. Trans., 12A, 965 (1981).
15. M. Rappaz and V. Voller, Metall. Trans., 21A, 749 (1990).
16. K. S. Yeum, V. Laxmanan, and D. R. Poirier, Metall. Trans., 20A, 2847 (1989).
17. H. Thresh, M. Bergeron, F. Weinberg and R. K. Buhr, Trans. Met. Soc. AIME, 242, 863 (1968).
18. F. Weinberg and E. Teghtsoonian, Metall. Trans., 3, 93 (1972).
19. F. Weinberg, Trans. Met. Soc., AIME, 221, 844 (1961).
20. A. Hellawell, The Solidification of Metals, Iron and Steel Institute, London, ISI Publication 110, 83 (1968).
21. R. M. Sharp and A. Hellawell, J. Crystal Growth, 11, 77 (1971).
22. A. Hellawell, Solidification and Casting of Metals, The Metals Soc., London, Book 192, p. 161 (1979).
23. B. Chalmers, Principles of Solidification, Wiley, New York, 1964.
24. K. A. Jackson, J. D. Hunt, D. R. Uhlmann, and T. P. Seward, Trans. Met. Soc. AIME, 236, 19 (1966).
25. M. E. Stern, Tellus, 12, 172 (1960).
26. G. B. McFadden, R. G. Rehm, S. R. Coriell, W. Chuck and K. A. Morrish, Metall. Trans., 15A, 2125 (1984).
27. S. R. Corriell, M. R. Cordes, W. J. Boettinger, and R. F. Sekerka, J. Crystal Growth, 49, 13 (1980).
28. R. W. Schmitt, Phys. Fluids, 26, 2373 (1983).
29. J. S. Turner, Rev. Fluid Mechanics, 17, 11 (1985).
30. H. Huppert, J. Fluid Mechanics, 212, 209 (1990).

31. N. Streat and F. Weinberg, Metall. Trans., 5, 2539 (1974).
32. R. J. MacDonald and J. D. Hunt, Trans. Met. Soc. AIME, 245, 1993 (1969).
33. S. M. Copley, A. F. Giamei, S. M. Johnson, and M. F. Hornbecker, Metall. Trans., 1, 2193 (1970).
34. H. E. Huppert, J. Fluid Mech., 173, 557 (1986).
35. G. J. Davies, Solidification and Casting, Applied Science, London, 1973.
36. F. Weinberg, J. Lait, and R. Pugh, Solidification and Casting of Metals, The Metals Society, London, Book 192, p. 334 (1979).
37. A. F. Gaimei and B. H. Kear, Metall. Trans., 1, 2185 (1970).
38. M. R. Bridge, M. P. Stephenson, and J. Beech, Metals Technology, 9, 429 (1982).
39. V. R. Voller, J. J. Moore, and N. A. Shah, Metals Technology, 10, 81 (1983).
40. M. Simpson, M. Yerebakan, and M. C. Flemings, Metall. Trans., 16A, 1687 (1985).
41. A. K. Sample and A. Hellawell, Metall. Trans., 13B, 495 (1982).
42. A. K. Sample and A. Hellawell, Metall. Trans., 15A, 2163 (1984).
43. J. R. Sarazin and A. Hellawell, Metall. Trans, 19A, 1861 (1988).
44. J. R. Sarazin, Ph.D. Thesis, Mich. Tech. Univ., 1990.
45. S. K. Morton and F. Weinberg, J. Iron and Steel Inst., 211, 13 (1973).
46. J. E. Lait, J. K. Brimacombe, and F. Weinberg, Iron and Steelmaking, 1, 36 (1974).
47. R. Bommeraju, J. K. Brimacombe, and I. V. Samarsekera, Trans. Iron and Steel. Soc., 5, 95 (1984).
48. F. Weinberg, Metall. Trans., 15B, 479 (1984).
49. S.G.R. Brown and J. A. Spittle, Metal. Sci. and Tech., 362 (1989).

Authors' Index

Subject Index